Regional Geology: A Case-Based Approach

Regional Geology: A Case-Based Approach

Calvin Graham

STATES
ACADEMIC PRESS
www.statesacademicpress.com

States Academic Press,
109 South 5th Street,
Brooklyn, NY 11249, USA

Visit us on the World Wide Web at:
www.statesacademicpress.com

ISBN: 978-1-63989-726-1

Cataloging-in-publication Data

Regional geology : a case-based approach / Calvin Graham.
p. cm.
Includes bibliographical references and index.
ISBN 978-1-63989-726-1
1. Geology. 2. Geology--Case studies. 3. Geology--Regional disparities.
I. Graham, Calvin.

QE26.3 .R44 2023
550--dc23

Table of Contents

Preface

The purpose of the book is to provide a glimpse into the dynamics and to present opinions and studies of some of the scientists engaged in the development of new ideas in the field from very different standpoints. This book will prove useful to students and researchers owing to its high content quality.

Regional geology is the geological study of large-scale regions. This area of study creates an important connection between local and global geology. The occurrence of geological processes and geologically significant boundaries help define the size and borders of each region. The objective of regional geology is to simplify local surface and subsurface data to a scale that helps in predicting and highlighting the generalizations that are implicit in continent-wide and global syntheses. Regional geology establishes the association among global plate tectonics, global climate changes, sea-level changes, and important studies in geoscience. It is applied in the search and conservation of natural resources such as water, ore deposits, and solid Earth energy resources. This book outlines the major concepts of regional geology in detail. Those in search of information to further their knowledge will be greatly assisted by this book.

At the end, I would like to appreciate all the efforts made by the authors in completing their chapters professionally. I express my deepest gratitude to all of them for contributing to this book by sharing their valuable works. A special thanks to my family and friends for their constant support in this journey.

Calvin Graham

1 Overview

1.1 Purpose and Task

The project named *Regional Geological and Mineral Survey of Hanggai (H50E009022), Xianxia (H50E010022), and Chuancun (H50E010023) Map Sheets, Zhejiang Province on a Scale of 1:50,000* (also referred to as the Project) with No. 1212011220527 is a part of *Geological and Mineral Resources Survey of Qingzhou–Hangzhou Metallogenic Belt*, a geological survey program initiated by China Geological Survey (also referred to as CGS). It lasted from 2012 to 2014 and the numbers of annual task documents issued by CGS during the three years are Ji[2012]02-013-021, Ji[2013]01-016-004, and Ji[2014]01-018-002, respectively.

1.1.1 Objective and Task

Based on systemic collection and comprehensive analysis of existing geological data and according to applicable technical specifications including *General Principles of Regional Geological Survey (1:50,000)*, *Technical Requirements for Regional Geological Survey (Scale: 1:50,000) (provisional)*, the specifications related to regional geochemical survey, and *the Notice on Strengthening the Regional Geological Survey of Metallogenic Belts (1:50,000)* issued by CGS, the following tasks are to be performed: (i) carrying out regional geological and mineral survey on a scale of 1:50,000, in order to determine the features of strata, rocks, and structures, establish the stratigraphic sequences of the areas to be surveyed (collectively referred to as the Area), and further break down stratigraphic units and highlight special geological bodies and informal stratigraphic units, (ii) strengthening the research into the relationship between mineralization and ore-bearing strata, magmatism, and tectonic activities, in order to systemically determine the geological conditions of mineralization in the Area, and (iii) verifying the anomalies and inspecting the ore occurrences of key parts of the Area, summarizing the metallogenic regularity, and putting forward key survey areas for geological prospecting in the Area. With an expected total area of the geological survey of 1334 km^2, the Project is designed to focus on the following work.

1. To carry out the research on multiple stratigraphic division of the Area; to conduct lithology mapping and lithofacies mapping by lithostratigraphic or tectonic stratum methods; to determine the stratigraphic distribution, lithologic association, era attribution, and ancient geographical environment of the Area; furthermore, to determine the stratigraphic sequences, further break down the stratigraphic units, and highlight special geological horizons and informal stratigraphic units.
2. To further determine the petrological, geochemical, and volcanic tectonic characteristics of the Mesozoic volcanic rocks in Tianmu Mountain based on the sequence division of volcanic strata, in order to ascertain the formation era of the volcanic rocks; to determine the characteristics of filling sequences and investigate the relationship between regional structure and the formation, development and closure of continental volcanic basins, in order to summarize the evolution law of the basins.
3. To determine the morphology, occurrence, rock types, contact relation, as well as the petrological and geochemical characteristics of various intrusives of Ma'anshan, Tangshe, Tonglizhuang, Wushanguan, Xianxia, Jiuliwan, and Dongling plutons and Songcun composite pluton within the Area; in addition, to further establish the rock units of the intrusive rocks of different eras and explore the relationship between mineralization and the magma activities of different eras.
4. To determine the characteristics of basic geological structures such as folds, fractures, and thrust (gliding) nappe structures in the Area, in order to establish the tectonic framework of the Area; to conduct correct division of the tectonic deformation periods; to focus on the survey of the characteristics of the composition, special distribution, and activities of Zaoxi-Moganshan fracture zone and Xuechuan-Xiaofeng fracture zone; and

to analyze the basin-controlling and rock-controlling mechanism of the fracture zones; to summarize the geological development history of the Area.

5. To carry out the study on geological conditions of mineralization and ascertain the stratum horizons, rock types and geological structures closely related to the mineralization; to inspect key ore occurrences and verify key geochemical anomalies, in order to determine prospecting clues and provide prospecting target areas; to summarize metallogenic regularity of the Area; and to propose the key survey areas for geological prospecting.
6. To systematically collect and collate existing data obtained from previous geochemical prospecting; to carry out stream sediment survey on a scale of 1:50,000 and delineate the geochemical anomalies; to verify and evaluate the key geochemical anomalies, preliminarily determine the causes of anomalies, and provide information for geological prospecting; to analyze more than 15 elements including Au, Ag, Cu, Pb, Zn, As, Sb, Bi, Cd, W, Mo, Sn, Cr, Co, and Ni.

1.1.2 Desired Production

1. A geological and mineral survey report of the joint survey areas; 1:50,000 geological maps based on map sheets and their explanations; 1:50,000 mineral resource maps based on map sheets; 1:100,000 ore-bearing tectonic maps base on map sheets and their explanations.
2. A report on stream sediment survey on a scale of 1:50,000 as well as related series of maps and databases attached.
3. Raw data (including the database of actual work material maps), the spatial database of the resulting maps, the geological survey report from the regional geological survey should all be submitted according to the requirements of the *Guideline on the Building of Spatial Databases of Geological Maps* and *Standard on Spatial Databases for Digital Geological Maps (2006)* issued by CGS.

1.2 Location, Transportation, Geography, and Economy

1.2.1 Overview of Location and Transportation of the Area

The Area is located at the junction of northern Zhejiang Province and southern Anhui Province, with east longitude of 119°15′00″–119°45′00″ and north latitude of 30°20′00″–30°40′00″. Administratively, it is a part of Lin'an City and Anji County of Zhejiang Province and Ningguo City and Guangde County of Anhui Province. It consists of three map sheets named Hanggai, Xianxia, and Chuancun, respectively, according to international sheet division on a scale of 1:50,000, covering a total area of about 1334 km^2.

The Area features a convenient road network, with provincial highways S11 and S13 directly passing through the Area as well as tarred roads connecting all the towns and villages of the Area (Fig. 1.1).

1.2.2 Overview of Natural Geography and Economy

Situated within the mid and low mountainous area of Tianmu Mountain in northern Zhejiang Province, the Area is high in the south and low in the north, with the average elevation of 500–1500 m. The highest elevation is 1587 m and is located in Longwang Mountain. Owing to dangerously steep and deep dissection as well as decades of afforestation, the Area is characterized by thick vegetation, sparse population, and rugged mountain roads, which comprise arduous conditions for geological survey in the Area.

The Area is featured by subtropical monsoon climate with four distinctive seasons, warm and humid. The annual average temperature is about 17 °C. The annual average precipitation is about 1602 mm, with rainy seasons lasting from March to June and from August to September mainly. The duration of the annual frost-free period is about 257 days. July and August constitute the midsummer of the Area, during which the temperature is usually 37–39 °C.

The Area boasts of the strong economy, rich and deep ecological culture, and an integrated economic pattern in which a number of animal and plant industries comprehensively develop including crops, oils, forestation, bamboo, tea, mulberry, fruits, vegetables, livestock, poultry, and fishery. Furthermore, the Area enjoys a sound market economy, individual economy with a long history, as well as agriculture and rural areas with high degrees of industrialization. Besides, the national forest reserves and resorts named Tianmu Mountain and Tianhuangping are also distributed in the Area.

1.3 The Past Geological Survey and Research

The systematic geological survey in the Area can be dated back to the 1960s. From then, the regional geological survey, mineral geological survey, hydrogeological survey, geophysical prospecting, and geochemical prospecting on a scale of 1:200,000 and part of geophysical prospecting and geochemical prospecting on a scale of 1:50,000 have been successively conducted in the Area. As of the beginning of

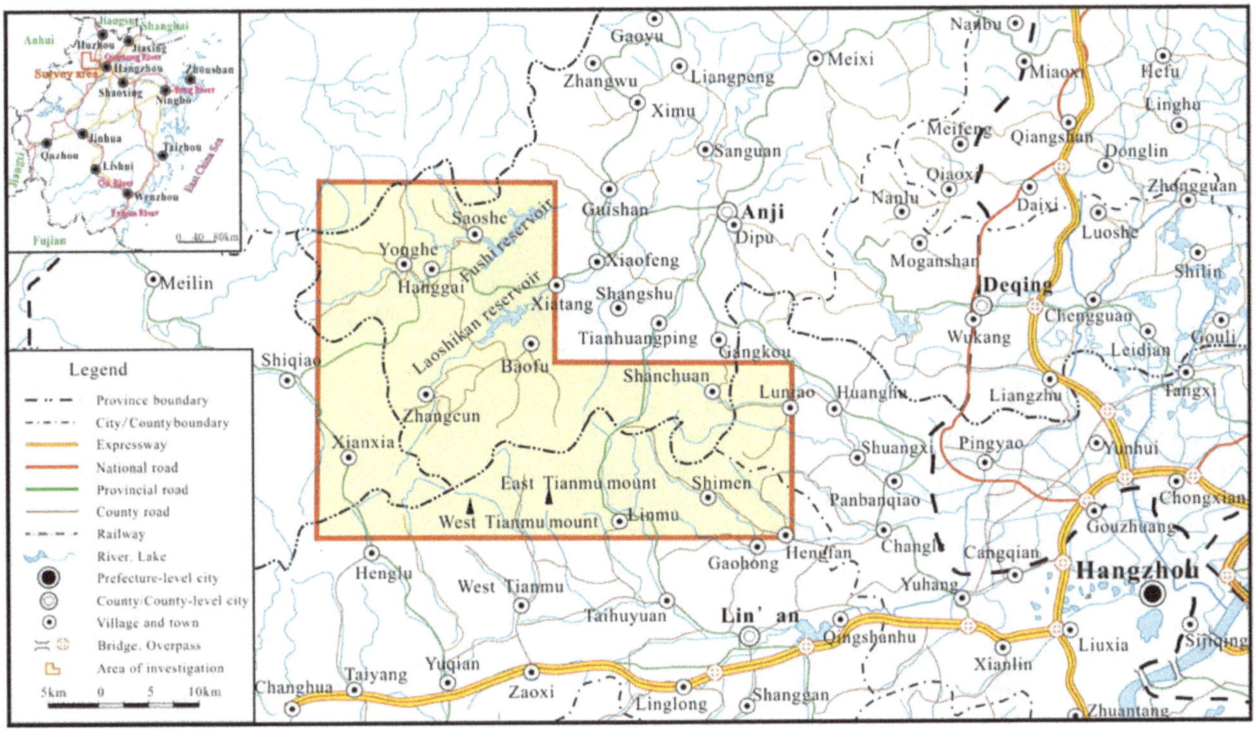

Fig. 1.1 Transportation and location of the area

this century, a regional geological survey of scale 1:250,000 had been carried out in the whole area. In addition, a number of mineral resource exploration, as well as special survey and research, have been performed in the Area.

1.3.1 Regional Geological Survey

In the 1950s, the Regional Geological Survey Team of Geological Exploration Bureau of Zhejiang Province conducted a regional geological survey on a scale of 1:200,000 in Lin'an map sheet (H–50–18). During this survey, the structures and magmatic rocks exposed in the Area were systematically investigated and were divided into 28 formations and 10 members. In the early 1960s, the Zhejiang Regional Survey Team of the Department of Geology, Nanjing University performed a comprehensive regional geological survey of the volcanic rock area of Tianmu Mountain and preliminarily investigated the lithology and lithofacies of the volcanic rocks. In the 1980s and 1990s, the Regional Geological Survey Team of Geological Exploration Bureau of Zhejiang Province and Zhejiang Bureau of Geology and Mineral Resources systematically summarized the lithostratigraphic condition of Zhejiang Province and the Area successively.

At the beginning of the twenty-first century, Anhui Institute of Geological Survey conducted a regional geological survey of Xuancheng map sheet (H50C002004) on a scale of 1:250,000, covering the work scope of this Project. The latest geological theory and methods were applied in the systematic study and summary of the strata, structures, and magmatic rocks in the Area, especially in the survey and comparative study on the movement history, association patterns, and nature of regional structures such as Jixi-Ningguodun Composite Anticline, Jixi-Ningguo fracture zone, and Xuechuan-Xiaofeng fracture zone as well as the rock association, petrology and intrusive sequence of Ma'anshan, Tangshe, Xianxia, and Wushanguan plutons.

1.3.2 Geophysical, Geochemical, and Mineral Exploration

(1) During 1955–1965

In the 1950s, the Regional Geological Survey Team of Geological Exploration Bureau of Zhejiang Province investigated the heavy concentrate and dispersed flow in Lin'an map sheet (H–50–18) while conducting the regional geological survey on a scale of 1:200,000 in the map sheet. As a result, 40 heavy concentrate anomalies and 65 metallic and dispersion flow anomalies were delineated; 25 mineral types and 94 ore occurrences were newly discovered; and 313 ore occurrences were inspected. Furthermore, 12 important metallogenic prospect areas were delineated, including Lujiashan tungsten-beryllium deposit, Zhangcun polymetallic scheelite, Baofu-Huanghu polymetallic magnetite deposit, Litali-Yucun fluorite deposit, Yujiakou-Tongguanshan boron-iron lead-zinc cassiterite deposit, Zhangcun-Henglutou fluorite deposit, Xixi polymetallic deposit, Xuechuan tungsten-beryllium-polymetallic-fluorite deposit, Machebu tinstone-polymetallic deposit, Xianlinbu-Wanshiqiao polymetallic-barite deposit, Fuyang lead-zinc-magnetite deposit, and the major deposits beyond these prospect areas.

In the early 1960s, the Fourth Group of the Geophysical Exploration Team of Geological Exploration Bureau of Zhejiang Province conducted a comprehensive reconnaissance survey of the geology, geophysical exploration, and geochemical exploration in Lin'an map sheet. The First Geological Team of Geological Exploration Bureau of Zhejiang Province conducted a comprehensive reconnaissance survey on a scale of 1:50,000 in the south of Anji (which was in the north of Tianmu Mountain). During this survey, 561 anomaly areas with various diffusion halos were delineated. Furthermore, 31 mineralized points and 21 mineral types were discovered. According to the comprehensive reconnaissance survey, it was believed that the widely distributed skarn and marble belts possibly contained the mineral resources (e.g., beryllium deposit) related to greisenization and albitization existing widely. However, systematic sampling and survey were not conducted. During the same period, the First Geological Team of Geological Exploration Bureau of Zhejiang Province conducted preliminary evaluation of the reconnaissance survey of Gaocun stibium deposit, Henglutou fluorite deposit, Zhujiashan-Yinshuidong boron deposit, while the Forth Geological Team of Geological Exploration Bureau of Zhejiang Province conducted preliminary evaluation of the reconnaissance survey of Gangkou-Zhangcun boron deposit and Gangkou Yujiawu fluorite deposit. The survey results are as follows: the fluorite deposits in Henglutou-Gangkou Yujiawu were all controlled by NW-trending faults; the NW-trending fracture zone near the pluton was the priority in the prospecting of the fluorite deposits; and skarnized breccia was an important prospecting indicator. The survey results also showed that the Zhujiashan-Yinshuidong boron mineralization was mainly born in the external contact zone between granodiorite and Sinian stratum, and the ore-bearing horizon was located in forsterite-bearing dolomitic marble that was about 10–20 cm to the stratum top. Besides, the boron mineralization was magnesium-bearing skarn ludwigite-magnetite deposit.

(2) During 1976–1996

In the late 1970s, the Geophysical Exploration Team of Geological Exploration Bureau of Zhejiang Province conducted an aeromagnetic survey in the north of Zhejiang Province on a scale of 1:50,000, covering the part in Zhejiang Province of the Area.

In the early 1980s, the Ninth Geological Team of Geological Exploration Bureau of Zhejiang Province carried out the reconnaissance survey of tungsten deposits in Tangshe and Ma'anshan areas. As a result, 12 various kinds of heavy concentrate anomalies were delineated including six anomalies of scheelite and wolframite, and one mineralized point of scheelite was newly discovered. The lardite vein was found in the thick stratiform dolomitic marble on the top of Xifengshi Formation during the reconnaissance survey. Besides, Yinshanjian scheelite-tinstone mineralized point, Tangsheling scheelite mineralized point, Baishawu scheelite-containing polymetallic mineralized point, and Shuangshekuling tungsten-beryllium ore occurrence was inspected by field reconnaissance. The First Geological Team of Geological Exploration Bureau of Zhejiang Province summarized the metallogenic regularity of the tungsten deposit in Anji-Chun'an area and made the preliminary prediction. According to the prediction, Ma'anshan pluton was featured by the successive distribution of tungsten-beryllium quartz-vein deposit (i.e., wolframite and beryl), fluorite deposit, skarn tungsten polymetallic deposit, polymetallic deposit, and stibium deposit from inside out; Tangshe granodiorite was featured by the distribution of disseminated scheelite mineralization, skarn scheelite polymetallic deposit, fluorite, and polymetallic deposit from inside out; and Xianxia pluton was characterized by the distribution of granodiorite, scheelite mineralization of skarn and quartz-vein types, and mineralization of chalcopyrite and pyritization from inside out. Based on this, Ma'anshan W–Be (Mo–Cu–Pb–Zn–CaF_2) and Tangshe Tungsten metallogenic prospect areas were delineated. In the late 1980s, the Ninth Geological Team of Geological Exploration Bureau of Zhejiang Province inspected the gold anomaly of Talishan, Tangshe area and accordingly determined that the gold mineralization degree was closely related to the alteration and broken degrees of the rocks as well as the development of tectonic cracks. However, the gold anomaly of Talishan, Tangshe area was characterized by weak mineralization and alteration, low development of tectonic cracks, unobvious anomaly concentration center or multiple anomaly concentration centers, low anomaly strength, and poor element association. Besides, the Ninth Geological Team of Geological

Exploration Bureau of Zhejiang Province verified the anomalies in the West Gangkou Road–Yujiawu area including Zn–Pb polymetallic anomalies of Yezhutang, Hongmiaoqiao, and Yujiawu, as well as gold anomalies of Yezhutang anomaly area. Furthermore, the team determined that Zn–Pb polymetallic mineralization occurred in diopside, garnet, and skarn in the lower part of Dachenling Formation, which lied in the external contact zone of granodiorite pluton, the fractured zone of the NW-trending fault, as well as the garnet, diopside, and skarn in the external contact zone of Xiyangshan Formation. The team believed that factors of polymetallic mineralization included horizon, lithology, crack fracture, and contact plane structure. During Grade 3 verification of polymetallic anomalies in Shangmei Village, Hanggai Town, it was determined that there were no visible anomalies of Cu, Pb, and Ag according to soil survey. According to geological reconnaissance survey and assessment of the deposits of Ag, Au, and Cu, it was believed that the deposits of meso-epithermal hydrothermal filling type were distributed in the Area, including dominant Ag deposits and associated polymetallic deposits of Cu, Pb, and Zn. Furthermore, the team believed that the chambered, moniliform, veined, lenticular, columnar, and stratiform mineral (mineralized) bodies were distributed in plane or profile in an echelon arrangement, with complex morphology and poor continuity. Besides, it was thought that the mineral (mineralized) bodies were closely related to fractured (caused by compression along with rock layers) and conjugated transtensional insequent faults. In addition, the mineral (mineralized) bodies occurred in the cores of synclines and microplissement development areas at the turning of synclines and anticlines, with extremely uneven distribution and very poor surface. However, the mineralization tended to be slightly more favorable downwards from the surface.

In the early 1990s, the Geophysical and Geochemical Exploration Team of Zhejiang Province conducted a stream sediment survey of Lin'an map sheet on a scale of 1:200,000. Consequently, the single-element geochemical maps and integrated anomaly maps were plotted, and several fourth-class geochemical prospecting areas were delineated including Ma'anshan-Tangshecun Ag–Cu–Pb–Zn–Mo–Sb–F–CaO, Shangshi-Gangkou Ag–Pb–Zn–Mo–Cu–Au, Sanqiaobu-Huanghu Au–Ag–Cu–Pb–Zn–F–CaO, Xixi Pb–Zn–Ag–Mo–Cu, Changchunqiao-Yuhang Pb–Zn–Cu–Ag–Mo and nonmetal, Heqiao-Wanshiqiao Sn–Mo–Bi–Pb–Zn–Au–Ag, and Fuyang Pb–Zn–Cu–Au–Ag prospecting areas. According to later aeromagnetic computation on a scale of 1:50,000 in Anji and Hanggai map sheets and the stream sediment survey on a scale of 1:200,000 by single-point sampling, the geochemical maps of 14 elements were plotted. Furthermore, the mid-deep and shallow geological bodies in the Area were inferred by the aeromagnetic calculation. It was pointed out that the middle part was the thickest area of the Lower Paleozoic Erathem with weak magmatic activities, while the eastern and southern parts were the active areas of magmatic rocks. According to the stream sediment survey, the remaining aeromagnetic anomalies were caused by andesite and stock-like granodiorite in Laocun Formation of Cretaceous Period, and the remaining aeromagnetic anomalies associated with granodiorite were closely related to metallic deposits. Besides, the metallogenic prospect areas were delineated during this survey (including Ag-bearing polymetallic metallogenic prospect areas of Yaocun, Hanggai-Shizhu area, Tongkengcun, Shangshi-Shishancun area, and Gangkou).

(3) During 2004–2014

In the twenty-first century, with the development of modern metallogenic theory and high-precision prospecting and test methods, various geological exploration institutions conducted further prospecting evaluation in the Area and its peripheral areas.

Zhejiang Geological Prospecting Institute of the China Chemical Geology and Mine Bureau conducted prospecting evaluation of the fluorite deposits in Changshan-Anji area of Zhejiang Province. It investigated and studied the metallogenic geological conditions and metallogenic model of fluorite, divided the deposits into the ones of meso-epithermal hydrothermal fissure filling type and the ones of hydrothermal filling and superposition type after contact metasomatism, conducted the prospective evaluation of fluorite resources in the NW-trending and SE-trending metallogenic zones, and evaluated the resources of important fluorite mining area. For example, according to the evaluation of the Jiantouwu (Yonghe) fluorite deposits within the Area, fluorite ore bodies occurred within the external contact zone between Ma'anshan pluton and keratinized silt sandstone of Yinzhubu Formation (O_1y). In addition, two fluorite ore bodies were found within the NW-trending tectonic fracture in the silicified zone with fluorite resources of 21,100 tonnes, and the fluorite resources in Tangshe-Zhangcun area were estimated to be 109,900 tonnes.

The institutions including the Ninth Geological Team of Geological Exploration Bureau of Zhejiang Province and Zhejiang Institute of Geological Survey conducted the resource reserve survey of Fe in the Gangkou mining area, fluorite in Yonghe mining area, and stibium in Gaocun mining area in the Area. As a result, the total accumulative identified resource reserve (122b + 2S22) of magnetite in the Gangkou Fe survey area was 689,000 tonnes, and the resource reserve (333) of stibium deposit in Gaocun was 57.87 tonnes.

According to the arrangement of the national mineral resource evaluation project, the Zhejiang Institute of Geological Survey carried out an evaluation of mineral resource potential in Zhejiang Province. According to the division of related metallogenic zones or belts, the Area was located in the Au–Ag–W–Mo–Cu–Pb–Zn–Sb–Fe–fluorite–Be–bentonite metallogenic subzone of Tianmu Mountain, which was located in the Cu–Pb–Zn–Ag–Au–W–Sn–Nb–Ta–Mn–sepiolite–fluorite–wollastonite metallogenic zone (III-71) in the northeastern part of Qinzhou Bay–Hangzhou Bay area. The Area was divided into the Yonghe (Anji County) fluorite–Sb–Cu, Zhangcun (Anji County)-Pingyao (Yuhang District) fluorite–S–Fe–Cu, and West Tianmu Mountain (Lin'an District) W-Mo-fluorite ore concentration areas, as well as Ma'anshan, Tonglizhuang, and Tanshe integrated pre-survey areas of Anji County.

In addition, 12 commercial survey projects are implemented in the Area currently, which involves pre-investigation, reconnaissance survey, and detailed survey. Most areas with exploration rights are situated in the peripheral parts of old mining areas or in known ore occurrences. According to exploration and reserve assessment, the submitted reserve (111b + 122b + 333) of fluorite resources in the Jianziwu (Yonghe) fluorite mining area was up to 743,000 tonnes (medium scale) as of 2012. According to a detail survey, the submitted resources (332 + 333) of lead and zinc, silver, and copper in the north of Gangkou Fe mining area were 22,500 tonnes, 46.78 tonnes, 5000 tonnes, respectively, as of 2014. As for the north contact zone of Tonglizhuang composite pluton, Tonglizhuang (Puluwu) fluorite deposit and Langcun (Tonglizhuang) W–Mo polymetallic deposit had been discovered as of 2014. The preliminarily estimated resources of fluorite and W–Mo deposit (WO_3) in the two deposits were 2,357,000 tonnes (large scale) and 18,500 tonnes (medium scale), respectively.

1.3.3 Scientific Research on Geology and Minerals

In 2007, Nanjing Institute of Geology and Paleontology, Chinese Academy of Sciences and Zhejiang Institute of Geological Survey discovered for the first time deepwater benthos (Evangelia–Dalmanitina (Songxites) combination) in Yuhang area, Hangzhou, which is 2000 m to the east of Chuancun map sheet. Accordingly, the institute considered that Yuhang was one of the refuges during the mass extinction of Late Ordovician organisms. This opened up a new prelude to the biostratigraphy study from Late Ordovician to Early Silurian and as well as offered a good opportunity for the study of biostratigraphy in the Area.

In 2012, the University of Science and Technology Beijing systematically studied the diagenesis and metallization of Wushanguan composite pluton in the east of the Area and published a series of papers. The university considered that the magma activities in the mining area occurred in 141–117 Ma, and the medium-grained biotite monzogranite, medium-grained syenogranite, medium-grained granodiorite, fine-grained granite, and bimodal vein-rock association successively constituted the emplacement sequence. All the rocks except for the bimodal vein-rock association constituted the Wushanguan composite pluton. The university also believed that two mineralization events took place in the Area. The first event was related to the emplacement of medium-grained granodiorite (137 Ma). During this period, skarn Cu–Fe mineralization featured by garnets and skarns took shape and was mainly distributed in the west of the Area. The second event was Pb–Zn (Ag) and Mo mineralization, which was related to the emplacement of fine-grained granite (135–134 Ma). During this period, the skarn Pb–Zn (Ag) mineralization formed with epidote skarn as the dominant mineralized result.

1.4 Overview of the Work

1.4.1 Work Stages of the Project

Based on full collection and utilization of existing outcomes obtained from the geological survey of Lin'an map sheet on a scale of 1:200,000, geological survey of Xuancheng map sheet on a scale of 1:250,000, geophysical and geochemical exploration on different scales, and mineral inspection, the Project conducted field survey and integrated research as required by the task documents issued by CGS. According to the task documents, the specific project stages are as follows:

From February to August in 2012, the Project team successively conducted the collection and research of existing information, preliminary remote-sensing interpretation, field reconnaissance, mapping of some sections, and survey of the master traverse. Furthermore, the Project team clearly determined various stratigraphic units and developed overall design.

From September 2012 to December 2014, according to the overall design and work deployment required by annual schemes, the Project team successively completed the digital geological survey on a scale of 1:50,000, stream sediment survey on a scale of 1:50,000 (Hanggai map sheet), the mapping of stratigraphic units and (volcanic) tectonic sections on different scales, and rough inspection of mineral resources. The Project team also collated various original data comprehensively and systemically, prepared various maps, and summarized fieldwork. Furthermore, the Project successfully passed the field acceptance organized by the east China project management office of CGS and was ranked excellent in terms of overall rating.

From January to August in 2015, the Project team resolved the problems proposed during the field acceptance and improved the original data. Furthermore, the Project team prepared the regional geological and mineral survey report as well as geological maps (including their explanations), mineral maps, tectonic maps of ore-bearing formations, and lithological and lithofacies tectonic maps of volcanic rocks. Besides, the Project team completed the spatial database of 1:50,000 digital geological maps.

During August 19–22th, 2015, the east China project management office of CGS organized competent experts to review the report of the outcomes obtained in the Project in Nanjing. The review expert group consisted of seven experts from Nanjing Institute of Geology and Paleontology, Chinese Academy of Sciences, Nanjing Center, China Geological Survey, Fujian Institute of Geological Survey, Jiangxi Institute of Geological Survey, Anhui Provincial Institute of Geological Survey, and Geological Exploration Technology Institute of Anhui Province. The expert group determined that the goals and tasks of the Project stipulated in the task documents were completed comprehensively, and the report featured rich content and detailed information. Therefore, the Project was ranked excellent in terms of overall rating.

During November 15–17th, 2015, All China Commission of Stratigraphy organized nearly 20 experts to conduct a field investigation and demonstration for the "Upper Ordovician Wenchang Formation Section of Hanggai Town, Anji County, Zhejiang Province" applied by the Project team, including Academician Xu Chen, Researcher Yuandong Zhang, and Researcher Huawei Cai from Nanjing Institute of Geology and Paleontology, Chinese Academy of Sciences, Researcher Zejiu Wang, Researcher Hongfei Hou, and Researcher Jianxin Yao from Chinese Academy of Geological Sciences, and Researcher Xiaofeng Wang from Nanjing Center, China Geological Survey. All the experts unanimously agreed to rank the section as the "Standard Cross Section of the Lower Yangtze Region in the Upper Ordovician Hirnantian." Besides, they recommended the Department of Natural Resources of Zhejiang Province, the People's Government of Anji County, and Zhejiang Institute of Geological Survey to include the protection and research of Hanggai sections into the geological environmental protection and ecological building planning, in order to further enhance the protection, utilization, and scientific research of the sections.

1.4.2 Workload Completed

According to relevant specifications and approved workload, the goals and tasks of the Project have been comprehensively completed. The main work includes 1334 km^2 of the regional geological survey on a scale of 1:50,000, 1334 km^2 of remote-sensing interpretation on a scale of 1:50,000, and 440 km^2 of stream sediment survey on a scale of 1:50,000. The detailed approved and completed physical workload is shown in Table 1.1.

Table 1.1 Main approved and completed physical workload

Work type		Unit	Approved workload				Actual workload completed	Completion percentage (%)	Remarks
			2012	2013	2014	Total			
Geological survey	Regional geological survey on a scale of 1:50,000	km^2	500	400	434	1334	1334	100.00	
	1:100 profile (field survey)	km					0.08		Replaced with the 1:2000 surveyed profiles as required by precision
	1:200 profile (field survey)	km					0.37		
	1:500 profile (field survey)	km					1.66		
	1:1000 profile (field survey)	km					9.97		
	1:2000 profile (field survey)	km	17	19		36	20.68	57.00	
	1:2000 profile (rough survey)	km	10.5			10.5	13.92	133.00	
	1:5000 profile (field survey)	km		43	19.5	62.5	63.92	102.00	
	1:5000 profile (rough survey)	km	6		12.5	18.5	22.70	123.00	

(continued)

Table 1.1 (continued)

Work type		Unit	Approved workload				Actual workload completed	Completion percentage (%)	Remarks
			2012	2013	2014	Total			
Geochemical exploration	Soil profile survey on a scale of 1:10,000	km			17.5	17.5	19.1	109.00	
	Stream sediment survey on a scale of 1:50,000	km^2		440		440	440	100.00	
Remote sensing	1:50,000 remote-sensing interpretation	km^2	1334			1334	1334	100.00	
Rock deposit tests	Silicate analysis	pcs	10	15	20	45	59	131.00	
	Trace element analysis	item	10	15	20	45	56	124.00	
	rare-earth element analysis	pcs	10	15	20	45	56	124.00	
	Spectral analysis	pcs	135	690	70	895	795	89.00	
	Chemical samples	pcs		10	5	15	16	106.00	
	Thin section identification	pcs	93	756	683	1532	1524	99.00	
	Stream sediment samples	pcs		1941		1941	1848	95.30	
	Soil samples	pcs		52	438	490	475	97.00	
	Microorganisms identification	pcs	50	50		100	101	101.00	
	Macrofossil palaeobios identification	pcs	150	150		300	300	100.00	
	Zircon LA-ICPMS U–Pb dating	dot	45	45		90	453	503.00	
	Zircon SHRIMP U–Pb dating	dot	45	45		90	190	211.00	
	C and S isotopes	pcs		80	30	110	81	74.00	
	Sr–Nd isotopes	pcs			30	30	21	70.00	

2

The Area and its Stratigraphic Division

The strata in the Area are the parts of Jiangshan–Lin'an stratigraphic subregion of the Jiangnan stratigraphic region. The exposed strata in the Area include the Nanhuan and Sinian of the Neoproterozoic; the Cambrian, Ordovician, and Silurian of the Paleozoic; the Cretaceous of the Mesozoic; and the Quaternary of the Cenozoic. Among these strata, the Paleozoic and Mesozoic developed the most. A small amount of Neoproterozoic strata outcrop in the middle west part and the east of the Area. The Paleozoic is mainly distributed in the northwest. The Mesozoic is mainly distributed in the southeast. The Quaternary of the Cenozoic is sporadically distributed in the hills in the west and east as well as the valleys in the central and northern part of the Area.

In this Project, 45 lithostratigraphic units were determined including 19 formations and 26 members in the Area based on the division scheme stated in *Lithostratigraphy of Zhejiang Province* (1996) according to lithologic association characteristics and sedimentary environment in the Area through the profile survey and the survey of geological observation traverse. They are described as follows from bottom to up. The Neoproterozoic Nanhuan is divided into Xiuning Formation (Nh_1x) and Nantuo Formation (Nh_2n). The Neoproterozoic Sinian is divided into the Lantian Formation ($Z_{1\text{-}2}l$) and Piyuancun Formation (Z_2p). The Cambrian, Ordovician, and Silurian of the Paleozoic constitute the major exposed strata in the Area. They are successively divided into 13 formations, namely Hetang ($Є_1h$), Dachenling ($Є_1d$), Yangliugang ($Є_2y$), Huayansi ($Є_3h$), Xiyangshan ($ЄOx$), Yinzhubu (O_1y), Ningguo ($O_{1\text{-}2}n$), Hule ($O_{2\text{-}3}h$), Yanwashan (O_3y), Huangnigang (O_3h), Changwu (O_3c), Wenchang (O_3w), and Xiaxiang (S_1x). The Mesozoic Cretaceous contains one formation named Huangjian (K_1h). The Cenozoic Quaternary contains one formation named Yinjiangqiao (Qhy). Some of these formations are further divided into lithological members according to the lithologic association. In detail, Lantian Formation is divided into four members ($Z_{1\text{-}2}l^1$, $Z_{1\text{-}2}l^2$, $Z_{1\text{-}2}l^3$, and $Z_{1\text{-}2}l^4$), Hetang Formation is divided into two members ($Є_1h^1$ and $Є_1h^2$), Yangliugang Formation is divided into two members ($Є_2y^1$ and $Є_2y^2$), Xiyangshan Formation is divided into three members ($ЄOx^1$, $ЄOx^2$, and $ЄOx^3$), Yinzhubu Formation is divided into three members (O_1y^1, O_1y^2, and O_1y^3), Hule Formation is divided into three members ($O_{2\text{-}3}h^1$, $O_{2\text{-}3}h^2$, and $O_{2\text{-}3}h^3$), Changwu Formation is divided into three members (O_3c^1, O_3c^2, and O_3c^3), Xiaxiang Formation is divided into two members (S_1x^1 and S_1x^2), and Huangjian Formation is divided into four members (K_1h^1, K_1h^2, K_1h^3, and K_1h^4). Therefore, there are 26 members in total in the Area (Table 2.1 and Fig. 2.1). For some lithological strata with special significance, such as potassium-bearing bentonite developing in Huangnigang Formation as well as the siltstone-bearing siliceous nodules and the black carbonaceous shale-bearing sponge and graptolite fossils in the middle part of Wenchang Formation, they were surveyed as informal stratigraphic units in the Project.

The sequence stratigraphic division in the Area was based on the observation data of the outcrops in the surveyed geological profiles. The subsequences and system tracts were divided into third-order sequences. The strata of the Nanhuan and Sinian with poor outcrops and low identification accuracy were divided into the second-order sequences (mesosequences) or sub-second-order sequences (orthosequences). The division criteria proposed by Wang and Shi (1998) were adopted to determine the age ranges of the second-order and third-order sequences. As a result, the age ranges of the second-order sequences (mesosequences), sub-second-order sequences (orthosequences), and third-order sequences (orthosequences) are 30–40 Ma, 9–12 Ma, and 2–5 Ma, respectively.

In terms of biostratigraphic division, the Project focused on the description of the biozones achieved through surveyed geological profiles due to the uneven distribution and the diverse species of the ancient organisms in the strata of the Area. Besides, other strata without available fossils were briefly described by utilizing previous achievements of the

Table 2.1 Evolution of formations in the area achieved from various lithostratigraphic divisions

Regional geological survey of Lin'an map sheet on a scale of 1:200,000 (1967)		Lithostratigraphy of Zhejiang Province (1996)			Regional geological survey of Xuancheng map sheet on a scale of 1:250,000 (2004)			This Project	
Q_4		Qh	Yinjiangqiao		Qh	Yinjiangqiao		Qh	Yinjiangqiao
J_3	Huangjian	J_3	Huangjian		J_3	Huangjian		K	Huangjian
	Laocun		Laocun			Laocun			
S_1	Anji	S_1	Xiaxiang		S_1	Xiaxiang		S_1	Xiaxiang
O_3	Zhangcun	O_3	Wenchang		O_3	Changwu		O_3	Wenchang
	Yuqian		Changwu						Changwu
	Huangnigang		Huangnigang			Huangnigang			Huangnigang
	Yanwashan	O_2	Yanwashan			Yanwashan			Yanwashan
O_2	Hule		Hule		O_2	Hule		O_2	Hule
	Niushang	O_1	Ningguo			Ningguo			Ningguo
O_1	Ningguo				O_1			O_1	
	Yinzhubu		Yinzhubu			Yinzhubu			Yinzhubu
$\in_3$	Xiyangshan	$\in_3$	Xiyangshan		$\in_3$	Xiyangshan		$\in_3$	Xiyangshan
	Shikongshan		Huayansi			Huayansi			Huayansi
$\in_2$	Yangliugang	$\in_2$	Yangliugang		$\in_2$	Yangliugang		$\in_2$	Yangliugang
$\in_1$	Hetang	$\in_1$	Dachenling		$\in_1$	Dachenling		$\in_1$	Dachenling
			Hetang			Hetang			Hetang
Z_2	Xijianshan	Z_2	Piyuancun		Z_2	Piyuancun		Z_2	Piyuancun
			Banqiaoshan	Lantian		Banqiaoshan	Lantian		Lantian
			Doushantuo		Z_1	Doushantuo		Z_1	
Z_1	Siliting	Z_1	Nantuo		Nh_2	Nantuo		Nh_2	Nantuo
	Zhitang		Xiuning		Nh_1	Xiuning		Nh_1	Xiuning

Area and its surrounding areas, in order to ensure the integrity of the biostratigraphic part in the report. In terms of chronostratigraphic division, the division criteria described in *the Stratigraphic Guide of China and Its Explanation (Revision)* (2001) and *Stratigraphic Chart of China* (2014) were referred to. Therefore, the Phanerozoic strata featuring high geological survey degree are divided into stages while other strata are divided into series. The Neoproterozoic strata and Mesozoic volcanic–sedimentary rocks were divided by indicators such as tectonic movements, lithofacies, and isotopes. The Paleozoic strata were divided according to the fossil zones of graptolites, trilobites, brachiopoda, and conchostracans. According to the existing stratigraphic survey of Zhejiang Province, the division standards stated in *Stratigraphic Chart of China* (2014) issued by the National Commission on Stratigraphy of China were adopted for the division of all strata in the Area except for the Cambrian which is still divided into upper, middle, and lower series.

Geochemical analysis (ore-bearing feature) of trace elements in strata is used to analyze the rock spectra collected from the stratigraphic units of different profiles. The statistics of 14 trace elements of various stratigraphic units were made by calculating the arithmetic mean values and concentration coefficients. Vinogradov mean value theorem (1972) was adopted to determine the background value of each trace element. The concentration coefficient (k), the ratio of the mean value to the background value of each element, was divided into three levels: $k < 0.5$, $1.5 > k \geq 0.5$, and k 1.5. Furthermore, in order to discover the regional enrichment and dilution of the elements in each stratigraphic unit,

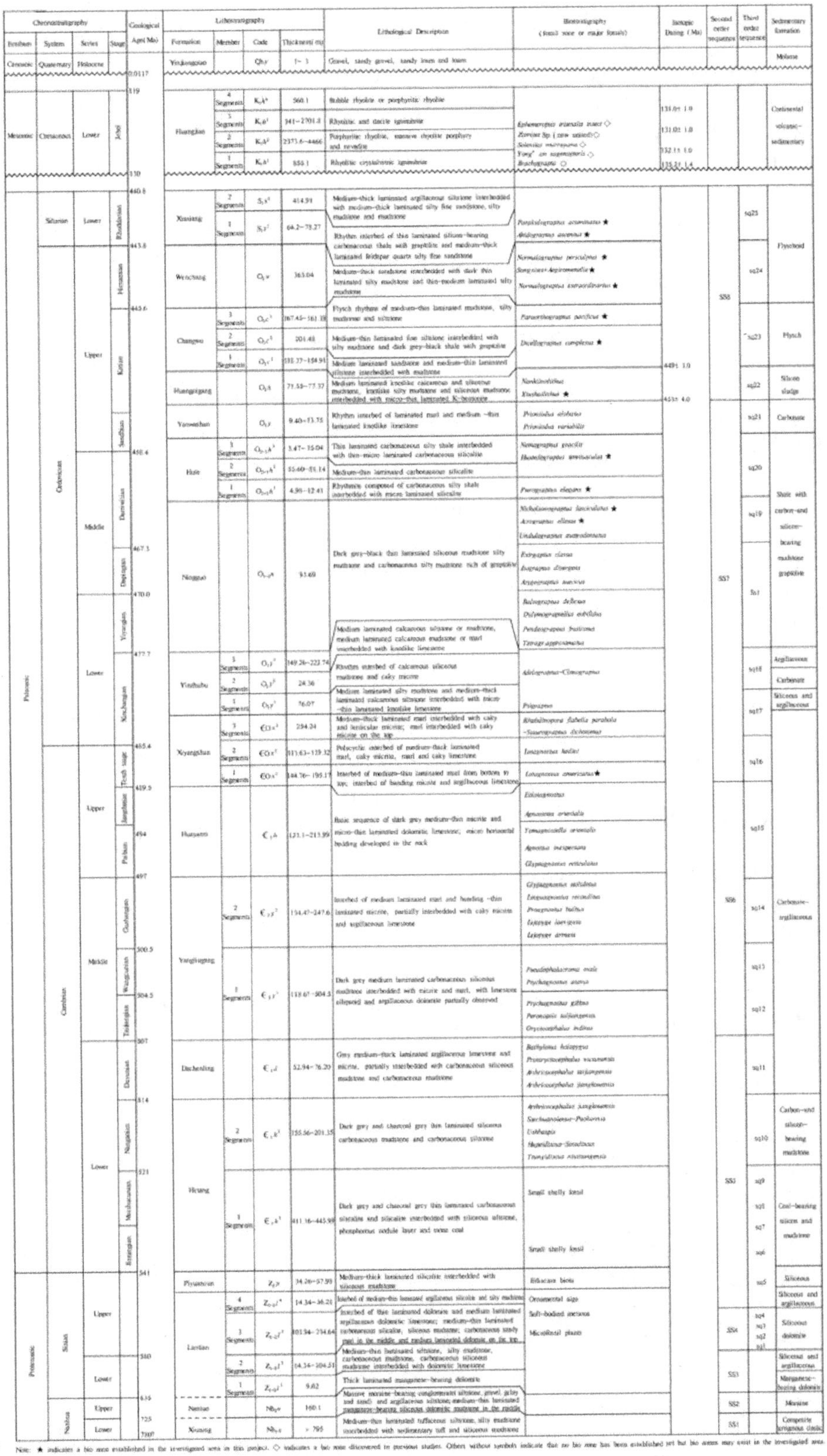

Note: ★ indicates a bio zone established in the investigated area in this project. ◇ indicates a bio zone discovered in previous studies. Others without symbols indicate that no bio zone has been established yet but bio zones may exist in the investigated area.

Fig. 2.1 Comprehensive stratigraphic histogram of the area the mean values of major trace elements in each stratigraphic unit were compared with the background value of strata in the northwest of Zhejiang Province provided in the *Research Report on Regional Stratigraphic Geophysical and Geochemical Parameters of Zhejiang Province* (1991). All enriched elements can pass the normal distribution test, indicating that the content of these trace elements was high in the early stage of sedimentation.

All stratigraphic units will be successively described in the order of lithostratigraphy, sequence stratigraphy, biostratigraphy and chronostratigraphy, event stratigraphy, sedimentary environment, and geochemistry of trace elements in strata (ore-bearing feature).

2.1 Nanhuan System

The Nanhuan System in the Area contains the Xiuning (Nh_1x) and Nantuo (Nh_2n) formations. They are mainly distributed in the external contact zone of Tangshe complex in the southwest corner of Hanggai map sheet, the external contact zone in the north of Xianxia complex in the west of Xianxia map sheet, and the northern part of Gaojiatang Village in the northeast of Chuancun map sheet. The outcrop area is about 23.12 km^2 in total, accounting for 1.82% of the bedrock area.

2.1.1 Xiuning Formation (Nh_1x)

Xiuning Formation was formerly known as Xiuning Sandstone. The name Xiuning Sandstone was created by Yuyao Li in Lantian Village, Xiuning County, Anhui Province, in 1936. Then, it was renamed Xiuning Formation by Yiyuan Qian in 1964 to mean a set of amaranthine and green sandstone occasionally interbedded with celadon argillaceous siltstone. When determining the lithostratigraphy of Zhejiang Province in 1996, Guohua Yu introduced the name Xiuning Formation into Zhejiang Province to be used in the northwestern part of Zhejiang Province to replace Zhitang Formation. In this Project, the name Xiuning Formation (Nh_1x) stated in *Lithostratigraphy of Zhejiang Province* was still adopted according to the characteristics of lithologic association of the Yaocun forest farm profile (PM044) of Hanggai map sheet and the geological observation traverse in the Area.

2.1.1.1 Lithostratigraphy

As the oldest formation in the Area, Xiuning Formation features poor outcrops and undiscovered bottom. The outcrop area is about 10.44 km^2, accounting for 0.82% of the bedrock area.

Regionally, the lithology of Xiuning Formation is characterized by two parts, i.e., the upper part and the lower part of the formation. The lower part consists of celadon tuffaceous silty-fine sandstone and amaranthine feldspar–quartz sandstone interbedded with a small amount of gravel-bearing coarse sandstone and sedimentary tuff. The upper part is composed of celadon and off-white tuffaceous siltstone as well as silty mudstone interbedded with sedimentary tuff. Horizontal, wavy, and veined bedding developed in the rocks. Owing to the Early Cretaceous magmatic intrusion and firing, hornfelsic alteration and lower horizon missing occurred in Xiuning Formation in the Area, leaving an outcrop thickness of 584.6 m.

1. **Stratigraphic section**

The lithology of Xiuning Formation is described by taking the example of Nanhuanian–Xiuning Formation (Nh_1x)–Nantuo Formation (Nh_2n) profile (rough survey) (Fig. 2.2) of Hanggai map sheet in the Yaocun forest farm, Anji County, Zhejiang Province. The details are as follows.

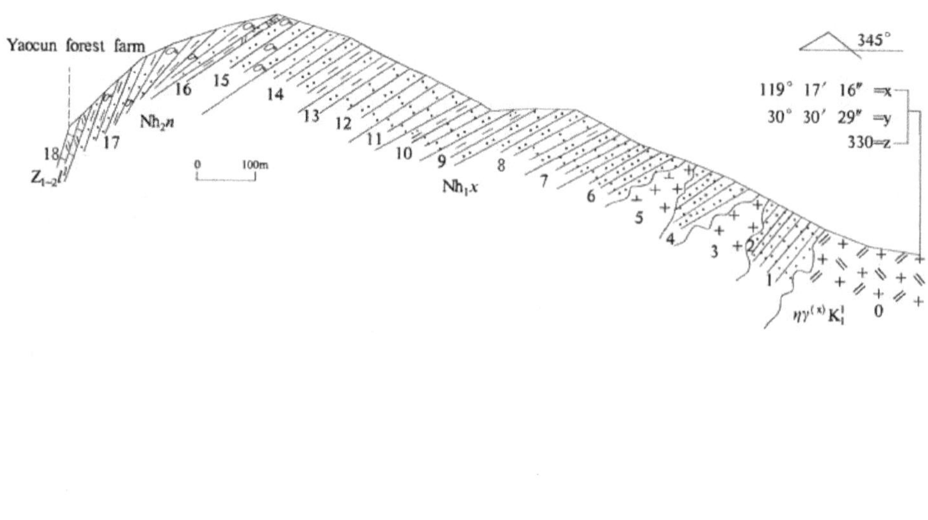

Fig. 2.2 Nanhuanian Xiuning (Nh_1x) Formation–Nantuo (Nh_2n) Formation profile in the Yaocun forest farm, Anji County, Zhejiang Province

Overlying stratum: celadon siltstone bearing moraine conglomerate of Nantuo Formation

-------------------------- Parallel unconformable contact --------------------------

Xiuning Formation Total thickness: 584.6 m

14. Celadon thin–medium laminated siltstone interbedded with silty mudstone, horizontal bedding developing; in the upper part: fine sandstone visible, wavy bedding and tabular cross bedding developing. 88.2 m

13. The upper and lower parts: celadon thin–medium laminated silty mudstone interbedded with a small amount of siltstone, horizontal bedding developing; the middle part: caesious thin–medium laminated argillaceous siltstone interbedded with banded silty mudstone, horizontal bedding developing. 21.1 m

12. Celadon medium laminated sedimentary tuff interbedded with thin banded siltstone, horizontal and wavy bedding developing. 68.5 m

11. Gray medium laminated sedimentary tuff, the thickness of a single layer: 10–30 cm. 49.3 m

10. Gray medium–thick lamellar siliceous mudstone in the upper part, rhythm interbeds of sedimentary tuff and siliceous siltstone in the middle part, tuffaceous siltstone in the lower part, horizontal and wavy bedding developing. 53.1 m

9. Celadon medium–thick laminated tuffaceous siltstone interbedded with thin laminated siliceous mudstone and sedimentary tuff; the sedimentary tuff: dot-shaped, oolite, etc.; horizontal and wavy bedding developing in the layers. 0.9 m

8. In the upper part: celadon medium laminated siltstone interbedded with tuffaceous siltstone; in the lower part: thin laminated argillaceous siltstone; small asymmetric wave ripples developing. 48.4 m

7. Celadon thin–medium laminated tuffaceous siltstone, interbedded with sedimentary tuff and thin laminated siliceous mudstone. 47.8 m

6. Celadon and gray medium laminated tuffaceous siltstone, the thickness of a single layer: 15–30 cm, horizontal bedding developing. 49.7 m

5. Off white fine-grained granodiorite dike.

4. Celadon–gray thick laminated tuffaceous siltstone, hornfelsic fuzzy bedding visible locally, appearing to be rhyolitic vitric tuff. 67.7 m

3. Flesh red fine-grained granite dike.

2. Celadon–gray medium laminated hornfelsic tuffaceous siltstone interbedded with thin laminated hornfelsic tuffaceous siltstone; the thickness of a single layer: 2 –20 cm; broken; silicification developing locally. 18.9 m

1. Rhythm interbeds of modena medium laminated hornfelsic tuffaceous fine sandstone and light celadon sedimentary tuff constitute, interbedded with celadon silty mudstone locally, horizontal and wavy bedding developing. 1.0 m

———————————— Bottom undiscovered ————————————

2. **Lithological Characteristics**

Tuffaceous siltstone, sedimentary tuff, and silty mudstone constitute the main rocks of Xiuning Formation (Nh_1x) in the Area.

The tuffaceous siltstone shows celadon–amaranthine. Felsic minerals are the main mineral components. They are equiaxed granular with a small grain size of less than 0.1 mm. Most of the siltstone is recrystallized due to thermal metamorphism, while there is a small amount still remaining the characteristics of the original rocks. The feldspar and quartz minerals remaining in the original rocks feature well-rounded shape and a slightly larger grain size of 0.1–0.2 mm.

The sedimentary tuff shows gray, mainly consisting of tuffaceous matter (75–80%) as well as siltstone and argillaceous matter (20–25%). The tuffaceous matter generally shows earthy and vitric and fine scaly locally. The siltstone and argillaceous matter have a grain size of 0.01–0.03 mm, mainly consisting of quartz.

The silty mudstone shows celadon, consisting of argillaceous matter (70%) and felsic silty matter (about 30%). The argillaceous matter has been recrystallized to be scaly mica due to thermal metamorphism. The felsic silty matter is distributed unevenly in the rocks, with a grain size of 0.02–0.04 mm.

3. **Basic Sequences**

Xiuning Formation in the Area features lower horizon missing due to rock erosion. It can be divided into two types of basic sequences according to the lithologic association exposed.

Basic sequences of type A: distributed in the lower part of Xiuning Formation, and composed of ① gray–celadon medium laminated tuffaceous (fine) siltstone and ② gray–celadon thin–medium laminated tuffaceous silty mudstone. The respective thickness of the two components is 10–40 cm and about 8–30 cm. Wavy and horizontal bedding developed in the first component, while horizontal bedding developed in the second component (Fig. 2.3a).

Basic sequences of type B: distributed in the middle and upper parts of Xiuning Formation, and composed of ① (fine) siltstone and ② (silty) mudstone. The basic sequences of this type in the middle part of Xiuning Formation are described as follows. They are about 1 cm thick, and the respective thickness of the (fine) siltstone and the (silty) mudstone is 2 mm and 5–8 mm generally. They become thinner or even wedge out sometimes. The grain sizes of the sands or silty sands in them remain unchanged generally, except that they significantly decrease near the contact interface with the underlying clay during gradual transition. Meanwhile, washing occurs on the interface while no sedimentary structures are observed in the argillaceous layer. As for the basic sequences of type B in the upper part of Xiuning Formation, the thickness is increased to be 3–5 mm

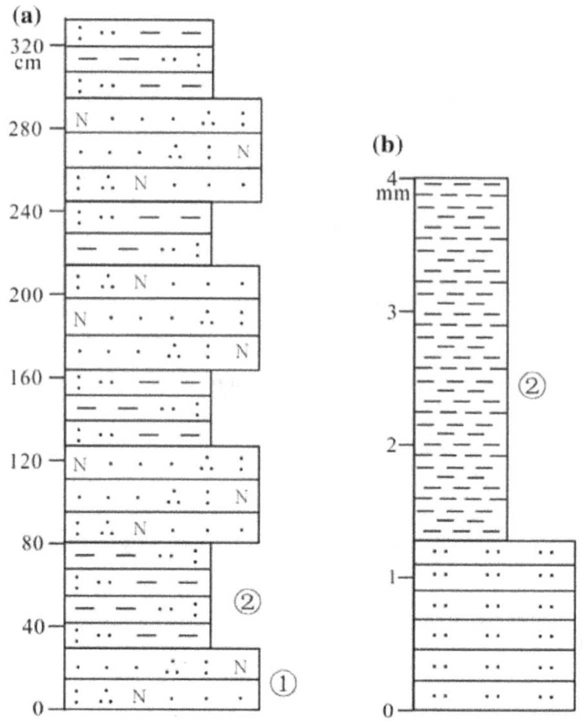

Fig. 2.3 Basic sequences of Xiuning Formation (Nh_1x)

(Fig. 2.3b) with the maximum and minimum values as 8 mm and 2 mm, respectively. Micro-fine horizontal bedding developed in the upper thicker argillaceous layer.

2.1.1.2 Sequence Stratigraphy

According to the characteristics of the lithology and lithofacies association of the profiles as well as the characteristics of the sequence boundary, there is one second-order sequence SS1 in Xiuning Formation.

Second-order sequence SS1

SS1 consists of transgressive systems tract (TST) and highstand systems tract (HST). The bottom of SS1 is missing due to the erosion of Tangshe complex. The top of SS1 is a parallel unconformity interface between silty mudstone and the overlying moraine conglomerate of Nantuo Formation, belonging to type-I sequence boundary. TST is located in the lower part of Xiuning Formation. The major sediments in TST include tuffaceous sandstone, feldspar–quartz sandstone, and tuffaceous silty mudstone, and minor sediments include siltstone, silty mudstone, and sedimentary tuff. Wavy, horizontal, and veined bedding developed in the rock layers of TST. Therefore, the sedimentary environment of TST is of littoral–tidal flat facies. HST is located in the middle and upper part of Xiuning Formation. The main sediments of HST include fine sandstone, siltstone, and mudstone. With micro-fine bedding developing in the layer, TST features the sedimentation of neritic facies. Generally, it is reflected from the sediments in SS1 that in Xiuning Formation, the sea level rose quickly in the lower part, reached the maximum in the middle part, and then gradually fell in the upper part during the period of Xiuning Formation. Therefore, the lower part of the Xiuning Formation is of retrogradation type, and the middle and upper parts of Xiuning Formation are of aggradation and progradation type.

2.1.1.3 Biostratigraphy and Chronostratigraphy

The Nanhuan paleontological research in Zhejiang Province was mainly carried out in the 1980s and focused on the profiles of the areas including Banqiao of Quzhou, Xiayabu of Jiande, Luocun of Fuyang, Heshang of Xiaoshan, Mengshan of Pujiang, and Xiaopokeng of Kaihua. No fossils were obtained in Xiuning Formation of the Area in this Project. According to the *Lithostratigraphy of Zhejiang Province* and the regional geological survey report of Changshan map sheet on a scale of 1:50,000, there are a total of 13 genera and 11 species micro-plant fossils of Xiuning Formation period (i.e., the Early Nanhuan) obtained in Xiayabu of Jiande, Shilonggang of Changshan, and Xiaopokeng of Kaihua. These micro-plant fossils are mainly of single-chlorella communities including *Leiominuscula minuta*, *Leiopsophosphaera densa*, *Lophominuscula prima*, *Lophosphaeridium* cf. *acielatum*, *Piyuanella irregutaris*, *Trachyminuscula* sp., *Trachysphaeridium* sp., *T. minor*, and *Trematosphaeridium* sp. There are also some new filamentous chlorella communities such as *Nostocomorpha prisca*, *Palaeolyngbya* cf. *barghoorniana*, and *Taeniatum verrucatum*. Among these communities, *Taeniatum verrucatum* is a characteristic molecule of the *Taeniatum-turuchanica* micro-plant association in the regions adjacent to the Area, and the geological age is the Early Nanhuan.

2.1.1.4 Analysis of Sedimentary Environment

The lithology of Xiuning Formation in the Area is mainly characterized by tuffaceous fine sandstone, siltstone, silty mudstone, mudstone, and sedimentary tuff. The sandstone in the lower part of Xiuning Formation experienced hornfelsic alteration, due to the thermal metamorphism caused by Mesozoic magma, etc. Horizontal and wavy bedding developed in the feldspar–quartz sandstone of the lower part of basic sequences of type A, indicating that water movement was strengthened by waves. Besides, the sands constitute the major rocks of the feldspar–quartz sandstone, reflecting the sedimentation of littoral neritic facies in the sand flats near and below the low water line under high-energy environment. Extremely thin and unstable argillaceous sediments tend to be deposited above the sandstone. They are low-energy sediments during slack water period, i.e., double-clay layer. The argillaceous layer with horizontal bedding developing in the upper part belongs to mudflat sedimentation of the supratidal zone during slack water period of high water. The thin silt layer was brought about by the extremely large tides. Therefore, the basic sequence of type B is of tidal flat sedimentation. Horizontal bedding is visible in the basic sequence, indicating that aggradation occurred. Besides, the sediments feature gradually smaller grain sizes and become silt, indicating the possible accumulation in low-lying areas on continental shelf.

2.1.2 Nantuo Formation (Nh_2n)

The name Nantuo Formation, formerly known as Nantuo Tillite, was created by Blackwelder et al. (1907) in Liantuo Town, Yichang City, Hubei Province, and used to mean the sandy and argillaceous tillite layer lying beneath a large set of limestone. It was introduced by Li and Zhao (1924). Yu (1996) introduced it to the northwestern part of Zhejiang Province to replace Leigongwu Formations. In this Project, the name Nantuo Formation (Nh_2n) stated in *Lithostratigraphy of Zhejiang Province* was still adopted according to the characteristics of lithologic association of the profile (PM044) of the Yaocun forest farm of the Hanggai map sheet in the Area and the geological observation traverse.

2.1.2.1 Lithostratigraphy

Nantuo Formation is associated with Xiuning Formation. The outcrop area of Nantuo Formation is about 12.68 km^2, accounting for 1% of the bedrock area.

Lithologic association of Nantuo Formation is characterized by the following three parts. The lower part of Nantuo Formation consists of caesious–celadon siltstone-bearing moraine conglomerate and silty mudstone. In this part, the grain size is 1–25 cm and the components mainly include granite and dolomite. Besides, there is a small amount of silicalite. The thickness of this part is 46.70 m. The middle part of Nantuo Formation is composed of gray thin–medium laminated siltstone and fine sandstone. In this part, horizontal bedding developed and the thickness of this part is 8.0 m. The upper part of Nantuo Formation consists of gray–celadon blocky siltstone-bearing moraine conglomerate clay. In this part, the grain size of the conglomerate is 0.5–15 cm or even 20 cm, and major and minor components of the conglomerate are dolomite and granite, respectively. The thickness of this part is 65.70 m.

1. **Stratigraphic section**

The lithology of Nantuo Formation is described by taking the example of Nanhuanian Xiuning Formation (Nh_1x)–Nantuo Formation (Nh_2n) profile (rough survey) (Fig. 2.2) of the Hanggai map sheet in the Yaocun forest farm, Anji County, Zhejiang Province. The details are as follows.

Overlying stratum: gray dolomitic limestone of the first member of Lantian Formation

-----------------------------Parallel unconformable contact-------------------------

Nantuo Formation Total thickness: 120.4 m

17. Celadon blocky siltstone bearing moraine conglomerate; the content of the conglomerate in the siltstone: 3%; the conglomerate: angular, grain size: 0.5–15 cm or even 20 cm, components: dolomite (major) and granite (minor); the dolomite: hollow due to weathering, with argillaceous cementation. 65.7 m

16. Gray thin laminated argillaceous siltstone and fine sandstone, horizontal bedding developing. 8.0 m

15. Celadon siltstone bearing moraine conglomerate; the content of the conglomerate in the siltstone: 5%–10%; the conglomerate: angular–subangular, grain size: 1–5 cm, components: granite and dolomite (major) and silicalite (minor). 6.7 m

-----------------------------Parallel unconformable contact-------------------------

Underlying stratum: celadon thin–medium laminated siltstone interbedded with silty mudstone of Xiuning Formation.

2. **Lithological characteristics**

The lithology of Nantuo Formation (Nh_2n) is mainly characterized by siltstone-bearing moraine conglomerate as well as silt and fine sandstone.

The siltstone-bearing moraine conglomerate: argillaceous cementation; components: sands (major) and moraine-bearing conglomerate (major); the sands: accounting for up to about 60–65%; grain size: silt and fine sand (major) and medium sands (a small amount); the components: quartz and feldspar; the argillaceous matter in the siltstone: concentrating locally, may be moraine-bearing conglomerate.

The silt and fine sandstone: subangular and subrounded; grain size: 0.01–0.25 mm; components: feldspar, quartz, and detritus; the detritus: rounded; grain size: less than 1 mm, contact and porous cementation available for argillaceous matter.

3. **Basic Sequences**

There are two types of basic sequences in Nantuo Formation according to the lithologic association of the profile (Fig. 2.4).

Basic sequences of type A: gray (silty) moraine-bearing sandstone; no sedimentary structures developing in the rock layers; the gravel: accounting for about 5% in the sandstone, angular–subangular; component: silicalite; grain size: 1–15 cm or even 25 cm. Therefore, the gravel features poor rounding, seriously different sizes, and random distribution. Furthermore, it can be observed from part outcrops that the gravel cuts along the bedding. Therefore, it can be concluded that the gravel came from falling from air but not movement and sedimentation with the detritus. Thus, very possibly, the gravel was carried to basin by ice rafting and then fell into fine-silty sediment after ice melted.

Basic sequences of type B: gray siltstone and fine sandstone; composition: silty-fine detritus; micro-fine horizontal bedding developing; no gravel visible; belonging to shallow shelf sedimentation of interglacial period.

2.1.2.2 Sequence Stratigraphy

According to the characteristics of the field survey of geological observation traverse as well as the lithology, lithofacies association, and sequence boundary of related profile (rough survey), there is one second-order sequence SS2 in Nantuo Formation in the Area. The internal structure of SS2 is described as follows:

Second-order sequence SS2

SS2 contains TST, starved section (CS), and HST. The bottom of SS2 is a parallel unconformity interface between the siltstone-bearing moraine conglomerate and the underlying silty mudstone of Xiuning Formation. The top of SS2 is a parallel unconformity interface between the conglomerate-bearing silty mudstone of Nantuo Formation and the overlying manganese-bearing dolomitic limestone of the first member of Lantian Formation. Both the top and the bottom are of type-I sequence boundary. TST is located in the lower part of the Nantuo Formation. The sediments of TST mainly consist of angular and subangular siltstone-bearing moraine conglomerate, and the main components are granite and dolomite. CS is located in the middle

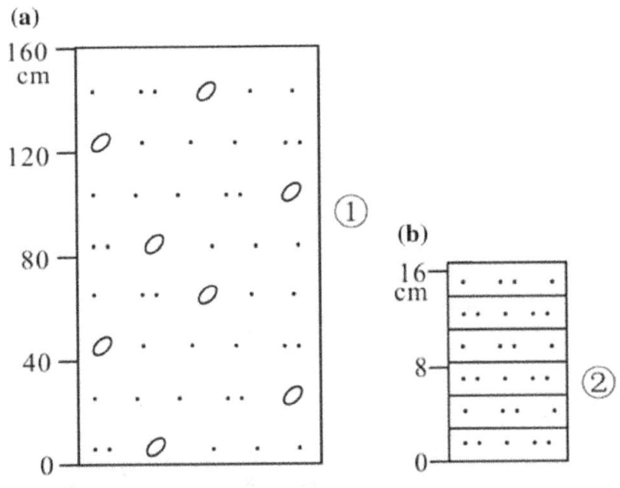

Fig. 2.4 Basic sequence of the Nantuo Formation (Nh_2n)

part of Nantuo Formation. After the early glacial period of the Nantuo Formation, the climate gradually became warm and the continental shelf sedimentary environment started in the Area. After glaciers melted, the water became deeper. As a result, the sediments decreased sharply, leaving thin sediments. HST is located in the upper part of Nantuo Formation. The siltstone-bearing moraine conglomerate constitutes the main rocks of HST. Compared with the bottom of SS2, HST features less and smaller conglomerate, indicating weaker glaciation. The sedimentary environment of this sequence successively experienced glacial period, interglacial period, and glacial period again. The age range of this sequence is 725–635 Ma, with a time span of 90 Ma. Therefore, the sedimentation time is relatively long.

2.1.2.3 Biostratigraphy and Chronostratigraphy

There are no fossils obtained in Nantuo Formation in the Area in this Project. According to previous data, there are 13 genera and 14 species of micro-plant fossils of the glacial period of Nanhuan Formation (i.e., the Early Nanhuan) obtained from Xiayabu of Jiande (Zhejiang Bureau of Geology 1965). These microfossil plants mainly include the following recurring single-chlorella communities such as *Trachysphaeridium minor*, *Trachyminuscula* sp., *Lophospaeridium* cf. *acielatum*, *Lophominuscula prima*, *Leiopsophosphaera densa*, *Leiominuscula minuta*, *Granomarginata prima*, *Asperatopsophosphaera* sp., *Nucellosphaeridum* sp., *Paleomorpha punctulata*, and *Bavlinella faveolata*. Besides, there are new spinous alga communities such as *Micrhystridium minimum*, *M. obtusum*, *M. odontodum*, and *Cymatiosphaera* sp. During the interglacial period of Nantuo Formation (i.e., the middle period of the Late Nanhuan), paleobiocenosis changed slightly and the microfossil plants obtained in the Xiayabu, Jiande, include *Leiominusculaminuta*, *Trachyminuscula* sp., *Lophominuscula prima*, *Leiopsophosphaera densa*, *Bavlinella faveolata*, *Polyporata obsoleta*, and *Paleamorpha puctulata*. In the second glacial period of Nantuo Formation (i.e., the late period of the Late Nanhuan), the microfossil plants obtained in Xiayabu, Jiande, include *Lophominuscula prima*, *Leiomarqinata* sp., *Margominuscuis* sp., *Granomarginata* cf. *prima*, *Leiopsophosphaera sp.*, *L. densa*, *Trachysphaeridium* sp., and *Polyporata obsolete*. In the Late Nanhuan, most microfossil plants were still single-chlorella communities, but some microfossil plants of new genera and species appeared including *Nucellsphaeridium*, *Granomarginata*, and *Bavliella*. Besides, spinous chlorella communities also appeared in this period such as *Gymatiosphaera* and *Micrhystridium*. Some genera and species such as *Bavlinella* are significant in comparison between regions and continents. In general, from the Early Nanhuan to the Late Nanhuan, the appearance of micro-plant association experienced slight changes which mainly include more species and more composite types.

Owing to the absence of Nanhuan fossils, stratigraphic division and comparison are mainly based on regional comparable geological events and absolute ages of related geological bodies. This requires that the division of the Nanhuan in the Area should be discussed within the northwest Zhejiang or even wider range. From bottom to top, the Nanhuan in the Area is divided into Xiuning Formation and Nantuo Formation. It is a set of highly maturated littoral neritic clastic rock–glacial rock system with small thickness (relative to the long duration of nearly 120 Ma). The weak differentiation of the lithofacies itself of the rock system in the Area indicate that the superficial part of the continental crust was at a loose and (sub-) stable state. The transgressional event in the Early Nanhuan is comparable in the whole of South China. According to previous studies, the ages obtained by zircon U-Pb isotopic dating in Xiuning Formation in Xiayabu, Jiande, the southwest Zhejiang, were 780 ± 10 Ma–776 ± 12 Ma (Yin et al. 2007), 759 ± 6 Ma (Gao et al. 2014), and 735 ± 14 Ma (Wang et al. 2015), respectively. That is, the transgressional event in the Early Nanhuan was the extensive transgression event around 750 Ma indicated by the bottom of Xiuning Formation. The glacial event in the Late Nanhuan (around 680 Ma) is prominent in the history of the earth, and it involves two glacial periods and one interglacial period. The moraine-bearing sediment in this period is comparable globally. According to the study on west Hunan Province–southwest Hubei Province area conducted by Yang et al. (1997), the Sm–Nd isochron age of the manganese carbonate deposit in the interglacial period of Nantuo Formation is 696 ± 52 Ma.

2.1.2.4 Analysis of Sedimentary Environment

In conclusion, Nantuo Formation of the Late Nantuo is of moraine formation. In the early and late periods of the Late Nantuo, the Area experienced two glacial events and thus

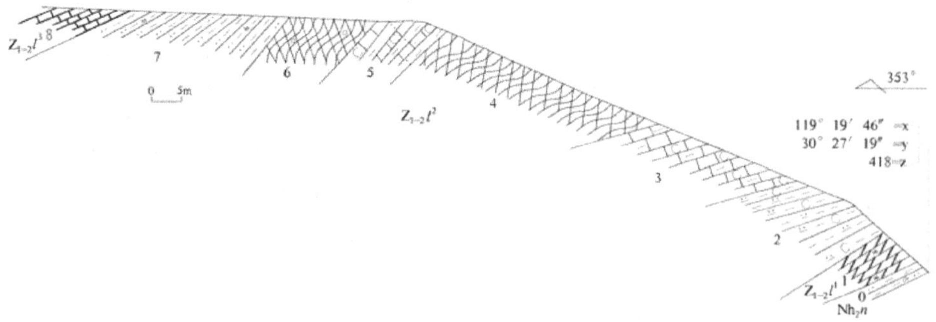

Fig. 2.5 First member ($Z_{1\text{-}2}l^1$)–second member ($Z_{1\text{-}2}l^2$) profile of Lantian Formation in Yeshanwu, Xianxia Town, Ningguo City, Anhui Province

can be divided into two glacial periods and one interglacial period. As a result, two different types of sedimentary rocks were deposited. The sedimentary rocks of the early and late periods were a set of caesious thick laminated and blocky formation of mudstone-bearing moraine conglomerate. With complex components as well as poor sorting and rounding, the conglomerate in the formation is the product of composite sedimentation formed when the moraine fell into the ocean due to the melting of the ice rafting or glaciers. This reflects a sedimentary environment of nearshore neritic transitional zone with weak hydrodynamic force. In the middle period of the Late Nantuo, the grayish silty-fine sandstone formation was deposited, indicating that the climate got warm, the glaciers melted, and the neritic deep shelf sedimentary environment had started.

2.2 Sinian System

The Sinian in the Area is mainly exposed in the west and distributed in the surrounding areas of Tangshe complex as well as the northern part of Xianxia complex including Zhangli, Shimen, Baishawu, Pingtoushan, and Kongfuguan–Taoshuwan–Waijianzi area. The outcrop area is about 29.49 km^2, accounting for 2% of the bedrock area. The middle and lower parts of the Sinian are classified under Lantian Formation ($Z_{1\text{-}2}l$), which is further divided into four members according to the lithologic association. The upper part of the Sinian is classified as Piyuancun Formation (Z_2p).

2.2.1 Lantian Formation ($Z_{1\text{-}2}l$)

The name Lantian Formation was created by Ding (1935) in Lantian Village, Xiuning County, Anhui Province, to mean a set of clastic and carbonate rock association below Shangchangyuan limestone (silicalite in Piyuan Village), which is above moraine layer. The Sinian in Zhejiang Province was highly researched in the 1950s. Xinfu Sheng established the Xifengsi Siliceous Shale in 1959, and Liu Hongyun and Sha Qing'an plotted the profile of the shale and named it Xifengsi Series in the same year. In 1965, Xifengsi Series was renamed Xifengsi Formation by Zhejiang Regional Geological Survey Team. In 1982, the Precambrian System Special Group of Zhejiang Regional Geological Survey Team introduced Dengying Formation and Doushantuo Formation, which were widely used in South China. In 1996, the Sinian in Jiangshan–Lin'an stratigraphic subregion was divided into Lantian Formation and Piyuancun Formation in *Lithostratigraphy of Zhejiang Province.* In this Project, the division scheme of Lantian Formation as described in *Lithostratigraphy of Zhejiang Province* was still adopted according to the characteristics of lithologic association of Nanchewu profile (PM010) of Hanggai map sheet, Dalingtou profile (PM037), Ligu'an profile (PM039), and Yeshanwu profile (PM031) in Xianxia map sheet, and the geological observation traverse within the Area.

2.2.1.1 Lithostratigraphy

The Lantian Formation in the Area mainly outcrops in the surrounding areas of Tangshe complex in the southwest corner of Hanggai map sheet and in the northwest part of Xianxia map sheet including Zhangli Village, Baishawu, Ligu'an, and Damiao–Zhaojia area discontinuously. It is distributed in "U" shape with an SW-trending opening in general. The outcrop area is about 29.49 km^2, accounting for 1.69% of the bedrock area.

The first Member of Lantian Formation ($Z_{1\text{-}2}l^1$): grayish black medium laminated argillaceous dolomitic limestone-bearing manganese and carbon. The rocks are heavily weathered and become brown wadite. The thickness of the member is 9.02 m.

The second member Lantian Formation ($Z_{1\text{-}2}l^2$): gray and dark gray thin–medium laminated argillaceous siltstone and silty mudstone, interbedded with thin–medium laminated and lenstoid carbonaceous limestone in the middle and lower parts. The thickness of the carbonaceous limestone layer varies greatly. In the silty mudstone, a small amount of fine-grained pyrite and horizontal bedding developed. The thickness of this member is 84.34–104.51 m.

The third member of Lantian Formation ($Z_{1-2}l^3$): The lower part is the interbed of grayish–off-white thin–medium laminated dolomite and gray argillaceous limestone, locally interbedded with medium laminated micrite. The medium part is gray and dark gray medium–thick laminated carbonaceous siliceous limestone and argillaceous limestone interbedded with black thin–medium laminated carbonaceous argillaceous silicalite. The upper part is gray–dark gray medium–thick laminated limy dolomite and dolomitic limestone interbedded with a small amount of dark gray thin laminated siliceous mudstone, and sandy dolomite can be observed occasionally. The thickness of the member is 103.94–234.64 m.

The fourth member of Lantian Formation ($Z_{1-2}l^4$): gray–dark gray thin laminated siliceous mudstone and carbonaceous siliceous mudstone, generally 2–10 cm and occasionally 20–30 cm thick per layer. Horizontal bedding develops. The thickness of this member is 14.34 m.

1. Stratigraphic section

The characters of the lithology and lithologic association of the first member and the second member of Lantian Formation are described by taking the example of the first member ($Z_{1-2}l^1$)–the second member ($Z_{1-2}l^2$) profile (Fig. 2.5) of Lantian Formation in Yeshanwu, Xianxia Town, Ningguo City, Anhui Province. The details are as follows:

The third member of Lantian Formation ($Z_{1-2}l^3$) Total thickness: >7.45 m

8. Off-white thin–medium laminated marbleized dolomitic limestone, very thin horizontal bedding developing, the thickness of single rock layer: 5–20 cm. 7.45 m

——— Conformable contact ———

The second member of Lantian Formation ($Z_{1-2}l^2$) Total thickness: 84.34 m

7. Dark gray hornfelsic silty mudstone containing fine-grained disseminated pyrite (1%–3%). They become speckled hornfels due to thermal alteration. 21.79 m

6. Off-white thin–medium laminated silty mudstone, containing fine-grained disseminated pyrite (5%–8%). They become speckled hornfels due to thermal alteration. 9.00 m

5. Interbed consisting of dark gray medium laminated siliceous mudstone and off-white carbonaceous limestone. The siliceous limestone contains a small amount of disseminated pyrit 8.27 m

4. Dark gray–grayish-black thin–medium laminated carbonaceous siliceous mudstone. They become speckled hornfels due to thermal alteration. 18.57 m

3. Grayish–off-white thin–medium laminated argillaceous limestone, containing sparse disseminated pyrite. Weakly tremolitized alteration occurs due to thermal action. The thickness of a single rock layer: 5–30 cm. 11.27 m

2. Dark gray–grayish black thin–medium laminated carbonaceous siliceous mudstone, containing a small amount of sparse fine-grained pyrite. 15.44 m

——— Conformable contact ———

The first member of Lantian Formation ($Z_{1-2}l^1$) 9.02 m

1. Grayish black medium laminated manganese-bearing dolomitic limestone, mainly containing calcite and some dolomite. A small amount of white net-veined quartz stringer of later period developed. A small amount of brown earthy weathered materials, i.e. wadite, are visible on the ground surface. 9.02 m

- - - - Parallel unconformable contact - - - -

Nantuo Formation of Later Nanhuan Epoch (Nh_2n) Total thickness: > 1.02 m

0. Gray, blocky-laminated conglomerate-bearing silty mudstone, mainly consisting of mud (70%), some silt (20%–25%) and gravel (5%–10%). The size of the conglomerates varies greatly, with grain size of 2–150 mm. The conglomerates are angular and sub-rounded and are unevenly distributed. 1.02 m

The characters of the lithology and lithologic association of the third member and the fourth member of Lantian Formation are described by taking the example of the third member ($Z_{1\text{-}2}l^3$)–the fourth member ($Z_{1\text{-}2}l^4$) profile (Fig. 2.6) of Lantian Formation of Hanggai map sheet in Nanchewu, Hanggai Town, Anji County, Zhejiang Province. The details are as follows:

Piyuancun Formation (Z_2p) Total thickness: > 1.28 m

26. Dark gray and grayish-black thick laminated–blocky argillaceous silicalite. 1.28 m

——————— Conformable contact ———————

The fourth member of Lantian Formation ($Z_{1\text{-}2}l^4$) Total thickness: 56.21 m

25. Gray thin laminated carbonaceous siliceous mudstone, locally interbedded with medium laminated silicalite. 1.78 m

24. Grayish black medium–thick laminated carbonaceous argillaceous silty silicalite. 7.81 m

23. Gray thin–medium-thickness silty silicalite. 19.54 m

22. Dark gray and grayish-black thin laminated carbonaceous silicalite. 15.76 m

21. Grayish black medium–thin laminated carbonaceous siliceous mudstone, interbedded with a layer of argillaceous limestone with the thickness of about 0.5m thick in the lower part. 11.32 m

——————— Conformable contact ———————

The third member of Lantian Formation ($Z_{1\text{-}2}l^3$) Total thickness: 234.64 m

20. Dark gray medium laminated limy dolomite. The thickness of a single layer is generally 10–30 cm. 17.94 m

19. Interbed consisting of gray medium laminated dolomitic limestone and thin laminated marlstone. 7.97 m

18. Dark gray medium laminated micrite dolomite. 13.89 m

17. Dark gray thick laminated blocky micrite dolomite interbedded with micro-thin laminated sandy dolomite. 9.77 m

16. Dark gray medium–thick laminated very thin grained dolomite. 8.42 m

15. Gray micro-thin laminated siliceous mudstone. 4.52 m

14. Dark gray medium–thick laminated argillaceous siliceous limestone interbedded with medium laminated limy dolomite. Horizontal bedding developed in all rock layers. 5.56 m

13. Dark gray thick laminated–blocky dolomite. 13.01 m

12. Interbeds constituting of gray thin laminated limy siliceous mudstone and micro-thin laminated dolomitic limestone. 1.97 m

11. Dark gray medium laminated carbonaceous marlstone. The thickness of a single rock layer is generally 20–40 cm. 2.10 m

10. Dark gray and grayish-black medium–thin laminated carbonaceous silicalite, interbedded with limestone lens in the middle part. The thickness of a single layer of carbonaceous silicalite is generally 5–30 cm. Micro horizontal bedding developed locally. 4.19 m

9. Gray and dark gray thick laminated–blocky carbonaceous sandy marlstone. 1.11 m

8. Grayish black medium–thin laminated carbonaceous silicalite, interbedded with a small amount of carbonaceous limestone lens. The thickness of a single rock layer is 5–30 cm. 16.85 m

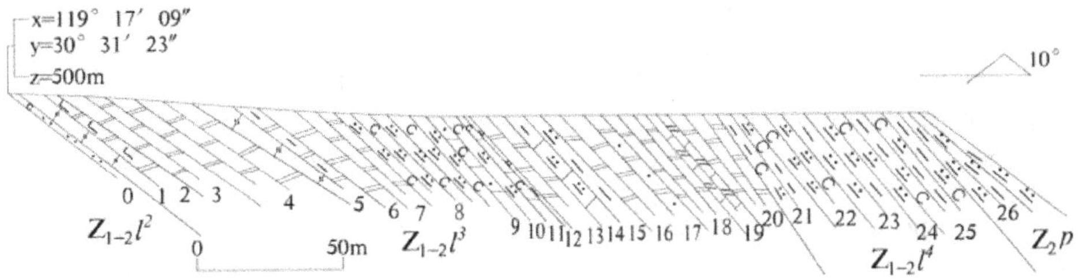

Fig. 2.6 Third member ($Z_{1\text{-}2}l^3$)–fourth member ($Z_{1\text{-}2}l^4$) profile of Lantian Formation in Nanchewu, Hanggai Town, Anji County, Zhejiang Province

7. Dark gray thick laminated silty–fine-crystalline dolomite. The thickness of a single rock layer is generally 50–100 cm. The middle part is medium laminated rocks. 11.73 m

6. Dark gray medium laminated fine-crystalline dolomite. The thickness of each layer is generally 15–30cm. 3.82 m

5. Interbed consisting of grayish medium–thin laminated micrite and micro-thin laminated argillaceous dolomite. 1.47 m

4. Dark gray and grayish-black thick laminated–blocky fine-crystalline dolomite. 44.98 m

3. Off-white thin laminated fine-crystalline dolomite. The thickness of each layer is generally 1–5 cm. 20.36 m

2. Interbed consisting of gray and off-white thin–medium laminated limy dolomite and micro-thin laminated argillaceous limestone. 5.91 m

1. Light off-white thin laminated micritic limestone. The thickness of each layer is generally 2–6 cm. 9.07 m

———————— Conformable contact ————————

Underlying strata: dark gray carbonaceous siltstone and silty mudstone of the second member of Lantian Formation.

2. **Lithological Characteristics**

The lithology of the first member of Lantian Formation ($Z_{1\text{-}2}l^1$) is mainly characterized by argillaceous dolomitic limestone-bearing manganese and carbon. The major component is cryptocrystalline calcite, accounting for 65–70%, and the grain size of it is about 0.01 mm. The argillaceous matter accounts for 20–25%. Amorphous carbonaceous matter accounts for 10% and is evenly distributed.

The main lithology of the second member of Lantian Formation ($Z_{1\text{-}2}l^2$) is characterized by carbonaceous siliceous mudstone, silty mudstone, and carbonaceous argillaceous limestone.

(1) Carbonaceous siliceous mudstone: composition: argillaceous matter (70%), cryptocrystalline siliceous matter (25%), and amorphous carbonaceous (5–10%). The argillaceous matter: fine scaly, unevenly distributed and concentrating locally; the cryptocrystalline siliceous matter: evenly distributed; amorphous carbonaceous: unevenly distributed and slightly concentrating locally.
(2) Silty mudstone: composition: argillaceous matter (75%), feldspar and quartz silt (20%), and carbonaceous matter (5%). The argillaceous matter becomes scaly mica partly due to thermal induced metamorphic recrystallization; the feldspar and quartz silt: grain size: 0.02–0.04 mm; the carbonaceous matter: unevenly distributed and concentrating locally.
(3) Carbon-bearing argillaceous limestone: composition: calcite (70%), argillaceous matter (20%), and amorphous carbonaceous matter (5–7%). The calcite:

particle size: 0.01–0.05 mm; the argillaceous matter: unevenly distributed and slightly concentrating locally; the amorphous carbonaceous matter: unevenly distributed and slightly concentrating locally; quartz sands: sparsely visible; particle size: 0.1–0.2 mm.

The lithology of the third member of Lantian Formation ($Z_{1-2}l^3$) is characterized by various types of rocks, mainly including dolomite, micrite, carbonaceous siliceous limestone, argillaceous limestone, argillaceous carbonaceous silicalite, limy dolomite, siliceous mudstone, and sandy dolomite.

(1) The dolomite: gray–dark gray; composition: dolomite (90–95%) and a small amount of calcite. The dolomite: granular or idiomorphic crystal generally; particle size: 0.1–0.2 mm; the calcite: cryptocrystalline generally and visible locally.
(2) The micrite: gray; composition: calcite (90%), argillaceous matter (< 10%), and occasionally visible silt. The calcite: particle size: < 0.01 mm, cryptocrystalline generally; the argillaceous matter: earthy.
(3) The carbonaceous siliceous limestone: gray–dark gray, composed of calcite (60–65%), siliceous matter (25%), and argillaceous matter (10–15%); grain size of the calcite: < 0.01 mm; the siliceous matter: cryptocrystalline, evenly distributed; the argillaceous matter: earthy.
(4) The argillaceous limestone: composition: calcite (60–65%), argillaceous matter (20–25%), and carbonaceous matter (5–10%). The calcite: particle size: < 0.01 mm; the argillaceous matter: earthy generally; the carbonaceous matter: black amorphous, concentrating locally.
(5) The argillaceous carbonaceous silicalite: black; composition: siliceous matter (45–50%), carbonaceous matter (25–30%), and argillaceous matter (25%). The siliceous matter: cryptocrystalline generally; the carbonaceous matter: black amorphous, evenly distributed; the argillaceous matter: horizontal bedding developing.
(6) The limy dolomite: gray–dark gray, composition: dolomite (75–80%) and calcite (20–25%). The dolomite: granular; particle size: 0.1–0.2 mm; the calcite: fine granular generally, unevenly distributed.
(7) The siliceous mudstone: dark gray; composition: argillaceous matter (60–75%), siliceous matter (25–30%), and silt (5–10%). The argillaceous matter: earthy generally; the siliceous matter: cryptocrystalline generally, evenly distributed; the silt: particle size: 0.01–0.03 mm, composed of quartz, etc.
(8) The sandy dolomite: composition: dolomite (85–90%) and sand grains (10–15%). The dolomite: particle size: 0.05–0.1 generally mm or > 0.1 mm occasionally, fine granular generally; the sand grains: composed of quartz, sub-well rounded; particle size: 0.2–0.25 mm, slightly unevenly distributed.

The main lithology of the fourth member of Lantian Formation Member ($Z_{1-2}l^4$) is characterized by siliceous mudstone and carbonaceous siliceous mudstone.

(1) The siliceous mudstone: composition: siliceous matter (60–70%) and argillaceous (35–40%). The siliceous matter: cryptocrystalline generally; the argillaceous matter: earthy.
(2) The carbonaceous siliceous mudstone: composition: argillaceous matter (50–65%), siliceous matter (25–30%), and carbonaceous matter (10%). The argillaceous matter: earthy generally; the siliceous matter: cryptocrystalline generally; the carbonaceous matter: black amorphous, unevenly distributed and concentrating locally.

3. **Basic Sequences**

There are eight types of basic sequences developing in the Lantian Formation, including one in the first member, three in the second member, three in the third member, and one in the fourth member (Fig. 2.7).

Basic sequences of type A: located in the first member of Lantian Formation ($Z_{1-2}l^1$), composed of single grayish black medium laminated argillaceous dolomitic limestone-bearing manganese and carbon, horizontal bedding developing. The basic sequences of this type belong to the sedimentation of platform facies and are of monotonous non-cyclic basic sequence.

Basic sequences of type B: located in the lower part of the second member of Lantian Formation ($Z_{1-2}l^2$), composed of gray–dark gray thin–medium laminated carbonaceous siliceous mudstone, containing a small amount of synsedimentary fine-grained pyrite, horizontal bedding developing. The basic sequences of this type belong to the sedimentation of bathyal basin facies and are of monotonous non-cyclic basic sequence.

Basic sequences of type C: located in the middle part of the second member of Lantian Formation ($Z_{1-2}l^2$), composed of ① grayish thin–medium laminated carbonaceous limestone and ② gray–dark gray thin–medium laminated carbonaceous siliceous mudstone. The carbonaceous limestone: 8–30 cm thick per layer, lensoid locally. The carbonaceous siliceous mudstone: 5–30 cm thick per layer, containing a small amount of synsedimentary fine-grained pyrite which concentrates along the bedding generally, horizontal bedding developing in the rock layers. The ratio between the two components is nearly 1:1. The carbonaceous siliceous mudstone is of deep shelf–bathyal basin facies (major) and

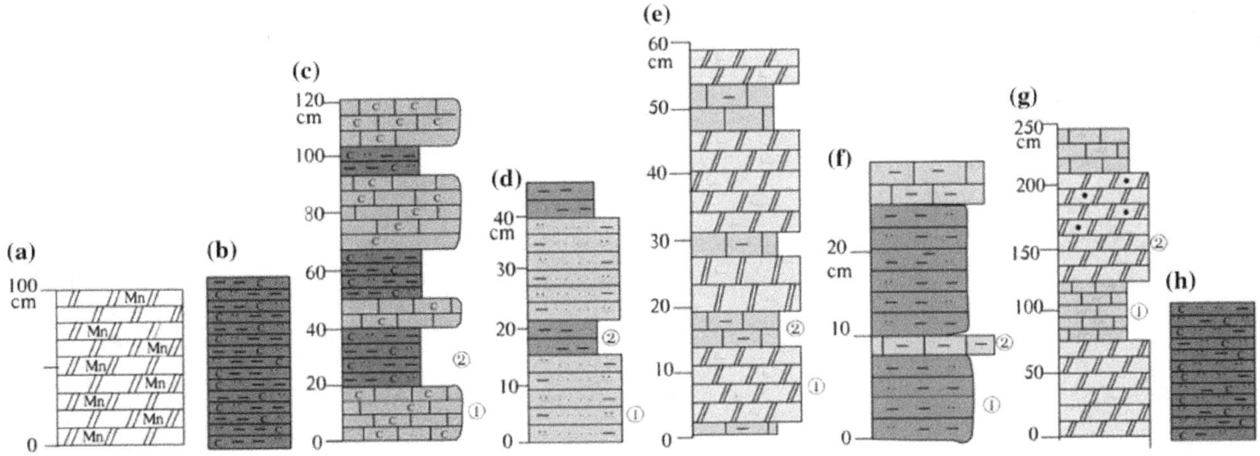

Fig. 2.7 Basic sequences of Lantian Formation ($Z_{1\text{-}2}l$)

the carbonaceous limestone is of shallow shelf facies (minor). The basic sequences of this type belong to non-cyclic basic sequence.

Basic sequences of type D: located in the upper part of the second member of Lantian Formation ($Z_{1\text{-}2}l^2$), composed of ① gray–dark thin–medium laminated silicon-bearing argillaceous siltstone and ② gray thin laminated silty mudstone. For the two components, the mineral composition is similar while the components vary greatly. Horizontal bedding developed in all the rock layers. Thus, the basic sequences of this type belong to shallow shelf–bathyal basin facies and are of non-cyclic basic sequence.

Lithologically, the second member of Lantian Formation gradually becomes carbonates and terrigenous silt from bottom to top, indicating that the sea level gradually rose.

Basic sequences of type E: located in the lower part of the third member of Lantian Formation ($Z_{1\text{-}2}l^3$); composed of ① grayish thin–medium laminated limy dolomite with no bedding developing and ② gray micro-thin laminated argillaceous limestone with horizontal bedding developing; belonging to shallow shelf–platform facies.

Basic sequences of type F: located in the middle part of the third member of Lantian Formation ($Z_{1\text{-}2}l^3$), composed of ① dark gray thin–medium laminated argillaceous silicalite and ② gray micro-thin laminated argillaceous limestone; micro-texture horizontal bedding developing; belonging to bathyal facies.

Basic sequences of type G: located in the upper part of the third member of Lantian Formation ($Z_{1\text{-}2}l^3$), composed of rhythm interbeds of ① gray–dark gray medium–thick laminated dolomitic limestone and ② gray–dark gray medium–thick laminated dolomite, with sandy dolomite visible occasionally; thick laminated structures visible in most dolomite; horizontal bedding developing in all rock layers; belonging to platform facies.

Lithologically, the third member of Lantian Formation has evolved from carbonate to carbonate-bearing argillaceous and siliceous matter, then to sand-bearing carbonate from bottom to top, reflecting that the sea level vibrated between rise and falling with rise as general trend.

Basic sequences of type H: located in the fourth member of Lantian Formation ($Z_{1\text{-}2}l^4$), composed of rhythmite of dark gray thin laminated and thin–medium laminated carbonaceous siliceous mudstone or dark gray thin laminated argillaceous silicalite, micro-texture horizontal bedding developing in rock layers. Thus, the basic sequences of this type belong to deep shelf–bathyal basin facies and are of monotonous non-cyclic basic sequence.

2.2.1.2 Sequence Stratigraphy

According to the characteristics of the lithology, lithofacies association, and sequence boundary of the profiles of Yeshangwu and Dalingtou of Xianxia map sheet, as well as Nanchewu profile of Hanggai map sheet, there is one second-order sequence SS3 in the first member–the second member of Lantian Formation, there is one second-order sequence SS4 in the third member of Lantian Formation, and there is one second-order sequence SS5 in the fourth member of Lantian Formation–Cambrian Dachenling Formation. Among these sequences, SS4 is further divided into four third-order sequences, i.e., Sq1–Sq4 (Fig. 2.8). The characteristics and internal composition of all these sequences from bottom to up are briefly described as follows:

Second-order sequence SS3

It is located in the first member–the second member of Lantian Formation. The bottom of it is a parallel unconformity interface between the argillaceous dolomitic

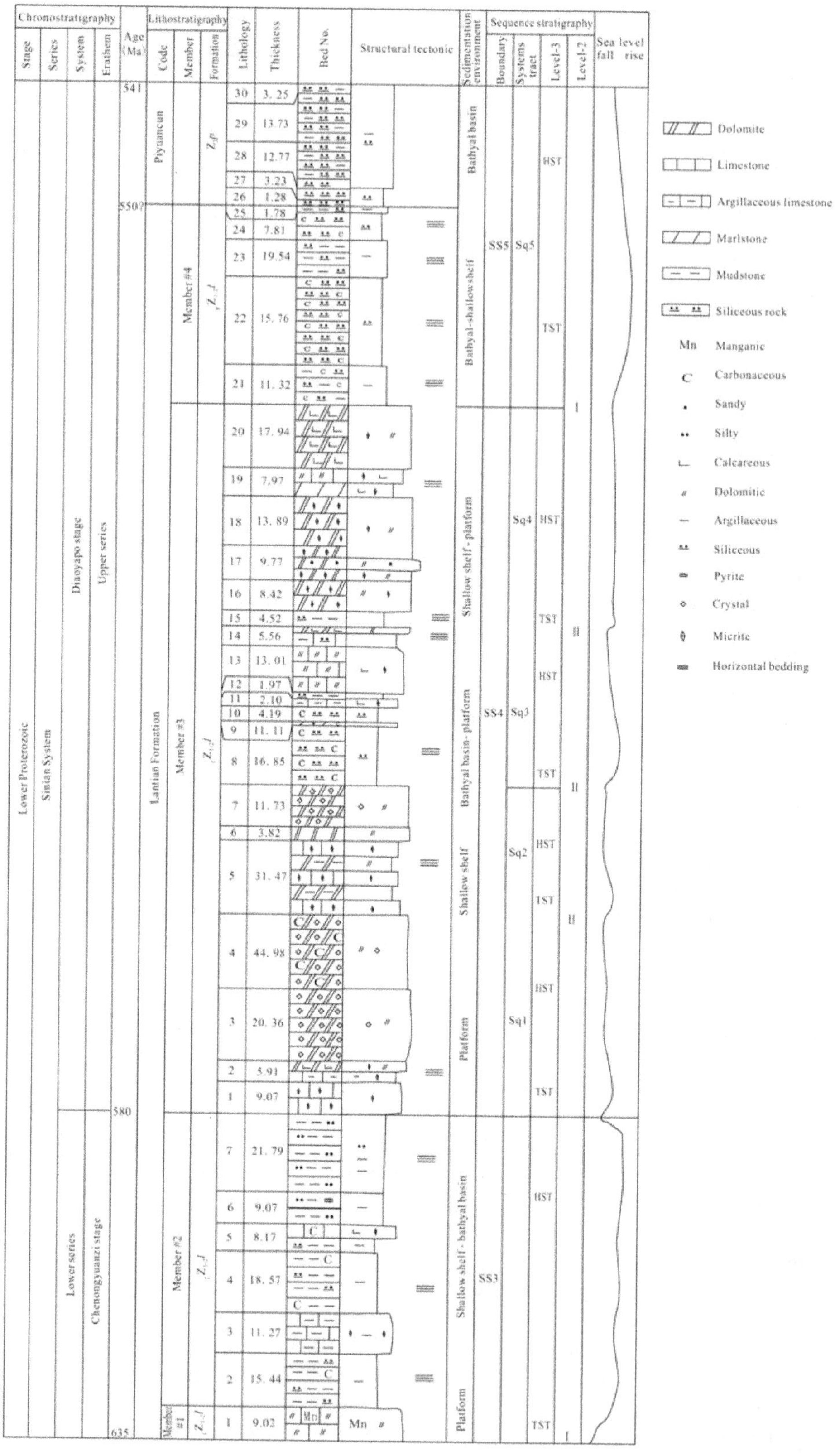

Fig. 2.8 Sequence stratigraphic framework of the Sinian in the area

limestone-bearing manganese and carbon in the first member of Lantian Formation and the silty mudstone-bearing moraine conglomerate in the underlying Nantuo Formation. It is of type-I sequence boundary. The top of SS3 is a conformity interface between the silty mudstone in the second member of Lantian Formation and the limestone in the overlying third member of Lantian Formation. It is of type-II sequence boundary. The sediments in the first member and the lower part of the second member mainly consist of calcium–magnesium carbonate rocks and argillaceous rocks bearing carbon and silicon, with horizontal bedding developing. Therefore, these sediments belong to shallow shelf facies–bathyal basin facies. The sediments in the upper part of the second member of Lantian Formation are silty mudstone and calcium–magnesium carbonate rocks, belonging to shallow shelf facies–platform facies and reflecting the process that sea level gradually fell after reaching the maximum value in the lower part of the second member of Lantian Formation. The lower part of SS3 is of retrogradation type, and the middle and upper parts of SS3 are of aggradation and progradation types, respectively. The age range of SS3 is 63–580 Ma, spanning 55 Ma.

Second-order sequence SS4

Third-order sequence Sq1: located in the lower and middle parts of the third member of Lantian Formation, including transgressive systems tract (TST) and highstand systems tract (HST). The bottom is of type-II sequence boundary. The top of it is the conformity interface between the dolomite in the lower part of the third member of Lantian Formation and the overlying carbonaceous siliceous shale; thus, the top is of type-II sequence boundary. Thin–medium laminated calcareous and dolomitic sediments are mainly distributed in TST, with horizontal bedding developing in rock layers, thus indicating shallow shelf facies. The thick blocky-laminated dolomitic sediments and a small amount of calcareous sediments are distributed in HST, indicating shallow shelf–platform facies. HST accounts for the overwhelming majority of Sq1. The siliceous-argillaceous sediments are interbedded and crossed with calcareous and dolomitic sediments in HST, reflecting the sea level rose in general. Thus, HST is of progradation–aggradation type.

Third-order sequence Sq2: located in the lower-middle part of the third member of Lantian Formation, including TST and HST. The top and bottom of it are both of type-II sequence boundaries. The carbonaceous siliceous sediments are mainly distributed in TST, with micro-texture horizontal bedding developing, thus indicating bathyal basin facies. Calcareous sediments (major), as well as argillaceous sediments and terrigenous quartz sands (minor), are distributed in HST, indicating shallow shelf–platform facies. Representing the regression process of seawater in general, Sq2 is a sequence of retrogradation type.

Third-order sequence Sq3: located in the upper-middle part of the third member of Lantian Formation, including TST and HST. The top and bottom are both of type-II sequence boundaries. Thin–medium laminated siliceous argillaceous sediments are mainly distributed in TST, thus indicating the bathyal basin facies. Medium–thick laminated calcareous and dolomitic sediments, as well as a small amount of thin–medium laminated siliceous argillaceous sediments, are distributed in HST, thus belonging to the bathyal basin–neritic platform facies. HST accounts for an overwhelming majority of Sq3. The siliceous-argillaceous sediments are interbedded and crossed with calcareous and dolomitic sediments in HST, reflecting that the sea level alternately rose and fell frequently and thus belonging to aggradation type.

Third-order sequence Sq4: located at the upper part of the third member of Lantian Formation, including TST and HST. The top and bottom of it are both of type-II sequence boundaries. Thin laminated siliceous argillaceous sediments are mainly distributed in TST besides a small amount of carbonaceous sediments, with micro-texture horizontal bedding developing in rock layers. Therefore, TST is of bathyal basin facies. Dominant medium–thick laminated dolomitic sediments, minor calcareous carbonate, and a small amount of terrigenous quartz sands are distributed in HST, indicating shallow shelf–platform facies. HST accounts for the overwhelming majority of Sq4, representing the process that sea level continually fell and thus belonging to progradation–aggradation type.

The age range of the sequences Sq1–Sq4 is 580–560 Ma, spanning 20 Ma. Each single third-order sequence spans 5 Ma on average.

2.2.1.3 Biostratigraphy and Chronostratigraphy

The renowned Lantian Biota is situated in Lantian area, Xiuning County, south Anhui. The fossils are preserved in the shale of the second member of Lantian Formation in the form of carbonaceous films. After the fossil biota was reported by Xing et al. (1989) for the first time, a number of articles on description and research of the macrofossils in the Area were published successively. Yan et al. (1992) described macro-algae fossils of 12 genera and 18 species and called this biotic association "Lantian Flora". Tang et al. (1997) mainly discussed a category of disk-shaped and ellipsoidal fossils in the fossil biota and considered them as the macrofossils evidence of sexual differentiation in metaphyte. In 1994–1999, Yuan et al. carried out systematic field collection and detailed research of the specimens of the biota and classified the 50 genera and species described previously under 12–15 species. Yuan et al. (2012) further researched the Lantian Flora that was discovered in 40-m-thick black shale of the second member of Lantian Formation and concluded that Lantian Flora was the earliest macrobiota with a highly rich species and complex biological structure so far. Therefore, they put forward the concept of Lantian Biota, which was mainly composed of multicellular macro-algae fossils and might contain some fossils related to metazoans.

Little survey and research have been conducted on paleontology of the Lantian Formation in Zhejiang. Previous survey and research on Sinian System in the northwest Zhejiang mainly focused on the areas near the Jiangshan Shaoxing fracture zone in the south, and much effort was paid to the profiles of Doushantuo Formation and Dengying Formation in Wujialing, Jiangshan City, the profiles of Doushantuo Formation and Dengying Formation in Shaojiashan, Zhuji City, the profile of Banqiaoshan Formation in Zhongjiazhuang, Fuyang City, and the profiles of Lantian Formation and Piyuancun Formation in Qiuyuan, Chun'an County. All of the profiles are contemporaneous heteropical sedimentation. In Doushantuo Formation and Dengying Formation of the Sinian in which rich palaeobios were obtained, the biota was mainly composed of microfossil plants, chitinozoans-like organism, and micro-vermes. The early Sinian biotic association was composed of microfossil plants. As for the biotic association of the Later Sinian, there were many stromatolites, chitinozoans-like animals, and micro-vermes besides massively flourishing microfossil plants. In addition to recurring microfossil plants, there were plenty of stromatolites, chitinozoans-like animals, and micro-vermes.

As a period when great changes happened in the biological world, the Sinian is an important stage of fauna origination. In the Dengying Formation of the Sinian, not only a lot of microfossil plants and stromatolites but also vermes and chitinozoans-like animals developed. The later species consist of the oldest marine fauna, which can be comparable to Ediacaran mollusk and trace fossil fauna. Some individuals of fauna are also found in similar strata in the USA, Sweden, UK, and southwest Africa. This fauna developed after the glacial period at the end of Proterozoic Era, and its age range is about 680–543 Ma. It provided an important biological basis for disintegration of the former Sinian in South China, which was then divided into then Nanhuan System and the Sinian System (Fig. 2.8).

2.2.1.4 Analysis of Sedimentary Environment

In the Early Sinian, with glacier melting and the sea level rising, the Lantian Formation in the Area experienced the following sedimentary environment. For the first member, about 8–10-cm-thick gray manganese-bearing dolomite was deposited, representing the carbonate sedimentation formation in the environment of an arid hot restricted marine platform facies. The second member of Lantian Formation mainly consists of dark gray–grayish black carbonaceous siliceous argillaceous sediments and a small amount of silty sediments, interbedded with a small amount of argillaceous carbonate rocks. Horizontal bedding developed in rock layers. Therefore, the second member represents a deep standing-water environment. Therefore, the second member belongs to the carbonate-bearing siliceous-argillaceous formation of shallow shelf outer margin–bathyal basin facies. For the third member of Lantian Formation, the sediments mainly consist of dark gray dolomite-bearing carbonate, interbedded with a small amount of carbonaceous siliceous sediments. The dolomite carbonate contains a large amount of terrigenous quartz sand grains locally. This indicates that this member was generally in an oxidization neritic environment and occasionally in a deep standing-water oxidation–reduction environment due to the sharp rise of the sea level. Therefore, the main part of the third member is a set of siliceous-argillaceous carbonate formation of shallow shelf–open platform facies. For the fourth member, dark gray–grayish black carbonaceous siliceous argillaceous sediments are dominant in the member and the silty sediments are available occasionally. Horizontal bedding developed in the rock layers. Therefore, the fourth member belongs to siliceous-argillaceous formation in the deep standing-water environment. As a result, Lantian Formation generally went through the change of the sedimentary environment, i.e., platform → bathyal basin → shallow shelf → open platform → bathyal basin successively.

As for the sedimentary environment of the black shale in Lantian Formation, very thin bedding structures developed in the shale where Lantian Biota fossils are preserved, and there are no signs of sedimentary structures and fossil transportation reflecting strong hydrodynamic environment. Thus, it is suggested that these macro-organisms should be buried in situ and they should live under the maximum wave

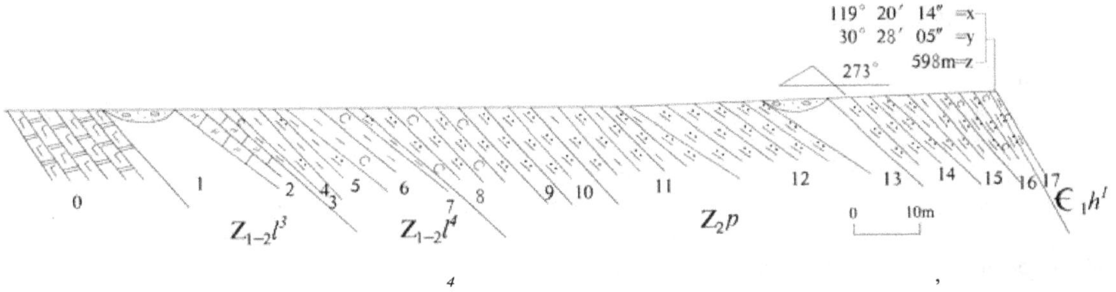

Fig. 2.9 Fourth member of Sinian Lantian Formation ($Z_{1-2}l^4$)–Piyuancun Formation (Z_2p) profile in Ligu an, Zhangcun Town, Anji County, Zhejiang Province

base and in the euphotic zone. According to the paleogeography back then and by reference to the standards on modern marine environment, the Lantian Biota should largely live in the margin of platform or local standing water of platform, with water depth ranging from 50 to 200 m. Preserved on shale surface in the form of carbonaceous film, the fossils are complete in morphology and sessile equipment is available for most species, suggesting that Lantian Biota is a complex benthic sessile macrobiota. This is very probably related to the intermittent anoxic event that happened universally in oceans during that period.

The Lantian Biota is a critical link in the evolvement of simple or micro-eukaryote into the organisms with complex physical conformation and diverse morphology, suggesting that the origination and early evolution of multicellular macro-organisms probably occurred in a deep standing-water environment. Regionally, the four lithological members of Lantian Formation vary slightly in lithologic association and stratum thickness, reflecting they were in the same stable sedimentary terrain within the Area. In Xianlin area, Yuhang, about 80 km to the south of the Area, the stratigraphic facies becomes the interbeds consisting of sandy dolomite and dolomite sandstones of Banqiaoshan Formation and the sedimentary environment is transitioned to neritic platform–tidal flat facies.

2.2.1.5 Trace Elements in Strata (Ore-Bearing)

Three rock spectra of Lantian Formation were systematically collected from the Nanchewu profile (PM010) of Hanggai map sheet. According to spectral analysis, the members of Lantian Formation vary slightly in terms of trace element enrichment. The first member is enriched in Sb, Bi, and Ag, and poor in Au, Be, Cu, Pb, Zn, W, Mo, Sn, and S. The second member is enriched in Sb, Bi, Ag, F, and S, and poor in Au, Be, Zn, and Mo. The third member is enriched in Sb, Bi, W, Mo, Ag, and S, and poor in Au, Be, Cu, Zn, and TFe. The fourth member of Lantian Formation is enriched in Au, Sb, Bi Pb, W, Mo, and Ag, and poor in Be and TFe. Therefore, all the members are enriched in Sb, Bi, and Ag, and poor in Au, Be, Cu, Zn, and Mo. Furthermore, the concentration coefficients of Au, Ag, Sb, and Bi of the fourth member are 10–25 times higher than those of other members. The trace element enrichment is related to rock type. The silicon-bearing carbonaceous mudstone is enriched in many types of trace elements with high concentration coefficients. The limestone ranks the second, and silicalite ranks the last in terms of trace element enrichment.

2.2.2 Piyuancun Formation (Z_2p)

The name Piyuancun Formation was created by Li and Li (1930) in Piyuancun Village, Lantian Town, Xiuning Country, Anhui Province. Qian et al. (1964) determined its meaning again and used it to refer to a set of silicalite in black and white between the parts under the Early Cambrian phosphorous and carbonaceous shale and Lantian Formation, with the horizon lying in the Sinian. Zhejiang Regional Geological Team (1990) introduced the Piyuancun Formation to Changhua–Anji area, and its horizon belonged to the Sinian. However, the horizon of the stratum in Kaihua–Lin'an and Jiangshan–Shaoxing areas which is equivalent to Piyuancun Formation was still located in the lower part of Hetang Formation. In this Project, the Piyuancun Formation stated in the list of *Lithostratigraphy of Zhejiang Province* was still adopted according to characteristics of lithologic association of Nanchewu (PM010) profile of Hanggai map sheet, the profiles of Xianxia map sheets including Meijiata (PM035), Dalingtou (PM0036), and Ligu'an (PM039), and the geological observation traverse in the Area.

2.2.2.1 Lithostratigraphy

The outcrops of Piyuancun Formation in the Area are largely associated with Lantian Formation and are mainly distributed in the surrounding areas of Tangshe complex in the southwest corner of Hanggai map sheet and in the northwestern part of Xianxia map sheet including Zhanglicun, Baishawu, Ligu'an, and Damiao–Zhaojia area discontinuously. Thus, the distribution in "U" shape is formed with SW-trending opening in general. The outcrop area is about 8.05 km^2, accounting for 0.63% of the bedrock area.

Lithologically, Piyuancun Formation (Z_2p) consists of thin–thick laminated silicalite and argillaceous silicalite, with very thin horizontal bedding developing in rock layers. Silicalite is distributed in the shape of a band in black and white locally. Compared with the overlying Hetang Formation, Piyuancun Formation is characterized by the rocks of lighter color and lower content of carbonaceous argillaceous sediments. The thickness of Piyuancun Formation is 22.47–57.98 m.

1. **Stratigraphic section**

The lithological characters of Piyuancun Formation are described by taking the example of the fourth member of Sinian Lantian Formation ($Z_{1\text{-}2}l^4$)–Piyuancun Formation (Z_2p) profile (Fig. 2.9) of Xianxia map sheet in Ligu'an, Zhangcun Town, Anji County, Xianxia, Zhejiang Province. The details are as follows:

The first member of Hetang Formation — Total thickness: > 0.66 m

17. Dark gray thin–medium laminated carbonaceous siliceous mudstone, 3–15 cm thick per layer, with well-developed bedding, micro-folds visible in rocks. 0.66 m

——— Conformable contact ———

Piyuancun Formation — Total thickness: 57.98 m

16. Gray medium laminated argillaceous silicalite, about 20 cm thick per layer, with bedding developing. 4.63 m

15. Gray thick laminated argillaceous silicalite, 50–100 cm thick per layer. 5.67 m

14. Gray thin laminated argillaceous silicalite, 2–10 cm thick per layer, with very thin horizontal bedding developing. 5.67 m

13. Covering. 6.31 m

12. Grayish black thin–medium laminated argillaceous silicalite, 3–25 cm thick per layer, with well-developed very thin horizontal bedding. 9.73 m

11. Hoary medium–thick laminated silicalite in black and white, 20–50 cm thick per layer, very thin horizontal bedding in black and white developing. 7.39 m

10. Dark gray blocky–laminated argillaceous silicalite, the thickness of a single layer: > 1m generally. 4.97 m

9. Hoary thin laminated silicalite, 2–10 cm thick per layer, with well-developed bedding. 2.84 m

8. Dark gray medium–thick laminated argillaceous silicalite, 20–70 cm thick per layer, with well-developed bedding. 10.77 m

——— Conformable contact ———

The fourth member of Lantian Formation Member — Total thickness: > 7.05 m

7. Hoary medium laminated carbonaceous siliceous mudstone, 20–40 cm thick per layer, with well-developed bedding. 1.33 m

6. Black thin laminated carbonaceous mudstone, 2–5 cm thick per layer, with well-developed bedding. 5.72 m

2. **Lithological Characteristics**

The rocks in Piyuancun Formation (Z_2p) are mainly of two types: silicalite and argillaceous silicalite.

Silicalite: dark gray, in black and white locally; cryptocrystalline texture; consisting of cryptocrystalline siliceous sediments (85–95%) and a small amount of argillaceous sediments (< 10%); medium–thick laminated generally and thin laminated partly; generally distributed at the lower part.

Argillaceous silicalite: gray and dark gray; cryptocrystalline texture; mainly consisting of cryptocrystalline siliceous sediment (50–75%); argillaceous sediment contained (10–25%, up to 30% locally), generally earthy; black non-crystalline carbonaceous sediment visible occasionally (< 10%), concentrated locally; generally thin laminated and partly medium laminated; generally distributed in the upper part.

3. **Basic Sequences**

There are two types of basic sequences developing in Piyuancun Formation according to rock association (Fig. 2.10).

Basic sequences of type A: distributed in the middle and lower parts of Piyuancun Formation, composed of dark gray, light gray–white cryptocrystalline silicalite, 0.2–2 mm thick, silicalite in sharply changed dark color and light color constituting very thin horizontal bedding, belonging to monotonous basic sequence.

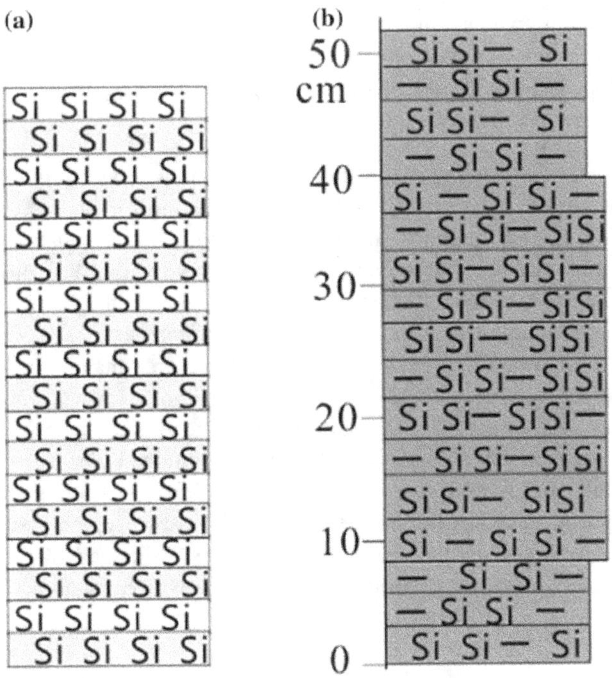

Fig. 2.10 Piyuancun Formation (Z_2p) basic sequence

Basic sequences of type B: distributed in the upper part of Piyuancun Formation, composed of thin–thick laminated argillaceous silicalite, very thin horizontal bedding developing in rock layers, belonging to monotonous basic sequence.

2.2.2.2 Sequence Stratigraphy

Piyuancun Formation is simple in terms of lithology and lithologic association. It is in conformable contact with its overlying and underlying strata, making it difficult to be further divided in terms of sequence stratigraphy and systems tract separately. According to the features of the lithology, lithofacies association, and sequence boundary of the profiles of Xianxia map sheet including Yeshanwu, Dalingtou, and Ligu'an as well as Nanchewu profile of Hanggai map sheet, there is one second-order sequence SS5 from the fourth member of Lantian Formation to Cambrian Dachenling Formation (Fig. 2.8), and there is one third-order sequence Sq5 from the fourth member of Lantian Formation to Piyuancun Formation.

Third-order sequence Sq5: located from the fourth member of Lantian Formation to Piyuancun Formation, including TST and HST. Its bottom and top belong to type-I and type-II sequence boundaries, respectively. TST is located in the fourth member of Lantian Formation, with argillaceous–siliceous sediment as well as a small amount of carbonaceous sediment constituting the sediments. Horizontal bedding developed in rock layers. Therefore, TST belongs to standing-water basin facies in bathyal weak reduction environment. HST is located in the upper part of Piyuancun Formation, and the sediments mainly consist of argillaceous–siliceous matter. Very thin horizontal bedding developed here. However, during the Piyuancun Formation period, Ediacaran Biota developed, which indicated oxidization environment and thus reflected that the sedimentary environment transitioned to bathyal–neritic weak oxidization environment. Therefore, from bottom to up, the sedimentary environment of Sq5 changed from bathyal basin facies to bathyal–neritic shelf facies, and is of progradational–aggradational type.

2.2.2.3 Biostratigraphy and Chronostratigraphy

Compared with adjacent southern Anhui area, less research has been carried out on palaeontology in Piyuancun Formation of Zhejiang Province. Dong et al. (2012) discovered Ediacaran fossils, macrofossils, unnamed fusiform fossils, and spherical fossils in the Piyuancun Formation of Lantian profile of Xiuning County, indicating that the silicalite of Piyuancun Formation was the sediment of Ediacaran Period. Among these fossils, *Palaeopaschnus* is distributed at the middle and upper parts of the silicalite in Piyuancun Formation and the distribution of *Horodyqkia* extends until near the boundary of Piyuancun Formation and Hetang

Formation, indicating the boundary between the Ediacaran and the Cambrian which is divided based on biostratigraphy largely coincides with the boundary between Piyuancun Formation and Hetang Formation in southern Anhui.

2.2.2.4 Analysis of Sedimentary Environment

The sediments of Piyuancun Formation mainly include gray–dark gray cryptocrystalline siliceous, with very thin horizontal bedding developing, indicating quiet water. There are very little carbon and sulfide in the sediment, suggesting weak oxidization environment. Compared with Lantian Formation, the Ediacaran biota and macro-algae indicate similar paleoclimate condition; however, the existence of metazoans indicates that oxygen was more sufficient in water.

In conclusion, the sedimentary environment of Piyuancun Formation is of shallow shelf facies. The boundary between the Sinian and the Cambrian is a very important turning point in geological history, since it represents a transformation from micro-bios-dominated Precambrian biosphere to animal-dominated Phanerozoic Eon biosphere and meanwhile reveals the beginning of oxygen-enriched surface environment (Anbar and Knoll 2002; Marshall 2006).

2.2.2.5 Trace Elements in Strata (Ore-Bearing)

Four rock spectra were collected from Piyuancun Formation in Nanchewu profile (PM010) of Hanggai map sheet. Piyuancun Formation is enriched in Sb, Bi, Mo, and Ag, and poor in Au, Be, Cu, Pb, Zn, W, F, S, and TFe. Compared with the four members of Lantian Formation, the content of Ag, Sb, and Bi is normal while the content of all other trace elements is all very lower.

2.2.3 Comparison of Regional Stratigraphy

The Sinian in Zhejiang Province is mainly distributed in the northwest of Jiangshan–Shaoxing fracture zone. With the change of paleotopography, rock layers of different thicknesses were deposited from northwest to southeast. According to the information on the Area and previous stratigraphic profiles, the northwest of Zhejiang can be divided into three zones of sedimentary facies, which include Hanggai–Kaihua hollow zone, Yuhang–Tonglu uplift zone, and Xiaoshan (Zhuji)–Jiangshan low-lying zone from northwest to southeast.

The Sinian in Zhejiang Province mainly consists of carbonate formation and siliceous argillaceous formation, followed by the formation of terrigenous clastic sedimentary rocks. In the Project, typical profiles in Hanggai area (the Area), Dashuping of Fuyang, Yuhang, and Fudun of Xiaoshan are selected for comparison of regional strata (Fig. 2.11).

The worldwide glacial period ended at the end of the Nanhuan. In the early period of the Early Sinian, a layer of carbonate rocks started to be deposited on tillite, which is called carbonatite cap. During this period, the strata of Hanggai area were about 9-m-thick manganese-bearing dolomite of the first member of Lantian Formation on the bottom of Sinian System, the strata of Fuyang area of Yuhang were about 28-m-thick dolomite of the first member of Doushantuo Formation, and the strata of Xiaoshan area were about 5-m-thick manganese-bearing dolomite of the first member of Doushantuo Formation.

In the middle and later period of the Early Sinian, the strata of Hanggai area consisted of carbonaceous siliceous silty mudstone of bathyal basin facies interbedded with a small amount of carbonaceous limestone of shallow shelf facies in the second member of Lantian Formation, and the thickness is about 85 m; the strata of Yuhang area consisted of dolomitic limestone interbedded with argillaceous limestone and a small amount of mudstone and dolomite in Doushantuo Formation; the strata of Xiaoshan area gradually changed from the interbed of caesious argillaceous dolomite and silty mudstone to the dolomite and amaranthine potassium-bearing siltstone of the second member of Doushantuo Formation from bottom to top, with terrigenous clasts increasing, indicating strong oxidization and high-energy environment.

In the early period of the Later Sinian, the strata of Hanggai area mainly consisted of the third member of Lantian Formation of shallow shelf facies. It changed from the interbed of thin laminated dolomite and argillaceous limestone to the thick laminated dolomite interbedded with a small amount of sand dolomite from bottom to top, and a thickness is 235 m; the strata of Yuhang area mainly consisted of dolomite and a small amount of carbonaceous calcareous mudstone, silty dolomite, and dolomitic quartz sandstone of the first member of Banqiaoshan Formation with terrigenous clasts increasing evidently, belonging to neritic platform facies; the strata of Xiaoshan region consisted of dolomite, siltstone, and potassic siltstone of the second member of Doushantuo Formation from bottom to

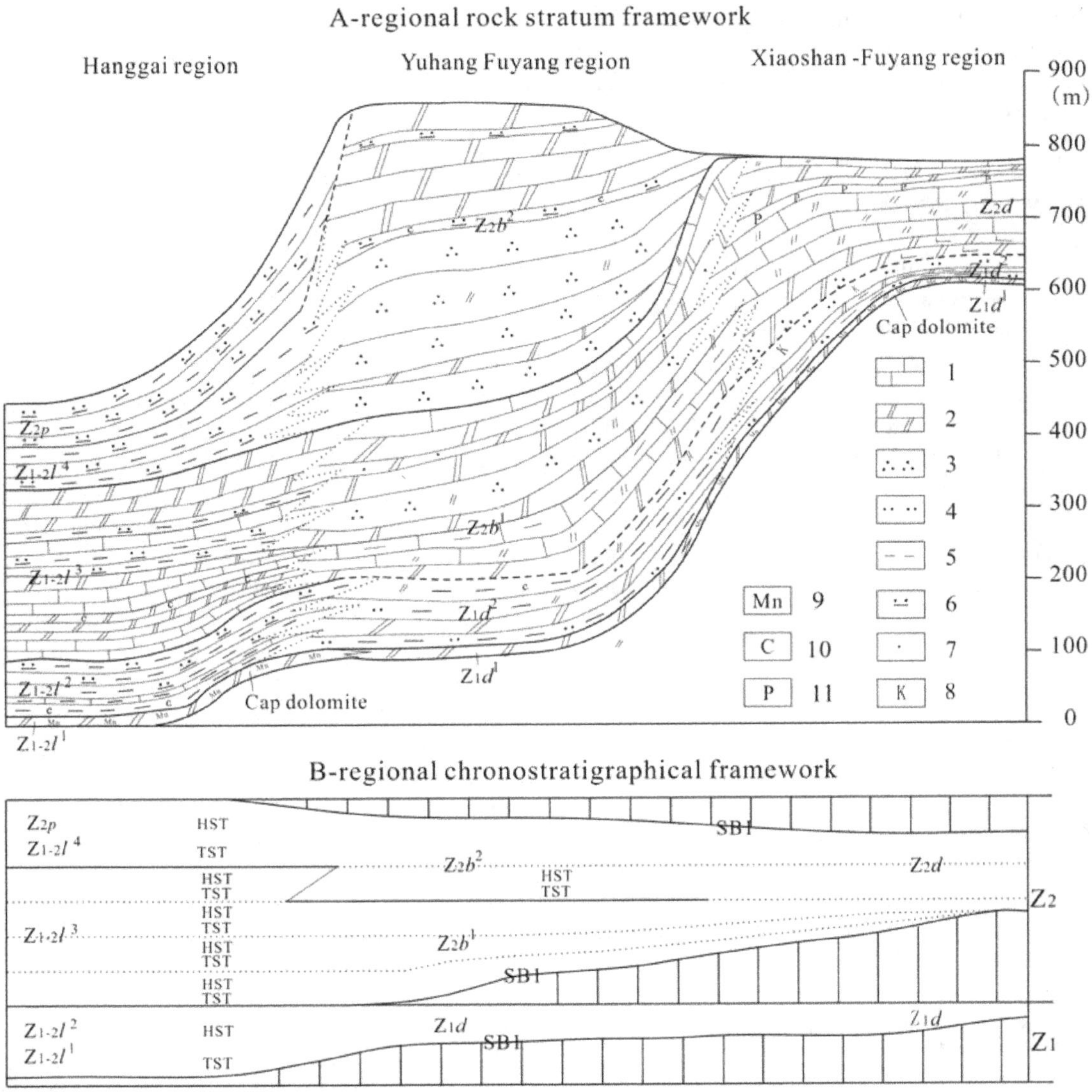

Fig. 2.11 Regional stratigraphic framework in Hanggai–Fuyang, Yuhang–Zhuji, Xiaoshang area of Sinian Period: 1. limestone; 2. dolomite; 3. quartz sandstone; 4. siltstone; 5. mudstone; 6. silicalite; 7. sandy sediment; 8. potassium-bearing sediment; 9. manganese-bearing sediment; 10. carbonaceous sediment; 11. collophanite; $Z_{1\text{-}2}l^1$—the first member of Lantian Formation; $Z_{1\text{-}2}l^2$—the second member of Lantian Formation; $Z_{1\text{-}2}l^3$—the third member of Lantian Formation; $Z_{1\text{-}2}l^4$—the fourth member of Lantian Formation; Z_2p—Piyuancun Formation; Z_2dy—Dengying Formation; Z_2b—Banqiaoshan Formation; Z_1d—Doushantuo Formation; SB1–type-I unconformity interface; TST—transgressive systems tract; HST—highstand systems tract

top. The dominated rocks of the strata in these areas were gradually changed from carbonate rocks to terrigenous clastic sedimentary rocks from northwest to southeast.

In the late period of the Later Sinian, the strata of Hanggai area consisted of the siliceous mudstone of bathyal–abyssal facies in the fourth member of Lantian Formation and the silicalite of Piyuancun Formation, and the thickness is about 110 m; the strata of Yuhang area mainly consisted of dolomitic sandstone and a small amount of gravel–clastic dolomite of platform facies in Banqiaoshan Formation, about 423 m thick, equivalent to neritic sandbank; the strata of Xiaoshan region consisted of dolomite, dolomite-bearing conglomerate clastic, and collophanite of neritic platform facies in Dengying Formation, about 135 m thick.

In conclusion, northwest Zhejiang in the Sinian consisted of a hollow basin in the northwest, uplift zone in the middle, and relatively low-lying areas in its southwest from northwest to southeast. The dominated rocks of northwest Zhejiang were gradually changed from carbonate rocks to terrigenous clastic sedimentary rocks from northwest to southeast.

2.3 Cambrian System

The Cambrian is widely distributed in the Area. It mainly outcrops in the southwestern part and the southeast corner of Hanggai map sheet as well as the western part and the southeast corner of Xianxia map sheet. Besides, it is sporadically distributed in the northeast and south of Chuancun map sheet. The total outcrop area is 170.60 km^2, accounting for 13.42% of the bedrock area. The Cambrian features complete and continuous outcrops and is divided into five formations from bottom to top, i.e., Hetang Formation ($\epsilon_1 h$), Dachenling Formation ($\epsilon_2 y$), Yangliugang Formation ($\epsilon_2 y$), Huayansi Formation ($\epsilon_3 h$), and Xiyangshan Formation ($\epsilon O x$). The adjacent formations are in conformable contact. According to the lithological association, Hetang Formation ($\epsilon_1 h$) and Yangliugang Formation ($\epsilon_2 y$) are, respectively, divided into two members, and Xiyangshan Formation is divided into three members.

2.3.1 Hetang Formation ($\epsilon_1 h$)

The name Hetang Formation was created by Lu et al. (1955) in Hetang Village, northeast of Dachen, Jiangshan City, Zhejiang Province. It was used to refer to black carbonaceous shale (stone coal layer) by Sheng (1951) and was subsequently used up to now by the preparation group of regional stratigraphic table of Zhejiang Province (1979) as well as in the literature, which includes *Stratigraphic Correlation Chart in China with Explanatory Text* (1982), *Regional Geology of Zhejiang Province* (1989), and *Lithostratigraphy of Zhejiang Province* (1995). In this Project, Hetang Formation was still adopted according to the lithological characteristics of the Nanche profile (PM010) of Hanggai map sheet, profiles of Xianxia map sheet including the Meijiata (PM035) and Ligu'an (PM040), and survey traverse in the Area.

2.3.1.1 Lithostratigraphy

The Hetang Formation in the Area mainly outcrops in the Gaocun forest farm–Tongcun forest farm–Tangkengwu area in the southwest of Hanggai map sheet and the Zhangli Village–Baishawu–Ligu'an–Damiao–Zhaojia area in the northwest of Xianxia map sheet. Hetang Formation is associated with Lantian Formation and Piyuancun Formation. Thus, the distribution in "U" shape is formed with SW-trending opening in general. Besides, it is sporadically distributed in Chenjiatang and the east of Tianhuangping in the northeast of Chuancun map sheet. The total outcrop area is about 49.18 km^2, accounting for 3.87% of the bedrock area.

The first member of Hetang Formation ($\epsilon_1 h^1$): The lithology is mainly characterized by the dark gray and grayish black thin laminated carbonaceous silicalite, the argillaceous silicalite, and the siliceous mudstone interbedded with carbonaceous mudstone. The lower-middle part is interbedded with medium laminated stone coal layer and multiple thin–medium laminated phosphate nodule layers. The thickness of the member is 146.50–445.99 m^2.

The second member of Hetang Formation ($\epsilon_1 h^2$): The lithology is mainly characterized by the dark gray and grayish black thin–medium laminated pyrite-bearing argillaceous carbonaceous silicalite, carbonaceous siliceous mudstone, interbedded with carbonaceous mudstone and partly with pyrite layers or pyrite nodule layers. The thickness of the member is 155.56–201.35 m.

1. **Stratigraphic section**

The lithology Hetang Formation is described by taking the example of the Cambrian Hetang Formation ($\epsilon_1 h$)–Dachenling Formation ($\epsilon_1 d$) profile (Fig. 2.12) of Hanggai map sheet in Nanchewu, Hanggai Town, Anji County, Zhejiang Province. The details are as follows:

Dachenling Formation Total thickness: >0.96 m

46. Gray medium–thin laminated argillaceous limestone, the thickness of a single layer: 5–20 cm generally. 0.96 m

———Conformable contact———

The second member of Hetang Formation Total thickness: 55. 56 m

45. Grayish black medium laminated carbonaceous silicalite, the thickness of a single layer: 20–60 cm. 1.57 m

44. Dark gray and gray medium laminated argillaceous carbonaceous silicalite, the thickness of a single layer: 20-40 cm. 16.25 m

43. Grayish black thin laminated carbonaceous argillaceous silicalite, the thickness of a single layer: 1–5 cm. 100.88 m

42. Grayish black thin laminated argillaceous carbonaceous silicalite, the thickness of a single layer: 3–10 cm. 15.16 m

41. Dark gray thin laminated silty mudstone. 11.60 m

40. Grayish black medium–thin laminated carbonaceous argillaceous silicalite. 1.54 m

39. Gray thin laminated carbonaceous silicalite, with the thickness of a single layer of 0.5 - 5 cm. 8.56m

———Conformable contact———

The first member of Hetang Formation Total thickness: 113.53 m

38. Gray medium laminated carbonaceous argillaceous silicalite, the thickness of a single layer: 20–50 cm. 15.78 m

37. Gray medium–thin laminated siliceous mudstone, very thin horizontal bedding developing. 10.35 m

36. Dark gray medium laminated siliceous mudstone. 10.77 m

35. Gray thin laminated carbonaceous argillaceous silicalite, the thickness of a single layer: 0.5–5 cm, very thin horizontal bedding developing. 45.59 m

34. Grayish black thin laminated argillaceous silicalite, the thickness of a single layer: 0.5–10 cm, very thin horizontal bedding developing. 6.06 m

33. Dark gray medium–thin laminated silicalite, the thickness of a single layer: 5–15 cm. 10.85 m

32. Grayish black thin laminated carbonaceous silicalite, the thickness of a single layer: 0.2–6 cm. 5.45 m

31. Dark gray and greyish-black thin laminated argillaceous silicalite, the thickness of a single layer: 2–8 cm, very thin horizontal bedding developing locally. 8.68 m

———Conformable contact———

Piyuancun Formation Total thickness: >3.25 m

30. Light gray thick laminated argillaceous silicalite, bands in black and white developing in the layer. 3.25m

The lithology Hetang Formation is described as follows by taking the sample of Hetang Formation ($\epsilon_1 h$)–Dachenling Formation ($\epsilon_1 h$) profile of the Cambrian of Xianxia map sheet in Meijiata, Xianxia Town, Ningguo City, Anhui Province (Fig. 2.13). The details are as follows:

Dachenling Formation Total thickness: >12.09 m

53. Black thick laminated micrite, the thickness of a single layer: 50–150 cm, very thin horizontal bedding developing. 12.09m

———————— Conformable contact ————————

The second member of Hetang Formation 201.35 m

52. Black thick laminated carbonaceous mudstone, the thickness of a single layer: 50–80 cm, very thin horizontal bedding developing. 22.93 m

51. Black medium laminated carbonaceous siliceous mudstone, the thickness of a single layer: 20–30 cm, horizontal bedding developing. 7.49 m

50. Black thin–medium laminated siliceous carbonaceous mudstone, the thickness of a single layer: 8–25 cm thick. 3.57 m

49. Black thin laminated carbonaceous siliceous mudstone interbedded with medium laminated carbonaceous siliceous mudstone, the thickness of a single-layer carbonaceous siliceous mudstone: 2–8 cm; very thin horizontal bedding developing, flaky; the thickness of a single layer of the carbonaceous mudstone: 15–30 cm. 8.10 m

48. Black medium laminated carbonaceous siliceous mudstone, the thickness of a single layer: 10–20 cm. 7.46 m

47. Black thin laminated carbonaceous siliceous mudstone interbedded with medium laminated siliceous mudstone; the thickness of a single layer of the carbonaceous siliceous mudstone: 4–8 cm, very thin horizontal bedding developing, flaky. 3.57 m

46. Black thin laminated carbonaceous siliceous mudstone, the thickness of a single-layer: 8–9 cm, horizontal bedding developing. 3.21 m

45. Black thin laminated carbonaceous siliceous mudstone, the thickness of a single-layer: 5–9 cm generally and >10cm partly. 5.00 m

44. Black medium laminated carbonaceous siliceous mudstone, micro-fine grained pyrite visible occasionally, the thickness of a single layer: 15–30 cm, horizontal bedding developing. 4.56 m

43. Black thin–medium laminated carbonaceous siliceous mudstone, the thickness of a single-layer: 6–11 cm, horizontal bedding developing. 23.93 m

42. Black thin laminated siliceous carbonaceous mudstone, the thickness of a single layer: 5–8 cm generally and >10 cm partly, very thin horizontal bedding developing. 9.68 m

41. Black medium laminated siliceous carbonaceous mudstone containing sparse micro-fine grained pyrite, the thickness of a single layer: 15–30 cm, very thin horizontal bedding developing. 5.58 m

40. Black thin laminated carbonaceous siliceous mudstone, disseminated pyrite sparsely distributed in the mudstone and concentrating locally and discontinuously along the bedding, the grain size of the pyrite: 0.1–0.5 mm, the thickness of a single layer: 5–9 cm generally and 10–12 cm partly. 7.93 m

39. Black medium laminated silicon-bearing carbonaceous mudstone interbedded with carbonaceous mudstone. The micro-fine vein consisting of aggregates of pyrite with a grain size of 0.1–1 mm is visible in the siliceous carbonaceous mudstone. The middle part of this bed is interbedded with the two layers of 3–5 cm thick carbonaceous mudstone. The thickness of a single layer of siliceous carbonaceous mudstone: 10–20 cm. 19.07 m

38. Black thin laminated siliceous carbonaceous mudstone, micro-fine grained pyrite developing along the bedding, the thickness of a single layer: 1–3 cm. 1.23 m

37. Black medium laminated carbonaceous siliceous mudstone containing disseminated pyrite with a grain size of 0.1–1 mm, the thickness of a single layer thickness: 10–30 cm, horizontal bedding developing. 4.12 m

36. Black thin laminated siliceous carbonaceous mudstone with pyrite being distributed occasionally, the thickness of a single-layer: 1–5 cm. 15.06 m

35. Black medium laminated carbonaceous silicalite containing disseminated pyrite, the thickness of a single layer: 10–30 cm, horizontal bedding developing. 0.67 m

34. Black thin laminated carbonaceous silicalite, micro-layer (disseminated) pyrite developing along the layers, the thickness of a single layer: 2–7 cm, very thin horizontal bedding developing; three layers of disseminated–densely disseminated pyrite visible. 4.53 m

33. Black medium laminated carbonaceous siliceous mudstone, micro-layer pyrite with a grain size of 0.1–0.2 mm visible in the layers, the micro-vein association consisting of 2 mm×5 mm pyrite nodule partly visible. 3.87 m

32. Dark gray–black thin laminated carbonaceous silicalite interbedded with medium laminated siliceous mudstone, the thickness of single-layer carbonaceous silicalite: 2–7 cm, the thickness of single-layer siliceous mudstone: 15–20 cm, the ratio of the carbonaceous silicalite to the siliceous mudstone: (2–5):1. 4.13 m

31. Dark gray medium laminated carbonaceous silicalite, the thickness of a single-layer: 10–20 cm, horizontal bedding developing. 4.54 m

30. Black thin laminated carbonaceous silicalite, the thickness of a single layer: 3–6 cm; aggregation of micro-fine grained pyrite visible along the bedding, micro interbedded, the thickness of single layer: 0.1–0.3 mm; 20 cm thick carbonaceous mudstone constitutes the top. 3.27 m

29. Black medium laminated carbonaceous silicalite; aggregation of micro-fine grained pyrite concentrating along the layers, forming the densely disseminated pyrite layer with a thickness of 0.1–1 mm; 3×10 cm pyrite nodule visible occasionally; the thickness of a single layer: 10–12 cm. 7.75 m

28. Black thin laminated carbonaceous silicalite; fine line consisting of 5% micro-fine grained pyrite visible along the layer; 3–4 layers of 1–3 cm × 3–7 cm pyrite nodules visible, with the major axis extending along the layers discontinuously. 10.10 m

——————— Conformable contact ———————

The first member of Hetang Formation Total thickness: 445.99 m

27. Black medium laminated carbonaceous silicalite, a small amount of micro-fine grained pyrite visible in the rock, the thickness of a single layer: 10–30 cm. 2.24 m

26. Black thin laminated siliceous carbonaceous mudstone, the micro-fine grained pyrite distributed along the bedding, interbedded with a layer of 1 cm thick pyrite. 2.18 m

25. Black medium laminated siliceous carbonaceous mudstone; the thickness of a single layer: 10–20 cm; the top: off-white clay layer, 0.1–2 cm thick, discontinuously distributed along the layers. 16.96 m

24. Covering. 31.23 m

23. Dark gray thin–medium laminated carbonaceous siliceous mudstone, the thickness of a single layer: 6–15 cm. 58.74 m

22. Covering. 30.76 m

21. Black medium laminated carbonaceous silicalite, the micro-fine grained disseminated pyrite visible occasionally in the rocks, the thickness of a single layer: 20–30 cm. 25.05 m

20. Black medium laminated carbonaceous silicalite, the thickness of a single layer: 10–20 cm, horizontal bedding developing, containing a small amount of disseminated pyrite. 6.21 m

19. Black thin laminated carbonaceous siliceous mudstone, containing a small amount of disseminated pyrite, the thickness of a single layer: 2–6 cm. 7.65 m

18. Black medium laminated carbonaceous siliceous mudstone, the micro-fine grained pyrite layer distributed along the bedding, 1 cm × 3 cm ellipsoid pyrite nodule partly visible, the covellite occasionally visible in cracks, the thickness of a single layer: 10–30 cm. 17.18 m

17. Covering. 22.46 m

16. Dark gray thin–medium laminated carbonaceous siliceous mudstone interbedded with micro–thin laminated carbonaceous mudstone, the thickness of a single layer of carbonaceous siliceous mudstone: 8–15 cm, horizontal bedding developing; the thickness of a single layer of carbonaceous mudstone: 0.5–1 cm, a small amount of micro-fine grained pyrite visible in the rocks. 7.72 m

15. Black medium laminated carbonaceous siliceous shale; the thickness of a single layer: 20–30 cm; horizontal bedding developing; interbedded with a layer of phosphate nodules with thickness of 30–40 cm near the upper part, the size of a single nodule: 1–3 cm, ellipsoid. 20.63 m

14. Black medium laminated carbonaceous mudstone, micro-fine grained disseminated pyrite concentrating along the bedding, the thickness of a single layer: 10–25 cm. 56.63 m

13. Dark gray thin laminated carbonaceous silicalite, the thickness of a single layer: 3–8 cm, horizontal bedding developing. 10.92 m

12. Dark gray thin–medium laminated siliceous carbonaceous mudstone interbedded with micro laminated carbonaceous mudstone, the thickness of a single layer of siliceous carbonaceous mudstone: 8–20 cm, very thin horizontal bedding developing; the thickness of a single layer of carbonaceous mudstone: 1cm. The ratio of the siliceous carbonaceous mudstone to the carbonaceous mudstone: (10–20):1. 21.72 m

11. Dark gray thin laminated carbonaceous mudstone, the thickness of a single layer: 1–8 cm, flaky. 17.46 m

10. Covering. 6.31 m

9. Dark gray thin laminated carbonaceous siliceous mudstone, the thickness of a single layer: 3–8 cm or partly up to 10cm. micro-fine grained pyrite visible locally. 14.48 m

8. Dark gray thin–medium laminated carbonaceous siliceous mudstone interbedded with micro laminated carbonaceous shale, horizontal bedding developing in carbonaceous siliceous mudstone, the thickness of a single layer: 8–20 cm, the thickness of a single layer of carbonaceous shale: 1–2 cm. 7.18 m

7. Dark gray thin laminated carbonaceous silicalite, the thickness of a single layer: 1–3 cm, horizontal bedding developing, prone to thin plate rocks after weathering. 3.85 m

6. Black medium laminated carbonaceous siliceous mudstone, a small amount of micro-fine grained pyrite distributed in the rocks in disseminated manner, the pyrite nodule with a grain size of 1–3 mm discontinuously distributed along the layers locally, the thickness of a single layer: 20–40 cm, very thin horizontal bedding developing. 4.67 m

5. Dark gray thin siliceous carbonaceous mudstone interbedded with thick laminated carbonaceous mudstone, the thickness of a single layer siliceous carbonaceous mudstone: 1–5 cm, horizontal bedding developing, the thickness of a single layer of carbonaceous mudstone: 0.5–1.5 cm, the ratio of the siliceous carbonaceous mudstone to the carbonaceous mudstone: (2–4):1. 21.60 m

4. Interbed consisting of dark gray medium laminated carbonaceous siliceous mudstone and thin siliceous mudstone, the thickness of a single layer of carbonaceous siliceous mudstone: 10–20 cm, the thickness of a single layer of siliceous mudstone: 1–3 cm. The ratio of carbonaceous siliceous mudstone to the siliceous mudstone: (2–4):1. 7.55 m

3. Dark gray medium laminated carbonaceous silicalite, very thin horizontal bedding in black and white developing, the thickness of a single layer: 30–45 cm. 1.32 m

2. Dark gray medium laminated carbonaceous silicalite, very thin horizontal bedding developing, flaky, the thickness of a single layer: 10–25 cm. 6.19 m

1. Dark gray thin laminated siliceous mudstone, the thickness of a single layer: 1–5 cm, very thin horizontal bedding developing. 17.1 m

———————— Conformable contact ————————

Piyuancun Formation (Z_2p) Total thickness:>13.16 m

0. Silicalite in black and white, very thin horizontal bedding developing, the thickness of a single layer of bedding: 1–3 mm. 13.16 m

2. Lithological Characteristics

The lithology of the first member of Hetang Formation is mainly characterized by carbonaceous silicalite, argillaceous silicalite, siliceous mudstone, carbonaceous mudstone, and phosphate nodule layer.

(1) The carbonaceous silicalite: dark gray–black, composited of cryptocrystalline siliceous matter (80–90%) and the amorphous carbonaceous matter (5–15%), horizontal bedding developing generally in the layers.
(2) The argillaceous silicalite: dark gray, argillaceous cryptocrystalline texture; mainly composed of cryptocrystalline siliceous matter (45–65%), argillaceous matter

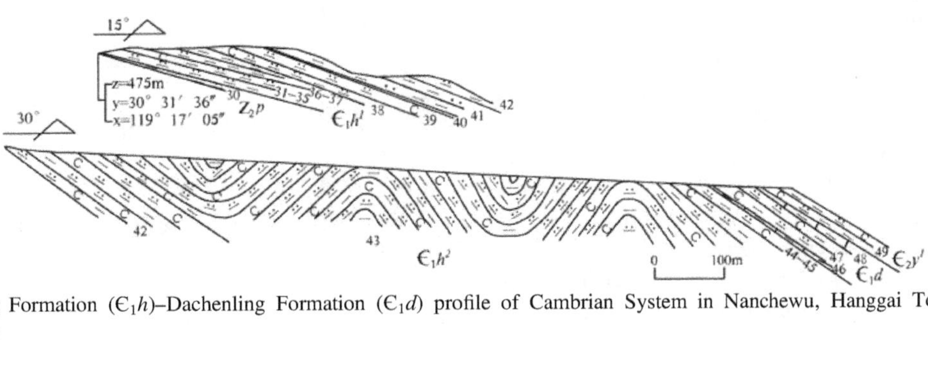

Fig. 2.12 Hetang Formation (€$_1h$)–Dachenling Formation (€$_1d$) profile of Cambrian System in Nanchewu, Hanggai Town, Anji County, Zhejiang Province

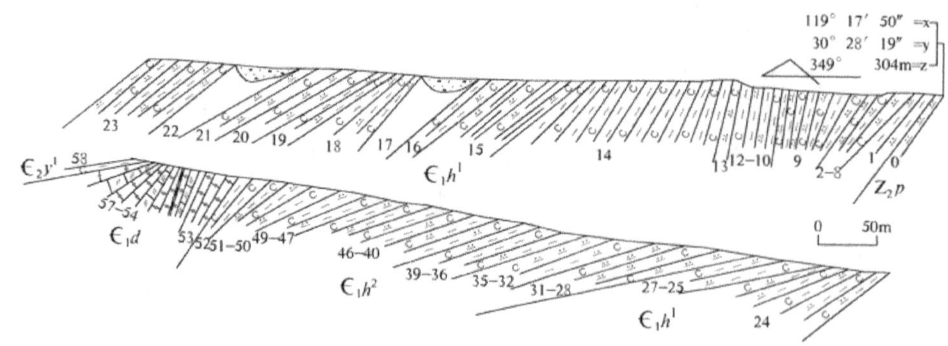

Fig. 2.13 Hetang Formation (Є$_1h$)–Dachenling Formation (Є$_1d$) profile of Cambrian Period in Meijiata, Xianxia Town, Ningguo City, Anhui Province

(20–35%), and amorphous carbonaceous (10–15%); micro-fine-grained pyrite visible occasionally.

(3) The siliceous mudstone: gray–dark gray, composition: argillaceous matter (65–75%) (major), siliceous (25–35%) (minor), micro-fine-grained pyrite visible occasionally, very thin horizontal bedding developing in the rock.

(4) The carbonaceous mudstone: black, mainly composed of carbonaceous matter (30–60%) and argillaceous (40–70%) matter, pyrite visible occasionally.

(5) The phosphate nodule layer: dark gray, in colloform texture, composited of black phosphorite-bearing siliceous spherulite of circularly shaped structure, nodule size: 2–30 mm, partly containing the small shelly fossils.

The lithology of the second member of Hetang Formation is mainly characterized by pyrite-bearing carbonaceous argillaceous silicalite and carbonaceous siliceous mudstone, interbedded with carbonaceous mudstone, a small amount of pyrite layer or pyrite nodule layer visible.

(1) The pyrite-bearing carbonaceous silicalite: dark gray; major component: cryptocrystalline siliceous matter (55–70%); minor components: amorphous carbonaceous matter (10–20%), argillaceous matter (15–25%), and pyrite (5–10%); silt visible occasionally (< 5%); aggregation of micro-fine-grained pyrite concentrating along the layers.

(2) The carbonaceous siliceous mudstone: grayish black; mainly composed of the argillaceous matter (45–65%), cryptocrystalline siliceous matter (15–25%), amorphous carbonaceous matter (10–15%), and pyrite (< 5%); micro-fine-grained pyrite concentrating along the layers.

(3) The carbonaceous mudstone: black, micro-laminated with the thickness of less than 2 cm generally, composed of carbonaceous (10–20%) and argillaceous (80–90%), pyrite visible occasionally.

(4) The pyrite layer or pyrite nodule layer, main component: micro-fine-grained pyrite (45–75%) with a grain size of less than 0.2 mm; minor component: argillaceous matter (15–25%), carbonaceous matter visible occasionally, disseminated pyrite concentrating along the layers, pyrite nodule with a size of 2–20 cm appearing locally; the thickness of the lithological layer formed: less than 3 cm.

3. **Basic Sequences**

There are three types of basic sequences developing in Hetang Formation (Є$_1h$), including three in the first member of Hetang Formation (Є$_1h^1$) and two in the second member of Hetang Formation (Є$_1h^2$). The two members share two same types of basic sequences (Fig. 2.14).

Basic sequences of type A: located in the lower parts of the first member and the second member, composed of ① grayish black thin–medium laminated carbonaceous argillaceous silicalite and ② black micro-laminated carbonaceous mudstone. For the first component: small pyrite lens commonly visible, micro-fine-grained pyrite distributed along the layers, horizontal bedding developing, spongy spicules as a result of pyrite metallization visible occasionally, thickness: 5–18 cm. For the second component: mainly

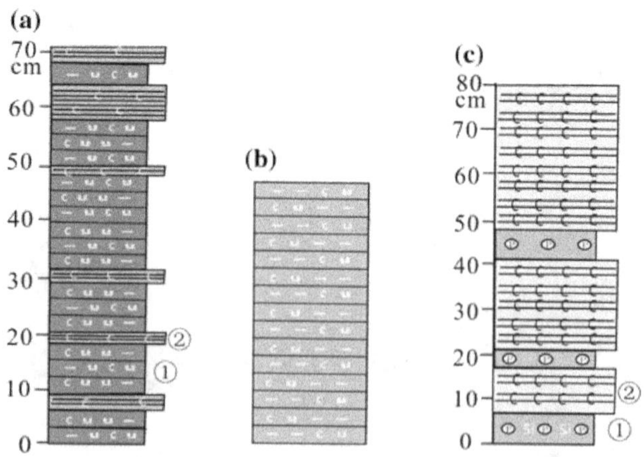

Fig. 2.14 Basic sequence map of Hetang Formation (Є$_1h$)

composed of carbonaceous matter, 2–20 mm thick, prone to deformation under compression and becoming graphitization films. The basic sequences of this type are composed of argillaceous–siliceous rocks and contain carbon and sulfide in reduction environment. Therefore, they belong to argillaceous–siliceous facies of abyssal basin and are of non-cyclic basic sequence.

Basic sequences of type B: located in the upper parts of the first member and the second member, composed of the dark gray thin–medium laminated carbonaceous siliceous mudstone with argillaceous matter as major component and siliceous matter as minor component, micro-fine-grained pyrite concentrating along the bedding. Therefore, they belong to the siliceous-argillaceous facies of abyssal standing-water reduction environment and are of monotonous basic sequence.

Basic sequences of type C: located in the lower part of the first member of Hetang Formation, composed of ① black thin laminated phosphorite-bearing siliceous spherulite and ② black thin laminated carbonaceous mudstone or stone coal layer. The siliceous spherulite in the phosphorite is of a circularly shaped structure, indicating high-energy water body. In the carbonaceous mudstone, there are high content of argillaceous matter (60–80%), as well as high content of carbonaceous and organic matter, indicating that the sedimentation is closely related to paleontological activities. The basic sequences of this type are of non-cyclic basic sequence.

2.3.1.2 Sequence Stratigraphy

According to the characteristics of lithologic association of the Cambrian as well as the types of tops and bottoms, there is a second-order sequence SS4 from the fourth member of Lantian Formation of the Late Sinian to the Dachenling Formation of the Early Cambrian, and there is a second-order sequence SS5 from the first member of Yangliugang Formation of the Middle Cambrian to the Huayansi Formation of the Late Cambrian. According to the changes in sedimentary facies of the second-order sequences, SS4 is further divided into seven third-order sequences (Sq5–Sq11), and SS5 is further divided into four third-order sequences (Sq12–Sq15).

There are a few types of rocks in Hetang Formation in the Area, and the main component is argillaceous–siliceous matter. Besides, there is a small amount of carbonaceous matter. This indicates monotonous standing-water reduction environment. The relative change in the content of argillaceous–siliceous matter reflects slight rise and falling of the sea level. Based on these, as well as the carbon and sulfide contained in the rocks of various periods, the transgression, the maximum flooding surface, and the regression are identified. Meanwhile, the division of Hetang Formation in terms of the transgressive systems tract (TST), the starved section, and the highstand systems tract (HST) is determined. According to the lithology, the petrofacies association, and the characteristics of sequence boundary of the Meijiata profile of Xianxia map sheet and Nanchewu profile of Hanggai map sheet, the first member of Hetang Formation ($Є_1h^1$) is divided into four third-order sequences (Sq6–Sq9) and the second member of Hetang Formation is divided into one third-order sequence. All of these third-order sequences are the parts of the second-order sequence SS5 (Fig. 2.15).

Third-order sequence Sq6: Located at the bottom of the first member of Hetang Formation, it includes TST and HST. The bottom of it is the conformity interface between the dark gray thin laminated siliceous mudstone and the silicalite or argillaceous silicalite of Piyuancun Formation. It is in regional parallel unconformable contact with Jiangshan–Changshan–Yushan area in the southwest. The top of it is the internal conformity interface of the first member of Hetang Formation. The top and the bottom are both of type-II sequence boundaries. TST is located in Bed 1–Bed 3, and the major and minor sediments in TST are siliceous matter and argillaceous matter, respectively. Owing to the very thin horizontal bedding developing partly, TST belongs to argillaceous–siliceous facies of abyssal basin. Located in Bed 4–Bed 6, HST contains the carbonaceous siliceous argillaceous matter bearing much micro-fine-grained pyrite as the main sediments. Furthermore, the horizontal bedding developed. Therefore, HST belongs to siliceous-argillaceous facies of bathyal basin. Therefore, the sedimentary environment in this sequence varies from abyssal basin facies to bathyal basin facies from bottom to top and this sequence is of progradation–aggradation type.

Third-order sequence Sq7: Located in the lower-middle part of the first member of Hetang Formation, it includes TST and HST, with the top and the bottom of type-II sequence boundary. TST is located in Bed 7–Bed 13. The sediments of it mainly include argillaceous–siliceous matter. Besides, there is a small amount of amorphous carbonaceous matter and associated synsedimentary micro-fine-grained sulfide locally. It can be indicated that the sedimentary environment of TST is standing-water reduction condition under wave base and TST belongs to abyssal siliceous-argillaceous facies. HST is located in Bed 14, and the major and minor sediments are the argillaceous and carbonaceous matters, respectively, associated with pyrite deposited. Thus, HST belongs to bathyal mudstone facies. Therefore, the sedimentary environment of this sequence gradually varies from the abyssal basin facies to the bathyal basin facies from bottom to top, and this sequence is of progradation–aggradation type.

Third-order sequence Sq8: Located in the upper-middle part of the first member of Hetang Formation, it includes TST and HST, with the top and the bottom of type-II sequence boundary. TST is located in Bed 15–Bed 16. The

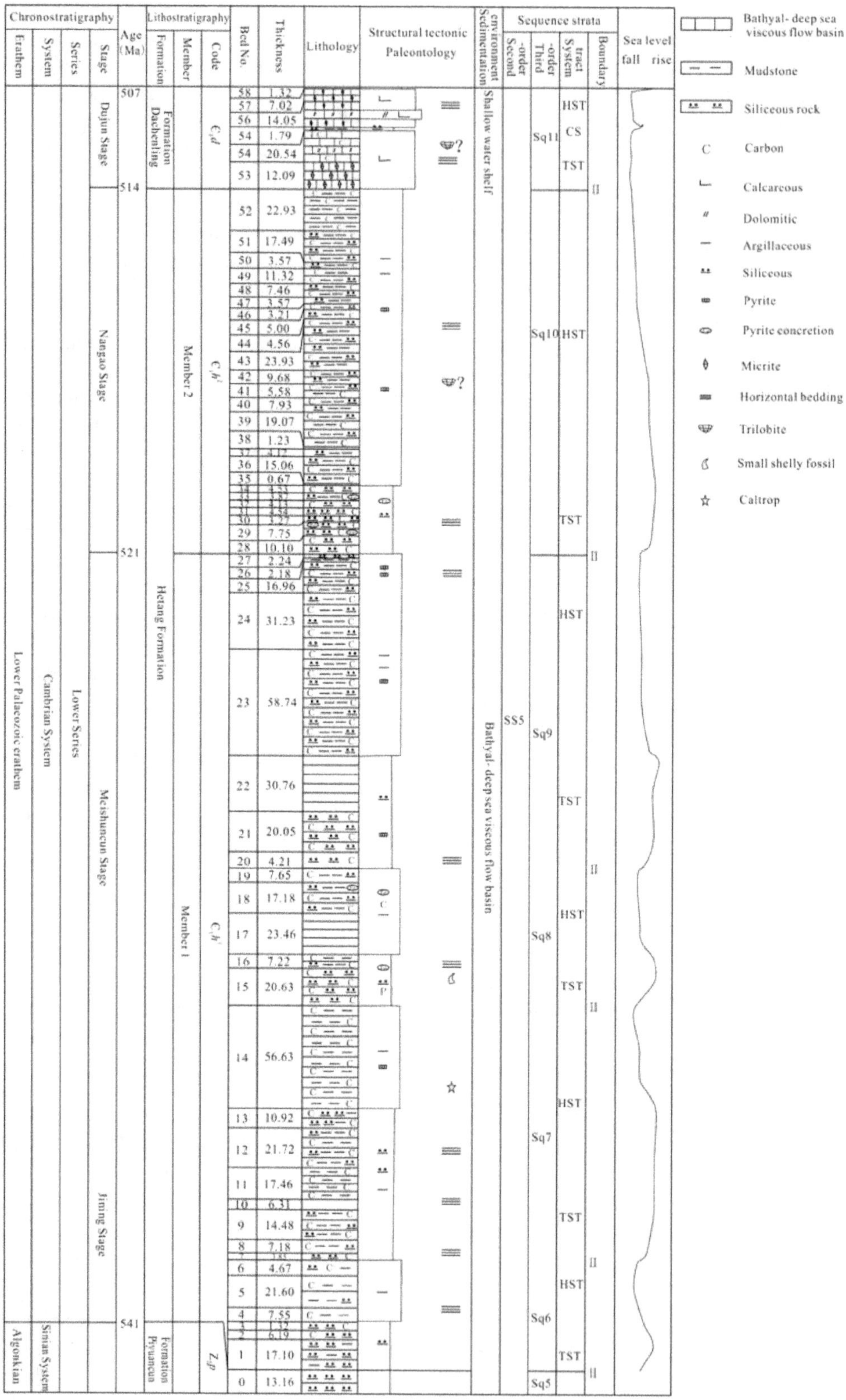

Fig. 2.15 Sequence stratigraphic framework of Hetang Formation–Dachenling Formation of Cambrian Period in the area

major and minor sediments of it are siliceous and argillaceous matters, respectively. Besides, there is a small amount of amorphous carbonaceous matter and associated micro-fine-grained sulfide deposited. Thus, TST belongs to abyssal argillaceous–siliceous facies in standing-water reduction environment under wave base. HST is located in Bed 17–Bed 19. The major and minor sediments of it are argillaceous and siliceous matters, respectively, with amorphous carbonaceous matter and pyrite commonly visible. Thus, HST belongs to bathyal mudstone facies under strong reduction environment. Therefore, the sedimentary environment in this sequence varies from the abyssal basin facies into the bathyal basin facies from bottom to top, and this sequence is of progradation–aggradation type.

Third-order sequence Sq9: Located in the upper part of the first member of Hetang Formation, it includes TST and HST, with the top and the bottom of type-II sequence boundary. TST is located in Bed 20–Bed 22. The major and minor sediments of it are siliceous and argillaceous matters, respectively. Besides, there is a small amount of amorphous carbonaceous matter and associated much micro-fine-grained sulfide deposited. Thus, TST belongs to abyssal argillaceous–siliceous facies in standing-water reduction environment under wave base. HST is located in Bed 23–Bed 27. The main and minor sediments of it are argillaceous and siliceous matters, respectively, with micro-layers of amorphous carbonaceous matter and pyrite commonly visible. Thus, TST belongs to bathyal mudstone facies under the strong reduction environment. Therefore, the sedimentary environment in this sequence varies from the abyssal basin facies into bathyal basin facies from bottom to top, and this sequence is of the progradation–aggradation type.

Third-order sequence Sq10: Located in the upper part of the second member of Hetang Formation, it includes TST and HST, with the top and the bottom of type-II sequence boundary. TST is located in Sq28–Sq34. The main sediment of it is siliceous matter. Besides, there is a small amount of argillaceous and amorphous carbonaceous matter as well as associated micro-fine-grained sulfide and nodules. Thus, TST belongs to abyssal argillaceous–siliceous facies in standing-water strong reduction environment under wave base. HST is located in Sq35–Sq52. The major and minor sediments are argillaceous and siliceous matters, respectively. Besides, there is a small amount of crystalline carbonaceous matter. There are commonly micro-fine-grained pyrite layers. Thus, HST belongs to bathyal mudstone facies under strong reduction environment. The sedimentary environment in this sequence changed from the abyssal basin facies into bathyal basin facies from bottom to top, and this sequence is of progradational–aggradational type.

The age range of the third-order sequences Sq6–Sq10 is 541–514 Ma, spanning 27 Ma. Each of the third-order sequences spans 5.4 Ma on average.

2.3.1.3 Biostratigraphy and Chronostratigraphy

No fossils were obtained in Hetang Formation of the Area in the Project. Ju (1983) reported that the fossils of *Hupeidiscus orientalis* (*Chang*) were discovered in Hetang Formation of Hanggai Town, Anji County. Much research on the palaeontology of Hetang Formation in Zhejiang Province was conducted in the areas such as Shangwan and Sanxikou of Fuyang, Hetang of Jiangshan, Duibian and Wujialing of Jiangshan, and Fenshui of Tonglu. Wanshi and Wuxi of Fuyang produced trilobite and peduncle, the first member of Hetang Formation in Duibian and Wujialing of Jiangshan, Dongxi of Tonglu, Shangwan of Fuyang, and Xinqiaotou of Xiaoshan produced small shelly animals, the second member of Hetang Formation in Duibian and Hetang of Jiangshan, Shangwan and Sanxikou of Fuyang, Xiayuqiao of Lin'an, and Fenshui of Tonglu produced trilobite and spongy caltrop.

The aforesaid organisms are the earliest metazoans since the Phanerozoic. Among these organisms, *Hupeidiscus orientalis* (*Chang*) is the earliest trilobite discovered in Zhejiang Province. According to *Stratigraphic Chart of China* (2014) of All China Commission of Stratigraphy, the small shelly animals in the first member of Hetang Formation belong to the Meishucunian, and the geological age range of them is the early–middle period of the Early Cambrian. The trilobites in the second member of Hetang Formation are parts of Hupeidiscus–Sinodiscus biozone. They belong to the Nangao, and the geological age range of them is the middle–late period of the Early Cambrian.

2.3.1.4 Analysis of Sedimentary Environment

The sediments of the Early Cambrian mainly consist of siliceous and argillaceous matter followed by carbonaceous matter. Besides, there is a small amount of terrigenous silt and sulfide. Regionally, a large amount of the literature indicates, based on the analysis of sediments, stable isotopes, and fossils, that the sedimentary environment of the Hetang Formation mainly is of deepwater basin–deepwater shelf facies, and the stone coal layer is mainly developed in the sagging part of the slope of the stagnant basin.

The composition of the sediment in Hetang Formation in the Area is evidently different from bottom to up. The content change of the pyrite, organic carbonaceous matter, and prehistoric fossils with facies significance indicates the change of the sedimentary environment from the weak oxidation environment to the reduction environment.

The sediments in the lower part of the first member of Hetang Formation mainly consist of major siliceous matter. Besides, there is a small amount of organic carbon and pyrite visible occasionally. The spongy spicules were well reserved in the rocks. All these indicate that the sedimentary environment in the lower part is still standing-water weak oxidation environment with weak hydrodynamic force. The stone coal layer in the middle part the first member of Hetang Formation mainly consists of organic carbon and contains multiple layers of phosphate nodules. Besides, most small shelly animals and the earliest trilobites were produced here. In this period, the highest content of oxygen in the water was beneficial to organism survival, and the first life explosion in earth history occurred during this period. The sediment in the upper part of the first member is mainly composed of the siliceous carbonaceous shale. However, it contains much sedimentary micro-fine-grained pyrite and pyrite layer with a thickness of 1–2 cm appears locally, indicating that the sedimentary environment was changed into a deep standing-water reduction environment.

The sediments of the second member of Hetang Formation mainly consist of argillaceous matter followed by carbonaceous and the siliceous matter. Horizontal bedding developed. Pyrite is generally distributed in the sediments, and the content of it is higher than that in the first member. Compared with the first member, the second member features more and evidently thicker (1–5 cm) pyrite interbeds. All these indicate a standing-water reduction environment.

Therefore, the sediments of the first member of Hetang Formation mainly include pelagic siliceous matter followed by a small amount of sulfide in the upper part. Therefore, the first member belongs to deepwater shelf–bathyal basin facies in weak oxidation–reduction environment with relatively scare terrigenous sediment sources. The main sediment in the second member of Hetang Formation is the siliceous argillaceous matter, containing much sulfide. Therefore, the second member belongs to bathyal–abyssal basin facies in standing-water reduction environment with scare terrigenous sediment sources.

Regionally, southwestward to Xianxia area of Anhui Province from the Area, the thickness of Hetang Formation increases gradually from nearly 300 to 650 m. It is inferred that the sedimentary basin became deeper accordingly. About 70 km southeastward to Yuhang area from the Area, the lithofacies of Hetang Formation changes into black carbonaceous dolomitic mudstone without silicalite, stone coal layer, and phosphorite, and the thickness is 25 m. This indicates a neritic platform environment.

2.3.1.5 Trace Elements in Strata (Ore-Bearing)

Eighteen rock spectra of the first member ($Є_1h^1$) and the second member ($Є_1h^2$) of Hetang Formation were systematically collected from the Nanchewu profile (PM010) of Hanggai map sheet in Hanggai Town, Anji County, Zhejiang Province. According to spectral analysis, the enriched elements of the first member of Hetang Formation are basically similar to those of the second member. The first member is enriched in Sb, Bi, Cu, Zn, Mo, and Ag, and poor in Au, Be, W, F, and the second member is enriched in Sb, Bi, Zn, Mo, and Ag, and poor in Au. These enriched and poor elements are related to rock type. The argillaceous rock bearing carbon and silicon are enriched in many types of trace elements with high concentration coefficients. The siliceous rock ranks the second.

According to the report of geological investigation of Lin'an map sheet on a scale of 1:200,000 (Zhejiang Bureau of Geology 1967), there are 4–15 layers of phosphate nodule or phosphorite in the first member of Hetang Formation. The thickness is 0.1–2.5 m, and the nodule size is 2–5 cm generally or even up to 10 cm × 25 cm. The content of P_2O_5 is 1–30%. The phosphate nodule layers mostly concentrate nearby stone coal bed. The first member of Hetang Formation in the Area is interbedded with 1–2 layers of stone coal. The ash content is 80–90%, and the thickness is 2–50 m. In the stone coal layers and the phosphate nodule layers, the content of vanadium is 0.3–2% or even up to 6% and content of Mo is 0.004–0.025%. Besides, the content of U and Ni is high, reaching the lowest production grade locally.

2.3.2 Dachenling Formation ($Є_1d$)

The name Dachenling Formation was created by Li and Yu (1965) in Dachenling Village in the southeast of Dachen Town, Jiangshan City, Zhejiang Province, when they discovered *Arthricocephalus* in the bottom of the former Yangliugang Formation. It is still used up to now by the preparation group of the regional stratigraphic table of Zhejiang Province (1979) as well as in the literature such as *Stratigraphic Correlation Chart in China with Explanatory Text* (1982), *Regional Geology of Zhejiang Province* (1989), and *Lithostratigraphy of Zhejiang Province* (1995). In this Project, the name Dachenling Formation is still adopted according to the characteristics of Wenguan profile (PM003),

Yekengwu profile (PM043), and Nanche profile (PM010) of Hanggai map sheet, Meijiata profile (PM035) of Xianxia map sheet, and the lithology along the survey traverse.

2.3.2.1 Lithostratigraphy

The distribution of the Dachenling Formation in the Area is basically the same as that of Hetang Formation. The outcrop area of Dachenling Formation is about 7.64 km^2, accounting for 0.60% of the bedrock area. The characteristics of the lithology and the lithologic association are as follows. The lower part of the formation consists of the interbeds of the gray–dark gray medium laminated dolomitic limestone and the micro-laminated calcareous dolomite. The middle part of the formation consists of the dark gray–black carbonaceous argillaceous or silty silicalite interbedded with dolomitic limestone. The upper part of the formation consists of the interbeds of dark gray thin–medium laminated micrite or dolomitic limestone and thick laminated marlstone or dolomitic limestone. This formation is in conformable contact with underlying Hetang Formation and overlying Yangliugang Formation. The thickness of Dachenling Formation is 52.94–76.20 m.

1. **Stratigraphic section**

The lithology of Dachenling Formation is described by taking the example of Cambrian Hetang Formation ($Є_1h$)–Dachenling Formation ($Є_1d$) profile (Fig. 2.13) of Xianxia map sheet in Meijiata, Xianxia Town, Ningguo City. The details are as follows:

The first member of Yangliugang Formation — Total thickness: >1.32 m

58. Dark gray thin–medium laminated carbonaceous mudstone, the thickness of a single layer: 6–20 cm, very thin horizontal bedding developing. 1.32 m

——————— Conformable contact ———————

Dachenling Formation — Total thickness: 55.48 m

57. Gray medium laminated micrite interbedded with micro laminated dolomitic limestone; the micrite: the thickness of a single layer: 10–25 cm, very thin horizontal bedding developing, pyrite nodules with a size of 1 cm visible occasionally; the dolomitic limestone: micro banded, fawn after weathering, the thickness of a single layer: 1–3cm; the ratio of the micrite to the dolomitic limestone is (3–8):1; the karst micro-landform with lithological difference is costate. 7.02 m

56. Gray thick laminated micrite interbedded with thick laminated dolomitic limestone; the micrite: the thickness of a single layer: 50–60 cm, very thin horizontal bedding developing; dolomitic limestone: the thickness of a single layer: 1–3cm, a part of the layers distributed in a banded and discontinuous manner; the ratio of the micrite to dolomitic limestone: (30–40):1. 14.05 m

55. Dark gray medium–thick laminated carbonaceous siliceous mudstone, the thickness of a single layer: 30–40 cm, very thin horizontal bedding developing. 1.79 m

54. Gray thick laminated carbonaceous micrite interbedded with the thick laminated dolomitic limestone; the thickness of a single layer of the carbonaceous micrite: 50–80 cm; the thickness of a single layer of the dolomitic limestone: 0.5–2 m; very thin horizontal bedding visible in the layers of both the carbonaceous micrite and the dolomitic limestone. 20.54 m

53. Black thick laminated micrite, the thickness of a single layer: 50–150 cm, very thin horizontal bedding developing. 12.09 m

——————— Conformable contact ———————

The second member of Hetang Formation — Total thickness: >40.42 m

52. Black thick laminated carbonaceous mudstone, the thickness of a single layer: 50–80 cm, very thin horizontal bedding developing. 22.93 m

51. Black medium laminated carbonaceous siliceous mudstone, the thickness of a single layer: 20–30 cm, horizontal bedding developing. 17.49 m

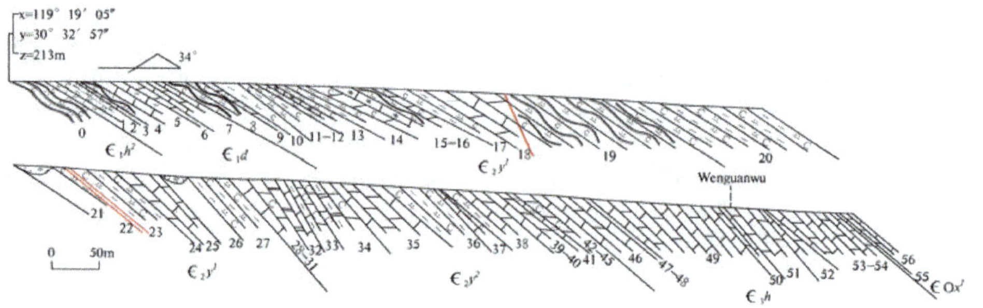

Fig. 2.16 Cambrian Dachenling Formation (Є$_1d$)–Huayansi Formation (Є$_3h$) profile in Wenguan Village, Hanggai Town, Anji County, Zhejiang Province

The lithology of Dachenling Formation is described by taking the example of Dachenling Formation (Є$_1d$)–Huayansi Formation (Є$_3h$) profile (Fig. 2.16) of Hanggai map sheet in Wenguan Village, Hanggai Town, Anji County, Zhejiang Province. The details are as follows:

The first member of Yangliugang Formation — Total thickness: >24.51 m

8. The middle part: poor and discontinuous outcrops, lithology: dark gray thin–medium laminated siliceous carbonaceous silt mudstone; the thickness of a single layer: 8–20 cm; very thin horizontal bedding developing; the upper part: thin laminated marlstone with a thickness of 2 cm. 24.51 m

——— Conformable contact ———

Dachenling Formation — Total thickness: 64.09 m

7. Interbed of dark gray thin micrite and argillaceous limestone; very thin horizontal bedding developing; the thickness of the micrite layers is not stable, and thickness of thick layers is even double that of thin layers. 6.4 m

6. Dark gray medium laminated marlstone, the thickness of a single layer: 15–30 cm, very thin horizontal bedding developing, prone to flake off along the bedding into 1–3 mm schistose. The upper part is interbedded with a small amount of medium laminated dolomitic limestone with the thickness of 10–15 cm. 4.13 m

5. Dark gray thin laminated micrite interbedded with thick laminated argillaceous limestone, very thin horizontal bedding developing in both the micrite and the argillaceous limestone. 25.45 m

4. Dark thin laminated carbonaceous silty siliceous mudstone, flaky. 10.60 m

3. Dark gray medium laminated dolomitic limestone, the thickness of a single layer: 15–30 cm, very thin horizontal bedding developing, spheroid partially due to weathering. 1.99 m

2. Black carbonaceous argillaceous silicalite, with a small amount of disseminated micro-grained pyrite (<1 mm), pyrite concentrating along the bedding locally. 10.21 m

1. Dark gray medium laminated dolomitic limestone, interbedded with carbonaceous siliceous mudstone with the thickness of 20 cm in the middle part; very thin horizontal bedding developing in the dolomitic limestone. 5.31 m

——— Conformable contact ———

The second member of Hetang Formation — Total thickness: >11. 36 m

0. Black carbonaceous silty siliceous mudstone, with a small amount of micro-fine grained pyrite concentrating along the bedding; the pyrite: content: <5%, grain size: <1 mm. 11.36 m

2. **Lithological Characteristics**

The lithology of Dachenling Formation is mainly characterized by dark gray dolomitic limestone, dark gray–black carbonaceous argillaceous silicalite, dark gray micrite, thick laminated argillaceous limestone, and a small amount of dark gray marlstone.

(1) The dolomitic limestone: dark gray, microcrystalline structure; composition: microcrystalline calcite (75–85%), dolomite (5–20%), amorphous carbonaceous matter (5–8%). The analysis results of the chemical samples obtained by chip-channel method from Yekengwu profile are provided in *Report on the Survey and Evaluation Results of Cambrian Limestone Resources of Zhejiang Province* (2007). According to the results, the content of CaO and MgO in the dolomitic limestone of Dachenling Formation is 45–49% and 2.8–4%, respectively.
(2) The carbonaceous argillaceous silicalite and pyrite-bearing carbonaceous silicalite: dark gray–black; composition: cryptocrystalline siliceous matter (55–70%), amorphous carbonaceous matter (10–20%), argillaceous matter (15–45%), and pyrite (5–10%); silt visible occasionally (< 5%); the pyrite: micro-fine-grained, concentrating along the layers.
(3) The micrite: dark gray, microcrystalline structure; main components: microcrystalline calcite (80–95%) and argillaceous matter, the respective content of CaO and MgO in the rock: 40–49% and < 1%.
(4) The argillaceous limestone: dark gray; microcrystalline structure; main components: microcrystalline calcite (70–80%), argillaceous matter (10–15%), and dolomite (5%); containing a small amount of carbonaceous; the content of CaO and MgO in the rock is 36–42% and 0.5–1.5%.
(5) The marlstone: dark gray, microcrystalline structure; composition: microcrystalline calcite (45–55%), argillaceous matter (25–30%), dolomite (5–10%), carbonaceous (5–10%), and micro-fine-grained pyrite (2–3%); the pyrite concentrating along the layer; the respective content of CaO and MgO in the marlstone: 36–38% and 2–5.8%.

3. **Basic sequences**

There are three kinds of basic sequences developing in Dachenling Formation ($€_1d$) (Fig. 2.17).

Basic sequences of type A: Located in the lower part generally, the lithology is characterized by: ① dark gray thin laminar dolomitic limestone: the thickness of a single layer: 0.5–2 cm generally or 3–5 cm occasionally, very thin horizontal bedding developing; and ② dark gray thin–medium laminated microcrystalline limestone: the thickness of a single layer: 2–10 cm, horizontal bedding developing. Since horizontal bedding is visible in both the dolomitic limestone and the microcrystalline limestone and the sediments are mainly composed of microcrystalline calcite, the basic sequences of this kind belong to carbonate facies of shallow shelf under standing-water sedimentary environment and are of non-cyclic basic sequence.

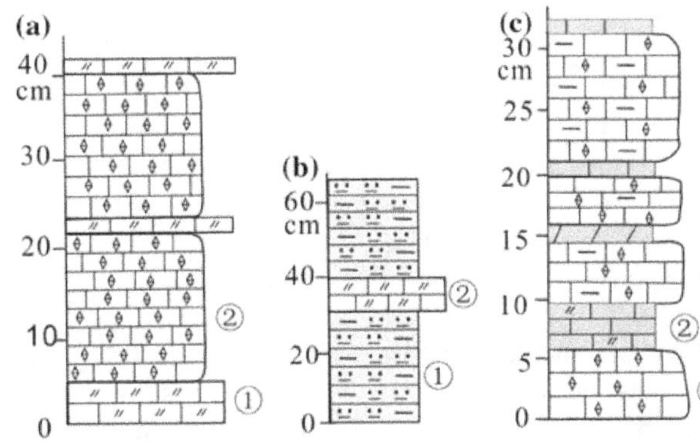

Fig. 2.17 Basic sequences of Dachenling Formation

Basic sequences of type B: Located in the middle part, the lithology is mainly characterized by dark gray–dark thin laminated carbonaceous argillaceous or silt silicalite; gray medium laminated dolomitic limestone is visible occasionally. The silicalite is mainly composed of cryptocrystalline siliceous matter, containing organic carbonaceous matter and pyrite. Therefore, the basic sequences of this kind belong to argillaceous–siliceous facies under bathyal–abyssal reduction environment. They consist of monotonous rocks; therefore, they are of monotonous basic sequence.

Basic sequences of type C: Located in the upper part, they are composed of the interbed of two kinds of rocks: ① dark gray thin laminated argillaceous micrite or micrite: mainly composed of microcrystalline calcite, with a small amount of dolomite of banded distribution, the thickness of a single layer: 2–13 cm, inconspicuous horizontal bedding developing; and ② dark gray micro-laminated limestone or dolomitic limestone: composed of microcrystalline calcite (major), argillaceous matter (minor), and dolomite (a small amount), the thickness of a single layer: 0.5–2 cm generally, very thin horizontal bedding developing. Stylolitic structures often developed between the two kinds of rocks. Compared with the basic sequences of type A, the basic sequences of type B contain more calcareous and argillaceous matter and feature evidently thicker rock layers, indicating that the water is relatively deeper and the basic sequences of type B belong to carbonate facies of the margin of shallow shelf.

According to the analysis of the lithology and sedimentary structure of the basic sequences in Dachenling Formation, the lower part is mainly composed of calcareous dolomite matter, indicating shallow sedimentary water; the middle part mainly consists of argillaceous–siliceous matter, indicating deep sedimentary water; the upper part mainly includes calcareous argillaceous matter, indicating that the water became shallower. Therefore, the sedimentary environment of Dachenling Formation can be described as: shallow-water weak oxidation → deepwater reduction → shallow-water weak oxidation.

2.3.2.2 Sequence Stratigraphy

According to the lithology and lithofacies association and the characteristics of sequence boundary of Meijiata Profile of Xianxia map sheet and Nanchewu profile of Hanggai map sheet, there is one third-order sequence Sq11 in Dachenling Formation. Sq11 belongs to the second-order sequence SS5 and is described as follows.

Third-order sequence Sq11 contains the transgressive systems tract (TST), starved section (CS), and the highstand systems tract (HST). The top and the bottom are conformity interfaces and belong to type-II sequence boundary. TST is located in the lower part of the sequence. The major and minor sediments of TST are calcareous matter and dolomitic matter, respectively. Besides, there is a small amount of argillaceous matter. Therefore, TST belongs to the sedimentary environment of the inner margin of shallow shelf. CS is located in the middle part of the sequence. The main sediment in CS is argillaceous–siliceous matter, resulting from scare chemical sedimentation of the precipitates under pelagic deepwater environment during the maximum marine flooding surface. HST is located in the upper part of the sequence. The major and minor sediments in it are calcareous matter and argillaceous matter, respectively. Besides, there is a small amount of dolomitic matter. Compared with the lower part of the sequence, HST contains more argillaceous matter; thus, HST belongs to oxidation–reduction sedimentary environment of the outer margin of shallow shelf. Therefore, the sedimentary environment in this sequence can be described as: shallow water → deepwater → shallow water. The water became deeper generally, and Sq11 is of aggradational–retrogradational type.

The age range of Sq11 is 514–507 Ma, spanning 7 Ma.

2.3.2.3 Biostratigraphy and Chronostratigraphy

No fossils were obtained in Dachenling Formation in the Project. Much research on the palaeontology of Hetang Formation in Zhejiang Province has been conducted in the areas such as Shangwan and Sanxikou of Fuyang, Hetang of Jiangshan, Duibian and Wujialing of Jiangshan, and Fenshui of Tonglu. According to the research, these areas produced trilobites, brachiopoda, hyoliths, worms, and paleopaly, which are parts of the Arthricocephalus biozone and belong to the Duyun according to *Stratigraphic Chart of China* (2014) prepared by All China Commission of Stratigraphy.

2.3.2.4 Characteristics of Stable Isotopes

As for the limestone of normal marine facies, the $\delta^{13}C$ value is 0 ± 2‰ and $\delta^{18}O$ value is from −13.0‰ to −1.30‰, with the average value of −9.76‰ (Zheng and Chen 2000); the $\delta^{34}S$ value is relatively stable in the Precambrian; the $\delta^{34}S$ value was the highest in the Early Paleozoic with the value of 26.0–35‰; $\delta^{34}S$ fluctuated in the Devonian, dropped to the lowest in the Permian, and increased to about 20‰ of the modern value irregularly after the Cretaceous Period $\delta^{34}S$ (Fig. 2.18).

Two samples were collected from the limestone of Dachenling Formation of Wenguan profile (PM003) for stable isotope analysis of $\delta^{13}C$, $\delta^{18}O$, and $\delta^{34}S$. The $\delta^{13}C$ values of the two limestone samples are −1.04 and 0.17‰, respectively, and the average is −0.44‰. The $\delta^{18}O$ values of the limestone samples are −16.66‰ and −14.55, respectively, and the average value is −15.61‰. The $\delta^{34}S$ values of the two samples are 23.1 and 35.3‰, respectively, and the average value is 29.2‰. Compared with the limestone of normal marine facies, the two limestone samples feature weak negative excursion of $\delta^{13}C$ and $\delta^{18}O$. This is consistent with the negative anomaly of $\delta^{13}C$ in the bottom of Dachenling Formation in Lin'an and Fuyang areas. The possible reason is the regional sea level rise in the

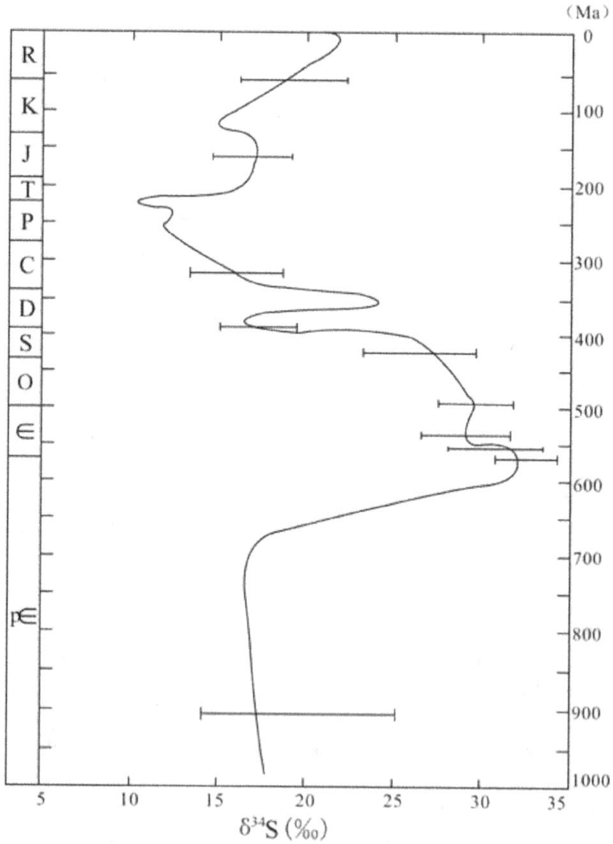

Fig. 2.18 Seawater sulfate $\delta^{34}S$ value change curve in the geological time (modified from Claypool et al. 1980)

Early Cambrian. Besides, this can prove that the Dachenling Formation in the Area belongs to the shallow shelf sedimentary environment with comparative deepwater. The $\delta^{34}S$ value is consistent with high $\delta^{34}S$ content range in the sulfate in the seawater of the Early Cambrian. The possible reason is as follows. With the organism explosion in the Early Cambrian, the sulfate-reducing bacteria reduced the sulfate, causing the fractionation of sulfur isotope. This further reflects that it is the shallow shelf environment with shallow water, moderate environment, and the rich nutrients that provided the condition for organism explosion in the Early Cambrian Period.

2.3.2.5 Analysis of Sedimentary Environment

The main lithology of the Dachenling Formation is characterized by limestone and argillaceous silicalite. As for the limestone, the major and minor components are microcrystalline calcite and dolomite, respectively; trilobites were produced in this formation and horizontal bedding developed, showing weak hydrodynamic force and shallow shelf environment with high oxygen content. There are 2–3 beds of argillaceous–siliceous sediment in the middle part of Dachenling Formation, with micro-fine horizontal bedding developing, indicating deep shelf–bathyal basin facies in a standing-water weak oxidation–reduction environment with scarce terrigenous sediment sources. Marl is absent due to the transition between the limestone and the argillaceous–siliceous sediment. Therefore, it can be inferred that the sea level once experienced a sharp rising and falling, and the sea level in the Dachenling Formation of the Area should experience 2–3 large-scale processes of rising and falling. Regionally, in Jiangshan–Zhuji area near Shaoxing–Jiangshan junction zone, a layer of micro-thin laminated carbonaceous mudstone can be identified in the middle part of the Dachenling Formation, indicating an inconspicuous sea level rise; in the Kaihua–Wangfu of Chun'an area, the Dachenling Formation is interbedded with 3–4 beds of argillaceous silicalite, indicating that sea level experienced four processes of rising and falling in this area.

2.3.2.6 Trace Elements in Strata (Ore-Bearing)

1. **Characteristics of trace elements in strata**

Ten rock spectra were collected in Dachenling Formation of Nanchewu profile (PM010) and Wenguan profile (PM003). According to spectral analysis, there is a big difference in trace element enrichment among a different lithology of Dachenling Formation. The argillaceous silicalite is enriched in Sb, Bi, Mo, Ag, and S, and poor in Au, Be, Zn, and TFe, while the limestone is enriched in Bi and Mo, and poor in Au, Be, Cu, Pb, Zn, W, and TFe. Therefore, argillaceous silicalite is obviously more enriched in trace elements than limestone. Furthermore, for the trace elements enriched in both argillaceous silicalite and limestone, the concentration coefficients in argillaceous silicalite are more than 5 times higher than those in limestone.

2. **CaO and MgO in main limestone and marl**

According to the *Report on the Survey and Evaluation Results of Cambrian Limestone Resources of Zhejiang Province* (2007), sampling by channel method and analysis was conducted for two beds of limestone in Dachenling Formation of Yekengwu profile, which is located in Hanggai Town, Anji County. According to the analysis results, the average content of CaO and MgO in the lower limestone (thickness: 23.05 m) is 47.56% and 2.04%, respectively. The average content of CaO and MgO in the upper limestone (thickness: 5.32 m) is 45.53% and 3.90%, respectively (Zhejiang Institute of Geological Survey 2007).

2.3.3 Yangliugang Formation (ϵ_2y)

The name Yangliugang Formation was created by Lu et al. (1955) in Yangliugang Village that is about 3.5 km to the northeast of Dachen, Jiangshan City, Zhejiang Province. According to the stratigraphic site conference data of western Zhejiang in 1963, Lu et al. (1955) believed that the Yangliugang Formation contained a part of Huayansi Formation and the stratum with the fossils of *Lejopyge* and *Oidalagnostus* constituted the top of Yangliugang Formation. Li and Yu (1965) divided the *Arthricocephalus* horizon from the bottom of the Yangliugang Formation and established Dachenling Formation accordingly. Since then, Yangliugang Formation is a part of the Middle Cambrian Series and thus the lithostratigraphy of it is consistent with its chronostratigraphy. Yangliugang Formation has been subsequently used up to now by the preparation group of regional stratigraphic table of Zhejiang Province (1979) as well as in the literature including *Stratigraphic Correlation Chart in China with Explanatory Text* (1982), *Regional Geology of Zhejiang Province* (1989), and *Lithostratigraphy of Zhejiang Province* (1995). In this Project, Yangliugang Formation was still adopted according to the lithological characteristics of Wenguancun profile (PM003) and Yekengwu profile (PM043) of Hanggai map sheet, Shiling profile (PM024) and Yutang profile (PM038) of Xianxia map sheet, and survey geological observation traverse in the Area.

2.3.3.1 Lithostratigraphy

The vast majority of the Yangliugang Formation (ϵ_2y) in the Area is distributed basically in the same way with Hetang Formation and Dachenling Formation. However, Yangliugang Formation features more extensive outcrops. The outcrop area is about 63.72 km^2, accounting for 5.01% of the bedrock area. In addition, Yangliugang Formation is distributed in a sporadic manner in Nanheling Village in the south of Xianxia intrusion in the southwest corner of Xianxia map sheet, Gaoling Village in the southeast corner of Xianxia map sheet, and Langling Village in the southern part of the Chuancun map sheet.

According to the characteristics of the lithology and lithologic association, Yangliugang Formation ($€_2y$) can be divided into the first member of Yangliugang Formation ($€_2y^1$)) and the second member of Yangliugang Formation ($€_2y^2$).

The first member of Yangliugang Formation ($€_2y^1$): grayish black thin laminated siliceous mudstone and silicalite, interbedded with argillaceous limestone; large lenticular limestone commonly visible; siliceous matter decreasing and calcareous argillaceous matter increasing upward; the bottom, consisting of thin laminated siliceous mudstone, in conformable contact with the dolomitic limestone of underlying Dachenling Formation; thickness: 118.05–279.13 m.

The second member of Yangliugang Formation ($€_2y^2$): The lower part consists of dark gray thin–medium laminated marl, interbedded with banded–thin laminated microcrystalline limestone generally and pie-strip shaped microcrystalline limestone locally; the upper part is composed of the interbed of dark gray thin–medium laminated marl and banded–thin laminated microcrystalline limestone; from bottom to up, argillaceous matter decreases, calcareous matter increases, marl gradually decreases, microcrystalline limestone gradually increases; the thickness of the member ranges 134.40–247.60 m.

1. **Stratigraphic section**

The lithology of Yangliugang Formation is described by taking the example of Dachenling Formation ($€_1d$)–Huayansi Formation ($€_3h$) profile (Fig. 2.16) of Hanggai map sheet in Wenguan Village, Hanggai Town, Anji County, Zhejiang Province. The details are as follows:

Huayansi Formation Total thickness: >15.76 m

46. Rhythm interbed of dark gray thin–medium laminated microcrystalline limestone and micro laminated marl. 15.76 m

——— Conformable contact ———

Yangliugang Formation Total thickness: 522.02 m

The second member of Yangliugang Formation Total thickness: 242.89 m

45. The lower part: interbed of dark gray banded microcrystalline limestone and thin laminated marl, the ratio between the two components: 1:1; the upper part: medium laminated marl interbedded with thin laminated microcrystalline limestone, the thickness of a single layer of the marl: 10–30 cm, the thickness of a single layer of microcrystalline limestone: 3–5 cm; the lower part: thinner marl with the thickness of a single layer of 8–15 cm, the microcrystalline limestone remaining the same compared with the middle part; very thin horizontal bedding visible in all rocks. 14.94 m

44. Dark gray medium laminated marl interbedded with medium laminated microcrystalline limestone; the marl: very thin horizontal bedding developing, prone to become thin laminated owing to weathering; the microcrystalline limestone: horizontal bedding developing, blocky owing to weathering, three layers in total, the thickness of a single layer: 30–35 cm. 2.74 m

43. Interbed of dark gray thin banded microcrystalline limestone and argillaceous limestone; the thickness of a single layer of the two components: 1–3 cm thick and 1–4 cm respectively. 5.42 m

42. Dark gray thin–medium laminated carbonaceous marl interbedded with a small amount of pie-strip shaped microcrystalline limestone; the thickness of a single layer of the carbonaceous marl: 8–15 cm; the size of a pie strip: (10–30) cm× (5–8) cm (length × width). 2.96 m

41. Interbed of dark gray thin banded microcrystalline limestone and dolomitic limestone; the banded microcrystalline limestone: 30–120 cm (length), very thin horizontal bedding developing; the thickness of the

single bed: 2–4 cm; the thickness of a single layer of limestone: 0.1–0.2 cm. 4.17 m

40. Dark gray thin–medium laminated argillaceous microcrystalline limestone, the thickness of a single layer: 8–20 cm, very thin horizontal bedding developing. 5.27 m

39. Interbed of dark gray thin laminated microcrystalline limestone and thick laminated marl; the thickness of a single layer of the two components: 4–8 cm and 1–3 mm respectively. 7.78 m

38. Interbed of dark gray thick laminated carbonaceous siliceous mudstone and rhythmic bed of thin laminated microcrystalline limestone and marl; the carbonaceous siliceous mudstone: very thin horizontal bedding developing, interbedded with a small amount of (5–10) cm × (5–30) cm microcrystalline limestone ellipsoid, the thickness of a single layer: 65–70 cm; the thin laminated microcrystalline limestone: very thin horizontal bedding developing, the thickness of a single layer: 2–8 cm; the thickness of a single layer of thin laminated marl: 1–3 cm. 20.08 m

37. Dark gray medium laminated microcrystalline limestone interbedded with black carbonaceous mudstone. For the microcrystalline limestone, the thickness of a single layer is 25–45 cm and gradually becomes 8–25 cm upwards. 5.26 m

36. Dark gray thin–medium laminated carbonaceous siliceous mudstone, interbedded with a small amount of microcrystalline limestone ellipsoid bed; the carbonaceous siliceous mudstone: very thin horizontal bedding developing, the thickness of a single layer: 6–15 cm; the size of microcrystalline limestone ellipsoid: 45 cm × 25 cm. 30.96 m

35. Interbed of dark gray thin–medium laminated marl and microcrystalline limestone; unclear horizontal bedding developing in the microcrystalline limestone; the marl: very thin horizontal bedding developing extremely, the thickness of a single layer: 8–20 cm; the ratio between the two components: about 1:1; the microcrystalline limestone gradually decreasing upwards. 43.09 m

34. Interbed of dark gray medium laminated carbonaceous marl and medium laminated argillaceous dolomite; the limestone: gradually decreasing upwards, the marl: very thin horizontal bedding developing extremely; the argillaceous dolomite: horizontal bedding developing, the thickness of a single layer: 20–45 cm, 4–5 layers in the lower part, the ratio to the argillaceous dolomite: 2:1 and 1:2 upwards, ellipsoid in the top. 25.77 m

33. Dark gray medium laminated argillaceous microcrystalline limestone, interbedded with black micro-thin bedded carbonaceous marl; the argillaceous microcrystalline limestone: unclear horizontal bedding developing, the thickness of a single layer: 12–70 cm; the carbonaceous marl: very thin horizontal bedding developing extremely, the thickness of a single layer: 1–8 cm generally and 30 cm locally. 13.75 m

32. Interbed of dark gray medium laminated carbonaceous dolomite and microcrystalline limestone; the microcrystalline limestone: vertical joints developing, thin segmented after being squeezed; the thickness of a single layer: 10–40 cm; the carbonaceous dolomite: the thickness of a single layer: 17–40 cm, the ratio to microcrystalline limestone: 2:1. 19.63 m

31. Dark gray medium–thick laminated carbonaceous marl, interbedded with five layers of microcrystalline limestone ellipsoid; the carbonaceous marl: containing micro-fine grained impregnated pyrite (<5%) that concentrate along the layers, the thickness of a single layer: 1–2 mm; the ellipsoid: (40–30) cm × (20–15) cm in size, micro horizontal bedding developing inside. 7.97 m

30. Interbed of dark gray thin–medium laminated microcrystalline limestone and marl; the thickness of a

single layer of microcrystalline limestone: 6—15 cm generally and 25—34 cm locally; the thickness of a single layer of marl: 2—13 cm. 11.81 m

29. Dark gray medium laminated carbonaceous argillaceous dolomite, the thickness of a single layer: 15—30 cm; very thin horizontal bedding developing extremely. 5.85 m

28. Interbed of dark gray medium laminated microcrystalline limestone and thin laminated carbonaceous marl; the microcrystalline limestone: very thin horizontal bedding developing, the thickness of a single layer: 15—20 cm; the ratio between the two components: about 1:1. 4.75 m

——————— Conformable contact ———————

The first member of Yangliugang Formation Total thickness: 79.13 m

27. Dark gray thin—medium laminated carbonaceous siliceous mudstone, interbedded with 10 layer of microcrystalline limestone ellipsoid, very thin horizontal bedding developing. 33.26 m

26. Covering. 13.10 m

25. Dark gray thin laminated microcrystalline limestone, interbedded with micro laminated marl; the microcrystalline limestone: very thin horizontal bedding developing, the thickness of a single layer: 2—5 cm generally and 25 cm locally; the thickness of the single bed of marl: 0.1—5 cm generally. 12.10 m

24. Dark gray medium laminated carbonaceous marl, interbedded with carbonaceous siliceous mudstone: the carbonaceous marl: finely disseminated pyrite visible occasionally, horizontal bedding developing, the thickness of a single layer: 20—30 cm. 8.47 m

23. Dark gray carbonaceous siliceous mudstone, fine-grained disseminated pyrite (<3%) visible. 19.16 m

22. Dark gray thin—medium laminated carbonaceous siliceous mudstone, very fine horizontal bedding developing, prone to flake off into flaky rocks with a thickness of 1—3 mm, a small amount of micro-fine disseminated pyrite (5%) visible in the mudstone that concentrates along the beds. 2.93 m

21. Covering. 8.60 m

20. Dark gray thin–medium laminated carbonaceous siliceous mudstone, horizontal bedding developing, occasionally interbedded with the limestone ellipsoid. 35.91 m

19. Dark gray thin—medium laminated carbonaceous siliceous mudstone, very thin horizontal bedding developing. 36.58 m

18. Dark gray medium laminated marl, interbedded with medium laminated microcrystalline limestone and thin laminated argillaceous dolomite; the marl: very thin horizontal bedding developing, the thickness of a single layer: 10—30 cm; the microcrystalline limestone: vertical joints developing, the thickness of a single layer: 10—30 cm, ellipsoidal after deformation; the thickness of a single layer of the argillaceous dolomite: 3–6 cm. 42.28 m

17. Dark gray thin laminated carbonaceous siliceous mudstone, interbedded with a microcrystalline limestone ellipsoid beds; the carbonaceous mudstone: very thin horizontal bedding extremely developing; the microcrystalline limestone ellipsoid: very thin horizontal bedding developing, size: 10–30 × 20–120 cm. 11.67 m

16. Dark gray medium laminated argillaceous limestone, very thin horizontal bedding developing. 0.50 m

15. Dark gray thick laminated carbonaceous limestone, interbedded with thin laminated calcareous mudstone and microcrystalline limestone ellipsoid; the thickness of a single layer of the carbonaceous limestone: 50—100 cm; the calcareous mudstone: prone to be platy, the thickness of a single layer: 3—5 cm; the microcrystalline limestone ellipsoid: 20 cm × 30 cm in size, horizontal bedding developing, long-axis parallel

bedding discontinuously distributed along the beds. 21.72 m

14. Dark gray medium laminated carbonaceous siliceous mudstone, interbedded with two layers of marl with the thickness of 25cm near the top; the siliceous mudstone: very thin horizontal bedding developing, the thickness of a single layer: 20–30 cm. 6.19 m

13. Dark gray medium laminated carbonaceous argillaceous silicalite, interbedded with three layers of microcrystalline limestone ellipsoids; the thickness of a single layer of carbonaceous argillaceous silicalite: 20–35 cm; the microcrystalline limestone ellipsoid: size (length × width): (20–120) × (15–50) cm approximately, long-axis parallel bedding developing; very thin horizontal bedding developing in all the layers and passing through the ellipsoid. 11.84 m

12. Covering. 7.02 m

11. Dark gray thin laminated carbonaceous argillaceous silicalite, the thickness of a single layer: 3–10 cm, very thin horizontal bedding developing, prone to flake off into platy rocks with a thickness of 3–5 mm. 5.74 m

10. Dark gray medium laminated carbonaceous calcareous mudstone, the thickness of a single layer: 10–40 cm, very thin horizontal bedding developing in all beds, microcrystalline limestone ellipsoid with a size of 10–30 cm × 20–50 cm visible locally along the beds; the long axis of the ellipsoid parallel to the bedding; the bedding in the mudstone parallel to the middle part of the ellipsoid; the bedding surrounding both sides of the ellipsoid visible. 14.38 m

9. Dark gray medium laminated carbonaceous marl, very thin horizontal bedding developing along the beds. 4.45 m

8. Dark gray thin–medium laminated silicon-bearing carbonaceous silty mudstone; the thickness of a single layer: 8–20 cm, very thin horizontal bedding developing. 20.03 m

———— Conformable contact ————

Dachenling Formation Total thickness: >2.84 m

7. Rhythm interbed of dark gray thin laminated microcrystalline limestone and micro laminated argillaceous limestone. 2.84m

2. Lithological characteristics

The lithology of Yangliugang Formation ($€_2y$) is mainly characterized by microcrystalline limestone, marl, siliceous mudstone, or argillaceous silicalite. Besides, there are a small amount of dolomitic limestone and carbonaceous dolomite. In the first member, siliceous mudstone or argillaceous silicalite is common in the lower part, and microcrystalline limestone and marl are more common in the upper part. In the second member of Yangliugang Formation ($€_2y^2$), microcrystalline limestone and marl are common, and the lower part is commonly interbedded with multilayers of dolomitic limestone and carbonaceous dolomite. Furthermore, a small amount of siliceous mudstone or argillaceous silicalite is interspersed in the dolomitic limestone and carbonaceous dolomite.

(1) The carbonaceous siliceous mudstone and silicalite: dark gray–black, argillaceous structure, mainly composed of argillaceous matter (10–25%), siliceous matter (60–75%), and silt (5–10%), additionally containing a small amount of carbon (5–10%).
(2) The dolomitic limestone: gray, micro-argillaceous phanerocrystalline structure, mainly composed of calcite (40–60%), dolomite (3–5%), and terrigenous clastic (such as quartz and sericite) (10–30%), the content of CaO and MgO: 30–40% and 1–9%, respectively.
(3) The marlstone: dark gray, micro-argillaceous phanerocrystalline, mainly composed of calcite (45–75%), argillaceous matter (25–30%), dolomite (5–18%), and carbonaceous (2–5%), the content of CaO and MgO: 29–35% and 1–4%, respectively.

(4) The micrite: gray, micro-argillaceous phanerocrystalline, composed of calcite (85–90%) and a small amount of argillaceous matter (10%), the content of CaO and MgO: 38–45% and 1–5%.
(5) The carbonaceous dolomite: gray, mainly composed of dolomite (75%), argillaceous matter (10–15%), and amorphous carbonaceous matter (5–10%).

3. **Basic sequences**

According to the rock types, association relationship, and sedimentary structure, there are five types of basic sequences developing in the Yangliugang Formation ($Є_2y$) from bottom to top (Fig. 2.19).

Basic sequences of type A: mainly distributed in the middle and lower parts of the first member, and consisting of thin–medium laminated siliceous mudstone. The components of the mudstone mainly include siliceous argillaceous matter besides a small amount of carbonaceous matter, silty matter, and pyrite. Therefore, the basic sequences of this type are of eupelagic siliceous argillaceous facies and belong to monotonic basic sequence.

Basic sequences of type B: distributed in the middle part and the lower part of the first member and the middle part of the second member, and consisting of ① the thin–medium laminated siliceous mudstone and ② ellipsoidal microcrystalline limestone. The siliceous mudstone: major component of the basic sequences of this type, the thickness of a single layer: 5–13 cm generally, fine horizontal bedding developing, containing a small amount of fine-grained pyrite, thus belonging to eupelagic argillaceous silicon facies. The microcrystalline limestone: ellipsoidal with a size of (5–15) cm × (20–50) cm, horizontal bedding developing, distributed discontinuously along the beds, therefore, belonging to carbonaceous facies of deep shelf. Since the limestone ellipsoid increases upward, the basic sequences of this type are of the basic sequence with carbonate rock increasing upward.

Basic sequences of type C: mainly distributed in the upper parts of the first member and the second member, and consisting of ① thin–medium laminated marl and ② microcrystalline limestone. The lower part of the basic sequences of this type is mainly composed of marl. The middle part is the marl-bearing microcrystalline limestone lenses, which gradually increase upward and become the main body of this type of basic sequences. The ratio of marl to microcrystalline limestone is about 1:1. Therefore, the basic sequences of this type are of the cyclic basic sequence with calcareous matter increasing and argillaceous matter decreasing gradually upward.

Basic sequences of type D: distributed in the lower part of the second member, consisting of ① medium laminated marl and ② the thin–medium laminated argillaceous dolomite or rhythmic layers of microcrystalline limestone. The main lithology is characterized by marl. The dolomite gradually decreases upward. Micro-fine horizontal bedding developed in the beds. This indicates that the sediment formed on a standing-water shallow shelf. Therefore, the

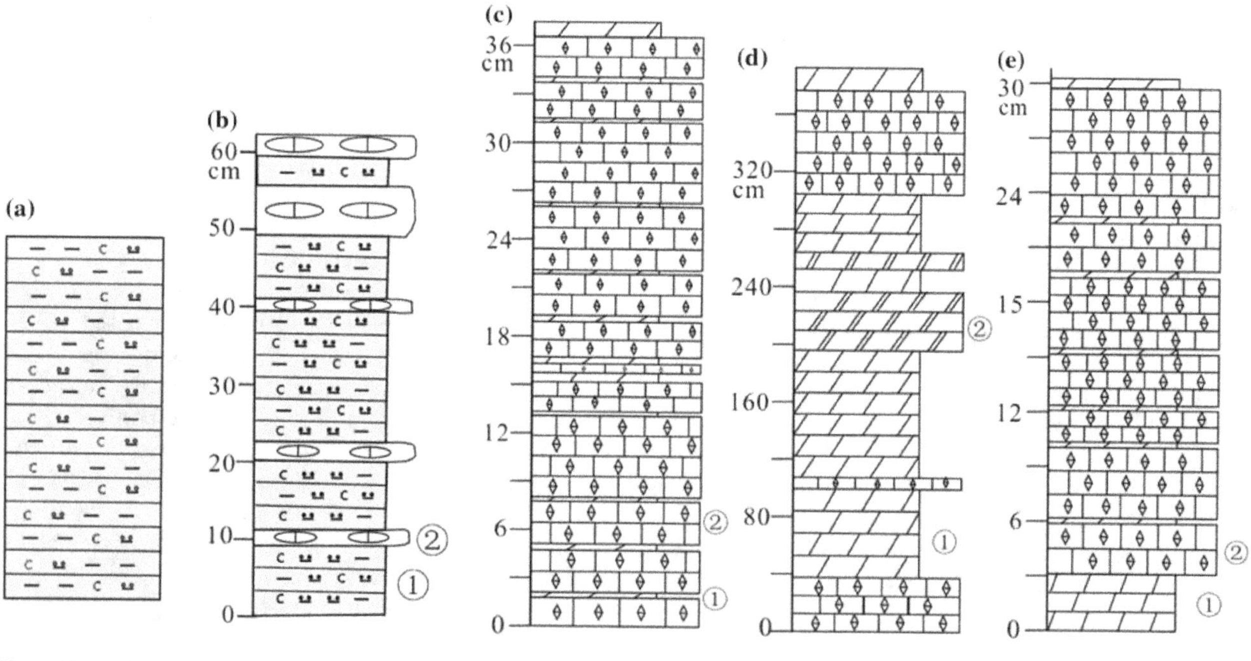

Fig. 2.19 Basic sequences of Yangliugang Formation ($Є_2y$)

basic sequences of this type are of a carbonate rock facies of shallow shelf. According to the relative changes of calcareous and magnesium carbonate, the basic sequences of this type are of the basic sequence with calcareous matter increasing and dolomite decreasing upward, generally reflecting the rise of the sea level.

Basic sequences of type E: distributed in the lower part and the middle-upper part of the second member, and consisting of ① medium laminated marl and ② thin–medium laminated argillaceous limestone or rhythmic layers of microcrystalline limestone. The marl constitutes the main lithology, and the amount of it is about 2–5 times that of microcrystalline limestone. Micro-fine horizontal bedding developed in the beds, indicating that the sediment formed on a relatively shallow shelf. Therefore, the basic sequences of this type are of carbonate rock facies of shallow shelf and belong to the basic sequence with calcareous matter increasing and argillaceous matter decreasing upward, reflecting the falling of the overall sea level.

In terms of the basic sequence association, the first member of the Yangliugang Formation mainly consists of the basic sequences of types A, B, and D from bottom to top, thus constituting a higher-order cyclic sequence. The second member of Yangliugang Formation is composed of the basic sequences of types D, B, and E from bottom to top, thus also constituting a higher-order cyclic sequence. Therefore, two progradational–aggradational sequences that become thicker upward are divided from Yangliugang Formation.

2.3.3.2 Sequence Stratigraphy

According to the association of lithology and lithofacies and the characteristics of sequence boundary of Wenguan profile of Hanggai map sheet, there is two third-order sequences (Sq12–Sq13) in the first member of Yangliugang Formation $Є_2y^1$) and one third-order sequence (Sq14) in the second member of Yangliugang Formation ($Є_2y^2$) (Fig. 2.20). All of these sequences belong to the second-order sequence SS6.

Third-order sequence Sq12: Located in the lower part of the Yangliugang Formation, it contains TST and HST. The bottom of Sq12 is the parallel unconformity interface between the siliceous argillaceous facies in bathyal basin sedimentary environment of Yangliugang Formation and the carbonate facies in shallow shelf sedimentary environment of Dachenling Formation. The sedimentation of intermediate transitional environment is missing. Therefore, the bottom is of type-I sequence boundary. The top of Sq12 is a conformity interface between the siliceous argillaceous facies in bathyal basin sedimentary environment and the argillaceous carbonate facies in deep shelf sedimentary environment, belonging to type-II sequence boundary. TST is located in Bed 8–Bed 14. The main sediment of TST is thin–medium laminated siliceous argillaceous matter. The upper part of TST is occasionally interbedded with microcrystalline limestone ellipsoid; thus, TST is of bathyal siliceous argillaceous facies in weak reduction sedimentary environment under wave base, in which sediment is lacking. HST is located in Bed 15–Bed 18. The sediment of it is calcareous carbonate mainly besides a small amount of pyrite-bearing siliceous and argillaceous matter. Horizontal bedding developed in HST. Therefore, HST belongs to the carbonate facies in the shallow shelf sedimentary environment. Since siliceous matter decreases and argillaceous calcareous matter increases from bottom to top in this sequence, Sq12 is a sequence of progradational–aggradational type with the sea level gradually falling.

Third-order sequence Sq13: Located in the upper part of the Yangliugang Formation, it contains TST and HST. The top and bottom of Sq13 are of type-II sequence boundary.

The TST is located in Bed 19–Bed 23. The sediment of TST is mainly siliceous argillaceous matter or argillaceous–siliceous matter, associated with sulfide and amorphous carbonaceous matter. Micro-fine horizontal bedding developed in TST. Therefore, TST is of bathyal siliceous argillaceous facies in reduction deep standing-water environment. HST is located in Bed 24–Bed 28. The sediment in HST is mainly calcareous carbonates and siliceous argillaceous matter, belonging to the facies of siliceous argillaceous matter-bearing carbonatite under deep shelf environment. Since siliceous argillaceous matter gradually decreases and carbonatite gradually increases from bottom to up in this sequence, Sq13 is of progradational–aggradational sequence with the sea level gradually falling.

Third-order sequence Sq14: Located in the second member of Yangliugang Formation, it includes TST, CS, and HST. The bottom of Sq14 is of type-I sequence boundary, and the top is of type-II sequence boundary. The TST is located on the Bed 29–Bed 35. The sediment in the bottom of TST is mainly dolomitic matter, and therefore, it is inferred that brief uplift and exposure once occurred. The sediment in the upper part of TST is mainly calcareous carbonate followed by argillaceous matter and dolomitic matter, belonging to the argillaceous carbonate rock facies under oxidization zone environment near wave base in the inner edge of shallow shelf. CS is located in Bed 36, and the sediment in it is siliceous argillaceous matter, belonging to bathyal siliceous argillaceous facies. HST is located in Bed 37–Bed 40, and the sediment in it is mainly calcareous carbonate, belonging to shallow shelf facies. Therefore, from

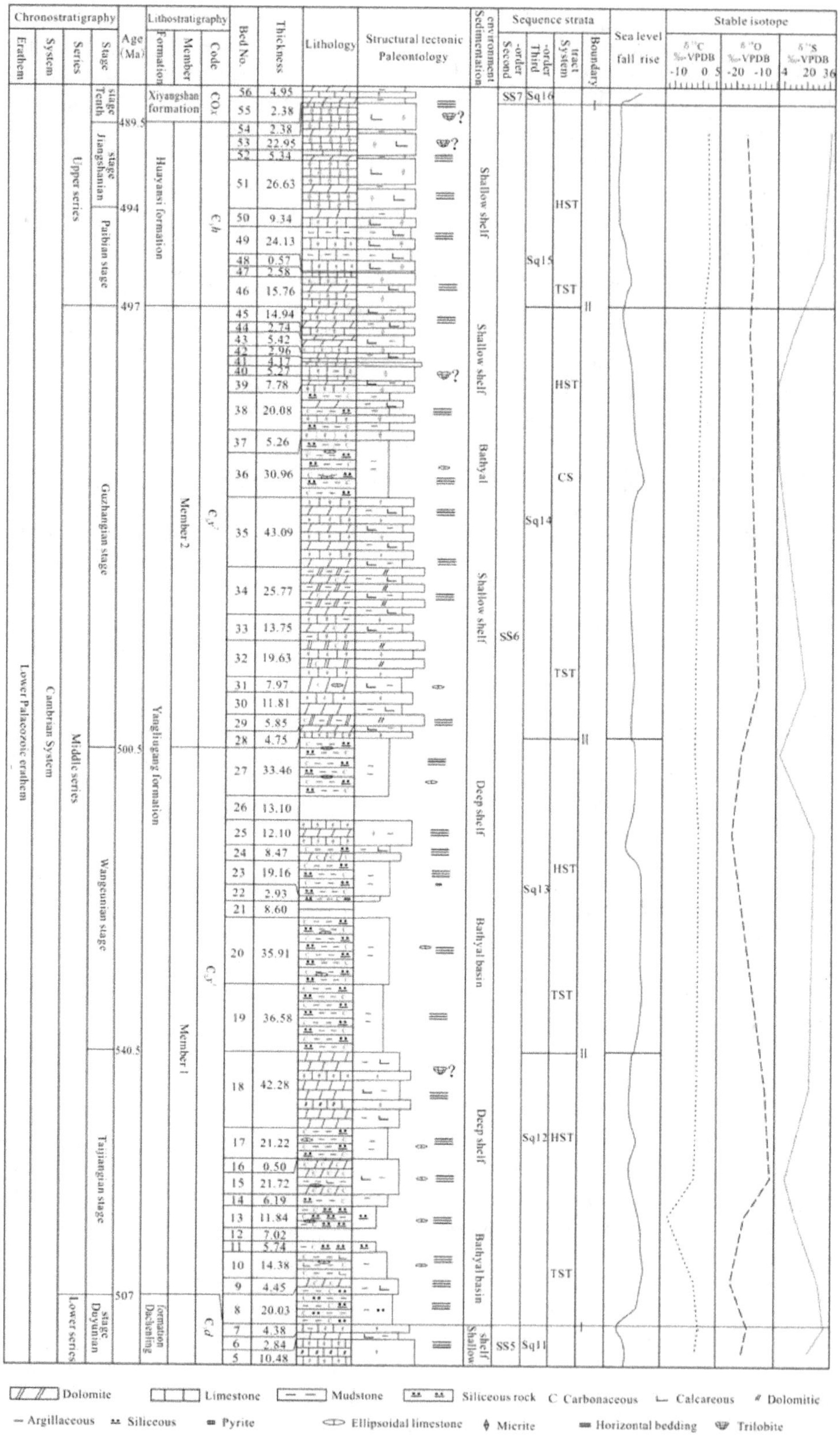

Fig. 2.20 Sequence stratigraphic framework of Yangliugang Formation–Huayansi Formation of the Cambrian Period in the area

bottom to top, the sediment in Sq14 is calcareous argillaceous matter, siliceous argillaceous mater, and argillaceous calcareous matter successively, indicating that the sea level experienced rise and falling successively. Therefore, Sq14 belongs to the sequence of progradation–retrogradation type.

In terms of time, the age range of Sq12, Sq13, and Sq14 sequences is 507 → 504.5 → 500.5 → 497 Ma, respectively. Therefore, the three sequences span 10 Ma in total. In detail, Sq12, Sq13, and Sq14 span 2.5 Ma, 4 Ma, and 3.5 Ma, respectively.

2.3.3.3 Biostratigraphy and Chronostratigraphy

In this Project, no fossils were obtained in Yangliugang Formation in the Area. Lu and Lin (1983) and Ju (1983) conducted detailed research on the palaeontology of Cambrian Period. In August 2011, Peng Shanchi et al. carried out systematic stratigraphic research on Dadoushan profile in Duibian, Jiangshan City, Zhejiang Province, and improved the materials obtained by Lu and Lin (1983). According to the improvement, five fossil zones are divided from the Yangliugang Formation from bottom to up, which are the fossil zones of *Ptychagnostus gibbus, Ptychagnostus atavus, Pseudophalacroma ovale, Lejopygelaevigata armata,* and *Proagnostus bulbus*, respectively, from bottom to up. Furthermore, according to the *Stratigraphic Chart of China* (2014), Yangliugang Formation belongs to the Taijiangian Stage, Wangcunian Stage, and Guzhangian Stage. In detail, the first member of Yangliugang Formation belongs to Taijiangian Stage and Wangcunian Stage, and the second member of Yangliugang Formation belongs to Guzhangian Stage.

2.3.3.4 Characteristics of Stable Isotopes

Five whole-rock stable isotope samples of $\delta^{13}C$, $\delta^{18}O$, and $\delta^{34}S$ of limestone were, respectively, collected from the first member and the second member of Yangliugang Formation in Wenguan profile (PM003).

As for the first member of Yangliugang Formation, the $\delta^{13}C$ value is from −10.69 to −0.39‰, with an average of −3.00‰; the $\delta^{18}O$ value is from −20.56 to −6.94‰ with an average of −14.48‰. As for the second member of Yangliugang Formation, the $\delta^{13}C$ value ranges from −1.84 to −0.02‰, with an average of –0.90‰; the $\delta^{18}O$ value is from −17.47 to −11.49‰, with an average of −14.48‰. Compared to the limestone of normal marine facies ($\delta^{13}C$ value: 0 ± 2‰, $\delta^{18}O$: from 13.0 to −1.30‰), the ten limestone samples experienced significant negative excursion of $\delta^{13}C$ value and $\delta^{18}O$ value. Consequently, it can be inferred that the sea level continued to rise after the period of Dachenling Formation and the Yangliugang Formation entered into the sedimentary environment such as bathyal basin and deep shelf. As a result, a large amount of organic matter with a large amount of $\delta^{13}C$ and $\delta^{18}O$ in the sediment were quickly buried, thus causing a negative excursion of $\delta^{13}C$ and $\delta^{18}O$ in the limestone.

The $\delta^{34}S$ value of the first member of Yangliugang Formation is 8.3–31.3‰, with an average of 21.34‰; the $\delta^{34}S$ value of the second member of Yangliugang Formation is 0.5–20.09‰, with an average of 7.46‰. The change of the $\delta^{34}S$ value is consistent with the variation range of $\delta^{34}S$ in the Cambrian Period. However, the $\delta^{34}S$ value tends to gradually decrease during the period of Dachenling Formation → the first member of Yangliugang Formation → the second member of Yangliugang Formation, indicating that after the organism explosion in Early Cambrian Period (premature), the degree of organism thriving gradually weakened, reducing the sulfur isotope fractionation of reducing bacteria.

2.3.3.5 Analysis of Sedimentary Environment

The sediment of the Yangliugang Formation in the Area is mainly siliceous argillaceous matter and argillaceous carbonate. There is a very little amount of terrestrial silt present. This is quite different from the Yangliugang Formation near the Jiangshan–Shaoxing fracture zone to the southeast and the Yangliugang Formation in Yuhang area, indicating a different palaeogeographic environment.

In the lower part of the first member of Yangliugang Formation, the sediment is mainly composed of siliceous argillaceous matter. Besides, a small amount of the synsedimentary micro-fine-grained pyrite is visible generally in the sediment. Horizontal bedding developed in the beds of the lower part. This indicates weak hydrodynamic force and that the sedimentary environment is the reproduction of the bathyal basin anoxic reduction environment in the Early and Late Cambrian Epoch. In the middle-lower part of the first member of Yangliugang Formation, the silicon argillaceous matter gradually decreases and the mud increases upward. The silicon argillaceous matter is interbedded with some carbonate layers or lens and becomes medium laminated marl upward in which micro-fine horizontal bedding developed. This indicates that the sedimentary environment had become standing-water deep shelf facies with weak oxidation after nearly 300 million years. The lithology and lithologic association of the middle-upper and the upper parts of the first member of Yangliugang Formation are similar to those of the lower-middle and lower parts of the first member of Yangliugang Formation, and the sedimentary environment change of the middle-upper and the upper parts also seems to be the reproduction of previous sedimentary environment.

In the second member of Yangliugang Formation, the main sediment is carbonated-bearing argillaceous matter interbedded with a small amount of siliceous argillaceous matter in the middle part. Micro-fine horizontal bedding developed in all sedimentary beds. This indicates a standing-water, low-energy, and deepwater body. There are small amounts of organic carbon and sulfide, indicating weak oxidation deep shelf sedimentary environment. The siliceous argillaceous matter means pelagic bathyal basin environment, reflecting the sea level experienced a large-scale rise and falling process in the Late Middle Cambrian Epoch.

Regionally, in the Yangliugang Formation of Yuhang area, which is 60–70 km away to the southeast of the Area and the Yangliugang Formation in the areas near Jiangshan–Shaoxing fracture zone including Zhuji, Quzhou, Jiangshan, and Changshan, the sediments are all limestone and marl, there is no siliceous matter, and very rare synsedimentary sulfides exist. This indicates that the sedimentary environment becomes shallow shelf facies.

2.3.3.6 Trace Elements in Strata (Ore-Bearing)

1. **Characteristics of Trace Elements in Strata**

Thirty-nine rock spectra were collected in Yangliugang Formation of Wenguan profile (PM003). The first member of Yangliugang Formation (Є_2y^1) and the second member of Yangliugang Formation (Є_2y^2) are similar in enriched elements. Argillaceous silicalite features the concentration coefficients of Sb, Bi, Mo, and Ag about two times those in limestone. The concentration coefficients of other trace elements in argillaceous silicalite are closely similar to those in the limestone.

2. **CaO and MgO in Main Limestone and Marl**

According to the *Report on the Survey and Evaluation Results of Cambrian Limestone Resources of Zhejiang Province* (2007), the average content of CaO and MgO in the limestone and marl with a thickness of 16 m on the top of the second member of Yangliugang Formation in the Yekengwu profile of Hanggai Town, Anji County, is 35.17% and 2.40%, respectively. The content of CaO and MgO in the rocks in other parts of the Yangliugang Formation is 10%–30% and 1%–5%, respectively.

2.3.4 Huayansi Formation (Є_3h)

Huayansi Formation, formerly known as Huayansi limestone, was renamed Huayansi Formation by Lu et al. (1955) in Huayan Temple (Tianma Mountain), Changshan County, Zhejiang Province. It was subsequently used up to now by the preparation group of regional stratigraphic table of Zhejiang Province (1979) as well as in the literature including *Stratigraphic Correlation Chart in China with Explanatory Text* (1982), *Regional Geology of Zhejiang Province* (1989), and *Lithostratigraphy of Zhejiang Province* (1995). In this Project, the name Huayansi Formation was still adopted according to the lithological characteristics of Wenguancun profile (PM003), Yekengwu profile (PM043), and Jinyindong profile (PM042) of Hanggai map sheet, Shiling profile (PM024) and Yutang profile (PM038) of Xianxia map sheet, and geological observation traverse in the Area.

2.3.4.1 Lithostratigraphy

Huayansi Formation (Є_3h) in the Area is distributed basically in the same way with Yangliugang Formation. However, Huayansi Formation features more extensive outcrops. The outcrop area is about 11.30 km^2, accounting for 0.89% of the bedrock area.

The lithology and lithologic association of Huayansi Formation (Є_3h) are characterized by dark gray thin–medium laminated dolomitic micrite interbedded with micro-laminated marl. Therefore, the lithologic association is monotonous. Besides, horizontal bedding developed and there is no obvious banded structure. Huayansi Formation is in conformable contact with the overlying Xiyangshan Formation and the underlying Yangliugang Formation, with a thickness of 109.68–213.99 m.

1. **Stratigraphic section**

The lithology of Huayansi Formation is described by taking the example of the Cambrian Dachenling Formation (Є_1d)–Huayansi Formation (Є_3h) profile (Fig. 2.16) of Hanggai

Xiyangshan Formation Total thickness: >2.38 m

55. Interbed consisting of dark gray thin–medium laminated marl and thin–medium laminated micrite. 2.38 m

——————————Conformable contact ——————————

Huayansi Formation Total thickness 109.68 m

54. Interbed consisting of dark gray thin–micro laminated argillaceous limestone and thin–medium laminated micrite; the argillaceous limestone: very thin horizontal bedding developing, the thickness of a single layer: 0.1–2 cm; the micrite: horizontal bedding developing, the thickness of a single layer: 5–15 cm. 2.38 m

53. Rhythm interbed consisting of dark gray thin laminated micrite and micro laminated marl; the respective thickness of a single layer of the micrite and the marl: 2–5 cm and 0.1–0.3 cm. 22.95 m

52. Rhythm interbed consisting of dark gray thin–medium laminated micrite and thin laminated marl; the respective thickness of a single layer of the micrite and the marl: 6–36 cm and 1–5 cm. 5.34 m

51. Dark gray thin–medium laminated micrite intercalated with marl; the thickness of a single layer of the micrite and the marl: 7–15 cm and 1–3 cm respectively. 26.63 m

50. Interbed consisting of dark gray thin laminated micrite and marl; the respective thickness of a single layer of the micrite and the marl: 2–6 cm and 1–4 cm. 9.34 m

49. Rhythm interbed consisting of dark gray thin laminated micrite interbedded with micro laminated argillaceous limestone; the respective thickness of a single layer of the micrite and the marl: 3–8 cm and 0.1–0.5 cm. 24.13 m

48. Dark gray thin–medium laminated micrite interbedded with micro laminated dolomitic limestone; the micrite: lotus-root-shaped owing to deform caused by compression; the dolomitic limestone: beige owing to weathering. 0.57 m

47. Rhythm interbed consisting of dark gray thin laminated micrite interbedded with by micro-texture marl; the respective thickness of a single layer of the micrite and the marl: 2–6 cm and 0.1–0.2 cm. 58 m

46. Rhythm interbed consisting of dark gray thin–medium laminated micrite interbedded with micro laminated marl; the micrite: very thin horizontal bedding developing, the thickness of single bed: 4–10 cm; the marl: protrude-shaped owing to weathering, the thickness of a single layer: 0.1–1 cm. 15.76 m

——————————Conformable contact ——————————

The second member of Yangliugang Formation ($Є_2y^2$) Total thickness: >14.94 m

45. The lower part: interbed consisting of dark gray banded micrite and thin laminated marl, the ratio between the two components: 1:1; the upper part: medium laminated marl interbedded with thin laminated micrite, the respective thickness of a single layer of the marl and the micrite: 10–30 cm and 3–5 cm; the lower part: thinner marl with thickness of 8–15 cm, the micrite remaining the same, very thin horizontal bedding visible in all layers. 14.94 m

map sheet in Wenguan Village, Hanggai Town, Anji County, Zhejiang Province. The details are as follows:

2. **Lithological characteristics**

The lithology of Huayansi Formation (Є$_3h$) is mainly characterized by dolomitic micrite and marl.

(1) The dolomitic micrite: sand–clastic microcrystalline structure; composition: calcite (80–90%), dolomite (3–10%), and carbonaceous matter (5–8%); the respective content of CaO and MgO: 37–50% and 2–5%, horizontal bedding developing.
(2) The marl: sand–clastic microcrystalline structure; composition: calcite (70–80%), argillaceous matter (10–20%), and amorphous carbonaceous matter (5–8%). The mixture of the argillaceous matter and dolomitic matter is distributed in micro-lamellar matter, with horizontal bedding developing extremely. The respective contents of CaO and MgO are 37–40% and 5–15%.

3. **Basic sequences**

Two types of basic sequences are developed in Huayansi Formation (Є$_3h$) (Fig. 2.21). All of them consist of ① dark gray thin–medium laminated micrite and ② micro-laminated argillaceous limestone. The amount of the micrite is about 5–10 times that of the marl. In the middle part of the sequences, the micrite is mainly medium laminated and accounts for a high proportion, indicating the process of rising and falling of the sea level. The sediments of this formation are mainly microcrystalline calcareous carbonates. The rocks are mostly medium laminated and partly thin laminated, with horizontal bedding developing. The argillaceous limestone appears in the interbed with a general thickness of less than 2–3 cm. Fine horizontal bedding extremely developed. Therefore, argillaceous limestone belongs to the carbonate facies of shallow shelf.

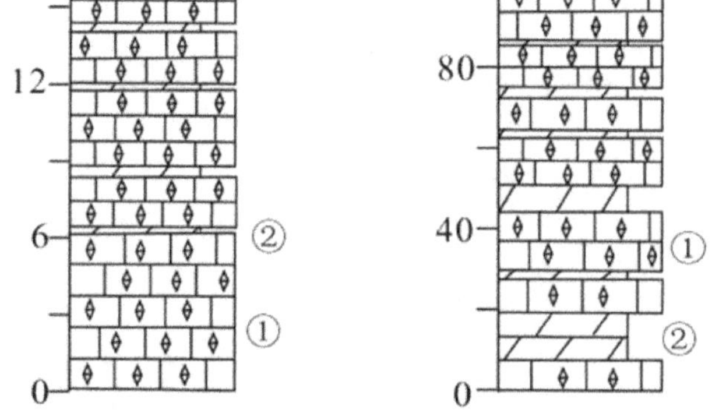

Fig. 2.21 Basic sequences of Huayansi Formation (Є$_3h$)

2.3.4.2 Sequence Stratigraphy

According to the lithology and lithofacies association and the characteristics of sequence boundary of the Wenguan profile of Hanggai map sheet, there is one third-order parasequence set Sq15, which belongs to the second-order sequence SS6 of Huayansi Formation (Є$_3h$).

Third-order parasequence set Sq15

The lithology of the Huayansi Formation is mainly characterized by micrite. Besides, there is a small amount of marl and occasionally visible dolomitic limestone. Among them, micrite accounts for more than 75%, and the thickness (structure) and proportion of it vary from bottom to up, reflecting the rise and falling of the sea level. Owing to this, it can be determined that Huayansi Formation is of transgression–regression sedimentary system, and the top and bottom of Huayansi Formation are both of type-II sequence boundaries. Three third-order parasequences can be identified in the Wenguan profile of Hanggai Town, Anji County. Among them, the first parasequence in the lower part belongs to TST and the second and the third parasequences in the upper belong to HST.

The first-order parasequence is located in Bed 46–Bed 47. The sediment of it is the interbed consisting of thin laminated microcrystalline mud interbedded with micro-laminated calcareous argillaceous matter, thus belonging to TST of argillaceous carbonate facies of shallow shelf sedimentation during transgression period.

The second and the third parasequences are, respectively, located in Bed 49–Bed 52 and Bed 53–Bed 54. The sediment is thin–medium laminated micrite interbedded with micro-laminated dolomitic limestone, with fine horizontal bedding developing in all layers, indicating that HST belongs to shallow shelf carbonate facies after the maximum flooding period.

In summary, the third-order parasequence set Sq15 in the Huayansi Formation reflects a comparatively stable rudimentary environment, in which the sea level experienced slight oscillation of rising and falling in shallow shelf sedimentary basin. Therefore, the third-order parasequence set belongs to the sequence of aggradation type. In terms of time, the age range of Sq15 is 497–489.5 Ma, spanning 8.5 Ma.

2.3.4.3 Biostratigraphy and Chronostratigraphy

In this Project, no fossils were obtained in Huayansi Formation in the Area. Peng (2011) conducted a systematic stratigraphic study on the Dadoushan profile in Duibian, Jiangshan City, Zhejiang Province, divided *Agnostotes orientalis* fossil zone from the profile, and established the "golden spikes" stratotype sections of Jiangshanian Stage. Furthermore, they divided the following seven fossil zones from Huayansi Formation. According to the *Stratigraphic Chart of China* (2014), the *Linguagnostus reconditus* zone belongs to the bottom of the Paibian Stage of early period of the Late Cambrian, the *Glyptagnostus stolidotus* zone belongs to the lower part of the Paibian Stage of the early period of the Late Cambrian, the Glyptagnostus reticulatus zone belongs to the middle part of the Paibian Stage of the early period of the Late Cambrian, the *Agnostus inexpectans* zone belongs to the middle and upper parts of Paibian Stage of the early period of the Late Cambrian, the *Corynexochus plumula* zone belongs to the upper part of Paibian Stage of the early period of the Late Cambrian, the *Agnostotes orientalis* zone belongs to the lower part of Jiangshanian Stage of the middle period of the Late Cambrian, and the *Eolotagnostus* zone belongs to the upper part of the Jiangshanian Stage of the middle period of the Late Cambrian.

2.3.4.4 Characteristics of Stable Isotopes

Two whole-rock stable isotope samples of $\delta^{13}C$, $\delta^{18}O$, and $\delta^{34}S$ of limestone were collected from the Huayansi Formation of the Wenguan profile (PM003).

The respective $\delta^{13}C$ values of the two samples are 2.66 and 2.09‰, with an average of 2.38‰. Therefore, the values feature slight positive excursion compared to the limestone of normal marine facies, reflecting that the sea level fell after the sedimentation of the Yangliugang Formation and the sedimentary environment is of shallow shelf facies. With the burial of the organic matter in the sediment, the consumption of $\delta^{13}C$ decreased, resulting in the positive excursion of the relative enrichment of $\delta^{13}C$ in limestone.

The respective $\delta^{18}O$ values of the two samples are −14.01 and −16.15‰, with an average of −15.08‰. The respective $\delta^{34}S$ values of the two samples are 31.7 and 37.5‰, with an average of 34.6‰. According to the analysis of the Dachenling and Yangliugang formations mentioned above, the $\delta^{18}O$ values should also have featured positive excursion but feature slight negative excursion according to the test results. As for the reasons, the $\delta^{18}O$ values were very possibly affected by the burial temperature of the sediment and temperature of the seawater (more than by the impact caused by the falling of the sea level) at that time. Qi (2006) believed that when the temperature rises and the seawater evaporates, ^{16}O is prone to escape from seawater into the atmosphere since it is light and highly active, while the fractionation coefficient of the isotope becomes weak with the rise of temperature since it is inversely proportional to temperature. Therefore, the difference in oxygen isotope content between seawater and vapor becomes smaller and the amount of ^{16}O that escapes from the sea decreases. As a result, in the seawater, the amount of ^{16}O and ^{18}O in seawater relatively increases and decreases, respectively (i.e., $\delta^{18}O$ decreases). The $\delta^{18}O$ values of the Huayansi Formation in the Area feature significant negative excursion compared to the limestone of normal marine facies. The reason may be that, with the continuous falling of the sea level in the shallow sedimentary environment, the burial temperature of the sediment rises. This promotes the short-term thriving of reducing bacteria followed by the increase of fractionation of the sulfur isotope. Consequently, the $\delta^{34}S$ value becomes abnormally high (average: 34.6‰) and positive anomaly occurs. However, since there were only two samples collected in this Project, the above results are poorly persuasive. Further efforts are required in the future in order to confirm whether the above explanation is reasonable.

2.3.4.5 Analysis of Sedimentary Environment

Huayansi Formation is the simplest stratum in the lithology and lithologic association of the Cambrian. The main sediment is thin–medium laminated microcrystalline calcite, interbedded with a small amount of micro-laminated mud or dolomitic lime mud occasionally. Fine horizontal bedding developed in the layers, but no obvious organic carbonaceous matter and sulfide are visible. The ancient organisms in Huayansi Formation are mainly planktonic Agnostus. It is indicated that the hydrodynamic force was weak in the environment and the oxygen content in the water was not favorable for the survival of benthic organisms. Therefore, the sedimentary environment of Huayansi Formation belongs to shallow shelf near the oxidation–reduction surface under the wave base.

Regionally, Huayansi Formation features in a similar way as the lithology and stratum thickness of the Area and is the main marker stratum of the regional geological survey of the Cambrian.

2.3.4.6 Trace Elements in Strata (Ore-Bearing)

1. **Characteristics of trace elements in strata**

Twelve rock spectra were collected in Huayansi Formation (ϵ_3h) of Wenguan profile (PM003). According to the results of the spectral analysis, Huayansi Formation (ϵ_3h) is enriched in Sb and Bi, and poor in Au, Be, Cu, Pb, Zn, W, Sn, and TFe. Compared with the lower Yangliugang Formation, Dachenling Formation, and Hetang Formation, Huayansi Formation features less enriched trace elements and the enrichment coefficients are 1 times smaller. This is related to the fact that microlithic limestone constitutes the main lithology of this formation.

2. **CaO and MgO content of main limestone and marl**

According to the *Report on the Survey and Evaluation Results of Cambrian Limestone Resources of Zhejiang Province* (2007), the limestone thickness of the Huayansi Formation in the Yekengwu profile of Hanggai Town, Anji County, is 135.95 m. Furthermore, among the 53 samples, the respective average content of CaO and MgO is 44.39% and 3.22%.

2.3.5 Xiyangshan Formation (ЄO$_x$)

The name Xiyangshan Formation, formerly known as Xiyangshan shale, was created by Lu et al. (1955) in Xiyang Mountain, which is 1.5 km to south of the town of Changshan County, Zhejiang Province. It belongs to the Late Cambrian and is a biostratigraphic unit. It was successively used in the *Qu County Map Sheet on a Scale of* 1:200,000 (1969), *Stratigraphic Correlation Chart in China with Explanatory Text* (1982), *Regional Geology of Zhejiang Province* (1989), etc. For this formation, the preparation group of regional stratigraphic table of Zhejiang Province (1979) adopted both division proposed by Lu et al. (1955) and lithostratigraphy division, while lithostratigraphy division was adopted in *Lithostratigraphy of Zhejiang Province* (1995). In this Project, Xiyangshan Formation (ЄOx) was still used according to the lithologic characteristics of Kaokengkou profile (PM004), Yekengwu profile (PM043), Jinyindong profile (PM042) in Hanggai map sheet, Shiling profile (PM024) and Yutang profile (PM038) in Xianxia map sheet, and survey traverse.

2.3.5.1 Lithostratigraphy

Xiyangshan Formation (ЄOx) is basically distributed in the same way as Yangliugang Formation in the Area. However, Xiyangshan Formation features more extensive outcrop. The outcrop area is about 38.76 km^2, accounting for 3.05% of the bedrock area.

According to the lithology and lithologic association, the Xiyangshan Formation (ЄOx) is divided into the first member of Xiyangshan Formation (ЄOx^1), the second member of Xiyangshan Formation (ЄOx^2), and the third member of Xiyangshan Formation (ЄOx^3). It is in conformable contact with the overlying Yinzhubu Formation and the underlying Huayansi Formation.

The first member of Xiyangshan Formation (ЄOx^1): The lower part is composed of the interbed consisting of deep gray thin–medium laminated marl and thin–medium laminated micrite, which are interbedded with carbonaceous and calcareous siliceous mudstones. Furthermore, the interbed gradually changes into the interbed consisting of dark gray vein-strip shaped, lenticular, and banded micrite and medium laminated argillaceous limestone from bottom to up. The lithologic association is mainly composed of argillaceous marl, followed by micrite. Besides, there is a small amount of siliceous mudstone. Very thin horizontal bedding developed in all layers. The marl is prone to be platy owing to weathering. The thickness of this member is 144.76–427.6 m.

The second member of Xiyangshan Formation (ЄOx^2): The middle and lower parts of the member consist of multicycle interbed of dark gray medium–thick laminated marl (micrite or dolomitic limestone locally) and marl-bearing vein-strip shaped micrite → small-vein-strip shaped micrite. The small-vein-strip shaped micrite is 1–2 cm thick and 5–100 cm long. Furthermore, the sparse small-vein-strip shaped micrite gradually becomes dense from bottom to top within the single circle. The upper part of the member is composed of the rhythmic interbed of dark gray medium laminated siliceous mudstone and marl and "vein-strip shaped micrite." Compared with the micrite strips of the first and the third member, the "vein-strip shaped micrite" in this member features a longer cross section and a greater thickness. The thickness of this member is 113.63 m.

The third member of Xiyangshan Formation (ЄOx^3): dark gray medium-thickness–thick argillaceous limestone, containing vein-strip shaped and lenticular micritic marl. The multicycle interbed of the vein-strip shaped micrite is mainly composed of argillaceous limestone. The part near the top of the member consists of dark gray medium laminated argillaceous limestone and marl, interbedded with the banded and thin laminated micrite. The thickness of this member is 254.24 m.

1. **Stratigraphic section**

The stratigraphic profile of Xiyangshan Formation is described by taking the example of Cambrian Xiyangshan Formation (ЄOx) profile (Fig. 2.22) of Hanggai map sheet in Kaokengkou Village, Hanggai Town, Anji County, Zhejiang Province. The details are as follows:

Yinzhubu Formation Total thickness: >10.99 m

55. Gray medium laminated calcareous argillaceous siltstone. 10.99 m

——————————Conformable contact ——————————

Xiyangshan Formation Total thickness 547.60 m

The third member of Xiyangshan Formation Total thickness 244.67 m

54. Two rhythmic interbed consisting of dark gray medium laminated silty mudstone and vein-strip shaped marl. The lower silty mudstone is 40 cm thick with horizontal bedding developing; the thickness of a single layer of upper vein-strip shaped marl is 40–50 cm; the vein-strip is (5–6) × (10–20) cm, accounting for 50%; the marl is 3–5 cm thick per vein-strip and gradually decreases upwards. 1.48 m

53. Interbed consisting of dark gray thin–medium laminated marl and vein-strip shaped micrite with the ratio between the two components of 1:1. As for the marl, the proportion in the lower part, middle part, and upper part of the layer is 100%, 50%, and 5% respectively, and the thickness of a single layer in the lower part, middle part, and upper part of the layer is 15–20 cm, 3–5 cm, and 0.5 cm respectively. As for the micrite, the proportion is contrary to that of the marl, it gradually becomes thin laminated from being vein-strip shaped from bottom to up, and the thickness of a single layer is 2–3 cm→ 6–15 cm from bottom to up. 4.02 m

52. Interbed consisting of dark gray small-vein-strip shaped marl and carbonaceous mudstone. The small-vein-strip shaped marl is 1 cm× (5–10) cm, accounting for 80%. The carbonaceous mudstone is 0.1–0.5 cm thick. 1.06 m

51. Dark gray thin laminated micrite; the thickness of a single layer: 3–10 cm. 1.06 m

50. Interbed consisting of dark gray medium laminated marl and thin laminated micrite. The thickness of a single layer of the marl: 20–30 cm; the thickness of a single layer of the micrite: 8–10 cm; the ratio between the two components: 1.5: 1. 2.47 m

49. Dark gray medium laminated marl, the thickness of a single layer: 20–50 cm. 4.90 m

48. Dark gray medium laminated micrite, the thickness of a single layer: 20–50 cm. 7.17 m

47. Dark gray medium laminated argillaceous limestone, the thickness of a single layer: 20–50 cm, horizontal bedding developing. 26.30 m

46. Dark gray banded micrite-bearing marl, the thickness of a single layer: 3–5 cm, horizontal bedding developing. 2.78 m

45. Dark gray medium –laminated argillaceous limestone bearing vein-strip shaped micrite. The thickness of a single layer of the argillaceous limestone: 40–100 cm; the thickness of a single layer of the vein-strip: (3–5) × (20–30) cm, the content of the vein-strip shaped micrite: <5%. 7.30 m

44. Dark gray medium laminated micrite interbedded with micro laminated argillaceous limestone. The micrite; horizontal bedding developing, the thickness of a single layer: 13–20 cm. The thickness of a single layer of argillaceous limestone: 0.5–1 cm. 2.91 m

43. Dark gray medium laminated carbonaceous marl, generally interbedded with vein-strip shaped micrite and locally interbedded with banded carbonaceous marl. The carbonaceous marl: very thin horizontal bedding developing, prone to flake off along the bedding. The vein-strip shaped micrite is (10–30) × (3-5) cm in size and is mostly visible in a single bed, accounting for 10%–15% of the layer. 46.83 m

42. Dark gray medium–thick laminated carbonaceous marl interbedded with a small amount of vein-strip shaped micrite. The carbonaceous limestone: very thin horizontal bedding developing, the thickness of a single layer: 30–60 cm. The vein-strip of micrite is 3–6 cm thick and 20–50 cm long and mostly visible in 1–3 layers; the micrite accounts for 5% of the layer. 29.70 m

41. Dark gray banded to thin laminated micrite, interbedded with micro-thin laminated marl. The micrite: the thickness of a single layer: 1–5cm, horizontal bedding developing. The marl: very thin horizontal bedding extremely developing, the thickness of a single layer varies greatly with a value of 0.1–1 cm. 2.91 m

40. Dark gray medium laminated marl, the thickness of a single layer: 20–50 cm, horizontal bedding developing. 27.47 m

39. Dark gray thin–medium laminated vein-strip shaped micrite and argillaceous limestone. The vein-strip of micrite is (1–3) × (5–20) cm thick, accounting for 10% of the layer. The argillaceous limestone: the thickness of a single layer: 8–20 cm, horizontal bedding developing. 23.49 m

38. Dark gray medium laminated argillaceous limestone, the thickness of a single layer: 20–50 cm, horizontal bedding developing. 19.39 m

37. Dark gray thin laminated argillaceous limestone bearing vein-strip shaped–banded micrite. The thickness of a single layer of argillaceous limestone is 3–10 cm. The vein-strip shaped or banded micrite is 2–5 cm thick per single bed and varies greatly in length. 23.41 m

36. Dark gray medium laminated argillaceous limestone, the thickness of a single layer: 30–50 cm, occasionally containing a small amount of the small vein-strip shaped micrite. 9.57 m

——————— Conformable contact ———————

The second member of Xiyangshan Formation Total thickness 101.60 m

35. Interbed consisting of dark gray thin laminated marl and vein-strip shaped, banded, and thin laminated micrite. The thickness of a single layer of the micrite is 1–5 cm. The marl accounts for 60% (in the upper part) and 40% (in the lower part) of the layer. 12.03 m

34. Dark gray thin–medium laminated siliceous mudstone and marl, interbedded with a small amount of vein-strip shaped–banded limestone. The thickness of a single layer of the marl is 5–30 cm; the thickness of a single vein-strip of the siliceous mudstone is similar to that of the marl; the thickness of the vein-strip shaped and banded limestone and vein-strip shaped limestone is 3–8 cm. 13.68 m

33. Interbed of dark gray thin laminated marl and vein-strip shaped–banded micrite. A single bed of the marl is 3–8 cm thick. The thickness of a single layer of the micrite is 1–3 cm. 3.35 m

32. Dark gray vein-strip shaped micrite and banded marl. The micrite is 1–3cm thick. The ratio between the two components is about 1:1. 1.01 m

31. Dark gray medium laminated siliceous mudstone, interbedded with small-vein-strip shaped micrite in the middle part. The single bed of siliceous mudstone is 10–30 cm thick. The size of a single vein-strip shaped micrite is (1–3) × (5–20) cm; the ratio between the two components is 1:1. 2.51 m

30. Dark gray vein-strip shaped limestone. 5.52 m

29. Dark gray medium laminated argillaceous limestone, very thin horizontal bedding developing in the layers. 1.19 m

28. Gray fine thick vein-strip shaped–banded micrite. 2.40 m

27. Gray medium laminated marl, horizontal bedding developing. 11.33 m

26. Interbed consisting of gray thin laminated vein-strip shaped micrite and marl. The marl is 1–10 mm thick, with horizontal bedding developing. The vein-strip shaped micrite is 1–10 mm thick and 3–40 cm long. The ratio of the two elements is about 1:1. 1.77 m

25. Interbed consisting of dark gray medium laminated marl and small-vein-strip shaped micrite. The thickness of a single layer of the marl is 10–20 cm; the micrite vein-strip is 2–5 cm thick and 10–30cm long. The single-rhythm vein-strip shaped micrite increases from 10% to 50% from bottom to top. Very thin horizontal bedding develops in all layers. 6.90 m

24. Dark gray medium laminated marl, the thickness of a single layer: >10 cm, very thin horizontal bedding developing, flaky after weathering. 9.83 m

23. Dark gray vein-strip shaped micrite and banded marl. The micrite: vein-strip shaped–banded, thickness: 1–3 cm, length: 10–100 cm, distributed discontinuously along the layer; the marl: very thin horizontal bedding developing, thickness: 1–3 cm, bypassing the strips. The ratio the marl to the micrite: 6:5 approximately. 2.61 m

22. Dark gray medium laminated marl interbedded with vein-strip shaped micrite, mainly consisting of the marl. As for the vein-strip shaped micrite, the thickness of a single layer is 3–6cm and the total thickness is 3–30 cm; the thickness of a single layer of the marl is 40–60 cm. 2.52 m

21. Dark gray apparently-thick laminated small-vein-strip shaped micrite, interbedded with a small amount of thin laminated argillaceous limestone; the vein-strip shaped micrite accounting for 70% of the layer. 7.67 m

20. Dark gray medium laminated dolomitic limestone, interbedded with a layer of micrite and a thickness of 5 cm in the lower part, very thin horizontal bedding developing in the layers. 1.45 m

19. Dark gray medium laminated small-vein-strip shaped micrite, the thickness of a strip: < 1cm, the micrite accounting for 50% of the layer, marl existing between the strips. 0.91 m

18. Dark gray medium laminated carbonaceous limestone, horizontal bedding developing. 1.36 m

17. Dark gray small-vein-strip shaped limestone. The small-vein-strip shaped micrite is 1–2 cm thick and 3–5 cm long, the micrite accounts for 70% of the layer. 1.12 m

16. Dark gray medium laminated micrite, very thin horizontal bedding developing. 3.32 m

15. Interbed of dark gray thin laminated marl and small-vein-strip micrite. The single vein-strip is 1–2 cm thick; the micrite accounts for 70% of the layer. 1.26 m

14. Dark gray medium–thick laminated marl, very thin horizontal bedding developing. 5.07 m

13. Two rhythmic interbed of dark gray vein-strip shaped marl and vein-strip limestone. The strips of limestone gradually increase and become thinner from bottom to top, accounting for 10%–30% in the lower part and 70% in the upper part; the overall ratio between the two components is 1:1. 7.61 m

———————— Conformable contact ————————

The first member of Xiyangshan Formation Total thickness: 223.5 m

12. Dark gray medium–thick laminated marl interbedded with a small amount of lenticular micrite. The marl:

the thickness of a single layer: 20–100 cm, interbedded with small-vein-strips in the middle part locally, very thin horizontal bedding developing in layers. The micrite lens: 5 × 30 cm, hollow owing to weathering. 26.00 m

11. Interbed consisting of dark gray vein-strip shaped–banded micrite and marl. The micrite: vein-strip shaped–banded, thickness: 1–2 m thick, length: 5–100 cm. The ratio of the vein-strip shaped micrite to the marl: 1:1 approximately. 2.49 m

10. Dark gray medium laminated carbonaceous silty marl, horizontal bedding developing, the thickness of a single layer: 10–30 cm. 3.56 m

9. Rhythm interbed consisting of dark gray medium laminated marl and apparently-thick laminated vein-strip shaped limestone. The lower marl: very thin horizontal bedding developing, the thickness: 2 m; the upper apparently-thick laminated vein-strip shaped limestone: the thickness of a strip: 1–5 cm; the ratio of the marl to the vein-strip shaped limestone: about 1:1. 15.87 m

8. Dark gray vein-strip shaped–banded micrite interbedded with thin laminated marl. The size of vein-strip shaped–banded micrite: 1–5 cm thick and 5–100 cm long; the marl: the thickness of a single layer: 1–2 cm, accounting for 10% of the bed. 3.81 m

7. Dark gray apparently-thick laminated vein-strip shaped micrite, interbedded with thin laminated argillaceous limestone. The thickness of a single layer of argillaceous limestone: 2–8 cm; the thickness of a vein-strip of micrite: 1–5 cm thick. 11.54 m

6. Dark gray medium laminated marl interbedded with lenticular carbonaceous marl. The marl: very thin horizontal bedding developing, the thickness of single bed: 20–50 cm; the carbonaceous limestone lens: size: (3–8) × (15–30) cm, appearing generally in the form of a single layer and partly in the form of 2–3 layers as a whole. 45.47 m

5. Dark gray medium laminated calcareous siliceous mudstone, the thickness of a single layer: 20–30 cm, prone to flake off, interbedded with a small amount of thin laminated micrite 10 meters upwards. 51.91 m

4. Interbed consisting of dark gray medium–thick laminated marl and thin–medium laminated micrite. The marl: very thin horizontal bedding developing, the thickness of single layer: 20–80 cm; the micrite: the thickness of a single layer: 10cm in the lower part of the layer and 30–50cm in the upper part of the layer. 11.79 m

3. Interbed consisting of dark gray thin laminated marl and vein-strip shaped micrite. The thickness of a single layer of the marl: 1–4 cm; the micrite: the thickness of a single layer: 2–4 cm, horizontal bedding developing. 1.32 m

2. Dark gray medium–thick laminated carbonaceous siliceous mudstone, interbedded with lenticular limestone. The carbonaceous siliceous mudstone: very thin horizontal bedding developing, prone to be peeled as platy, the thickness of a single layer: 30–60 cm. The lenticular limestone: 5 × 20 cm in size, distributed discontinuously along the layer. 15.25 m

1. Interbed consisting of dark gray medium laminated carbonaceous marl and thin–medium laminated micrite. The carbonaceous marl: very thin horizontal bedding developing, platy owing to weathering, the thickness of a single layer: 12–30 cm. The micrite: horizontal bedding developing, the thickness of a single layer: 6–30 cm generally. There is slightly more marl than the micrite. 6.16 m

———— Conformable contact ————

Huayansi Formation Total thickness: >11.09 m

0. Dark gray medium laminated micrite, micro-fine horizontal bedding developing. 11.09 m

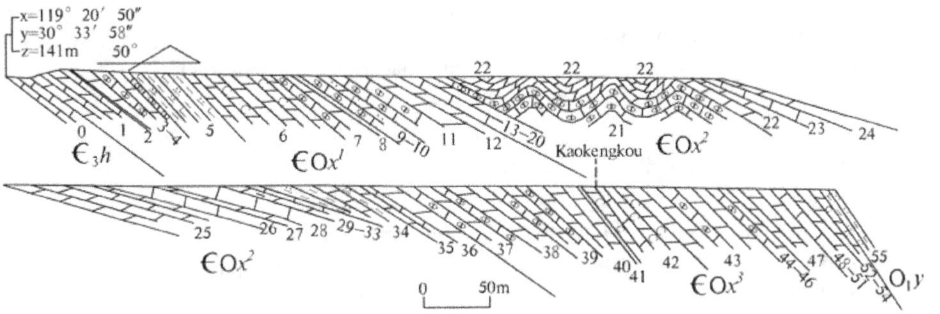

Fig. 2.22 Surveyed stratigraphic profile of the Cambrian Xiyangshan Formation (ЄO*x*) in Kaokengkou Village, Hanggai Town, Anji County, Zhejiang Province

2. **Lithological characteristics**

The lithology of the Xiyangshan Formation (ЄO*x*) is mainly characterized by dark gray micrite, dolomitic limestone, argillaceous limestone, marl, and siliceous mudstone. Among these rocks, siliceous mudstone is only distributed in the lower part of the first member and the upper part of the second member, and other rocks are generally distributed in all members of the Xiyangshan Formation.

(1) The micrite: gray and dark gray, microcrystalline structure, major component: calcite (75–90%), minor components: argillaceous matter (10–20%) and amorphous carbonaceous matter (0–10%). The respective content of CaO and MgO: 30–42% and 2–4%.
(2) The dolomitic limestone: gray, micro- to argillaceous–crystalline structure, major components: calcite (40–60%), dolomite (3–5%), and terrigenous clastic matter (quartz, sericite, etc.) (10–30%). The respective content of CaO and MgO: 30–40% and 1–9%.
(3) The argillaceous limestone: dark gray, microcrystalline structure, major component: calcite (55–70%), minor components: argillaceous matter (10–30%) and amorphous carbon (5–10%). The respective content of CaO and MgO: 15–30% and 2–4.8%.
(4) The marl: dark gray, argillaceous structure, major component: argillaceous matter (50%), minor components: calcite (15–25%), sericite (5%), and quartz (25–30%). The respective content of CaO and MgO: 9–13% and 3–5%.
(5) The siliceous mudstone: dark gray, mainly composed of argillaceous matter (60%), cryptocrystalline siliceous matter (30%), and amorphous carbonaceous or calcareous matter (10%); very thin horizontal bedding developing; prone to flake-off owing to weathering.

3. **Basic sequences**

There are three types of basic sequences in the first member of Xiyangshan Formation (ЄO*x*^{1}) (Fig. 2.23).

Basic sequences of type A: distributed in the lower part of the first member of Xiyangshan Formation, and consisting of rhythmic interbed of ① medium laminated marl and ② thin–medium laminated micrite. From bottom to top, the component ① gradually increases and component ②

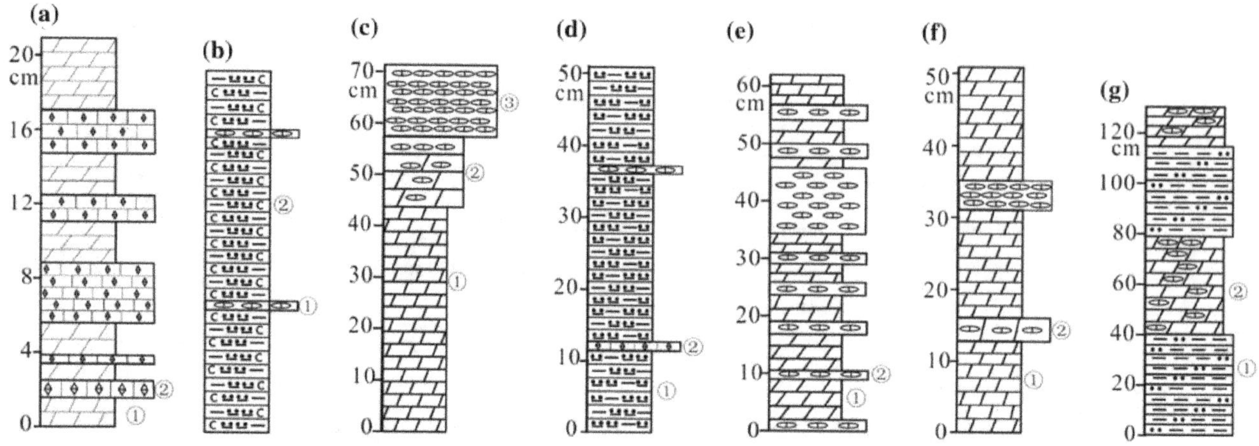

Fig. 2.23 Basic sequences of Xiyangshan Formation (€Ox)

gradually decreases. The basic sequences of this type are of non-cyclic basic sequence with argillaceous matter gradually increasing upward.

Basic sequences of type B: distributed in the middle-lower part of the first member of Xiyangshan Formation, and composed of ① a small amount of micritic marl occasionally interbedded with ② medium laminated carbonaceous siliceous mudstone or the calciferous siliceous mudstone. The basic sequences of this type are of non-cyclic basic sequence with monotonous lithology.

Basic sequences of type C: distributed in the upper part of the first member of Xiyangshan Formation, and consisting of ① thin–medium laminated marl, ② cake-striped micritic marl, and ③ lenticular and cake-striped micrite. The lithological boundary is completely different, and the monocyclic lithologic association is characterized by the gradual increase of the cake-striped micrite from bottom to top, as well as the gradual increase of micrite. It is indicated that the sea level once became shallower. In the basic sequences of this type, the proportion of the cake-striped micrite increases from bottom to top. Generally, the basic sequences of this type are of polycyclic basic sequence with argillaceous matter decreasing and microcrystalline calcite increasing.

The sediments in the lower part of this member are mainly carbonates followed by a small amount of siliceous argillaceous matter, with very thin horizontal bedding developing in the layers. It is indicated that the sedimentary environment is of deep shelf siliceous mud facies–shallow shelf calcareous carbonate facies. The sediment in the upper part is mainly calcareous carbonate followed by argillaceous matter, with horizontal bedding developing. It is indicated that the rudimentary environment is of shallow shelf argillaceous carbonate facies.

There are two types of basic sequences in the second member of Xiyangshan Formation (€Ox^2).

Basic sequences of type D: distributed in the upper part of the second member of Xiyangshan Formation, and consisting of ① medium laminated carbonaceous siliceous mudstone or calcareous siliceous mudstone interbedded with ② a small amount of micritic marl. The lithologic association is relatively monotonous, and the basic sequences of this type are of non-cyclic basic sequence.

Basic sequences of type E: the main components of the second member of Xiyangshan Formation, and composed of ① thin to medium laminated marl, ② cake-striped micritic marl, ③ cake-striped micrite. Micro-fine horizontal bedding developed. In the lithologic association consisting of ① and ③, the cake-striped micrite gradually increases and the marl gradually decreases from bottom to top, indicating that the sea level once became shallower. Therefore, the basic sequences of this type are of cyclic basic sequence with the sea level gradually rising and calcareous matter increasing.

The sediment in this member is mainly calcareous matter followed by a small amount of argillaceous matter. The basic sequences of this type are of shallow shelf carbonate facies.

There are two basic sequences in the third member of Xiyangshan Formation (€Ox^3).

Basic sequences of type F: distributed in the middle-lower part of the third member of Xiyangshan Formation and composed of ① medium–thick laminated marl and ② lenticular, cake-shaped micrite. Among them, the component ① is 2–3 times of the component ②. Since micro-fine horizontal bedding developed, the marl is prone to flake off. The cake-striped limestone decreases upward. Therefore, the basic sequences of this type are of polycyclic basic sequence with calcareous matter gradually decreasing and argillaceous matter gradually increasing upward.

Basic sequences of type G: distributed in the upper part of the third member of Xiyangshan Formation, and consisting of ① silty mudstone and ② cake-striped micritic marl. Micro-fine horizontal bedding developed in all layers, the silty mudstone increases, and the cake-striped mudstone decreases. Therefore, the basic sequences of this type are of a non-cyclic basic sequence with argillaceous matter increasing upward.

The sediment in the lower part of this member is mainly calcareous followed by a small amount of argillaceous matter, with micro-fine horizontal bedding developing and thus reflecting that the main body is of carbonate facies of deep shelf with weak hydrodynamic force. The sediment in the upper part is argillaceous and calcareous carbonate-bearing terrigenous clasts, indicating carbonate facies of shallow shelf.

2.3.5.2 Sequence Stratigraphy

According to the characteristics of rocks and the types of the top and bottom, a second-order sequence SS7 is divided from the Late Cambrian Xiyangshan Formation to the Late Ordovician Yanwashan Formation. Furthermore, the SS7 is further divided into six third-order sequences (Sq16–Sq21) and one sub-second-order orthosequence set Ss1. According to the association of lithology and lithofacies as well as the characteristics of the boundary of Kaokengkou profile of Hanggai map sheet, there are two third-order sequences Sq16 and Sq17 in the Xiyangshan Formation (€O*x*) (Fig. 2.24).

Third-order sequence Sq16: Located in the first member and the second member of Xiyangshan Formation, it includes TST and HST. The bottom of Sq16 is composed of medium laminated argillaceous limestone and is in conformable contact with the thin laminated micrite of the underlying Huayansi Formation. The top consists of small-vein-strip shaped micritic marl and in conformable contact with the medium laminated argillaceous mudstone of the overlying third member. The top and bottom are all of type-II sequence boundary. TST is located in Bed 1–Bed 5 in the middle-lower part of the first member of Xiyangshan Formation. The sediment of TST is carbonaceous siliceous argillaceous matter and calcareous carbonate. Therefore, TST belongs to weak reduction environment of deep shelf during the maximum flooding period–weak oxidization environment of the outer margin of shallow shelf under the wave base. HST is located in Bed 6–Bed 35 in the upper part of the first member and the second member of Xiyangshan Formation. The sediment is mainly microcrystalline calcite followed by a small amount of argillaceous matter. Accordingly, HST belongs to shallow shelf oxidation environment after the maximum flooding period. Therefore, this sequence reflects a complete transgression–regression process, generally indicating the decline of the sedimentary basin after Huayansi Formation period and belonging to the sequence of progradational–aggradational type. In terms of time, the age range of Sq16 is 489.5–485.4 Ma, spanning 4.1 Ma.

Third-order sequence Sq17: Located in the area from the third member of Xiyangshan Formation to the first member of Yinzhubu Formation, it includes TST and HST. The bottom is an interface with a lithological abrupt change between the marl of the third member of Xiyangshan Formation and the caky limestone of the overlying second member of Xiyangshan Formation. Regionally, graptolite is discovered in the third member of Xiyangshan Formation and planktonic Agnostus is the main palaeobios in the second member. Furthermore, the values of oxygen and sulfur stable isotopes fluctuate near the bottom. This is presumably caused by global tectonic events. Therefore, the bottom belongs to the type-I sequence boundary. The top is a conformity interface and a type-II sequence boundary. TST is located in Bed 36–Bed 54. The main sediment in it includes carbonate and argillaceous matter, but horizontal bedding developed. Therefore, TST should be under weak oxidization sedimentary environment with weak hydrodynamic force and belongs to deep shelf–shallow shelf carbonate rock facies. HST is located in the first member of Yinzhubu Formation. The main sediment in it is siliceous argillaceous matter, belonging to the siliceous argillaceous facies under deep shelf weak oxidization environment during the maximum flooding period. Therefore, this sequence represents the continuous transgression process and is of retrogradation–aggradation type. In terms of time, the age range of Sq17 is inferred to be 485.4–481.00 Ma, spanning about 4.4 Ma.

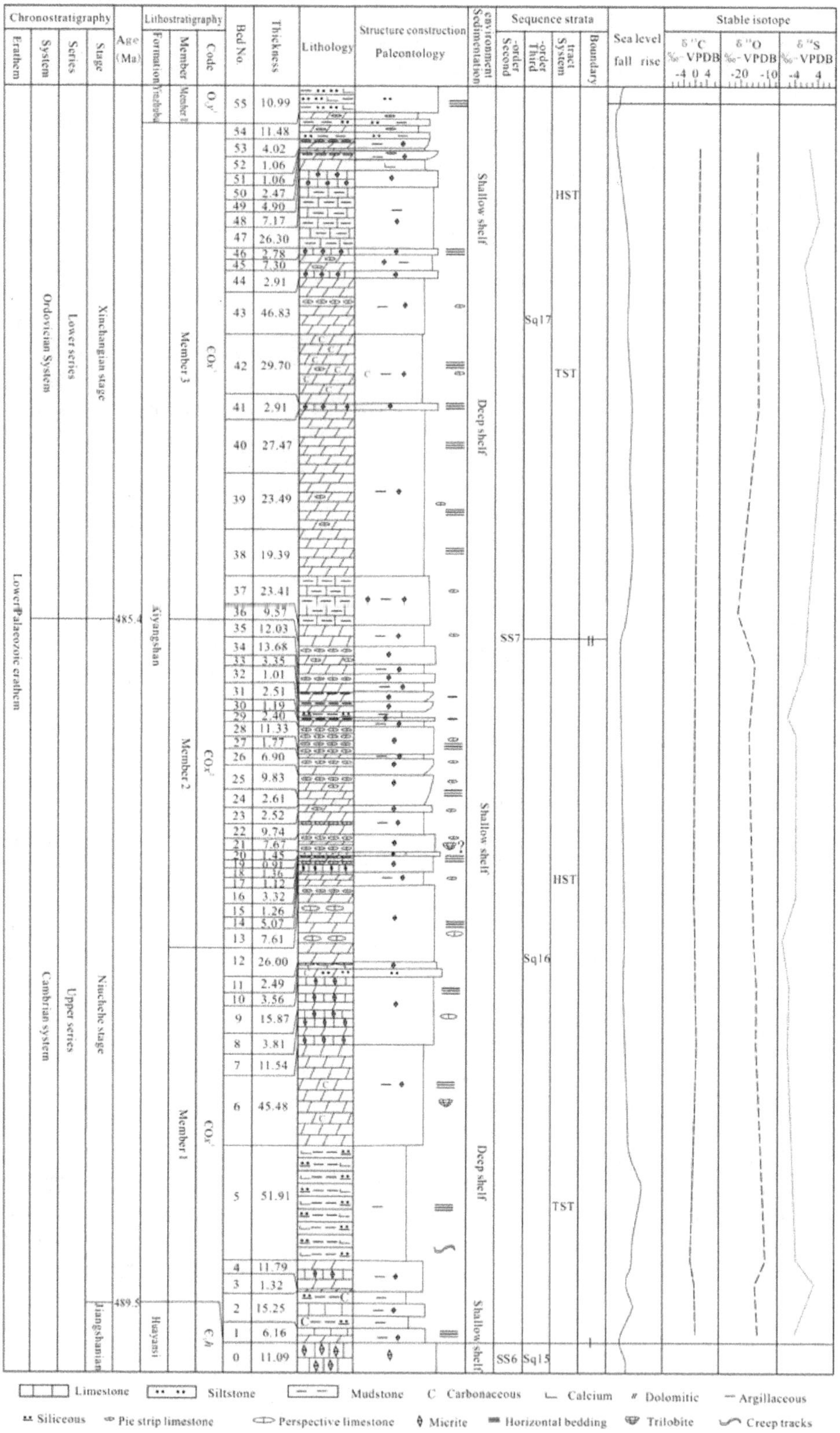

Fig. 2.24 Sequence stratigraphic framework of Cambrian Xiyangshan Formation in the area

Fig. 2.25 *Lotagnostus americanus* in the first member of Xiyangshan Formation

2.3.5.3 Biostratigraphy and Chronostratigraphy

In this Project, *Lotagnostus americanus* fossils, the standardized fossils of the *Lotagnostus americanus* zone, are initially found in Bed 6 of the first member of Xiyangshan Formation (ЄOx^1) of Kaokengkou profile in Hanggai Town, Anji County (Fig. 2.25). They are also the first ones discovered in the north Zhejiang Province. Besides, the trackway of grid-like creeps is found in the marl of the Bed 2.

According to the *Stratigraphic Chart of China* (2014), the fossil zones in the Xiyangshan Formation are as follows: The *Lotagnostus americanus* zone occurred in the medium laminated marl of the first member of Xiyangshan Formation, the *Lotagnostus hedini–Cordylodus proavus* zone produced in the small cake-striped rocks and the thin–medium laminated marl in the second member of Xiyangshan Formation according to speculation, and the *Hysterolenus–Staurograptus dichotomus* zone produced in the marl and calcareous mudstone in the third member of Xiyangshan Formation. Xiyang Formation can be divided into the Tenth Stage and the Xinchangian Stage from bottom to top. The *Lotagnostus americanus* zone in the first member of Xiyangshan Formation belongs to the Tenth Stage of the Cambrian. The *Lotagnostus hedini* zone in the second member of Xiyang Formation also belongs to the Tenth Stage of the Cambrian. The *Hysterolenus–Staurograptus dichotomus* zone in the third member of Xiyangshan Formation belongs to the Ordovician Xinchang Stage.

2.3.5.4 Characteristics of Stable Isotopes

Six, five, and five limestone samples for whole-rock stable isotopes $\delta^{13}C$, $\delta^{18}O$, $\delta^{34}S$ analysis were, respectively, collected from the first member, the second member, and the third member of Xiyangshan Formation in Kaokengkou Profile (PM004).

The $\delta^{13}C$ value of the first member of Xiyangshan Formation is from −1.35 to 0.71‰, with an average of 0.108‰; the $\delta^{13}C$ value of the second member is from −0.06 to −0.37‰, with an average of 0.13‰; the $\delta^{13}C$ value of the third member is 0.53–1.61‰, with an average of 1.20‰. The $\delta^{13}C$ value of the whole Xiyangshan Formation is consistent with the $\delta^{13}C$ value of the limestone of normal marine facies (0 ± 2‰), reflecting that the sea level was comparatively calm at that time. This also indicates deep shelf sedimentation and that the organic matter in the sediment was buried and deposited at a normal rate.

The $\delta^{18}O$ value of the first member of Xiyangshan Formation is from −14.6 to –0.76‰, with an average value of −13.46‰; the $\delta^{18}O$ value of the second member is from −16.54 to −14.89‰, with an average of −15.476‰; the $\delta^{18}O$ value of the third member is from −20.87 to −13.62‰, with an average of −15.44‰. Compared with the $\delta^{18}O$ value of the limestone of normal marine facies (from −13.0 to −1.30‰), the $\delta^{18}O$ value of the Xiyangshan Formation features negative excursion. The reason may be that the Xiyangshan Formation inherited the temperature of the seawater from the Huayansi Formation. Therefore, the burial temperature of the sediment increased with time, and the temperature of the seawater was comparatively high and remained stable in the late period of the Xiyangshan Formation.

The $\delta^{34}S$ value of the first member of Xiyangshan Formation is −6.1‰–3.8‰, with an average value of −1.833‰; the $\delta^{34}S$ value of the second member is from −8.0‰ to −0.1‰, with an average of −4.52‰; the $\delta^{34}S$ value of the third member is −0.6‰–6.6‰, with an average of 2.46‰. Compared with the sulfate in seawater during the Xiyangshan Formation period in the geological history, the Xiyangshan Formation features significant negative anomaly of $\delta^{34}S$ value, which is abnormally low. This may be related to the sedimentary–diagenetic environment. The Xiyangshan Formation was in the alternation period of Cambrian and Ordovician periods, and the Area entered the sedimentary period in an open environment (deep shelf sedimentation → shallow shelf sedimentation). The rate of sulfate reduction is much lower than the supply rate of SO_4^{2-}. As a result, the $\delta^{34}S$ value of the sulfide formed owing to the reduction of sulfate-reducing bacteria in the seawater is 40–60‰, lower than the $\delta^{34}S$ value of the seawater, with an average of 50‰ (Zhang 1989).

2.3.5.5 Analysis of Sedimentary Environment

The Xiyangshan Formation is located in the middle-upper part of the Cambrian. In the lower part of the first member, the main sediment is dark gray siliceous argillaceous matter and marl and the trackway of mollusk creeps are visible along the layers. It is indicated that the sedimentary environment is near the redox interface, which is unfavorable for the survival of benthic shellfish and belongs to a deep shelf environment. In the upper part of the first member, the sediment is argillaceous carbonate interbedded with a small amount of lenticular and banded microcrystalline calcite. In the sediment, there are Agnostus and deepwater trilobite with comparatively complex planktonic structure. Besides, horizontal bedding developed. It is indicated that the hydrodynamic force was weak and oxidation was comparatively enough at that time. Therefore, the sedimentary environment in this part belongs to a shallow shelf sedimentary environment. In the second member, vein-strips of microcrystalline calcite increase, indicating that the sea level continued to fall and the seawater became shallow. Therefore, the second member belongs to a stable shallow shelf environment. In the lower part of the third member, the main sediment is thin–medium laminated sand mud interbedded with a small amount of microcrystalline calcite, with micro-fine horizontal bedding developing. It is indicated that the sea level rose and the hydrodynamic force was weak then, and the sedimentary environment belongs to deep shelf environment. In the part toward the top of Xiyangshan Formation, the sediment is carbonate, indicating that sea level fall occurred and the sedimentary environment is turned into a shallow shelf environment.

Regionally, the lithology of the Xiyangshan Formation in the Area is basically the same as that in the Area. That is, the lithology of the Xiyangshan Formation is mainly characterized by carbonate rocks followed by a small amount of calcareous mudstone. In the areas such as Duibian of Jiangshan City and Fuyang-Dongqiao of Tonglu County, the lithofacies of the upper part or the top of the Xiyangshan Formation is transformed into a sparry limestone, with mudstone lacking and conglomerate and landslide tectonic limestone interspersing. It is indicated that the sedimentary environment was converted into a slope. In Yuhang area, the lithofacies of Xiyangshan Formation turned into dolomite and the sedimentary environment transformed into a carbonate platform environment.

2.3.5.6 Trace Elements in Strata (Ore-Bearing)

1. **Characteristics of Trace Elements in Strata**

Sixty-five rock spectra were collected from three lithological members of Xiyangshan Formation of Kongkengkou profile (PM004) in Hanggai Town, Anji County. According to spectral analysis, the first member of Xiyangshan Formation is enriched in Bi, Sb, Ag, F, S, and TFe, the second member is enriched in Bi, F, S, and TFe, and the third member is enriched in Bi, F, S, and TFe. The content of S decreases upward, indicating the sea level rise and the oxygen content increase at that period.

2. **Content of CaO and MgO in Main Limestone and Marl**

According to the *Report on the Survey and Evaluation Results of Cambrian Limestone Resources of Zhejiang Province* (2007), the analysis results of the limestone and marl samples obtained by chip-channel method from the first member and the second member of Xiyangshan Formation of Yekengwu Profile in Hanggai Town, Anji County, are as follows. The respective average content of CaO and MgO of the six samples at the bottom of the first member is 38.83 and 3.32%. The respective average content of CaO and MgO of the four samples in the middle-lower part of the first member is 43.97 and 2.40%. The content of CaO and MgO of the rocks in other parts of Xiyangshan Formation is 15%–30% and 2%–5.64%, respectively.

2.3.6 Comparison of Regional Stratigraphy

In Zhejiang Province, the Cambrian topographic pattern was inherited from the Sinian. The sediments of it are mainly composed of siliceous argillaceous matter and carbonates from early to late period. The lithology of the sediment varies in different sedimentary zones as the sea level rises and falls. The regional stratigraphy was compared by taking the samples of the profile in the Area in Hanggai area, Chaoshan profile in Yuhang area, and Shima profile in Zhuji, Xiaoshan area from the northwest to the southeast.

The regional stratigraphy in the Early Cambrian is as follows. The strata of the early stage in Hanggai area mainly consist of the rocks in the first member of the Hetang Formation, including carbonaceous, argillaceous, and siliceous rocks of abyssal basin facies, intercalated with stone coal layers and phosphorite nodule layers. There are 1–3 layers of stone coal in the siliceous rocks and sedimentary pyrite contained generally. The thickness of the strata is 411 m. The strata of the middle stage in Hanggai area mainly consist of the rocks in the second member of Hetang Formation, which mainly includes carbonaceous siliceous mudstone of bathyal facies followed by carbonaceous argillaceous–siliceous rocks. The thickness of the strata is 200 m. The strata of the later stage in Hanggai area mainly consist of the rocks in Dachenling Formation, including the interbed consisting of carbonate and siliceous argillaceous matter of deep shelf facies–bathyal basin facies. The thickness of the strata is 52 m. In Yuhang–Fuyang area, the strata of the early stage consist of carbonaceous dolomitic mudstone of deep shelf facies in Chaoshan Formation, with a thickness of 25 m; the strata of the middle and late stage consist of the rocks of Dachenling Formation including the dolomite of platform facies containing a small amount of chert, with a thickness of 98–103 m. In Xiaoshan–Zhuji area in the southeast, the strata of the early stage consist of carbonaceous mudstone and stone coal layers of semi-closed estuarine facies in Hetang Formation, with a thickness of less than 30 m; the strata of the middle stage consist of carbonaceous siliceous mudstone and argillaceous–siliceous rocks of deep shelf facies in the middle-upper part of Hetang Formation, with a thickness of 120 m; the strata of the late period consist of the dolomitic limestone and micrite of neritic platform facies in Dachenling Formation, with a thickness of 120 m.

The regional stratigraphy in the Middle Cambrian is as follows. In Hanggai area, the strata consist of the rocks in Yangliugang Formation, which gradually change from the siliceous argillaceous matter of deep shelf facies–bathyal basin facies from bottom to top, indicating the falling process of the seawater. In Yuhang–Fuyang area, the strata consist of dolomitic limestone of shallow shelf facies–platform facies in Yangliugang Formation, with a thickness of 100 m. In Xiaoshan–Zhuji area, the strata are composed of banded marl and limestone of shallow shelf facies in Yangliugang Formation, with a thickness of 230 m.

The regional stratigraphy in the Later Cambrian is as follows. In Hanggai area, the strata of the early stage consist of the micrite of shallow shelf facies in Huayansi Formation with fine horizontal bedding developing, and the thickness of the strata is 123 m; the strata of the middle stage consist of the argillaceous limestone of deep shelf facies interbedded with lenticular micrite in the first member of Xiyangshan Formation, with a thickness of 145 m; the strata of the late stage consist of the interbed of argillaceous limestone and lenticular micrite of shallow shelf facies in the second member of Xiyangshan Formation, with a thickness of 113 m. In Yuhang–Fuyang area in the central area, the strata consist of dolomite and psephitic dolomite of platform facies in Chaofeng Formation, with a thickness of 175 m. In Xiaoshan–Zhuji area, the strata of the early stage consist of carbonate of shallow shelf facies in Huayansi Formation, with a thickness of 130 m; the strata of the middle stage consist of marl and micrite of shallow shelf facies in the lower part of Xiyangshan Formation, with a thickness of 60 m; the strata of the later stage consist of knotlike and lenticular argillaceous limestone of shallow shelf facies–platform facies in the middle-upper part of Xiyangshan Formation, which is the sediment in environment with high-energy oxidization and 200 m thick.

In northwest Zhejiang, the sediment gradually changes from siliceous mud and calcareous carbonate in Hanggai area to magnesium carbonate in Yuhang area and argillaceous calcareous carbonate in Xiaoshan area from northwest to southeast. In the central and southern areas with high terrain, some sediments are missing owing to the falling of the sea level in the Early and Later Cambrian (Fig. 2.26).

2.4 Ordovician System–Silurian System

Widely distributed in the Area, the Ordovician and the Silurian are exposed mainly in the eastern half of Hanggai map sheet and secondarily in the southwest of Xianxia map sheet. Besides, they are sparsely distributed in the southeast corner of Xianxia map sheet as well as the northwest, southeast, and northwest corners of Chuancun map sheet. The total outcrop area is 206.11 km^2, accounting for 16.21% of the bedrock area.

The Ordovician in the Area outcrops completely and continuously, and is divided into seven formations from bottom to top, namely: Yinzhubu Formation (O_1y), Ningguo Formation ($O_{1\text{-}2}n$), Hule Formation ($O_{2\text{-}3}h$), Yanwashan Formation (O_3y), Huangnigang Formation (O_3h), Changwu Formation (O_3c), and Wenchang Formation (O_3w). The Silurian in the Area mainly outcrops in Xiaxiang Formation (S_1x) and is in conformable contact with the underlying Wenchang Formation. According to the lithologic association, Yinzhubu Formation (O_1y), Hule Formation ($O_{2\text{-}3}h$), and Changwu Formation (O_3c) can be further divided into three members, respectively, and Xiaxiang Formation (S_1x) into two members.

2.4.1 Yinzhubu Formation (O_1y)

The name Yinzhubu Formation was established by Tinghu Zhu in 1924 in Yinzhubu, Fenshui Town, Tonglu County, Zhejiang. The stratotype profile of it was not yet determined then. The Yinzhubu Formation called by Lu et al. (1955) referred specifically to the stratum producing trilobite fossils in the former Yinzhubu System. It was subsequently used in the literature such as *Qu County Map Sheet on a Scale of*

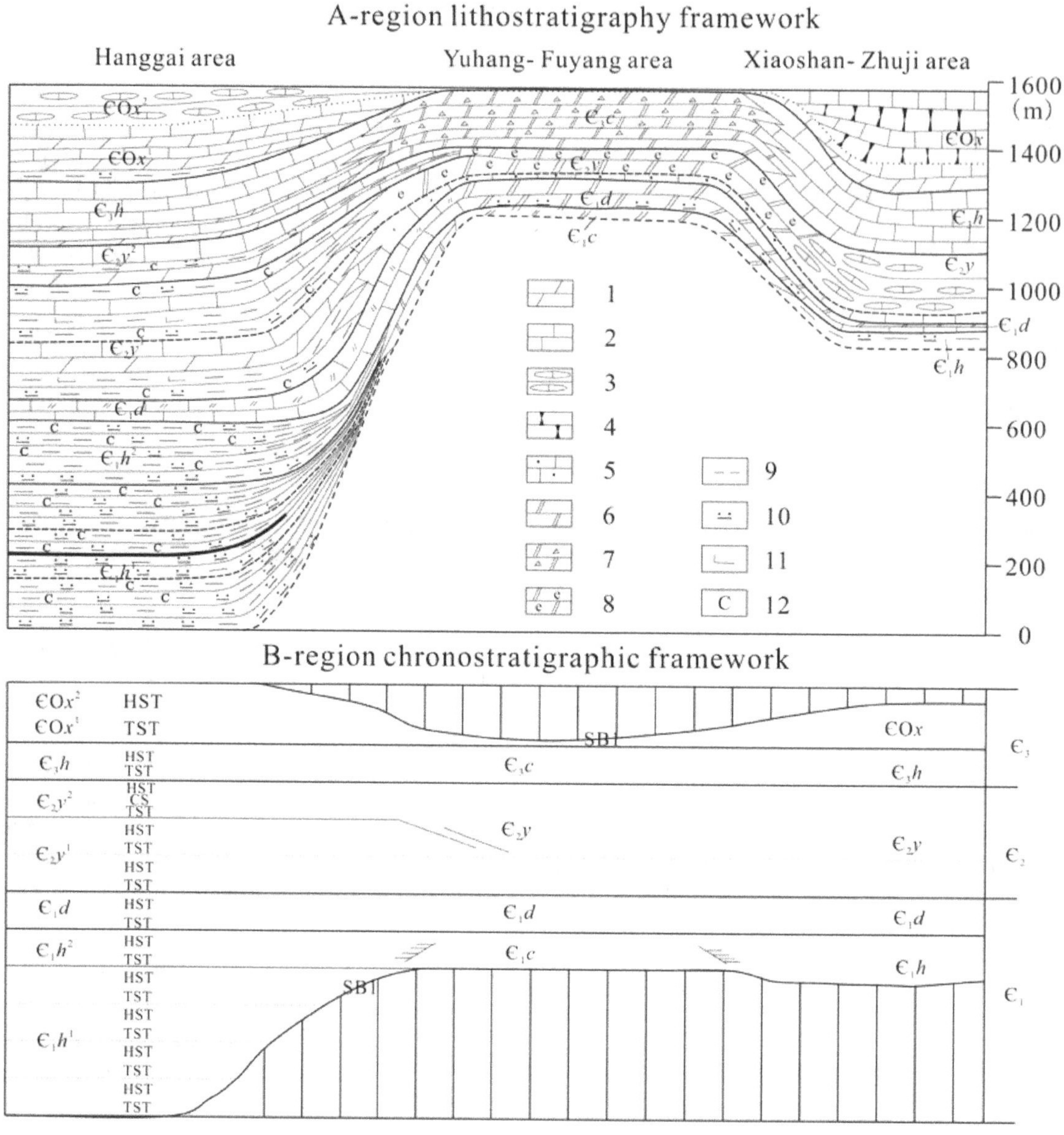

Fig. 2.26 Stratigraphic framework of Cambrian Hanggai–Yuhang–Fuyang–Xiaoshan–Zhuji area: 1. marl; 2. limestone; 3. banded limestone; 4. knotlike limestone; 5. sandy limestone; 6. dolomite; 7. brecciaous dolomite; 8. bioclastic dolomite; 9. argillaceous rock; 10. siliceous rock; 11. calcareous matter; 12. carbonaceous matter; $\in_1h^1$—the first member of Hetang Formation; $\in_1d$—Dachenling Formation; $\in_2y^1$—the first member of Yangliugang Formation; $\in_2y^2$—the second member of Yangliugang Formation; $\in_3h$—Huayan Temple Formation; $\in Ox^1$—the first member of Xiyangshan Formation; $\in Ox^2$—the second member of Xiyangshan Formation; $\in_1c$—Chaoshan Formation; $\in_3c$—Chaofeng Formation; SB1—type-I unconformity interface; TST—transgressive systems tract; CS—starved section; HST—highstand systems tract

1:200,000 (1969), *Stratigraphic Correlation Chart in China with Explanatory Text* (1982), and *Regional Geology of Zhejiang Province* (1989). In *Lithostratigraphy of Zhejiang Province* (1995), Yinzhubu Formation refers to a set of strata above caesious–olive thick laminated calcareous mudstone. This is different from the Yinzhubu Formation called by Lu et al. (1955). In this Project, the name Yinzhubu Formation (O_1y) is still adopted according to the lithological characteristics of Huangdouwu profile (PM005) and Lijiabian profile (PM011) in Hanggai map sheet, and the geological observation traverse within the Area.

2.4.1.1 Lithostratigraphy

Yinzhubu Formation (O_1y) in the Area, the most widespread stratum in the Lower Paleozoic, outcrops basically in the same way with Late Cambrian Xiyangshan Formation. The outcrop area is about 58.38 km^2, accounting for 4.59% of the bedrock area.

According to the lithologic association, Yinzhubu Formation (O_1y) can be divided into the first member of Yinzhubu Formation (O_1y^1), the second member of Yinzhubu Formation (O_1y^2), and the third member of Yinzhubu Formation (O_1y^3). Yinzhubu Formation is in conformable contact with its overlying Ningguo Formation and underlying Xiyangshan Formation.

The first member of Yinzhubu Formation (O_1y^1): consisting of gray medium laminated siliceous mudstone, medium–thick laminated silty siliceous mudstone interbedded with a small amount of micro-thin laminated knotlike micrite, and carbonaceous silty mudstone or argillaceous silt interbedded with calcareous siliceous mudstone from bottom to top; very thin horizontal bedding developing in the layers; the thickness of a single layer: 10–50 cm. The size of a piece of knotlike micrite: 2–4 cm generally. The thickness of this member: 76.07–510.00 m.

The second member of Yinzhubu Formation (O_1y^2): consisting of rhythm interbeds of dark gray calcareous siliceous mudstone or calcareous sandy mudstone, silty mudstone-bearing small-vein-strip shaped micrite; locally interbedded with argillaceous silt and marl. The small-vein-strip shaped micrite: no bedding developing, length × width: about (1–10) × (0.5–2) cm, small-vein-strip shaped micrite in a single rhythm increasing gradually from bottom to top. The thickness of this member: 24.36 m.

The third member of Yinzhubu Formation (O_1y^3): the lower part: gray medium laminated calcareous mudstone or interbeds consisting of medium laminated calcareous mudstone and knotlike micrite, locally interbedded with small-vein-strip shaped micrite; the middle part: dark gray medium laminated marl interbedded with vein-strip shaped micrite visible occasionally; the middle and upper part: caesious and gray medium laminated silty siliceous mudstone, bioturbation relics developing. The thickness of this member: 149.79–222.74 m.

1. **Stratigraphic section**

The lithology of Yinzhubu Formation is described by taking the example of Ordovician Yinzhubu Formation (O_1y) profile (Fig. 2.27) of Hanggai map sheet in Huangdouwu–Qiaotou area, Hanggai Town, Anji County, Zhejiang Province. The details are as follows:

Ningguo Formation ($O_{1\text{-}2}n$) Total thickness: > 37.00 m

25. Gray–dark gray siliceous silty mudstone. 37.00 m

——— Conformable contact ———

Yinzhubu Formation Total thickness: 323.17 m

The third member of Yinzhubu Formation Total thickness: 222.74 m

24. Interbed consisting of caesious and gray medium laminated siliceous mudstone and medium laminated argillaceous siltstone, the main component: siliceous mudstone, horizontal bedding developing. 81.34 m

23. Gray medium laminated silty siliceous mudstone, locally interbedded with medium laminated calcareous sandstone and calcareous mudstone. Very thin horizontal bedding developing, organism burrows and creep trackway locally visible, carbonaceous matter with a size of 3 mm visible. 92.46 m

22. Gray medium laminated marl interbedded with vein-strip shaped limestone, horizontal bedding developing in the vein-strip shaped limestone. 4.85 m

21. Gray medium laminated silty siliceous mudstone, locally interbedded with thin laminated silty mudstone. The thickness of a single layer of the silty mudstone: 2–6 cm. Silty siliceous mudstone: horizontal bedding developing, the thickness of a single layer: 10–50 cm. 34.34 m

20. Gray thin laminated siliceous mudstone, interbedded with knotlike limestone. The siliceous mudstone: very thin horizontal bedding developing, the thickness of a single layer: 2–4 cm. The knotlike limestone:

spreading along the layer discontinuously, the size of a knot: (2–3) cm× (2–5) cm, the thickness of a single layer: 25–35 cm. 9.11 m

19. Interbed consisting of dark gray medium laminated silty siliceous mudstone and knotlike limestone and thin laminated silty mudstone bearing small-vein-strip shaped limestone. The size of vein-strip shaped limestone: 1–3 cm thick and 5–30 cm long. The thickness of a single layer of knotlike limestone: 3–15 cm; the thickness of a single of silty siliceous mudstone: 13–70 cm; the thickness of a single layer of silty mudstone bearing small-vein-strip shaped limestone: 15–140 cm. 7.12 m

18. Dark gray–gray medium laminated siliceous mudstone, horizontal bedding developing, the thickness of a single layer: 10–30 cm. 4.40 m

17. Dark gray thin laminated silty siliceous mudstone bearing vein-strip shaped limestone, silty strips (< 1mm) developing along the mudstone bedding, the thickness of a single layer: 2–5 cm. A vein strip: (1–2) cm × (1–25) cm, distributed along the layer discontinuously. 2.35 m

16. Gray medium laminated siliceous mudstone, fine-texture horizontal bedding developing, containing a small amount of carbonaceous clastics. Bioturbation structure locally visible. 21.37 m

15. Gray thick laminated calcareous mudstone interbedded with knotlike limestone. The calcareous mudstone: horizontal bedding developing, the thickness of a single layer: 55–82 cm. The knotlike limestone: distributed discontinuously along the layer. The diameter and thickness of a node of the knotlike limestone: 1–15 cm and 1–2 cm respectively. 11.36 m

——— Conformable contact ———

The second member of Yinzhubu Formation Total thickness: 24.36 m

14. The lower part: dark gray calciferous siliceous mudstone, the thickness: 70 cm. The top: dark gray, calcareous siliceous mudstone bearing dense small-vein-strip shaped limestone, thickness: 40 cm. Very thin horizontal bedding developing in the calciferous siliceous mudstone. 2.77 m

13. The lower part: dark gray calcareous silty mudstone interbedded with small-vein-strip shaped argillaceous limestone; the calcareous silty mudstone: no evident horizontal bedding developing, the thickness: 8–15 cm; the size of vein-strip shaped argillaceous limestone: (1–1.5) cm × (8–12) cm. The upper part: calcareous siltstone interbedded with a layer of vein-strip shaped limestone, very thin horizontal bedding developing.

3.39 m

12. The lower part: dark gray calcareous mudstone bearing vein-strip shaped marl, the upper part: calcareous siltstone. The proportion of vein-strip marl is the same as that of the calcareous mudstone. The size of vein-strip: 1–2 cm in thickness and 1–50 cm in length. The calcareous mudstone: bioturbation structure developing, the thickness of a single layer: 60–110 cm. The ratio of the siltstone to the mudstone: approximately (3–5):1.

2.03 m

11. Dark gray medium–thick laminated silty calcareous mudstone, horizontal bedding developing. 2.64 m

10. Gray thin–medium laminated calcareous silty mudstone interbedded with thin laminated knotlike limestone. The thickness of a single layer of the calcareous silty mudstone: 6–33 cm generally and 60 cm locally. The thickness of a single layer of knotlike limestone: 1–3 cm generally and 6–12 cm locally. Deformed lotus-root shaped structure visible. 3.39 m

9. The lower part: dark gray calciferous siliceous mudstone, containing vein-strip shaped micrite. The top: calciferous fine-sand siltstone (15 cm). Very thin horizontal bedding developing in all layers. 3.74 m

8. The lower part: dark gray calciferous siliceous mudstone bearing small-vein-strip shaped marl; the vein-strip: 30 × 1–3 cm, distributed discontinuously along the layers with an interval of 5–10 cm; the calciferous siliceous mudstone gradually changing to calcareous fine sandstone upwards. Very thin horizontal bedding developing in the fine sandstone. The ratio between the calciferous siliceous mudstone and the calcareous fine sandstone: 1:1. 1.72 m

7. Dark gray marl bearing vein-strip shaped limestone. The marl: very thin horizontal bedding developing, the bedding surrounding the vein-strip. The vein-strip shaped limestone: no bedding developing, the length × thickness: approximately (1–10) × (0.5–2) cm, distributed discontinuously along the layer. The interval between the vein-strips: 1–2 cm in the lower part, gradually increasing to 5–10 cm upwards. The upper part: calcareous mudstone or marl bearing vein-strip shaped limestone, the thickness: 78 cm. 4.68 m

——————— Conformable contact ———————

The first member of Yinzhubu Formation Total thickness: 76.07 m

6. Dark gray thick-blocky and laminated calcareous argillaceous siltstone, interbedded with a small amount of calcareous siliceous mudstone. The calcareous argillaceous siltstone: very thin horizontal bedding developing, fine-bar shaped after weathering. 3.85 m

5. Dark gray silty mudstone interbedded with calcareous siliceous mudstone. The calcareous siliceous mudstone: very thin horizontal bedding developing, silt stripes visible along the bedding. The width of a single strip: 0.1–0.3 cm. The argillaceous matter in the lower part is interbedded with much small-vein-strip shaped limestone, which decreases upwards. A single of the vein-strip: 1–2 cm × 5–20 cm. The marl in the middle part of the layer is 60 cm thick, with bioturbation structure developing. 3.39 m

4. Gray–dark gray medium laminated siliceous mudstone interbedded with thin laminated knotlike limestone. A layer of vein-strip shaped limestone interspersed in the middle part of the layer. The siliceous mudstone: very thin horizontal bedding and bioturbation structure developing, a small amount of small pyrite nodule visible. 4.93 m

3. Gray thick laminated siliceous mudstone, very thin horizontal bedding developing in the layers, creep trackway and many charcoal flats visible, a small amount of pyrite nodule with a particle size of 1–3 mm discovered. 6.95 m

2. Gray thick laminated to blocky silty siliceous mudstone, interbedded with micro–thin laminated knotlike marl. The silty siliceous mudstone: very thin horizontal bedding developing, the thickness of a single layer: 75–220 cm. The knotlike marl: accounting for 1%–2% of this layer, the size of a single node: 2–4 cm generally. Small pyrite nodules visible occasionally. 6.51 m

1. Gray medium laminated siliceous mudstone, very thin horizontal bedding developing in the layers. 50.44 m

——————— Conformable contact ———————

Xiyangshan Formation (ЄO*x*) Total thickness: > 14.10 m

0. Dark gray medium laminated marl interbedded with a small amount of medium laminated micrite. A layer of pie-stip shaped limestone interspersed in the top. The marl: horizontal bedding developing, the thickness of a single layer: 20–30 cm. The size of vein-strip shaped limestone: (5–20) cm × (3–8) cm. 14.10 m

The lithology of Yinzhubu Formation is described by taking the example of the profile of the third member of Yinzhubu Formation (O_1y^3) (Fig. 2.28) in Lijiabian Village, Hanggai Town, Anji County, Zhejiang. The details are as follows:

Ningguo Formation Total thickness: >72.33 m

15. Black medium–thin laminated siliceous silty carbonaceous shale, containing a small amount of pyrite, bedding developing extremely, graptolite visible. 10.1 m

14. Grayish black–modena medium laminated hornfelsic siliceous silty mudstone, containing a small amount of dark fine specks, bedding developing extremely, graptolite visible. 19.3 m

13. Black micro–thin laminated carbonaceous siliceous silty mudstone, bedding extremely developing, the thickness of a single layer: 3–10 mm, graptolite visible. 28.26 m

12. Grayish black–modena hornfelsic siliceous silty mudstone, horizontal bedding extremely developing, graptolite visible. 10.97 m

11. Black micro–thin laminated carbonaceous silty siliceous mudstone, bedding extremely developing, the thickness of a single layer: 1–3 mm. 3.69 m

——— Conformable contact ———

The third member of Yinzhubu Formation Total thickness 149.26 m

10. Gray-black hornfelsic mudstone, horizontal bedding developing extremely, much pyrite concentrating along the bedding. 2.81 m

9. Grayish black hornfelsic mudstone, micro-fine horizontal bedding developing. 43.05 m

8. Grayish black pyrite-bearing hornfelsic mudstone, the content of the pyrite: about 5%, the pyrite occurring in lamellar shape along the horizontal bedding. 8.18 m

7. Grayish black pyrite-bearing hornfelsic mudstone, the content of the pyrite: about 5%, and the pyrite distributed in the fine-grained disseminated form. 29.52 m

6. Grayish purple speckled hornfelsic mudstone, horizontal bedding developing comparatively in the layer. 7.92 m

5. Grayish yellow hornfelsic mudstone, horizontal bedding developing comparatively in the layer. 2.50 m

4. Gray speckled altered mudstone, the particle size of the specks: about 1 mm, pyrite sparsely distributed, horizontal bedding visible clearly in the layer. 10.95 m

3. Gray hornfelsic mudstone, horizontal bedding developing in the layer. 3.79 m

2. Gray–dark gray hornfelsic mudstone, sporadic pyrite of small particle size visible (2–3%), horizontal bedding developing in the layers. 29.54 m

1. Gray medium laminated mudstone interbedded with a small amount of knotlike limestone, the node diameter of the knotlike limestone: 4–6 cm, horizontal bedding developing in the mudstone. 11.00 m

——— Conformable contact ———

The second member of Yinzhubu Formation Total thickness: > 8.67 m

0. Gray calcareous mudstone bearing vein-strip shaped limestone. The length and width of a vein-strip of Limestone: 7–10 cm and about 1–2 cm generally. 8.67 m

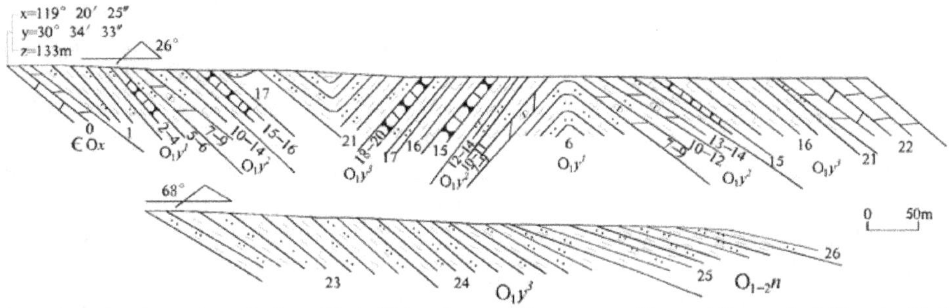

Fig. 2.27 Ordovician Yinzhubu Formation (O_1y) profile in Huangdouwu–Qiaotou area, Hanggai Town, Anji County, Zhejiang Province

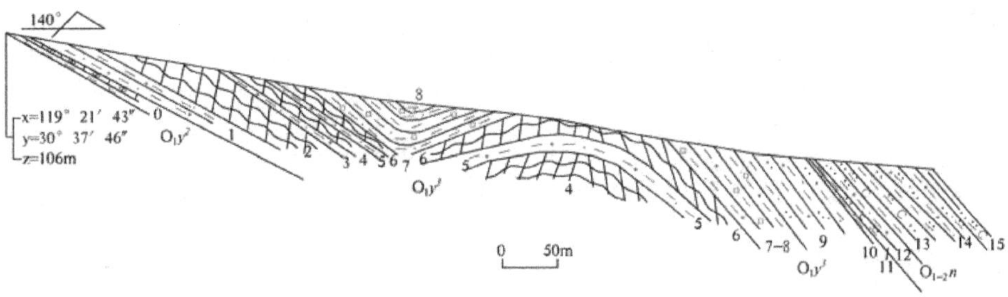

Fig. 2.28 Profile of the third member of Yinzhubu Formation (O_1y^3) in Lijiabian Village, Hanggai Town, Anji County, Zhejiang Province

2. **Lithological characteristics**

The lithology of the first member of Yinzhubu Formation (O_1y^1) is characterized by calcareous argillaceous silts and silty siliceous mudstone, siliceous mudstone, and knotlike marl.

(1) The calcareous argillaceous silts: gray; medium–thick laminated structure; composition: quartz silt (60%), argillaceous matter (25–30%), and micritic calcite (15–20%); very thin horizontal bedding developing; distributed in the upper part of the first member of Yinzhubu Formation.
(2) The silty siliceous mudstone: gray; medium–thick laminated structure; composition: argillaceous (60%), cryptocrystalline siliceous matter (20–25%), and quartz silt (10–15%); very thin horizontal bedding developing; the thickness of a single layer: 45–220 cm; distributed in the lower part of the first member of Yinzhubu Formation.
(3) The siliceous mudstone: gray–dark gray; medium laminated structure; composition: argillaceous (70%), cryptocrystalline siliceous matter (25%), silt occasionally visible (5%); very thin horizontal bedding developing; creep trackway and many charcoal flats distributed; distributed in the middle and lower parts of the first member of Yinzhubu Formation.
(4) The knotlike marl: gray; composition: micritic calcite (55–60%) and argillaceous matter (45–50%); diameter of a node of knotlike marl: 2–4 cm generally; mostly distributed discontinuously in mudstone along the bed; small pyrite nodules occasionally visible; distributed in the middle and lower parts of the first member of Yinzhubu Formation.

The lithology of the second member of Yinzhubu Formation (O_1y^2) is mainly characterized by calciferous siliceous mudstone and small-vein-strip shaped argillaceous limestone followed by calcareous silty mudstone.

(1) The calciferous siliceous mudstone: gray–dark gray; medium laminated structure; composition: argillaceous matter (55–75%), siliceous matter (25–30%), a small amount of micritic calcite (10–15%), and silt occasionally visible (5–10%); distributed in the upper and lower parts of the second member of Yinzhubu Formation.
(2) The small-vein-strip shaped argillaceous limestone: gray; size of vein-strip: about 30 × (1–3) cm; composition: microcrystalline calcite (50–65%), argillaceous matter (20–25%), and a small amount of silt; vein-strips distributed discontinuously along the layer at the interval of 3–10 cm; consisting of the main lithology of the second member of Yinzhubu Formation.

(3) The calcareous sandy mudstone: gray; medium laminated structure; composition: argillaceous (60%), micritic calcite (25–30%), and a small amount of silt (10–15%); the thickness of a single layer: 6–33 cm, up to 60 cm locally; distributed in the middle part of the second member of Yinzhubu Formation.

The lithology of the third member of Yinzhubu Formation (O_1y^3) is mainly characterized by silty siliceous mudstone, calcareous mudstone, knotlike marl, and silicon-bearing mudstone.

(1) The silty siliceous mudstone: gray–caesious; thick to medium laminated; composition: argillaceous matter (60–65%), cryptocrystalline siliceous matter (20–25%), and a small amount of silt (10–15%); silty stripes with a thickness of < 1 mm developing along bedding; consisting of the main lithology of the third member of Yinzhubu Formation.
(2) The calcareous mudstone: gray; medium laminated structure; composition: argillaceous matter (60–65%), microcrystalline calcite (20–30%), and a small amount of silt and carbonaceous clastics; fine-texture horizontal bedding developing; distributed in the bottom of the third member of Yinzhubu Formation.
(3) The marlstone: gray; composition: micritic calcite (65–70%) and argillaceous (30–5%); mostly knotlike; size: 2–4 cm generally; small-vein-strip partly shaped; distributed discontinuously along the layers in the mudstone; distributed in the middle and lower parts of the third member of Yinzhubu Formation.
(4) The siliceous mudstone: gray; medium laminated structure; composition: argillaceous (70–75%), cryptocrystalline siliceous (20–25%), and silt (5%); horizontal bedding developing; the thickness of a single layer: 10–30 cm; distributed in the middle part of the third member of Yinzhubu Formation.

3. **Basic sequences**

There are three types of basic sequences developing in the first member of Yinzhubu Formation (O_1y^1) (Fig. 2.29).

Basic sequences of type A: distributed in the middle and lower parts of the first member; composed of medium to thick laminated siliceous mudstone; the thickness of a single layer: 20–60 cm; horizontal bedding developing; belonging to low-energy siliceous argillaceous facies near the oxidation–reduction zone. Therefore, the basic sequences of this type belong to the monotonous basic sequence.

Basic sequences of type B: distributed in the middle part of the first member; composed of ① gray medium to thick laminated silty siliceous mudstone or siliceous mudstone and ② micro-thin laminated mudstone-bearing knotlike limestone; the mudstone: horizontal bedding developing, many bioturbation structure visible; the knotlike limestone: generally 2–4 cm in diameter, distributed discontinuously along the layers; from bottom to top, the thickness of the layers decreasing, knotlike limestone increasing, and siliceous argillaceous matter decreasing, reflecting the rise of the sea level. Therefore, the basic sequences of this type belong to non-cyclic basic sequence with siliceous argillaceous matter decreasing and calcareous matter increasing upward.

Basic sequences of type C: distributed in the upper part of the first member; composed of ① dark gray, medium to thick laminated carbonaceous silty mudstone or thick laminated argillaceous silt, interbedded with ② calcareous siliceous mudstone; micro-fine horizontal bedding developing in the layers; bioturbation structure commonly visible; from bottom to top, silt and the layer thickness increasing. Therefore, the basic sequences of this type belong to non-cyclic basic sequence with layer thickness and grain size increasing upward.

The sediments in the first member mainly include argillaceous–siliceous matter followed by a small amount of calcareous and terrigenous silt. Horizontal bedding is generally visible. Bioturbation structures are also commonly visible. The siliceous argillaceous matter comparatively decreases upward, reflecting siliceous argillaceous facies near oxidation–reduction interface under the wave base in deep shelf sedimentary environment. There are terrigenous clastics and silt in the sediments upward, indicating that there is a gradual sea level fall.

There is one type of basic sequence developing in the second member of Yinzhubu Formation (O_1y^2).

Basic sequences of type D: composed of ① dark gray calcareous siliceous mudstone or calcareous silty mudstone, ② siliceous mudstone-bearing small-vein-strip shaped limestone, and ③ small-vein-strip shaped limestone interbedded with mudstone. From bottom to top, the silt and siliceous argillaceous matter decrease gradually, the calcareous matter increases gradually, and the thickness of the layers decreases. Therefore, the basic sequences of this type belong to cyclic basic sequence with grain size increasing and layer thickness decreasing upward.

The sediments mainly contain calcareous argillaceous matter and sparse fossils, and therefore the second member belongs to carbonatite facies of deep–shallow shelf sedimentary environment near the oxidation–reduction interface.

There are two types of basic sequences developing in the third member of Yinzhubu Formation (O_1y^3).

Basic sequences of type E: distributed in the middle and lower parts of the third member, composed of ① gray

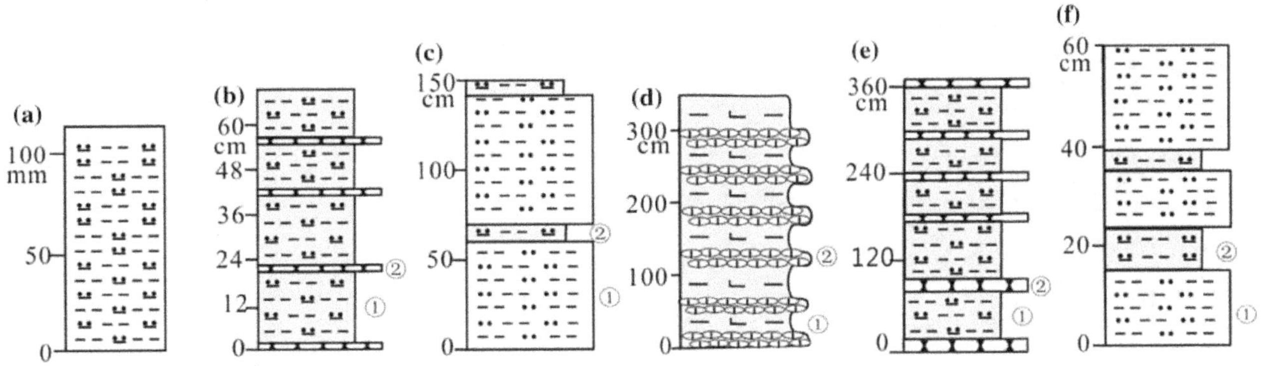

Fig. 2.29 Basic sequences of Yinzhubu Formation (O_1y)

medium to thick laminated silty siliceous mudstone or siliceous mudstone and ② mudstone-bearing micro-thin laminated knotlike limestone. The knotlike limestone is 2–4 cm in size generally, distributed discontinuously along the layers; from bottom to top, the thickness of the layers decreasing, the silt and knotlike limestone decreasing, and the siliceous argillaceous matter increasing, reflecting the sea level rising. Therefore, the basic sequences of this type belong to a non-cyclic basic sequence with layer thickness and grain size decreasing upward.

Basic sequences of type F: distributed in the upper part of the third member, composed of the interbeds of ① dark gray, medium laminated, silty mudstone or argillaceous siltstone and ② thin–medium laminated calcareous siliceous mudstone. Silt content and the thickness of the layers increase upward. Therefore, the basic sequences of this type belong to the basic sequence with layer thickness and grain size increasing upward.

In this member, sediments are mainly argillaceous–siliceous, a small amount of calcareous and terrigenous silt, and rock strata generally have horizontal bedding and sparse biological fossils, and it is the deepwater shelf carbonate-bearing silicon mud facies.

According to the lithological column of Yinzhubu Formation, from the second member, the calcareous matter decreases and siliceous argillaceous increases gradually upward and downward. Therefore, the sediments are characterized by symmetric distribution vertically.

2.4.1.2 Sequence Stratigraphy

According to rock features and types of top and bottom of Ordovician–Silurian systems, a second-order sequence SS7 is divided from Late Cambrian Xiyangshan Formation to Late Ordovician Yanwashan Formation and one second-order sequence SS8 is divided from Late Ordovician Huangnigang Formation to Early Silurian Xiaxiang Formation. According to the changes of sedimentary facies in the second-order sequences, SS7 is further divided into six third-order sequences (Sq16–Sq21) and one sub-second-order orthosequence set Ss1 while SS8 is further divided into four third-order sequences (Sq22–Sq 25) (Fig. 2.30).

According to the features of the lithologic association in Yinzhubu Formation (O_1y) profile of Huangdouwu–Qiaotou area in Hanggai Town, Anji County, Zhejiang Province, Yinzhubu Formation can be divided into two third-order sequences (Sq17–Sq18), both of which belong to second-order sequence SS7. Among these sequences, the third-order sequence Sq17 is composed of the third member of Xiyangshan Formation and the first member of Yinzhubu Formation as described in the previous section.

Third-order sequence Sq18: Located from the second member to the third member of Yinzhubu Formation, it consists of transgressive systems tract (TST) and highstand systems tract (HST). The top and bottom of Sq18 belong to type-II sequence boundary. TST is situated from Bed 7 to Bed 14, and the main sediments include thin–medium laminated microcrystalline calcite and siliceous argillaceous matter. Micro-fine horizontal bedding developed in the siliceous argillaceous matter, indicating deep shelf siliceous-argillaceous facies. The vein-strip shaped limestone in the sediments is shallow shelf carbonate facies, indicating frequent vibration and alternate change of the sea level during a new round of transgression. HST is situated from Bed 15 to Bed 24, and the main sediment is siliceous argillaceous matter. Besides, the lower part is interbedded with a small amount of knotlike limestone. Bioturbation structures and charcoal flakes are visible in mudstone of the upper part of HST. Very thin horizontal bedding developed in the rocks, indicating deep shelf siliceous-argillaceous facies in a weak oxidation environment. The sequence represents continual transgression and is a sequence of retrogradation–aggradation type.

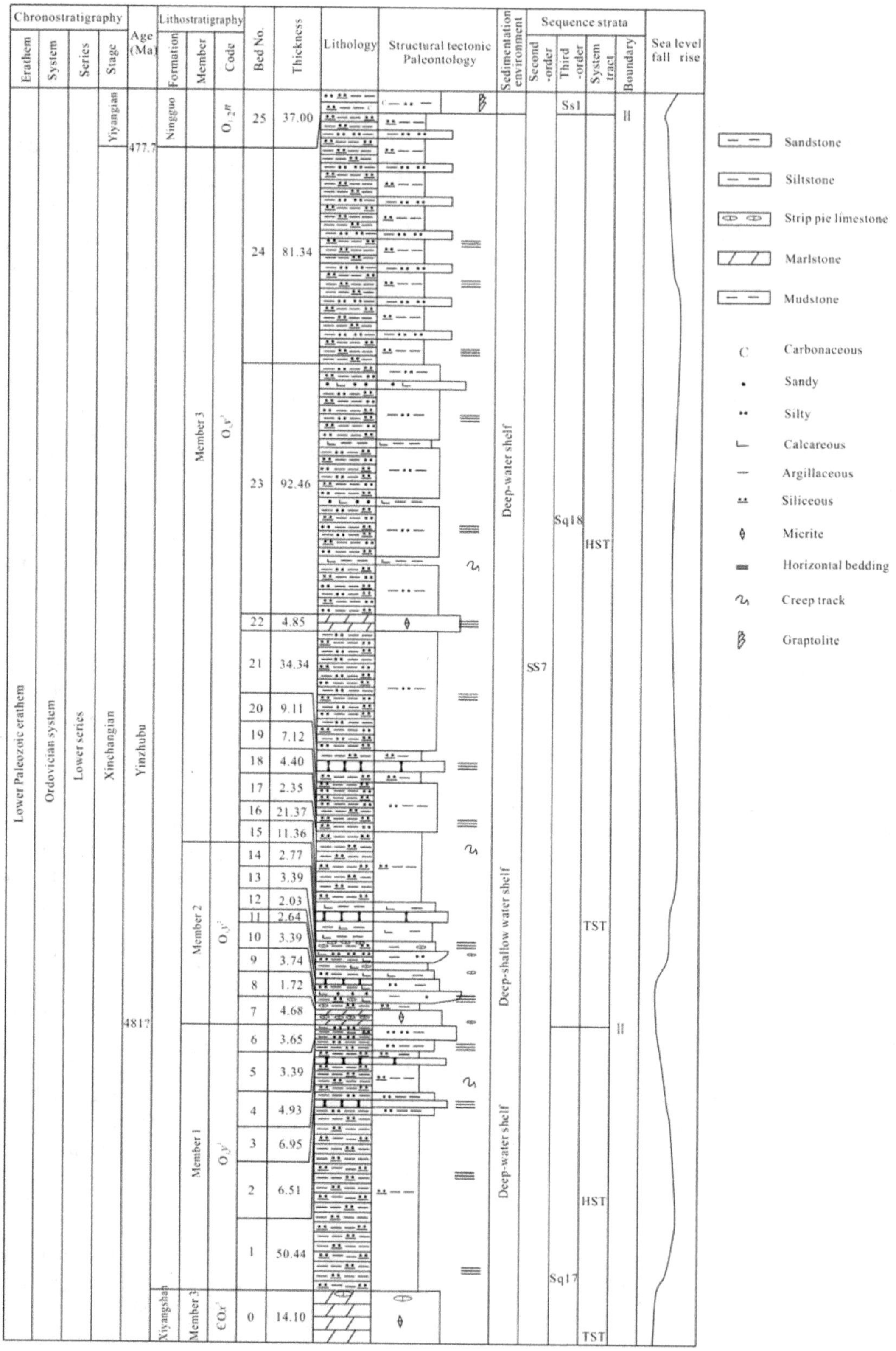

Fig. 2.30 Sequence stratigraphic framework of Ordovician Yinzhubu Formation in the area

2.4.1.3 Biostratigraphy and Chronostratigraphy

Yinzhubu Formation of the Area features sparse fossils. In this Project, no fossils were obtained and only bioturbation structure of Mollusca was found in the mudstone. According to *Regional Geology of Zhejiang Province* (1989), trilobite, brachiopod, and graptolite fossils are produced in the mudstone-bearing knotlike limestone in the upper profile of Yinzhubu Formation in Zhitang Village, Longyou County, and in the lower part of Yinzhubu Formation in Duibian Village, Jiangshan City. They belong to *Anisograptus–Clonograptus* zone. According to *Stratigraphic Chart of China* (2014), they are parts of Xinchangian Stage of the later period of the Early Ordovician. Therefore, Yinzhubu Formation is Early Ordovician strata.

2.4.1.4 Analysis of Sedimentary Environment

After the Cambrian when the sediments were dominated by carbonatite, great changes took place in the Ordovician sedimentary environment in northwest Zhejiang, and the sediments became dominated by argillaceous and terrigenous clasts and interbedded with a small amount of carbonate.

The sedimentary environment of Yinzhubu Formation of the early period of the Early Ordovician is as follows. In the first member, the sediments mainly include argillaceous matter followed by a small amount of siliceous matter, and very thin horizontal bedding developed, indicating a deep shelf environment under the oxidation–reduction interface. Terrigenous silt increases upward, the interbed of knotlike microcrystalline calcite appears, and Mollusca trackway of soft body developed in the siliceous mud, reflecting the rise of sea level and the increase of oxygen content in seawater and indicating the sedimentary environment above deep shelf oxidation zone.

The sedimentary environment of Yinzhubu Formation of the middle period of the Early Ordovician is as follows. In the second member, the sediments mainly include siliceous argillaceous matter and carbonate, which appears alternately in a rhythmic way, reflecting the frequent change of the seal level and indicating the sedimentary environment with alternate vibration of the deep shelf and shallow shelf.

The sedimentary environment of Yinzhubu Formation of the middle–late period of the Early Ordovician is as follows. In the third member, the sediments in this sequence are contrary to that of the early period. In detail, from bottom to top, the sediments gradually change from carbonate-bearing siliceous argillaceous matter into silty siliceous argillaceous matter, indicating a deep shelf sedimentary environment with sea level rise.

In conclusion, the sediments in Yinzhubu Formation change symmetrically upward and downward from the second member, generally reflecting the falling → rise process of the sea level and indicating the change process consisting of deep shelf → shallow shelf → deep shelf in the ancient geographic environment.

There is no considerable change in lithology and stratum thickness of Yinzhubu Formation in the Area, suggesting that Yinzhubu Formation is located in the same stable sedimentary basin. Regionally, in Dongqiao, Fuyang–Jingshanling, Hangzhou area, the lithofacies is transformed into knot-strip shaped marl, reticulate limestone, and banded limestone in Lunshan Formation (O_1l) and limestone with a polygonal reticulate structure in Honghuayuan Formation (O_1h), indicating platform sedimentary environment. In Jiangshan–Shaoxing area, the lithology is changed into caesious mudstone and silty mudstone, which contain sporadical calcareous nodule. In Changshan–Tonglu area, the lithology is characterized by amaranthine, caesious, and olive calcareous mudstone that contains calcareous concretion universally. The mudstone is 160–280 m thick and contains trilobite fossils, indicating a shallow shelf oxidation zone sedimentary environment.

2.4.1.5 Trace Elements in Strata (Ore-Bearing)

Sixty-five rock spectrums were systemically collected in the Yinzhubu Formation (O_1y) profile (PM005, 6) in Huangdouwu and Qiaotou, Hanggai Town, Anji County, Zhejiang Province, and statistics were made on the arithmetic mean values and concentration coefficients of 14 main trace elements of the three members of Yinzhubu Formation. According to the analysis and calculation results, the rocks of the three members are mainly enriched in Bi and F followed by Be, Sb, Cu, Pb, Zn, W, Ag, Sn, S, and TFe. This suggests that after long-distance transformation and sorting of terrigenous clasts, the content of most trace elements in pelagic siliceous argillaceous sediment is unchanged and only a small amount of trace elements get enriched and diluted.

2.4.2 Ningguo Formation ($O_{1\text{-}2}n$)

Ningguo Formation ($O_{1\text{-}2}n$) was established and named by Jie Xu in 1934 in Hulesi Town, Ningguo City, Anhui Province. Lu et al. (1955) introduced it to Zhejiang and named a section of the upper part (producing Early Ordovician *Didymograptushirundo*, etc.) in the former Yinzhubu System (established by Liu and Zhao in 1927) in northwest Zhejiang as Ningguo Formation, since the section features the same lithology and fossils as the Ningguo Formation in south Anhui. It was subsequently used by the preparation group of the regional stratigraphic table of Zhejiang Province (1979) as well as in the literature including *Stratigraphic Correlation Chart in China with Explanatory Text*

(1982) and *Regional Geology of Zhejiang Province* (1989). In this Project, the name Ningguo Formation ($O_{1-2}n$) was still adopted according to the lithological characteristics of Jiumulong profile (PM007) of Hanggai map sheet and the survey traverse.

2.4.2.1 Lithostratigraphy

The Ningguo Formation ($O_{1-2}n$) in the Area is mainly distributed in east Hanggai map sheet and secondarily in south Xianxia map sheet. The total outcrop area is about 4.79 km^2, accounting for 0.38% of the bedrock area.

The lithology and lithologic association of Ningguo Formation ($O_{1-2}n$) are characterized by dark gray thin to medium laminated silty mudstone, carbonaceous siliceous silty mudstone, and grayish black thin–medium laminated carbonaceous siliceous mudstone. In addition, very thin horizontal bedding developed and rich graptolite is present. It is in conformable contact with overlying Hule Formation and underlying Yinzhubu Formation. The thickness of the formation is 93.69 m.

1. **Stratigraphic section**

The stratigraphic profile of Ningguo Formation is described by taking the example of the Ordovician Ningguo Formation ($O_{1-2}n$)–the first member of Hule Formation ($O_{2-3}h^1$) profile (Fig. 2.31) in Jiumulong, Hanggai Town, Anji County, Zhejiang Province. The details are as follows:

The first member of Hule Formation Total thickness: >3.89 m

23. Grayish black thin–medium laminated carbonaceous silicalites interbedded with a small amount of carbonaceous mudstone. The carbonaceous silicalites: 6–20 cm per single layer, horizontal bedding developing. The carbonaceous siltstone: 1–5 cm thick per layer, accounting for 10% of the layer, horizontal bedding developing in the layers, and graptolite fragments visible. 3.89 m

———————— Conformable contact ————————

The first member of Hule Formation Total thickness: 5.24 m

22. Grayish black medium laminated silty carbonaceous siliceous mudstone, interbedded with a small amount of thin laminated carbonaceous silicalite, the thickness of a single layer: 10–15 cm, horizontal bedding developed. The thickness of a single layer of thin laminated carbonaceous silicalite: 1 cm and gradually increasing to 6–10 cm upwards. In the part with a distance of about 0–70 cm from the bottom of the layer, plenty of fossils of the following ancient organisms were obtained: *Pterograptuselegans* Holm, 1881, *Tetragraptus erectus* Mu, Geh et Yin, *dichograptid gen.* et sp., *Haddingograptuseurystoma* (Jaanusson, 1960) *Archiclimacograptus* sp. indet., *Kalpinograptusovatus* (Hall, 1902) *Haddingograptusoliveri* (Bouček), *Proclimacograptusangustatus* (Ekström), *Glossograptus* sp. Indet., *Archiclimacograptus*cf. *Caelatus* (Lapworth), and *Archiclimacograptusangulatus* (Bulman). 2.86 m

21. Grayish black thin laminated silty carbonaceous mudstone, occasionally interbedded with micro laminated silicalite. The silty carbonaceous mudstone: 5–9 cm thick per single layer, horizontal bedding developing. The silicalite: 3–0 mm thick per single layer, evenly distributed. Many fossils of *Pterograptus elegans* Holm 1881 visible in the carbonaceous silty carbonaceous mudstone 120 cm and 180–210 cm away from the bottom. 2.38 m

———————— Conformable contact ————————

Ningguo Formation ($O_{1-2}n$) Total thickness: 93.69 m

20. Grayish black thin–medium laminated silty carbonaceous siliceous mudstone, the thickness of a single layer: 7–20 cm, very thin horizontal bedding developing. Many fossils of *Nicholsonograptusfasciculatus* visible 200 cm away from the top (true thickness), plenty of graptolite visible 100–140 cm to the top, *Nicholsonograptusfasciculatus* visible 60–70 cm away from the top. 10.06 m

19. Covering. 1.64 m

18. Grayish black thin–medium laminated siliceous silty mudstone, the thickness of a single layer: 7–20 cm, very thin horizontal bedding developing. 2.31 m

17. Grayish black medium laminated carbonaceous siliceous mudstone, interbedded with mudstone and a thickness of 1–3cm; the siliceous carbonaceous mixture: banded, the thickness of a single layer: 10–20 cm; very thin horizontal bedding developing. 4.46 m

16. Grayish black thin laminated carbonaceous siliceous mudstone, the carbonaceous matter in the mudstone distributed in the shape of micro-fine stripe along the layers, very fine horizontal bedding developing along the layers. 3.26 m

15. Covering. 5.25 m

14. Dark gray thin–medium laminated carbonaceous siliceous mudstone, the thickness of a single layer: 6–13 cm, very thin horizontal bedding developing. 4.69 m

13. Grayish black thin laminated carbonaceous siliceous mudstone, the thickness of a single layer: 3–8 cm, very thin horizontal bedding developing. 3.38 m

12. Dark gray carbonaceous siliceous mudstone, the thickness of a single layer: 10 cm, very thin horizontal bedding developing, graptolite fossil visible. 1.92 m

11. Grayish black thin–medium laminated banded carbonaceous siliceous mudstone, the thickness of a single layer: 7–20 cm, very thin horizontal bedding developing, the following graptolite fossils obtained: *Cryptograptustrieornis*, *Tylograptusintermedsis*Mu, and *Phyllograptusilicifolios* Holl. 9.18 m

10. Dark gray thin laminated carbonaceous siliceous silty mudstone, off-white striates occurring along the bedding owing to weathering, the thickness of a single layer: < 10 cm, very thin horizontal bedding developing, carbonaceous matter concentrating along the bedding. 6.00 m

9. Grayish black thin laminated carbonaceous siliceous mudstone, the thickness of a single layer: < 10cm, very thin horizontal bedding developing. No graptolite fossil visible in the mudstone. 1.64 m

8. Dark gray–grayish-black thin–medium laminated silty fine-sandy mudstone, the thickness of a single layer: 8–15 cm, very thin horizontal bedding developing, *Dichograptid* gen. & sp. acquired. 5.98 m

7. Grayish black medium laminated carbonaceous siliceous silty mudstone, the thickness of a single layer: 10–30 mm. 1.83 m

6. Dark gray thin–medium laminated carbonaceous mudstone, the thickness of a single layer: 10 cm generally and 15 cm locally, very thin horizontal bedding developing. *Tylograptus* sp., *Allograptus* sp., and *Archiclimacograptus* sp. acquired. 15.00 m

5. Covering. 6.80 m

4. Dark gray thin–medium laminated siliceous silty mudstone, the silt distributed in the shape of micro-fine strips along the layer, the thickness of a single layer: 6–20 cm, very thin horizontal bedding developing. *Diplograptid* & sp. acquired 1.48 m

3. Covering. 4.32 m

2. Dark gray thin–medium laminated siliceous silty mudstone, the thickness of a single layer: mostly 8–15 cm and >10 cm, horizontal bedding developing, off-white after weathering. 1.08 m

1. Dark gray–grayish-black thin–medium laminated silty mudstone, grayish-black mostly owing to high carbon content, the thickness of a single layer: mostly 8–20 cm and >10 cm, horizontal bedding developing. 3.96 m

——— Conformable contact ———

Yinzhubu Formation Total thickness: > 1.44 m

0. Gray thin laminated silty mudstone, the thickness of a single layer: 3–10 cm, very thin horizontal bedding composed of grayish–dark gray matter developing. 1.44 m

2. **Lithological characteristics**

The lithology of Ningguo Formation ($O_{1\text{-}2}n$) is mainly characterized by siliceous silty mudstone and carbonaceous siliceous mudstone.

(1) The siliceous silty mudstone: dark gray, thin–medium laminated, composition: argillaceous matter (50–55%), silt (15–35%), and a small amount of carbonaceous matter and cryptocrystalline siliceous matter (10–20%), containing rich graptolite.

(2) The carbonaceous siliceous mudstone: grayish black, thin–medium laminated, banded, composition: argillaceous matter (50–65%), cryptocrystalline siliceous matter (10–25%), and amorphous carbonaceous matter (10–15%), very thin horizontal bedding developing, containing rich graptolite.

3. **Basic sequences**

Two types of basic sequences (Fig. 2.32) developed in Ningguo Formation ($O_{1\text{-}2}n$), which is, respectively, composed of (carbonaceous) siliceous silty mudstone and carbonaceous siliceous mudstone. Therefore, the basic sequences of the two types are both characterized by a single lithology and are both of non-cyclic basic sequence composed with monotonous lithology. The siliceous mudstone constitutes the main body of Ningguo Formation. In the siliceous mudstone, there are rich graptolite as well as small amounts of terrigenous silt and carbonaceous, and micro-fine horizontal bedding developed extremely, indicating the sedimentation of abyssal graptolite shale facies.

2.4.2.2 Sequence Stratigraphy

Ningguo Formation in the Area features simple rock types, which are mainly argillaceous–siliceous matter followed by a small amount of carbonaceous matter and terrigenous silt, indicating that the sedimentary environment is a standing-water deepwater reduction environment. Therefore, only the changes in relative contents of siliceous matter, argillaceous matter, and terrigenous silt can be used to reflect the rise and falling of the sea level, identify transgression, maximum flooding surface, and regression, and determine the division of the systems tract.

On the basis of features of lithology, palaeontology, lithofacies association, and sequence boundary of Jiumulong profile of Hanggai map sheet, one sub-second-order orthosequence set Ss1 and one third-order sequence Sq19 were divided from Ningguo Formation ($O_{1\text{-}2}n$). Both Ss1 and Sq19 belong to the second-order sequence SS7 (Fig. 2.33).

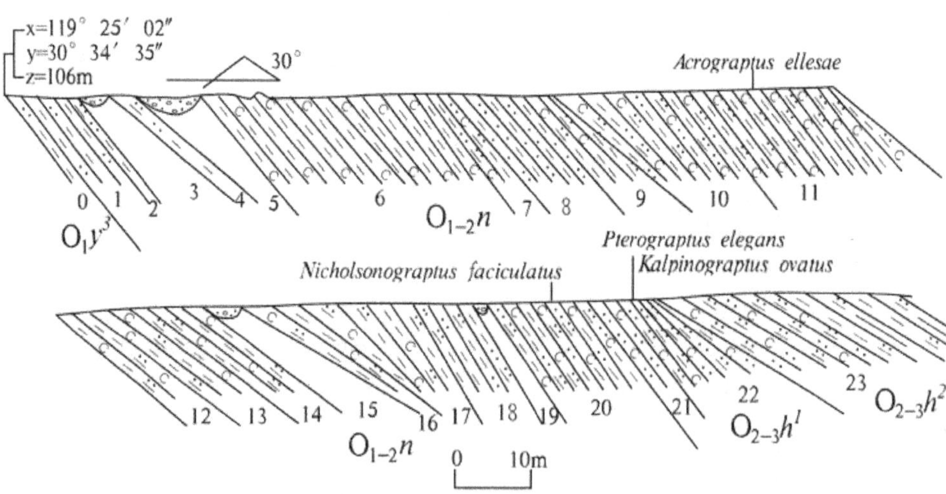

Fig. 2.31 Ordovician Ningguo Formation ($O_{1\text{-}2}n$)–the first member of Hule Formation ($O_{2\text{-}3}h^1$) profile in Jiumulong, Hanggai Town, Anji County, Zhejiang, Province

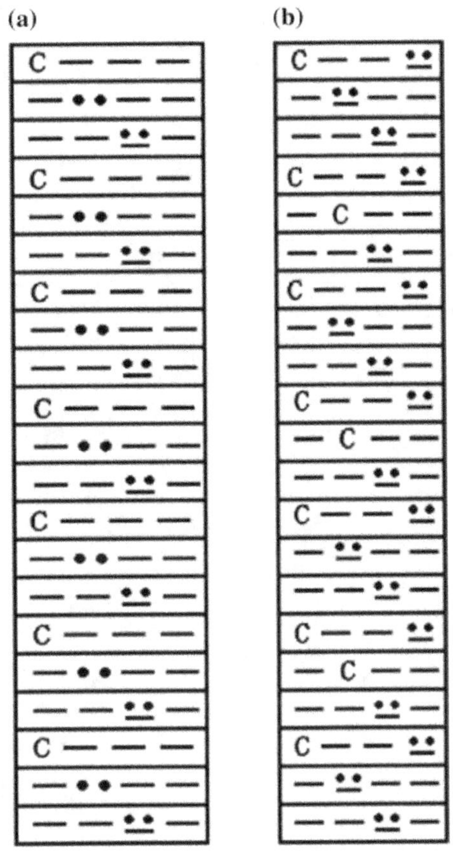

Fig. 2.32 Basic sequences of Ningguo Formation ($O_{1-2}n$)

Sub-second-order orthosequence set Ss1: Located in the lower part of Ningguo Formation, it consists of TST and HST. The top and bottom of Ss1 are of type-II sequence boundary. TST is located from Bed 1 to Bed 5. The sediments in the bottom of TST mainly include siliceous silty argillaceous matter followed by some terrigenous clasts, indicating the siliceous-argillaceous facies of bathyal subcompensational basin. Till Bed 6, Ss1 entered the maximal flooding period, and the sediments in this bed are thick lamellar graptolite-bearing carbonaceous shale. HST is located from Bed 6 to Bed 9. The sediments in it mainly include argillaceous matter, containing rich siliceous carbonaceous matter and producing graptolite fossils, thus indicating the sedimentation in an abyssal basin reduction environment. The sequence set represents a weak process of transgression–retrogradation and is a sequence of aggradation type. Since the sequence set stands approximately in the Yiyangian Stage and the Dapingian Stage, the age range of it is 477.7–467.3 Ma, spanning about 10.4 Ma.

Third-order sequence Sq19: Located in the middle and upper parts of Ningguo Formation, it consists of TST and HST. The top and bottom of Sq19 are of type-II sequence boundary. TST is located from Bed 9 to Bed 14. The sediments in it mainly include argillaceous matter and occasionally contain terrigenous fine sand and silt. The terrigenous clasts decrease upward, indicating the rise of the sea level. In Bed 15–Bed 17, Sq19 entered the maximal flooding period, and the sediments in these beds are siliceous argillaceous matter with very thin horizontal bedding developing, indicating that the sedimentary environment is still a standing-water reduction environment. HST is located in Bed 18–Bed 20. The sediments in it mainly include argillaceous matter and a small amount of carbonaceous siliceous matter. The terrigenous clasts gradually increase upward, indicating the gradual falling of the sea level. Since the sediments mainly include carbonaceous siliceous argillaceous matter-bearing planktonic graptolite, HST shall be of siliceous-argillaceous facies in an abyssal reduction environment. Sq19 represents the sedimentation in the slight vibration of the sea level under continuous abyssal environment and is a sequence of aggradation type. The sequence starts from the bottom of Darriwilian Stage, and its top is located in the *Nicholsonograptusfasciculatus* zone, which is in the middle part of Darriwilian Stage. The age range of the sequence is 458.4–467.3 Ma, spanning about 8.9 Ma.

2.4.2.3 Biostratigraphy and Chronostratigraphy

During the period of Ningguo Formation, the Area was in basin sedimentation in closed reduction environment and featured most flourishing planktonic graptolite and other scarce ancient organisms. According to *Stratigraphic Chart of China* (2014), there are 10 graptolite zones in Ningguo Formation, and from bottom to top, they are: *Tetragrapproximatus* zone, *Pendeograptusfruticosus* zone, *Didymograptelluseobifidus* zone, *Baltograptus* (*Corymbograptus*) *deflexus* zone, *Azygograptussuecicus* zone, *Isograptuscaduceuimitatus* zone, *Exirgaptusclavus* zone, *Undulograptusaustrodentatus* zone, *Acrograptusellesae* zone, and *Nicholsonograptusfasciculatus* zone. In this Project, many graptolite fossils were obtained from Ningguo Formation and Hule Formation, and two graptolite zones were identified in Ningguo Formation. The distribution of the graptolite is shown in Fig. 2.34.

1. ***Acrograptusellesae* zone**

It is located in Bed 10–Bed 18 in the middle and upper parts of Ningguo Formation of Jiumulong profile. The graptolite obtained in the zone mainly includes *Acrograptusellesae* Rudemann (Fig. 2.35a) and other associated graptolites including *Tylograptusgeniculiformis* Mu, *Tylograptus intermedius* Mu, *Allograptus* sp. graptolite, *Archiclimacograptus* sp. graptolite, *lossograptus* sp., *Archiclimacograptus* sp., *Cryptograptustricornis* (Carruthers),

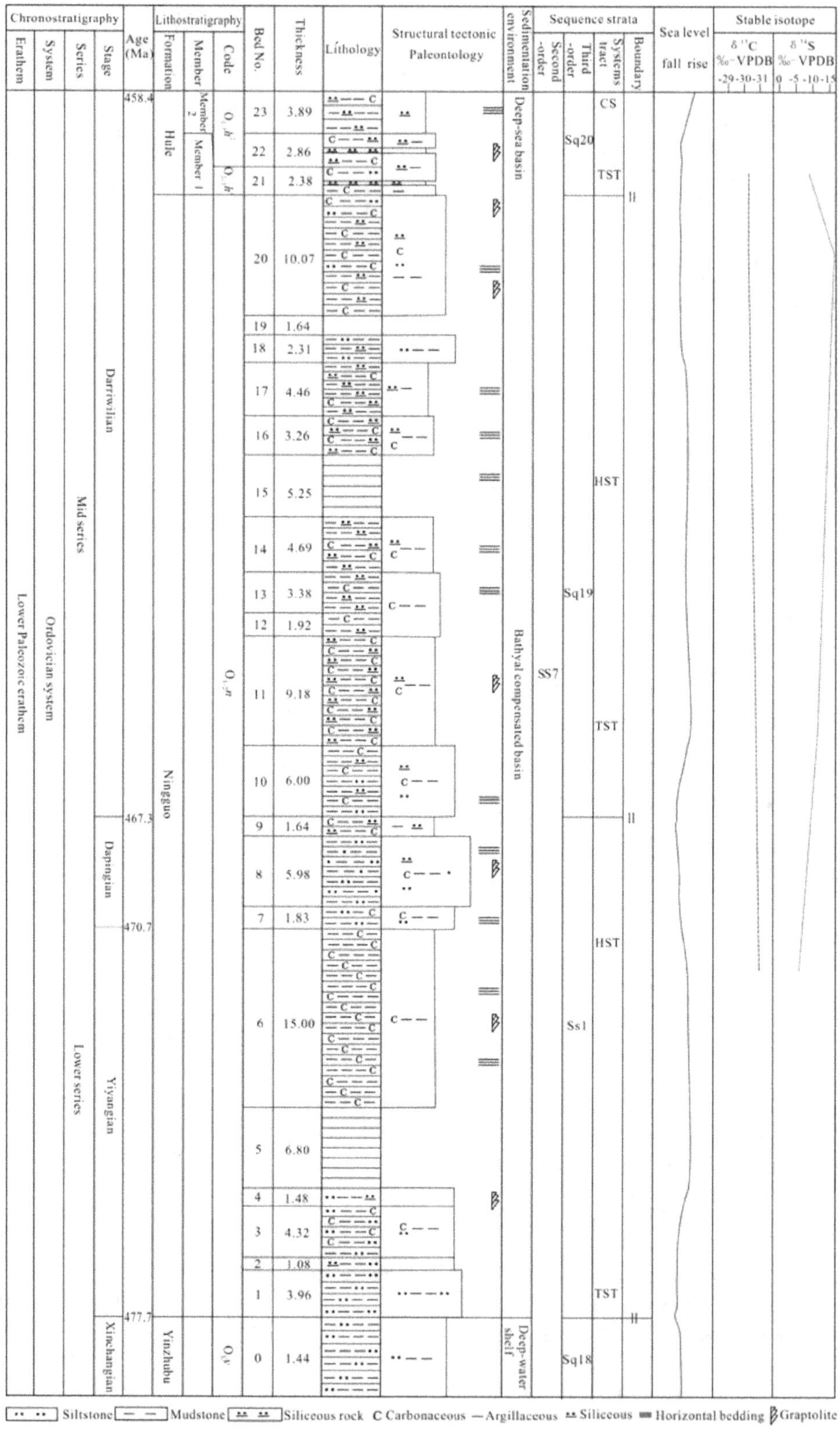

Fig. 2.33 Sequence stratigraphic framework of Ordovician Ningguo Formation in the area

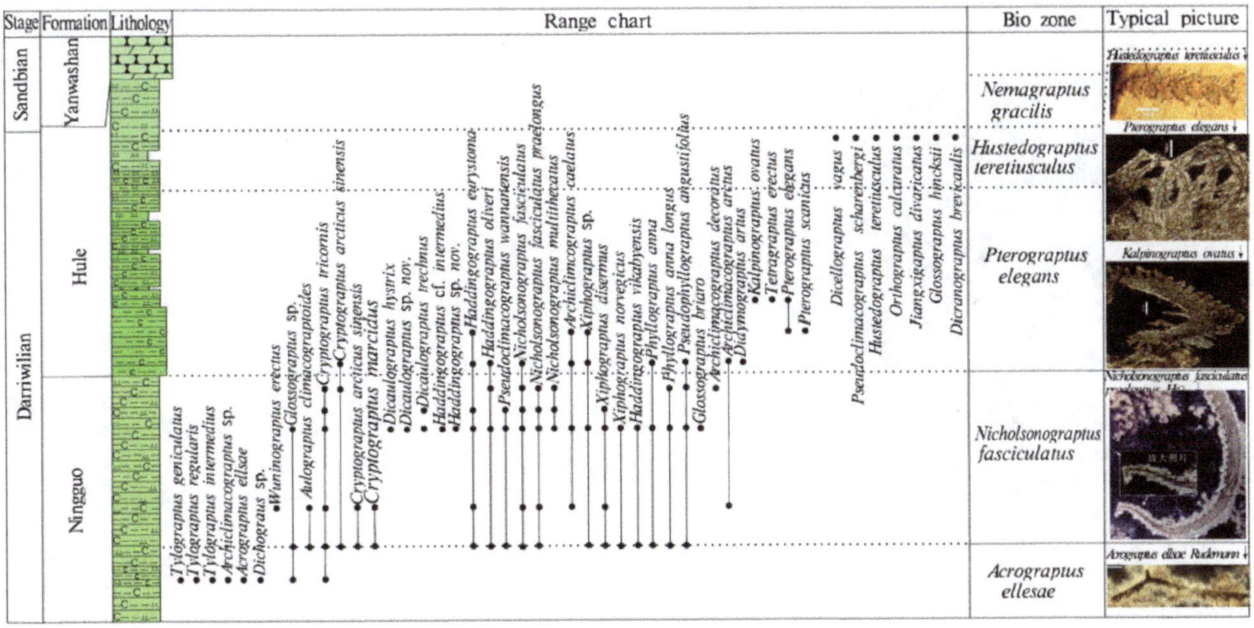

Fig. 2.34 Range chart of part fossils found in Ningguo Formation–Hule Formation in the area

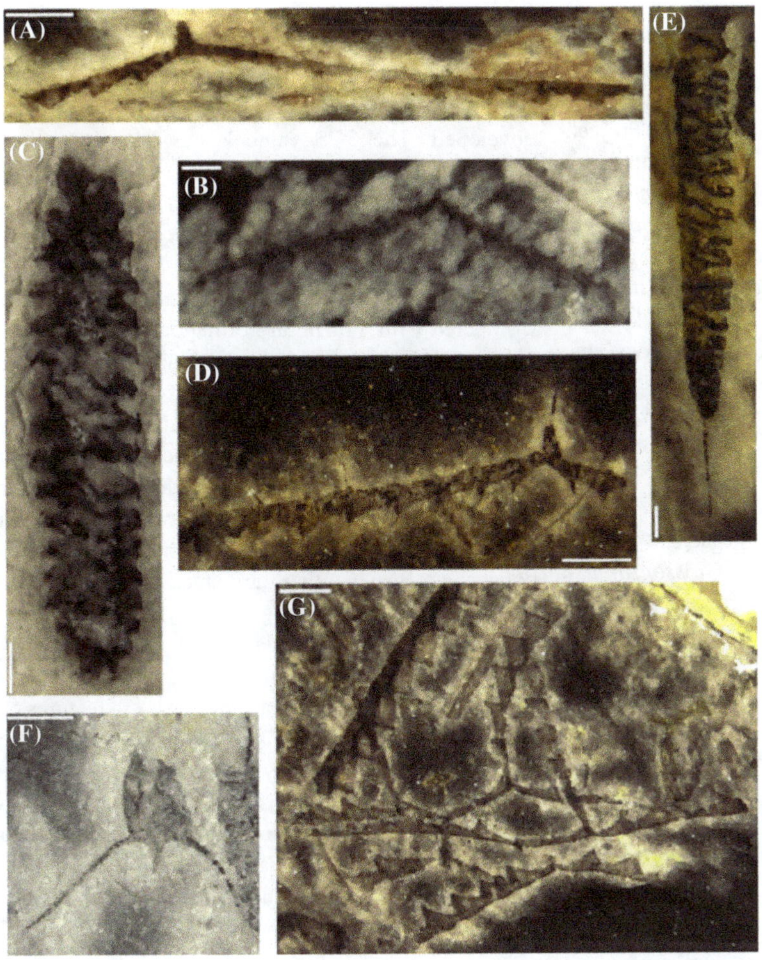

(A) *Acrograptusellsae* Rudemann (PM007-11-4).

(B) *Tylograptus geniculiformis* Mu (PM007-11-3).

(C) *Glossograptus* sp. (PM007-11-1-28).

(D) *Tylograptus intermedius* Mu (PM007-11-1-22).

(E) *Archiclimacograptus* sp. (PM007-11-1-35).

(F) Cryptograptustricornis (Carruthers) (PM007-11-1-14).

(G) *Dichograptus* sp. (PM007-11-1-37).

Scale length = 1mm.

(The sample No. consists of Profile No. – Bed No. – specimen location – specimen No.)

Fig. 2.35 Graptolite in *Acrograptusellsae* zone in Bed 11 of Ningguo Formation ($O_{1\text{-}2}n$) in Jiumulong profile (PM007)

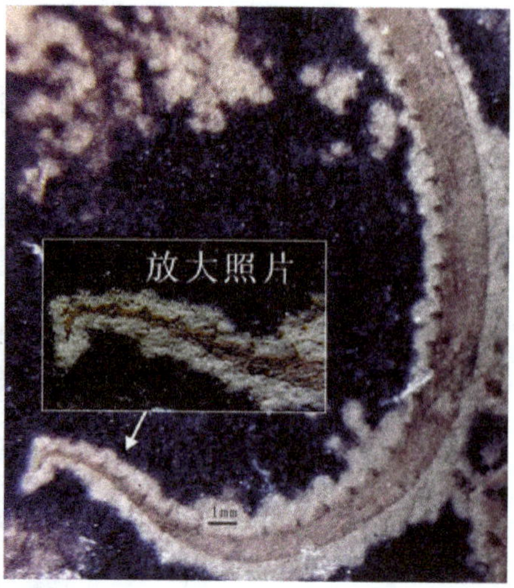

Fig. 2.36 *Nicholsonograptus fasciculatus praelongus* Hsü at the top of Ningguo Formation

Tylograptusintermedsis Mu, *Phyllograptusilicifolios* Holl, *Diplograptid*gen. & sp., *Dichograptid* gen. & sp. (Figure 2.35).

2. ***Nicholsonograptusfaciculatus* zone**

It is located at the top of Ningguo Formation. In this Project, the graptolite obtained from the four collection points located from bottom to top of Bed 20 in Jiumulong profile (PM007) mainly includes exquisite *Nicholsonograptus fasciculatus praelongus* Hsü (Fig. 2.36) and other paragenetic graptolite including *Nicholsonograptusfasciculatus* (Nicholson), *Nicholsonograptusmultithecatus* Ge, *Cryptograptusarcticussinensis* Ni, *Archiclimacograptuscaelatus* (Lapworth), *Archiclimacograptusdecoratus*, *Pseudoclimacograptuswannanensis* Li, *Archiclimacograptusarctus* (Elles and Wood), *Haddingograptusoliveri* (Bouček), *Dicaulograptus*sp. Nov., *Cryptograptustricornis* (Carruthers), *Xiphograptusnorvegicus* (Berry), *Aulograptusclimacograptoides*, *Wuninograptus erectus* Ni, *Pseudophyllograptus angustifolius* (Hall), *Glossograptusbriaros* Ni, *Dicaulograptushytrix* (Bulman), and Dicaulograptustrechnus Ni (Figs. 2.37, 2.38 and 2.39).

The above-mentioned is the typical paleontologic condition of Ningguo Formation in the Area. According to Stratigraphic Chart of China (2014), Ningguo Formation involves Yiyangian Stage, Dapingian Stage, and part of Darriwilian Stage. Among these 10 graptolite zones, Zone 1–Zone 4 belong to Yiyangian Stage of the Early Ordovician, Zone 5–Zone 7 belong to Dapingian Stage of the lower stage of the Middle Ordovician, and Zone 8–Zone 10 belong to Darriwillian Stage of the middle stage of the Middle Ordovician. Ningguo Formation spans from the lower and middle parts of the Ordovician.

2.4.2.4 Characteristics of Stable Isotopes

Three whole-rock samples of stable isotopes $\delta^{13}C$ and $\delta^{34}S$ of carbonaceous mudstone were collected from Ningguo Formation of Jiumulong profile (PM007). The $\delta^{13}C$ value ranges between −30.99 and −30.24‰, with an average of −30.66‰. The $\delta^{34}S$ value ranges between −15.40 and −5.20‰, with an average of −10.33‰. According to Zheng and Chen (2000), the content of organic carbon $\delta^{13}C$ in sedimentary rocks of different periods in China ranges between −35 and −20‰, with an average of −24.4‰, and the content of $\delta^{34}S$ ranges between −10 and 10‰; the sedimentary rocks here mainly refer to fossil fuels (such as coal), most of which are humic coal and sapropelic coal with the $\delta^{13}C$ content mostly between −35 and −30‰. Therefore, it can be thought that the normal content of organic carbon $\delta^{13}C$ in sedimentary rocks ranges between −35 and −30‰. Obviously, the content of organic carbon $\delta^{13}C$ of Ningguo Formation in the Area is within the normal range. Therefore, it can be inferred that the sedimentary environment of Ningguo Formation in the Area is similar to the coal-forming environment, i.e., bathyal subcompensational basin sedimentation in which chitin lipoid organisms relatively developed. The $\delta^{34}S$ value is consistent with the normal content of $\delta^{34}S$ in shale. This further verifies that the Area of the Ningguo Formation period was in a reduction environment that was pretty favorable for coal forming.

2.4.2.5 Analysis of Sedimentary Environment

Compared with Yinzhubu Formation period, the Ordovician Ningguo Formation period experienced a qualitative change in sedimentary environments. That is, the open circulated oxidation environment in the Yinzhubu Formation period was transformed into the stagnant closed reduction environment Ningguo Formation period. This can be reflected by sediments. The sediments in Ningguo Formation are mainly dark gray–black siliceous argillaceous matter followed by carbonaceous. Besides, there are small amounts of terrigenous clastic silt and micro-fine-grained pyrite. The sediments contain rich well-preserved planktonic graptolite fossils and no demersal shellfish, with micro-fine horizontal bedding developing. All these indicate that the basin features extremely short sources of terrigenous clasts, very weak hydrodynamic force, and insufficient oxygen in the bottom. This is unfavorable for organism survival and belongs to bathyal subcompensational basin sedimentation with strong reduction.

Regionally, large lithofacies transition takes place in this formation. In the area along Banqiao (Lin'an)–Jingshanling

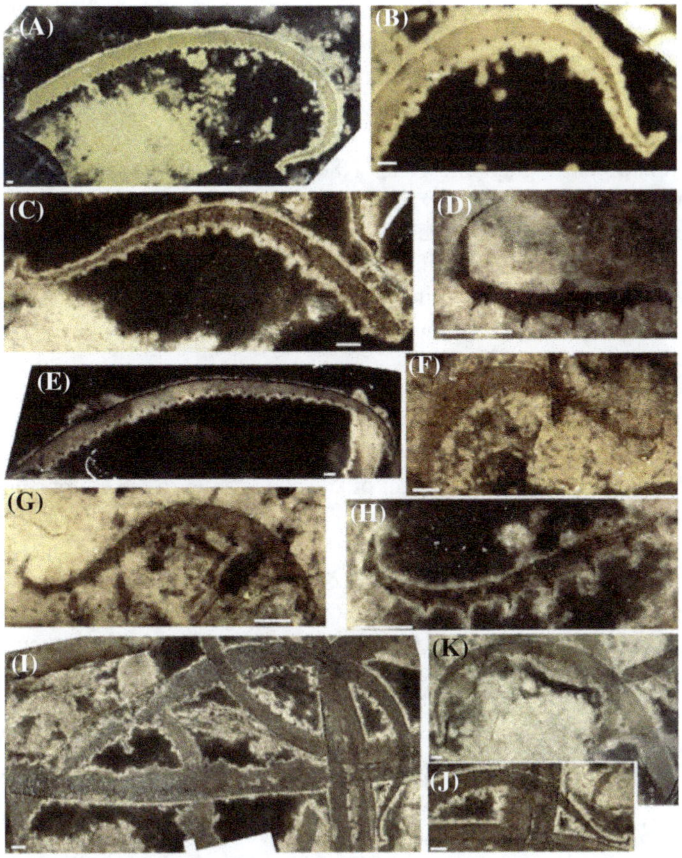

A-B. *Nicholsonograptus fasciculatus praelongus* Hsü (PM007-20).
C-E.*Nicholsonograptus fasciculatus* (Nicholson) (PM007-20-1-16).
D,H.*Nicholsonograptus* sp. (PM007-20-2-11,PM007-20-2-38). Two Juveniles.
F-G,I-K *Nicholsonograptus multithecatus* Ge (PM007-20-1-13, PM007-20-1-55, PM007-20-1, PM007-20-n-8).
J is the amplified start point of I.
Scale length = 1mm.
(The sample No. consists of Profile No. – Bed No. – specimen location – specimen No.)

Fig. 2.37 Graptolite in *Nicholsonograptusfasciculatus* zone in Bed 20 of Ningguo Formation ($O_{1\text{-}2}n$) in Jiumulong profile (PM007)

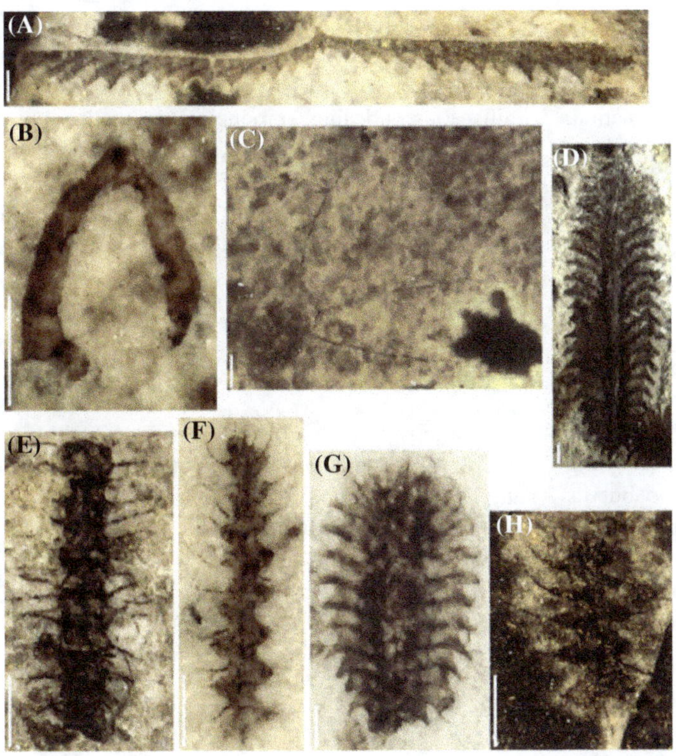

A.*Xiphograptus Norvegicus* (Berry)(PM007-20-1-53).
B. *Aulograptus climacograptoides* (Bulman) (PM007-20-a-71).
C. *Wuninograptus erectus* Ni (PM007-20-b-32).
D. *Pseudophyllograptus angustifolius* (Hall) (PM007-20-a-89).
E. *Glossograptus briaros* Ni (PM007-20-1-14).
F. *Dicaulograptus hytrix* (Bulman) (PM007-20-1-58).
G. *Phyllograptus anna ultimus* Rudemann (PM007-21-53).
H. Dicaulograptus trechnus Ni (PM007-20-1-14)
Scale length = 1mm.
(The sample No. consists of Profile No. – Bed No. – specimen location – specimen No.)

Fig. 2.38 Other graptolite fossils in *Nicholsonograptusfasciculatus* zone in Bed 20 of Ningguo Formation ($O_{1\text{-}2}n$) in Jiumulong profile (PM007)

(Hangzhou), the lithofacies of Ningguo Formation is characterized by node-bearing calcareous mudstone interbedded with limestone lenticels, and demersal organisms and exotic conodont (the main ancient organisms), indicating a neritic platform oxidation environment. In Jiangshan area, the bottom of Ningguo Formation consists of calcarenite, and Huangnitang area of Changshan is interbedded with several layers of calcarenite and calcisiltite. In addition to graptolite, there are a small amount of pleopod and deepwater Brachiopoda in the formation of this area. This indicates that the Ningguo Formation of this area belongs to the sedimentary environment of slope transitional zone. In conclusion, from southeast to northwest, the lithology of the Ningguo Formation in Zhejiang Province gradually turns from interbeds of shale and intraclastic limestone into siliceous mudstone. Furthermore, the thickness of the formation gradually increases. In detail, it is 44.8–29.4 m in Jiangshan–Shaoxing area, 147.8 m in Liujia of Tonglu, 93.69 m in the Hanggai of Anji, and 175 m in Changhua–Anji area, indicating the paleogeographic landscape of lithofacies featuring gradually increasing depth of sedimentary basin, during decreasing hydrodynamic force, and sharply increasing oxygen content in seawater from southeast to northwest.

2.4.2.6 Trace Elements in Strata (Ore-Bearing)

Fifteen rock spectra were systemically collected in the Ningguo Formation of Jiumulong profile (PM007), and statistics were made on the arithmetic mean values and concentration coefficients of 14 main trace elements. According to spectral analysis and statistics, Ningguo Formation is mainly enriched in Sb, Bi, Pb, and Mo, and the enrichment coefficient of Bi is up to 46.13 times. Furthermore, most of the enriched elements are consistent with those in black rock series of Cambrian Hetang Formation. Therefore, it can be inferred that some materials from abyssal volcanic eruption and hot springs flowed and penetrated into the sediments.

2.4.3 Hule Formation ($O_{2\text{-}3}h$)

The name Hule Formation ($O_{2\text{-}3}h$) was created by Jie Xu in 1934 for Hulesi Town, Ningguo County, Anhui Province. The Hule Formation in Zhejiang refers to the shale in the lower part called Yanwashan System by Jichen Liu and Yazeng Zhao in 1927 since its biota is similar to but its lithology is a little different from Hule Formation in south Anhui. Lu et al. (1955) referred to the shale as Hule Formation. It was subsequently used by the preparation group of regional stratigraphic table of Zhejiang Province (1979) as well as in the literature including *Stratigraphic Correlation Chart in China with Explanatory Text* (1982), *Regional Geology of Zhejiang Province* (1989), and *Lithostratigraphy of Zhejiang Province* (1995). This formation includes graptolite zones N8–N9 in the former Ningguo Formation (previously called Niushang Formation). In this project, the name Hule Formation ($O_{2\text{-}3}h$) is still adopted given the lithological features of Xinqiao profile (PM001) and Lijiabian profile (PM012) in Hanggai map sheet, Xiajia profile (PM030) in Xianxia map sheet, and the geological observation traverse.

2.4.3.1 Lithostratigraphy

Hule Formation ($O_{2\text{-}3}h$) in the Area is a stratum with the highest content of siliceous matter in the Ordovician system of Lower Paleozoic Erathem. The outcrop range is basically the same as that of Ningguo Formation. The outcrop area is about 3.64 km^2, accounting for 0.29% of the bedrock area.

Based on the lithology and lithologic association, Hule Formation ($O_{2\text{-}3}h$) can be divided into three members: the first member ($O_{2\text{-}3}h^1$), the second member ($O_{2\text{-}3}h^2$), and the third member ($O_{2\text{-}3}h^3$). They are in conformable contact with its underlying Ningguo Formation and in disconformable contact with its overlying Yanwashan Formation.

The first member of Hule Formation ($O_{2\text{-}3}h^1$): rhythm beds consisting of grayish black thin–medium laminated carbonaceous silty shale interspersed with a small amount of micro-laminated silicalite, presence of graptolite. The thickness of this member: 4.98–12.41 m.

The second member of Hule Formation ($O_{2\text{-}3}h^2$): interbeds consisting of dark gray–grayish black thin laminated carbonaceous silicalite and thin–medium laminated silty argillaceous silicalite, producing graptolite locally. The thickness of this member: 55.60–81.14 m.

The third member of Hule Formation ($O_{2\text{-}3}h^3$): black thin laminated siliceous mudstone interspersed with micro-thin laminated carbonaceous silicalite, the components turning into siliceous mudstone upward, rich in graptolite. The thickness of this member: 3.47–15.04 m.

1. **Stratigraphic section**

The lithology of Hule Formation is described by taking the example of Ordovician Hule Formation ($O_{2\text{-}3}h$)–Huangnigang Formation (O_3h) profile (Fig. 2.40) in Xinqiao Village, Hanggai Town, Anji County, Zhejiang. The details are as follows:

Yanwashan Formation Total thickness: 13.75 m

10. Interbeds consisting of gray thin laminated marl and knotlike limestone. The marl: horizontal bedding developing, the thickness of a single layer: 3–7 cm; the node in the limestone: 2–7 cm in diameter, discontinuously distributed in moniliform shape and laminated form. The bottom of this bed: knotlike limestone with a thickness of 10 cm, the occurrence is consistent with that of its underlying stratum. 1.94 m

——— Conformable contact ———

The third member of Hule Formation Total thickness: 3.47 m

9. Dark gray siliceous carbonaceous shale, shale bedding developing. 1.93 m

8. Black thin laminated siliceous mudstone, interspersed with micro-thin laminated siliceous shale. The siliceous mudstone: very thin horizontal bedding developing; the thickness of a single layer of siliceous shale: 1–3 cm. Producing the following ancient organisms: *Dicellograptusvagus* Elles& Wood, 1904; *Pseudoclimacograptusscharenbergi*; *Hustedograptusteretiusculus* (Hisinger); *diplograptid* gen. &sp. Indet.; *Corynoides* sp.; *Orthograptuscalcaratus* Lapworth. 1.54 m

——— Conformable contact ———

The second member of Hule Formation Total thickness: 66.80 m

7. Black thin laminated argillaceous silicalite, containing silt (5–10%), the thickness of a single layer: 2–5 cm, very thin horizontal bedding developing, producing graptolite. 11.41 m

6. Dark thin laminated silty argillaceous silicalite, the thickness of a single layer: generally 2–8 cm and rarely 10–11 cm. 76 m

5. Dark gray thin laminated carbonaceous silicalite, the thickness of a single layer: 2–7 cm, pretty broken. 3.60 m

4. Black thin laminated silty argillaceous silicalite, the thickness of a single layer: 3–7 cm, horizontal bedding developing. 17.64 m

3. Grayish black thin laminated argillaceous siliceous shale, the thickness of a single layer: 2–10 cm, rich in graptolite. 7.29 m

2. Covering. 10.10 m

——— Conformable contact ———

The first member of Hule Formation Total thickness: 4.98 m

1. Rhythmic layer consisting of grayish-black thin–medium laminated carbonaceous siliceous shale interspersed with micro laminated carbonaceous silicalite. The carbonaceous siliceous shale: mostly 19 cm thick per single layer, very thin horizontal bedding developing, graptolite visible; the silicalite: no bedding developing, hard, the thickness of a single layer: 0.3–0.5 cm. 4.98 m

——— Conformable contact ———

Ningguo Formation Total thickness: > 17.05 m

0. Black carbonaceous siliceous shale, very thin horizontal bedding developing along the layer, in black and white after weathering, white bands containing no carbonaceous matter. 17.05 m

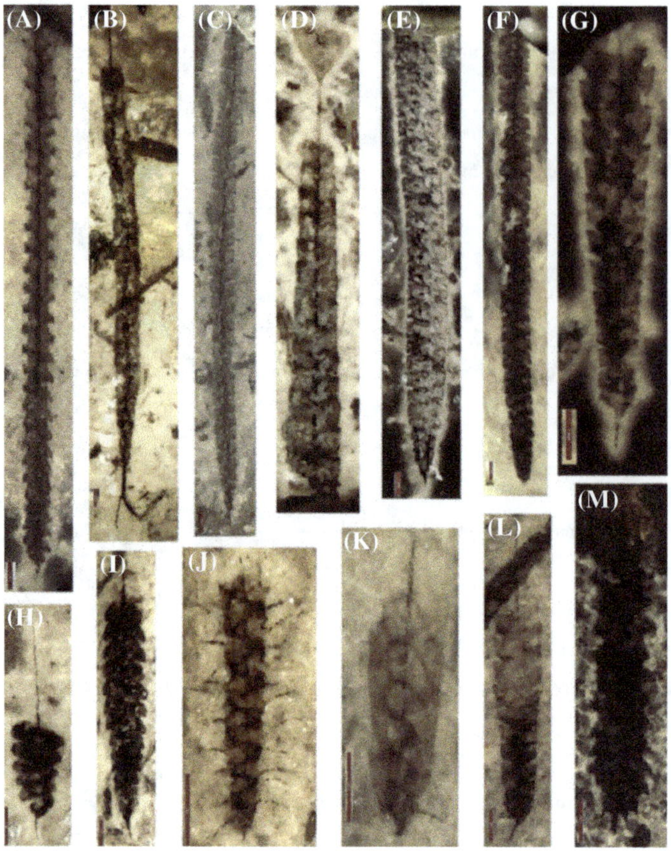

A. Cryptograptusarcticussinensis Ni (PM007-20-3-87).
B-C. Archiclimacograptuscaelatus (Lapworth) (PM007-20-1-54, PM007-21-2-63).
D. Archiclimacograptusdecoratus (PM007-20-3-9).
E. Haddingograptus sp. (PM007-21-2-6).
F. Pseudoclimacograptuswannanensis Li (PM007-20-2-2).
G, L. Archiclimacograptusarctus Elles & Wood (PM007-21-84, PM007-20-b-14).
H-I. Haddingograptusoliveri (Bouček) (PM007-20-3-38, PM007-20-1-61).
J. Dicaulograptussp. nov.(PM007-20-1-47).
K. Cryptograptustricornis (Carruthers) (PM007-20-2-4).
M. Cryptograptus sp. (PM007-20-a-70).
(The sample No. consists of Profile No. - Bed No. - specimen location - specimen No.)

Fig. 2.39 Other graptolite fossils at Beds 20–21 of Ningguo Formation ($O_{1-2}n$)–Hule Formation ($O_{2-3}h$) of Jiumulong profile (PM007)

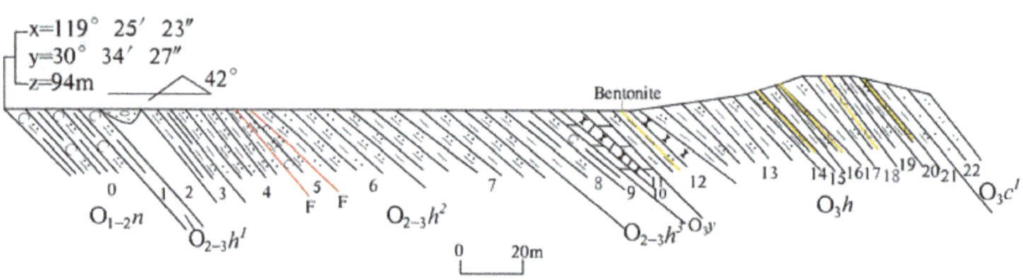

Fig. 2.40 Ordovician Hule Formation ($O_{2-3}h$)–Huangnigang Formation (O_3h) profile in Xinqiao Village, Hanggai Town, Anji County, Zhejiang

2. **Lithological characteristics**

The lithology in the first member of Hule Formation ($O_{2-3}h^1$) is mainly characterized by grayish black thin–medium laminated carbonaceous silty siliceous shale and micro-laminated silicalite.

(1) The carbonaceous silty siliceous shale: dark gray–black, composition: argillaceous matter (80–85%), a small amount of carbonaceous and siliceous matter (10–15%), and occasionally visible silt (5%), the thickness of a single layer: mostly about 19 cm, very thin horizontal bedding developing, rich in graptolite.
(2) The silicalite: dark gray, mainly composed of siliceous matter, no bedding developing, hard, the thickness of a single layer: 0.3–0.5 cm, projecting after weathering.

The lithology of the second member of Hule Formation ($O_{2-3}h^2$) is mainly characterized by dark gray–grayish black thin laminated carbonaceous silicalite and thin–medium laminated silty argillaceous silicalite.

(1) The carbonaceous silicalite: dark gray, composition: cryptocrystalline siliceous matter (50–60%) and carbonaceous matter (25–40%), hard.
(2) The silty argillaceous silicalite: dark gray, composition: cryptocrystalline siliceous (50%), argillaceous (25–30%), silt (15–20%), and micro-fine-grained pyrite (< 5%); silt is mainly composed of terrigenous quartz clasts.

The lithology of the third member of Hule Formation ($O_{2\text{-}3}h^3$) is mainly characterized by black thin laminated siliceous mudstone and thin–medium laminated carbonaceous silicalite.

(1) The siliceous mudstone: dark gray, composition: argillaceous matter (75%) and a small amount of siliceous matter (20–25%), very thin horizontal bedding developing, rich in graptolite.
(2) The charcoal-bearing silicalite: dark gray–black, composition: siliceous matter (80%–90%) and a small amount of carbonaceous matter (10–20%).

3. **Basic sequences**

One type of basic sequences developed in the first member of Hule Formation ($O_{2\text{-}3}h^1$) (Fig. 2.41).

Basic sequences of type A: composed of grayish black carbonaceous silty shale and micro-laminated silicalite; the thickness of a single of the micro-laminated silicalite: 1–2 mm in the lower part of the sequence and gradually increasing upward as 3–5 cm; belonging to non-cyclic basic sequence with layer thickness increasing and grain size decreasing upward.

In this member, the major and minor sediments are argillaceous matter and siliceous, respectively. The planktonic graptolite constitutes the main fossils. Therefore, this member belongs to graptolite-bearing mudstone facies of bathyal subcompensational basin.

Two types of basic sequences developed in the second member of Hule Formation ($O_{2\text{-}3}h^2$).

Basic sequences of type B: distributed in the middle part of the second member of Hule Formation; composed of dark gray–grayish black thin laminated carbonaceous siliceous; belonging to monotonous basic sequence.

Basic sequences of type C: main basic sequences in the second member of Hule Formation, composed of thin–medium laminated silty argillaceous silicalite, belonging to the non-cyclic basic sequence composed of monotonous lithology.

In this member, the major and minor sediments are siliceous matter and argillaceous matter, respectively. A small amount of planktonic graptolite is visible. Therefore, this member belongs to argillaceous–siliceous facies of abyssal subcompensational basin.

One type of basic sequences developed in the third member of Hule Formation ($O_{2\text{-}3}h^3$).

Basic sequences of type D: composed of rhythm interbed of ① black thin laminated siliceous mudstone and ② micro-thin laminated carbonaceous silicalite. From bottom to top, the micro-laminated silicalite in a small amount

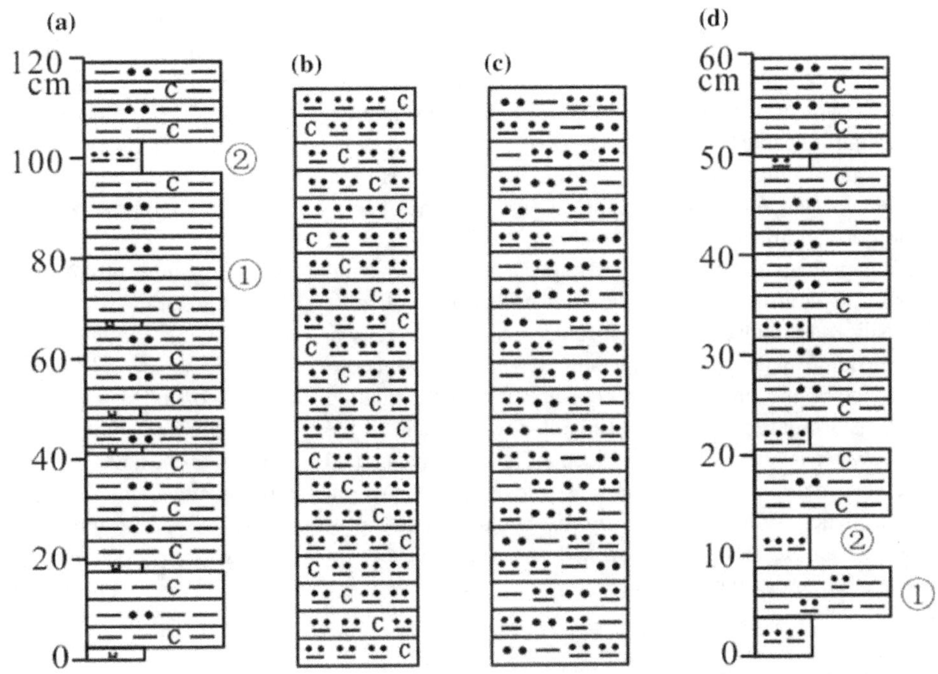

Fig. 2.41 Basic sequences of Hule Formation ($O_{2\text{-}3}h$)

gradually becomes visible occasionally and the thickness gradually decreases. This trend is contrary to that of the basic sequences of type A. Therefore, the sequences of this type are of non-cyclic basic sequence with argillaceous matter increasing and thickening upward.

In this member, the sediments are mainly argillaceous matter followed by siliceous matter, and planktonic graptolite is rich in the rocks. Therefore, this member belongs to graptolite-bearing argillaceous–siliceous facies of bathyal subcompensational basin.

2.4.3.2 Sequence Stratigraphy

According to the features of lithology, palaeontology, lithofacies association, and sequence boundary of Xinqiao profile in Hanggai map sheet, there is one third-order sequence (Sq20) in Hule Formation ($O_{2-3}h$), which belongs to second-order sequence SS7 (Fig. 2.42).

***Third-order sequence Sq*20**

Located in Hule Formation, Sq20 consists of TST, CS, and HST. The top and bottom of it are of type-II sequence boundary. TST is located in Bed 1–Bed 4, i.e., the first member of Hule Formation and the upper part of the second member. The lower part of TST is grayish black silty carbonaceous siliceous mudstone interspersed with micro-laminated silicalite, which turns to dark gray–black thin- to medium-blocky siliceous mudstone and argillaceous silicalite upward with siliceous matter increasing and argillaceous matter decreasing. Very thin bedding and graptolite developed, indicating a siliceous-argillaceous facies of bathyal subcompensational basin–abyssal subcompensational basin and reflecting transgression process with the sea level rising gradually. CS is located in the thin laminated carbonaceous silicalite of Bed 5, which is in the middle part of the second member of Hule Formation, indicating silicalite facies in the abyssal stagnant reduction environment during the maximum flooding period. HST is located in Bed 6–Bed 8 involving the second member of Hule Formation and the third member. From bottom to top, the components of HST change from dark gray thin laminated silty argillaceous silicalite and argillaceous silicalite into siliceous mudstone. The silicalite gradually decreases, argillaceous matter increases gradually, and the siliceous mudstone in the upper part is rich in graptolite, indicating siliceous-argillaceous facies of bathyal–abyssal basin in the high water-level stagnant reduction environment and reflecting the falling process of the sea level. In conclusion, this sequence, from bottom to top, reflects a complete process of transgression–regression and is a sequence of retrogradation–progradation type.

2.4.3.3 Biostratigraphy and Chronostratigraphy

The Hule Formation in the Area inherited the sedimentary environment of Ningguo Formation period. The ancient organisms in this formation are mainly planktonic graptolite and can be divided into three graptolite zones, i.e., *Pterograptus elegans* zone, *Hustedograptus teretiusculus* zone, and *Nemagraptus gracilis* zone. In this Project, *Pterograptus elegans* zone was found in the first member of Hule Formation, *Hustedograptus teretiusculus* zone was found in the lower part of the third member of Hule Formation, and no fossils were found in the black carbonaceous shale, which is 2 m thick at the top of the third member of Hule Formation. But there is *Nemagraptus gracilis* zone in similar horizons of other regions in Zhejiang Province. The features of ancient organism association in *Pterograptus elegans* zone and *Hustedograptus teretiusculus* zone in Hule Formation are described as below:

1. ***Pterograptus elegans zone***

Pterograptus elegans zone is roughly located in Bed 21–Bed 23 in the first member of Hule Formation in Jiumulong profile (PM007) in Hanggai Town, Anji County, and the main graptolite includes *Pterograptus elegans* Holm 1881 (Fig. 2.43) and associated graptolites including *Tetragraptus erectus* Mu, Geh et Yin, dichograptid gen. et sp., *Haddingograptuseurystoma* (Jaanusson 1960), *Archiclimacograptus* sp. Indet., *Kalpinograptus ovatus* (Hall 1902, Fig. 2.44c, d), *Haddingograptusoliveri* (Bouček), *Proclimacograptusangustatus* (Ekström), *Glossograptus* sp. Indet., *Archiclimacograptus cf. caelatus* (Lapworth), and *Archiclimacograptus angulatus* (Bulman).

Among these graptolite, *Kalpinograptus ovatus* (Hall 1902) (Fig. 2.44) is a typical fossil in Keping region, Xinjiang, and is also the one found for the first time in Lower Yangtze area, thus providing the basis for stratigraphic comparison between South China and North China.

2. ***Hustedograptus teretiusculus*** zone

Hustedograptus teretiusculus zone is in the lower part of the third member of Hule Formation of Lijiabian profiles PM012 and D2025. The fossils in the zone from bottom to top mainly include *Hustedograptus teretiusculus* (Hisinger, Fig. 2.45) and associated fossils including *Dicellograptus vagus* (Elles and Wood 1907), *Pseudoclimacograptus scharenbergi*, diplograptid gen. &sp. Indet., *Corynoides* sp., *Orthograptus calcaratus* (Lapworth), *Dicellograptus geniculatus*, *Dicranograptus brevicaulis*, and *Climacograptus* sp.

Hustedograptus teretiusculus zone belongs to Darriwilian Stage and also represents the graptolite zone at the top.

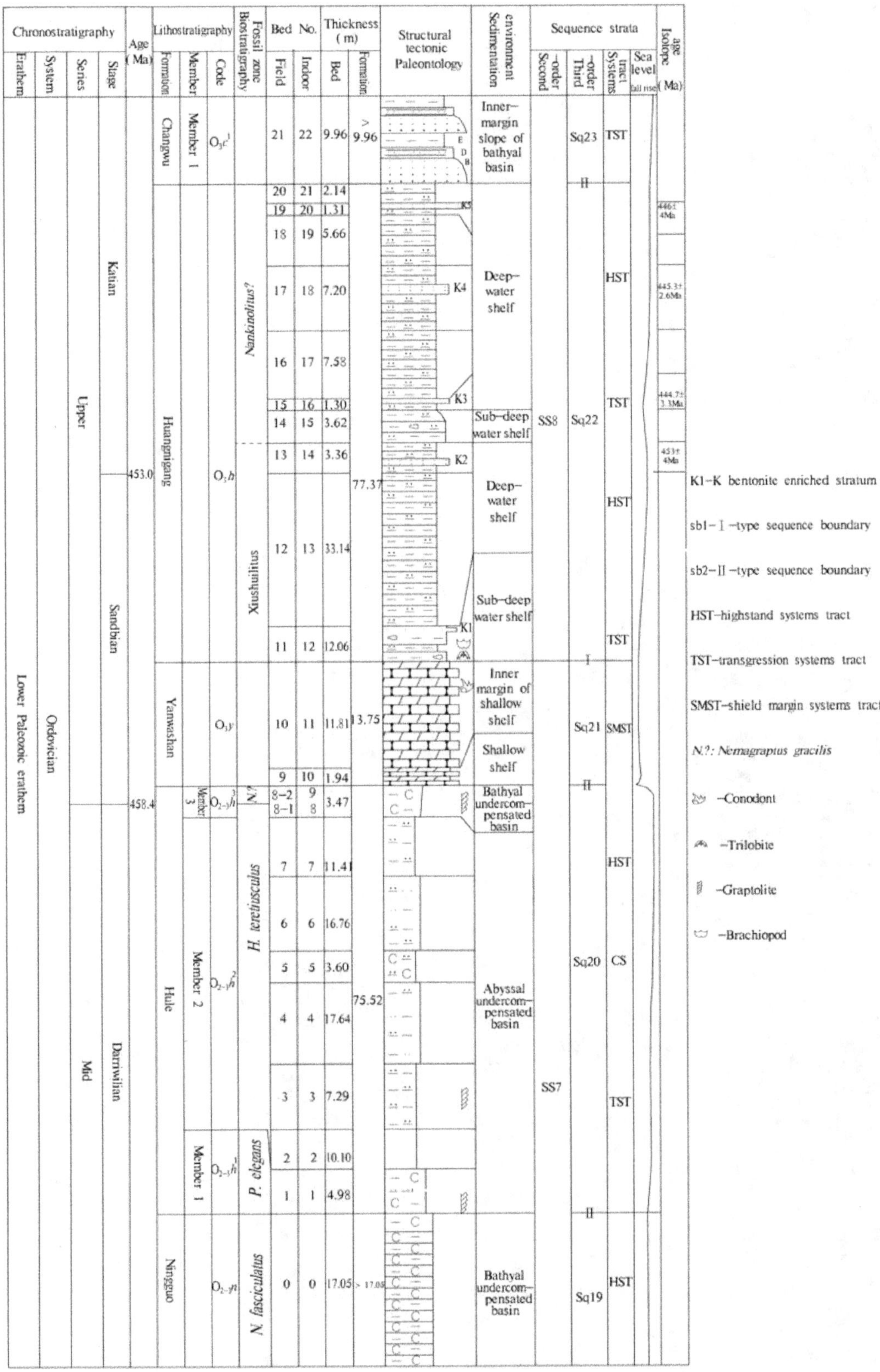

Fig. 2.42 Sequences of stratigraphic framework of Ordovician Hule Formation–Huangnigang Formation in the Area

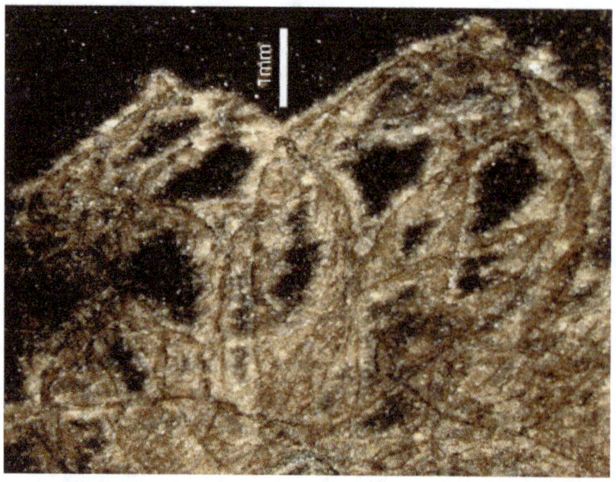

Fig. 2.43 *Pterograptus elegans* in the bottom of the first member of Hule Formation

No fossil was obtained in the black shale, which is about 2 m thick at the top of the third member of Hule Formation. By comparison with Hule Formation in other regions of Zhejiang, it is inferred that there may be *Nemagraptus gracilis zone* in the shale.

According to *Stratigraphic Chart of China* (2014), the *Pterograptus elegans* zone and the *Hustedograptus teretiusculus* zone in the lower part of Hule Formation belong to Darriwilian Stage of the Middle Ordovician. The *Nemagraptus gracilis zone* in its upper part of Hule Formation belongs to Sandbian Stage of Late Ordovician Epoch. Hule Formation spans Middle and Lower Ordovician Series.

2.4.3.4 Features of Stable Isotopes

Four carbonaceous mudstone samples, respectively, are collected from Hule Formation, including one collected in

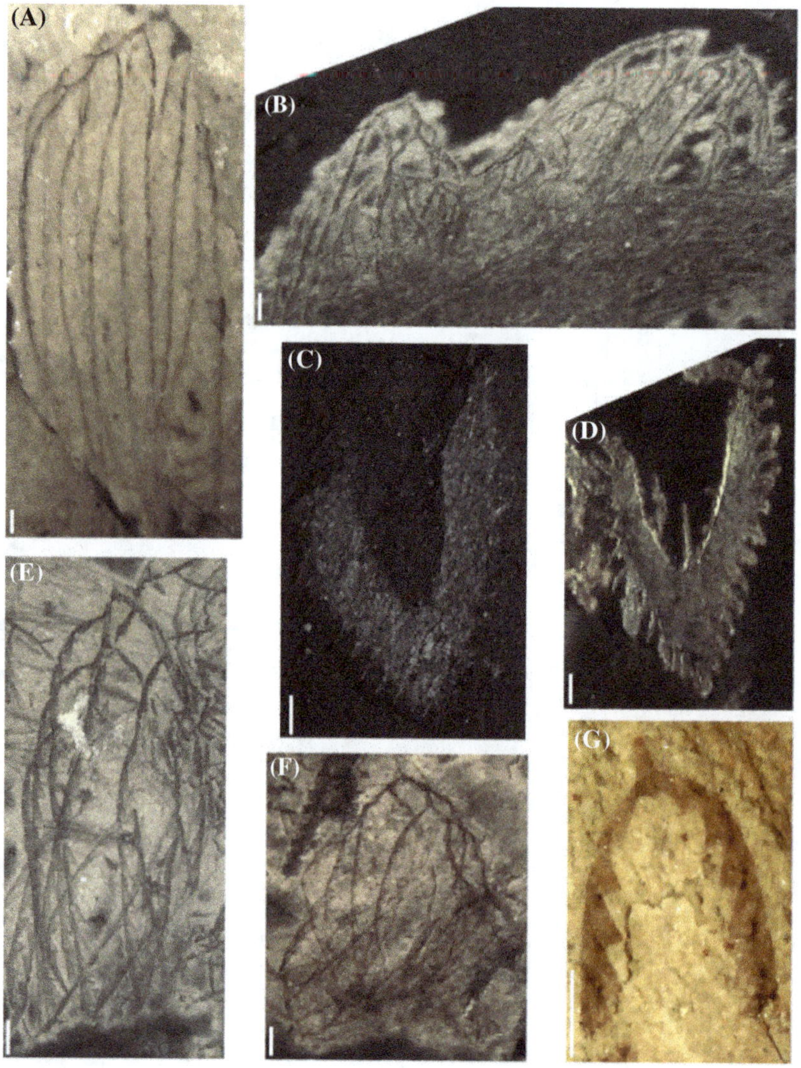

A-B, E-F. *Pterograptus elegans* Holm (PM007-22-1-24, PM007-22-1-49, PM007-21-2-69, PM007-22-1-58).

C-D. *Kalpinograptus ovatus* (T.S.Hall) (PM007-22-1-35, PM007-22-1-65).

G. *Didymograptus artus* Elles and Wood (PM007-21-3-4). Scale length = 1 mm. (The sample No. consists of Profile No. – Bed No. – specimen location – specimen No.)

Fig. 2.44 Graptolite fossil in *Pterograptus elegans* zone of Hule Formation ($O_{2\text{-}3}h$) in Bed 21–Bed 22 of Jiumulong profile (PM007)

Fig. 2.45 *Hustedograptus teretiusculus* in the third member of Hule Formation

the first member of Hule Formation of Jiumulong profile (PM007) and three collected in the third member of Hule Formation of Lijiabian profile (PM012) to conduct whole-rock stable isotope test and analysis of $\delta^{13}C$ and $\delta^{34}S$. The $\delta^{13}C$ value of Hule Formation ranges between −30.31 and −27.93‰ with the mean of −28.78‰, approximate to those of Ningguo Formation and still within the content of organic carbon $\delta^{13}C$ in coal. Hule Formation inherited the sedimentary environment of Ningguo Formation, i.e., bathyal subcompensational basin sedimentation. The $\delta^{34}S$ value of Hule Formation is from −8.90 to 23.90‰ with an average of 13.60‰, indicating an obvious positive anomaly. However, it is still within the range of $\delta^{34}S$ in global coal (−30‰–24‰) (Zheng and Chen 2000). The possible cause of the anomaly is that the marine organisms were thriving relatively owing to the warm temperature in the Area and thus the isotope fractionation of the sulfate-reducing bacteria was comparatively strong.

2.4.3.5 Analysis of Sedimentary Environment

Hule Formation inherited the sedimentary environment of Ningguo Formation, i.e., the stagnant closed reduction environment. This can be reflected as follows. In the top and bottom of Hule Formation, the sediments are mainly dark gray–black argillaceous matter followed by siliceous matter. Besides, there are small amounts of terrigenous clastic silt, micro-fine pyrite, and carbonaceous matter. The top and bottom are rich in well-preserved planktonic graptolite fossils, without any benthic shellfish. Micro-fine horizontal bedding developed. All these indicate that the sedimentary environment of the top and bottom of Hule Formation features weak hydrodynamic force, short sources of terrigenous clasts in basin, and insufficient oxygen in bottom basin. Therefore, the sedimentary environment is not favorable for the survival of organisms and belongs to bathyal subcompensational basin environment with strong reduction. In the middle part of Hule Formation, the sediments are mainly siliceous matter followed by argillaceous matter. Besides, there are small amounts of carbonaceous matter and sulfide. All these indicate that the sedimentary environment of the middle part of Hule Formation features extremely weak hydrodynamic force, extremely short sources of terrigenous clasts in basin, and extremely insufficient oxygen in the bottom basin, thus belonging to closed bathyal subcompensational basin environment with strong reduction. Regionally, this formation features slight lithofacies transition and the lithology of it is mainly characterized by siliceous matter followed by argillaceous matter. In Banqiao (in Lin'an)–Jingshanling (in Hangzhou) area, the thickness of the formation decreases to be merely 7 m. In Jiangshan and Changshan areas, the bottom of Hule Formation is interspersed with a small amount of calcarenite and belongs to slope outer margin sedimentary environment.

In conclusion, the Hule formations all through Zhejiang Province are the products of the stable closed abyssal reduction sedimentary environment and are critical marker strata for the geological survey of the Ordovician.

2.4.3.6 Trace Elements in Strata (Ore-Bearing)

Eight rock spectrum samples were systematically collected from Hule Formation of Jiumulong profile (PM007) in this project. Among these samples, one was collected from the first member, five from the second member, and two from the third member. Furthermore, three spectrum samples of carbon mudstone were collected from the third member of Hule Formation of Lijiabian profile (PM012). Analysis was made on 14 main trace elements, and statistics were made on the arithmetic mean values and concentration coefficients.

The first member of Hule Formation features similar enriched trace elements to the second member of Hule Formation and Ningguo Formation. They are all enriched in Sb, Bi, Pb, and Mo. The enrichment of Ag and F is slightly different among them. The third member of Hule Formation features different enriched trace elements compared with the first and the second members. In this member, the Vinogradov values of all the trace elements except Au are greater than 1; the enriched elements include Sb, Bi, Cu, Pb, Zn, W, Mo, Ag, and S, among which Bi, Cu, Zn, Ag, and S are extremely rich. Therefore, it can be inferred that materials from abyssal volcanic eruption and hot springs possibly penetrated this member.

2.4.4 Yanwashan Formation (O_3y)

The name Yanwashan Formation (O_3y) was created by Jichen Liu and Yazeng Zhao in 1927 in Yanwashan, Changshan County, Zhejiang Province. This formation roughly refers to the limestone in the central part of Yanwashan System created by Liu and Zhao (1924), which was renamed Yanwashan Formation by Lu et al. (1955). It was subsequently used up to now by the preparation group of regional stratigraphic table of Zhejiang Province (1979) as well as in the literature including *Stratigraphic Correlation Chart in China with Explanatory Text* (1982), *Regional Geology of Zhejiang Province* (1989), and *Lithostratigraphy of Zhejiang Province* (1995). In this Project, the name Yanwashan Formation (O_3y) was still adopted according to the lithological characteristics of Xinqiao profile (PM001) and Lijiabian profile (PM011) in Hanggai map sheet, Xianxia profile (PM030) of Xianxia map sheet, and the survey traverse.

2.4.4.1 Lithostratigraphy

The Yanwashan Formation (O_3y) in the Area is mainly distributed in the east of Hanggai map sheet, followed by the south of Xianxia map sheet. The outcrop area is about 1.15 km^2, accounting for 0.09% of the bedrock area.

The lithology and lithologic association of Yanwashan Formation (O_3y) are characterized by the rhythm interbed consisting of gray thin laminated marl and thin–medium laminated knotlike micrite. In the lower part, the main component is marl and horizontal bedding developed. The upper part is dominated by knotlike limestone. The single rhythm features gradual transition from micrite to marl from bottom to top. Yanwashan Formation (O_3y) is in conformable contact with the overlying Huangnigang Formation (O_3h) and in disconformable contact with the underlying third member of Hule Formation ($O_{2\text{-}3}h^3$) (Fig. 2.46). The thickness of Yanwashan Formation (O_3y) is 9.40–13.75 m.

1. **Stratigraphic section**

The lithology of the Yanwashan Formation (O_3y) is described by taking the sample of Ordovician Hule Formation ($O_{2\text{-}3}h$)–Huangnigang Formation (O_3h) profile (Fig. 2.40) in Xinqiao Village, Hanggai Town, Anji County, Zhejiang Province, as an example. The details are as follows:

Huangnigang Formation — Total thickness: > 12.06 m

12. Gray medium laminated siliceous mudstone interspersed with knotlike limestone or interbeds consisting of the two components. 12.06 m

———— Conformable contact ————

Yanwashan Formation — Total thickness: 13.75 m

11. Gray apparently-thick-laminated knotlike micrite interspersed with a small amount of dark gray thin laminated marl. The thickness of a single apparent thick layer: 40–100 cm generally; the thickness of a single layer of knotlike micrite: 2–5cm generally and up to 10–18 cm individually. the thickness of a single layer of marl: 2–8 cm. Gradual change from knotlike micrite to marl occurring from bottom to top. Conodont fossils in the upper part of the layer: *Protopanderodus liripipus*, *p. varicostatus*, *Spinodus spinatus*, *Periodon* sp., *Baltoniodus* sp., and *Yaoxianognathsus* sp., etc. 11.81 m

10. Interbeds consisting of gray thin laminated marl and knotlike limestone. The marl: horizontal bedding developing, the thickness of a single layer: 3–7cm. The notes of knotlike limestone: 2–7cm in diameter, distributed in moniliform shape and laminated form. The bottom of this bed: knotlike limestone with a thickness of 10 cm, flat, the occurrence is consistent with the overlying strata. 1.94 m

———— Conformable contact ————

The third member of Hule Formation — Total thickness: > 1.93 m

9. Dark gray siliceous carbonaceous shale, shale bedding developing. 1.93 m

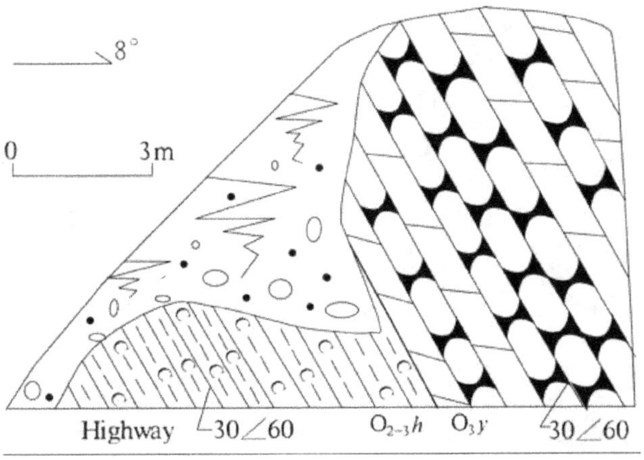

Fig. 2.46 Geological sketch map of conformable contact between Yanwashan Formation and Hule Formation

2. **Lithological characteristics**

The lithology of the Yanwashan Formation (O_3y) is mainly characterized by knotlike limestone and marl.

(1) The nodular limestone: grayish–gray, composed of cryptocrystalline calcite (> 90%), the thickness of a single layer: 2–5 cm generally, up to 10–18 cm individually, part of the knotlike limestone nodule containing trilobite fossils.

(2) The marl: gray; composition: microcrystalline calcite (55–60%) and argillaceous (40–45%); calcite: particle size: less than 0.01 mm generally, cryptocrystalline.

3. **Basic sequences**

Two types of basic sequences developed in the Yanwashan Formation (O_3y) (Fig. 2.47).

Basic sequences of type A: composed of knotlike limestone and thin laminated marl. The marl accounts for a large proportion in the lower part. Upward, the content of knotlike limestone gradually increases and the thickness of a single layer increases. Therefore, the basic sequences of this type belong to the non-cyclic basic sequence with thickness and calcareous matter increasing upward.

Basic sequences of type B: composed of thin laminated micrite, knotlike limestone, and thin laminated marl.

The sediments in this formation are mainly calcareous matter followed by argillaceous matter with conodont fossils in the rocks, belonging to the argillaceous carbonate facies of inner edge of continental shelf → the shallow shelf.

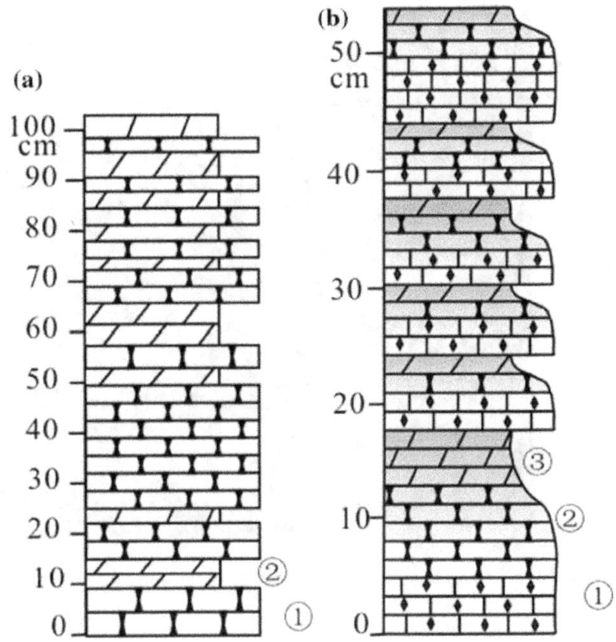

Fig. 2.47 Basic sequences of Yanwashan Formation (O_3y)

2.4.4.2 Sequence Stratigraphy

According to the characteristics of the lithology, palaeontology, lithofacies association, and sequence boundary of Xinqiao profile in Hanggai map sheet, there is one third-order sequence (Sq21) in the Yanwashan Formation (O_3y), which belongs to the second-order sequence SS7.

Third-order sequence Sq21

Located in the Yanwashan Formation, it contains shelf margin systems tract (SMST), transgressive systems tract (TST), and highstand systems tract (HST). The bottom of Sq21 is the interface between the thin laminated marl or knotlike limestone of the Yanwashan Formation and the black siliceous carbonaceous mudstone of the third member of the Hule Formation. From the graptolite shale of the Hule Formation to the Yanwashan Formation, the bathyal sedimentation abruptly changed into the shallow shelf sedimentation. As a result, a large number of sediments in transitional environments are missing, while no weathering

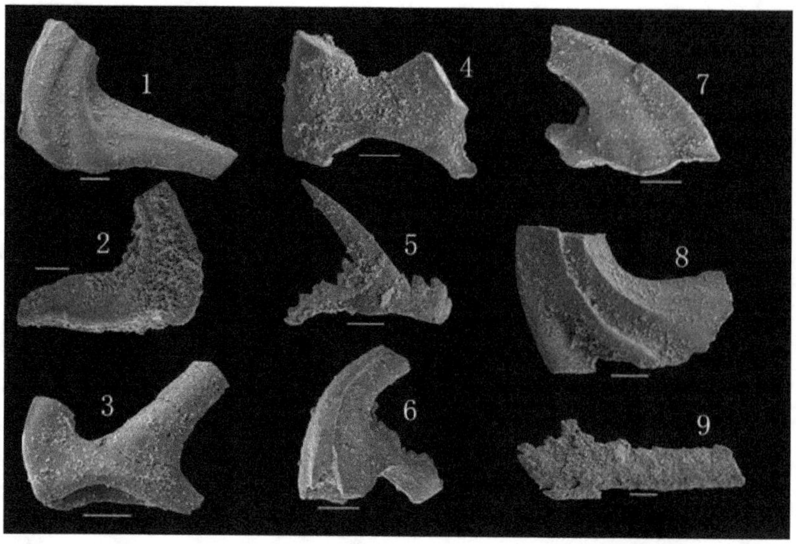

1-2: *Protopanderodus liripipus* Kennedy, Barnes et Uyeno, 1979 Lateral views of S elements;
3-4: *Spinodus spinatus* (Hadding, 1913) Lateral views of S elements;
5-6: *Periodon* sp. Lateral views of P and S elements;
7: *Baltoniodus?* sp. *Lateral view of S element;*
8: *Protopanderodus varicostatus* (Sweet et Bergström, 1962) Lateral view of S element;
9: *Yaoxianognathus* sp. Lateral view of S element

Fig. 2.48 Conodont fossils in knotlike limestone in the upper part of Yanwashan Formation

and erosion are visible. The top of Sq21 is the conformity interface between the knotlike micrite interspersed with a small amount of marl of Yanwashan Formation and siliceous mudstone of Huangnigang Formation. The bottom is of type-II boundary, and the top is type-I sequence boundary (Fig. 2.42). SMST is located in the Yanwashan Formation. The lower part of it consists of thin laminated marl and knotlike micrite. A small number of shellfish fragments are visible. Therefore, SMST belongs to argillaceous carbonate facies of shallow shelf. TST is located in the lower part of the Huangnigang Formation where the siliceous mudstone part interspersed with knotlike limestone is distributed. The knotlike limestone contains a large number of benthic trilobite and brachiopoda fossils and a small number of planktonic Pseudagnostus idalis. Therefore, TST belongs to siliceous mudstone facies near the oxidation zone. HST is located in the lower part of the Huangnigang Formation where the siliceous mudstone is distributed. Horizontal bedding developed, and there is no presence of biological fossils. Therefore, HST belongs to deep shelf siliceous mudstone facies with weak reduction during the maximum flooding period.

The sequence generally reflects the process of sea level rise from bottom to top and is a sequence of retrogradation type.

2.4.4.3 Biostratigraphy and Chronostratigraphy

Yanwashan Formation experienced a transformation of sedimentary environment. Affected by Guangxi movement, the sedimentary environment of the Area changed from closed deep sea to open shallow sea, carbonate began to be deposited, and conodonts constitute the main ancient organism fossils.

A total of five conodont samples were collected in Yanwashan Formation. Only a small amount of conodonts was found in the PM002YX4 sample of knotlike limestone in the upper part of this formation, including *Protopanderodus liripipus, P. varicostatus, Spinodus spinatus, Periodon* sp., *Baltoniodus* sp., and *Yaoxianognathsus* sp. (Figure 2.48). They are of the Late Ordovician, but the fossil zones in which they belong are undetermined.

According to the *Stratigraphic Chart of China* (2014), there are two conodont zones in Yanwashan Formation, which belong to the middle part of the Sandbian Stage; Yanwashan formation is of the Upper Ordovician Series.

2.4.4.4 Analysis of Sedimentary Environment

From Ordovician Yanwashan Formation, closed stagnant reduction sedimentary environment of the Area ended, and a set of argillaceous carbonate formation was deposited. Owing to the absence of organic carbon and sulfide in the sediments, the production in reduction environment, the absence of benthic shellfish fossils, and the product in the oxidation environment, it is inferred that the sedimentary environment is an open basin. Furthermore, the Area should be in deepwater environment under redox zone of shallow shelf, which is unfavorable for the survival of benthos. Judging from the incomplete preservation of conodonts, the sedimentary area should be close to the high-energy shallow sea, resulting in the defect of conodonts. Regionally, Yanwashan Formation features slight lithofacies transition and is an important marker stratum. However, the thickness of the formation changes, indicating that there is a stable and consistent shallow shelf environment in Zhejiang Province.

2.4.4.5 Trace Elements in Strata (Ore-Bearing)

Three rock spectra were collected in the Yanwashan Formation of Xinqiao map (PM001), and statistics of the arithmetic mean values and concentration coefficients of the analysis results were made. According to the statistics, Yanwashan Formation is enriched in Sb, Bi, Pb, W, and Mo, among which Bi is extremely rich.

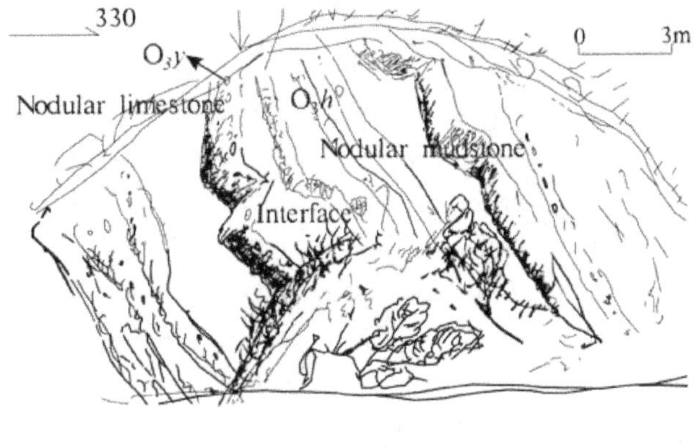

Fig. 2.49 Geological sketch map of the contact relation of Huangnigang Formation with Yanwashan Formation

2.4.5 Huangnigang Formation (O_3h)–Late Ordovician Volcanic Event Stratum

The name Huangnigang Formation (O_3h) was created by Lu et al. (1955) in Huangnigang Village, Jiangshan City, Zhejiang Province, and was used to refer to the shale in the upper part of the Yanwashan created by Liu and Zhao (1924). It was subsequently used up to now by the preparation group of regional stratigraphic table of Zhejiang Province (1979) as well as in the literature including *Stratigraphic Correlation Chart in China with Explanatory Text* (1982) and *Regional Geology of Zhejiang Province* (1989). In this Project, the name Huangnigang Formation (O_3h) was still adopted according to the lithological characteristics of Xinqiao profile (PM001) in Hanggai map sheet, Xianxia profile (PM030) in Xianxia map sheet, and geological observation traverse.

2.4.5.1 Lithostratigraphy

Huangnigang Formation (O_3h) in the Area is mainly distributed in the east of Hanggai map sheet followed by the south of Xianxia map sheet. The outcrop area is about 3.62 km^2, accounting for 0.28% of the bedrock area.

According to the lithology and lithologic association, Huangnigang Formation (O_3h) can be divided into four parts, i.e., the bottom, the lower part, the middle part, and the upper part. The bottom consists of grayish yellow medium laminated siliceous mudstone interbedded with knotlike limestone or the interbeds of the two components. It is about 1.5 m to the top. Three layers of micro-laminated light grayish yellow K-bentonite developed in the bottom. Generally, there are rich fossils of trilobite and brachiopoda in the knotlike limestone.

The lower part is composed of gray thin–medium laminated siliceous mudstone and interbedded with nine layers of light grayish yellow micro-laminated K-bentonite near the top. The maximum thickness of the K-bentonite is 2 cm. The middle part consists of grayish yellow medium laminated siliceous mudstone interbedded with three layers of knotlike limestone. The upper part is composed of gray thin–medium laminated siliceous mudstone interbedded with four eruption cycles of buff K-bentonite, which are, respectively, located in Layer 14, Bed 16, Bed 18, and Bed 20 from bottom to top. The maximum thickness of the K-bentonite layer in each layer is 5.5 cm, 3.5 cm, 2 cm, and 9 cm, respectively. From bottom to top, Huangnigang Formation can be further induced into two large rhythm beds, namely siliceous mudstone interbedded with knotlike limestone rhythm bed and siliceous mudstone rhythm bed. A total of 42 layers of K-bentonite are visible with thickness varying from 0.2 to 9 cm. This formation is in conformable contact with the overlying Changwu Formation and the underlying Yanwashan Formation (Fig. 2.49). The thickness of the formation is 71.55–77.37 m.

1. **Stratigraphic section**

The lithology of Huangnigang Formation is described by taking the example of Ordovician Hule Formation ($O_{2-3}h$)–Huangnigang Formation (O_3h) profile (Fig. 2.40) in Xinqiao Village, Hanggai Town, Anji County, Zhejiang Province. The details are as follows:

The first member of Changwu Formation Total thickness: > 9.96 m

22. Interbeds consisting of caesious sandstone, silty mudstone, and mudstone, BDE Bouma sequences developing. The grayish–gray sandstone: fine sand gradually turning into silt from bottom to top, the thickness of a single layer: 10–30 cm; the gray silty mudstone: horizontal bedding developing, the thickness of a single layer: 5–10 cm; the gray mudstone: no bedding developing, the thickness of a single layer: 10–20 cm. 9.96 m

——————— Conformable contact ———————

Huangnigang Formation (O_3h) Total thickness: 77.37 m

21. Caesious medium laminated siliceous mudstone, no bedding developing, the thickness of a single layer: 30–50 cm generally. 2.14 m

20. Gray medium laminated siliceous mudstone interbedded with six layers of light grayish-yellow K-bentonite, silt visible occasionally in the siliceous mudstone; the respective thickness of the six layers of K-bentonite from bottom to top: 0.2 cm, 9 cm, 3 cm, 0.5 cm, 0.2 cm, and 0.5 cm. 1.31 m

19. Gray medium laminated siliceous mudstone, silt visible occasional in the mudstone, the thickness of a single layer: 10–30 cm generally. 5.66 m

18. Gray thin–medium laminated siliceous mudstone interbedded with 17 layers of light grayish-yellow K-bentonite. Silt visible occasionally in the siliceous mudstone; the respective thickness of the 17 layers of K-bentonite from bottom to top: 2 cm, 0.5 cm, 1 cm, 0.2 cm, 0.6 cm, 0.2 cm, 2 cm, 0.2 cm, 0.5 cm, 0.5 cm, 0.2 cm, 0.2 cm, 0.2 cm, and 1 cm. 7.2 m

17. Gray medium laminated siliceous mudstone, no bedding developing, the thickness of a single layer: 35 cm generally and 4–8 cm locally. 7.58 m

16. Gray medium laminated siliceous mudstone interbedded with seven layers of light grayish-yellow K-bentonite; the respective thickness of the seven layers of K-bentonite from bottom to top: 0.5 cm, 3.5 cm, 0.5 cm, 0.2 cm, 0.2 cm, 0.2 cm, and 1 cm. 1.30 m

15. Gray medium laminated siliceous mudstone; no bedding developing; the thickness of a single layer: 25–50 cm generally and 5–10 cm in the bottom; interbedded with three layers of knotlike micrite with a width of 2–3 cm. 62 m

14. Gray thin–medium laminated siliceous mudstone interbedded with nine layers of light grayish-yellow K-bentonite; main components of the K-bentonite: illite, quartz, montmorillonite, and kaolinite; the respective thickness of the nine layers of K-bentonite from bottom to top: 1 cm, 0.5 cm, 0.5 cm, 0.5 cm, 0.2 cm, 0.5 cm, 5.5 cm, 1 cm, 2 cm; gray siliceous mudstone existing between the layers of K-bentonite. 3.36 m

13. Gray medium laminated siliceous mudstone, the thickness of a single layer: 30–50 cm general and 10–20 cm locally. 33.14 m

12. Gray thin–medium laminated siliceous mudstone interbedded with knotlike micrite or interbeds

consisting of the two components. The siliceous mudstone: 8–18 cm thick per layer, no bedding developing. The knotlike micrite: discontinuously distributed along the bed in the shape of an ellipsoid with the size of 1–5 cm, long axis parallel to the bedding. Three layers of K-bentonite visible about 1.5 m to the top, the respective thickness of the three layers: 1.5 cm, 1 cm, 1 cm; mudstone with the thickness of 5cm existing between the first layer and the second layer; mudstone with a thickness of 1.5 cm existing between the second layer and the third layer. A large number of fossils of trilobite obtained in the knotlike micrite 3–4 cm away from the bottom including *Alceste sinensis* (Sheng) the assesse insect + tail, chest *Corrugatagnostus* sp., wrinkled face ball subgroups in the head, *Cyclopyge recurva* Lu, *Dionide* sp., *Dionidella subquadrata* Han & Ju, *Ovalocephalus tetrasulcatus* (Kielan), *Saukia* cf. *Superioris* Zhou, and *Xiushuilithus* sp. 12.06 m

——— Conformable contact ———

Yanwashan Formation Total thickness: > 11.81 m

11. Gray apparently-thick laminated knotlike micrite interbedded with a small amount of dark gray thin laminated marl. The thickness of apparently thick bed: 40–100 cm generally; the thickness of a single layer of knotlike limestone in the apparently thick layer: 2–5 cm generally and up to 10–18 cm individually. The thickness of a single layer of marl: 2–8 cm. From bottom to top, the knotlike micrite gradually changes into marl. Fossils of the following conodonts in the upper part: *Protopanderodus liripipus*, *P. varicostatus*, *Spinodus spinatus*, *Periodon* sp., *Baltoniodus* sp. and *Yaoxianognathsus* sp. 11.81 m

1. **Lithological characteristics**

The lithology of the Huangnigang Formation (O_3h) is mainly characterized by gray–caesious thin–medium laminated silicon-bearing mudstone, siliceous mudstone, knotlike micrite, and K-bentonite.

(1) The knotlike micrite: gray–grayish, composed of cryptocrystalline calcite (> 90%), the diameter of a single node: 2–5 cm in general, fossils and fragments visible in the middle part.
(2) The siliceous mudstone: generally gray and locally gray–caesious, composition: argillaceous matter (70–80%) and cryptocrystalline siliceous matter (20–30%), no bedding developing, the thickness of a single layer: 10–35 cm generally.
(3) The silicon-bearing mudstone: gray–caesious, composition: argillaceous matter (85–90%) and a small amount of cryptocrystalline siliceous matter (10%), silt visible occasionally, the thickness of a single layer: 8–50 cm, no bedding developing.
(4) The K-bentonite: grayish (celadon), tuffaceous structure, blocky tectonics, weak bedding structure, composition: volcanic ash and a small amount of fine-grained subangular feldspar (5–10%), minerals mostly turning into kaolin owing to compression and hydrolysis, with weak waxy luster and soap-like slippery feeling, distinctly different from the upper and lower gray silicon-bearing mudstone and silty mudstone.

3. **Basic sequences**

Three types of basic sequences developed in the Huangnigang Formation (O_3h) (Fig. 2.50).

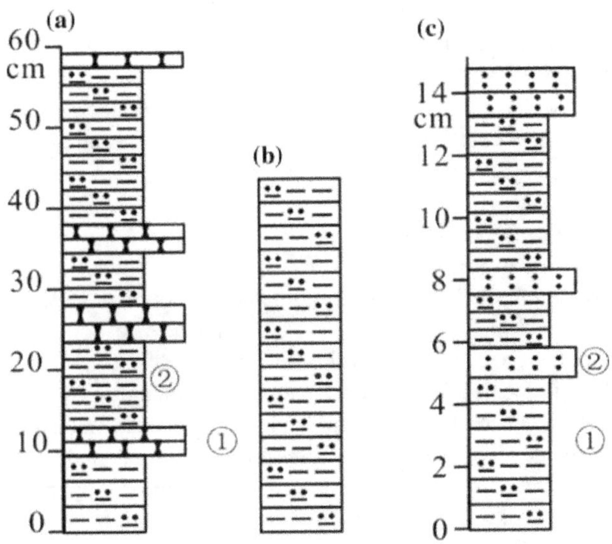

Fig. 2.50 Basic sequences of Huangnigang Formation (O_3h)

Basic sequences of type A: distributed in the lower parts of the two large rhythm beds that are located in the lower and middle parts of Huangnigang Formation, consisting of silicon-bearing mudstone-bearing knotlike limestone and siliceous mudstone; the knotlike limestone: 2–5 cm thick, ellipsoid, discontinuously distributed along the siliceous mudstone, concentrating in Bed 2–Bed 5 and gradually decreasing upward; belonging to the cyclic basic sequence with argillaceous–siliceous matter increasing upward.

Basic sequences of type B: distributed in the middle-lower and middle-upper parts of Huangnigang Formation, consisting of thin–medium laminated siliceous mudstone with medium laminated siliceous mudstone as main component, belonging to monotonic basic sequence.

Basic sequences of type C: distributed in the middle-upper part of the Huangnigang Formation; consisting of thin–medium laminated siliceous mudstone and micro-thin laminated K-bentonite; the K-bentonite: the product of volcanic eruption, the thickness of a single layer: mostly less than 1 cm and up to 9 cm as the maximum; belonging to the monotonic basic sequence.

The sediments in this formation are dominated by siliceous argillaceous matter, followed by knotlike marl. The siliceous mudstone: no fossils of ancient organisms discovered, micro-fine horizontal bedding developing, belonging to siliceous-argillaceous facies under the oxidation zone of deep shelf. The knotlike limestone: containing the fossils of benthic trilobites, gastropod, brachiopoda, crinoid, and planktonic Pseudagnostus idalis. Therefore, this member belongs to carbonate facies of basin on the margin of shallow shelf.

2.4.5.2 Sequence Stratigraphy

According to the characteristics of the lithology, ancient organisms, lithofacies association, and sequence boundary of Xinqiao profile in Hanggai map sheet, there is one third-order sequence Sq22 in the upper part of the Huangnigang Formation, which belongs to the second-order sequence SS8.

Third-order sequence Sq22

Located in the Huangnigang Formation, it contains TST and HST. The bottom of Sq22 is of type-I sequence boundary, and the top of it is of type-II sequence boundary (Fig. 2.42). TST is located in the lower part of the Huangnigang Formation where thin–medium laminated siliceous mudstone interbedded with knotlike limestone is distributed. The sediments of TST are dominated by silicon-bearing mudstone. Besides, there is a small amount of micritic calcite. Nankinolithus was ever obtained from the knotlike limestone, indicating that seawater was rich in oxygen. All these indicate that TST belongs to the sedimentation of siliceous-argillaceous facies of the outer edge of shallow shelf. HST is located in the upper part of the Huangnigang Formation. The sediments of it are composed of silicon-bearing argillaceous matter. Furthermore, HST features scarce palaeontology and little organic matter. Besides, horizontal bedding developed. All these indicate that HST belongs to the siliceous-argillaceous facies of deep shelf in weak oxidation–reduction environment under wave base. In conclusion, the sequence experienced continuous transgression process and is a sequence of retrogradation type.

2.4.5.3 Biostratigraphy and Chronostratigraphy

The benthic paleontological fossils dominated by trilobites developed in the knotlike micrite or calcareous mudstone in the Huangnigang Formation. According to the *Stratigraphic Chart of China* (2014), two trilobite zones developed in the Huangnigang Formation, i.e., *Xiushuilithus* zone in the lower part and *Nankinolithus* zone in the upper part.

In this Project, a total of 7–8 fossils of trilobite genus were obtained in the knotlike limestone in the bottom of the Huangnigang Formation (O_3h) of Xinqiao profile (PM001) in the Area: *Xiushuilithus* sp., *Alceste sinensis* (Sheng), *Corrugatagnostus* sp., *Cyclopyge recurva* Lu, *Dionide* sp., *Dionidella subquadrata* Han & Ju, *Ovalocephalus tetrasulcatus* (Kielan), and *Saukia* cf. *Superioris* Zhou. All these trilobites belong to the *Xiushuilithus* zone (Fig. 2.51). Owing to the poor outcrop, only trilobite fragments were found in the rhythm layers of knotlike limestone in Bed 2 in the middle part of the Huangnigang Formation in the Area. According to the material of Longtan profile in Tonglu County in the south, which is in the middle part of the profile, the caesious calcareous mudstone-bearing knotlike limestone occasionally produces the fossils of *Nankinolithus, Birmanites,* and *Hammatocnemis* which belong to the *Nankinolithus* zone.

The Xiushuilithus and *Nankinolithus* zones are located above the conodonts in the Yanwashan Formation, and the SHRIMP isotope age (D006TW) of the K-bentonite collected from the bottom of the siliceous mudstone-bearing knotlike limestone in Bed 2 of Xinqiao map (PM001) is 453 ± 4 Ma. According to the *Stratigraphic Chart of China* (2014), the two trilobite zones in the Huangnigang Formation belong to the Late Ordovician, among which the *Hustedograptus teretiusculus* zone belongs to Sandbian and *Xiushuilithus* zone is of Katian.

2.4.5.4 Event Stratigraphy

A total of 47 K-bentonite interbeds were discovered in the strata of Huangnigang Formation and Changwu Formation in Xinqiao map (PM001), which can be divided into eight cycles (K_1–K_8) as per the distribution density. The

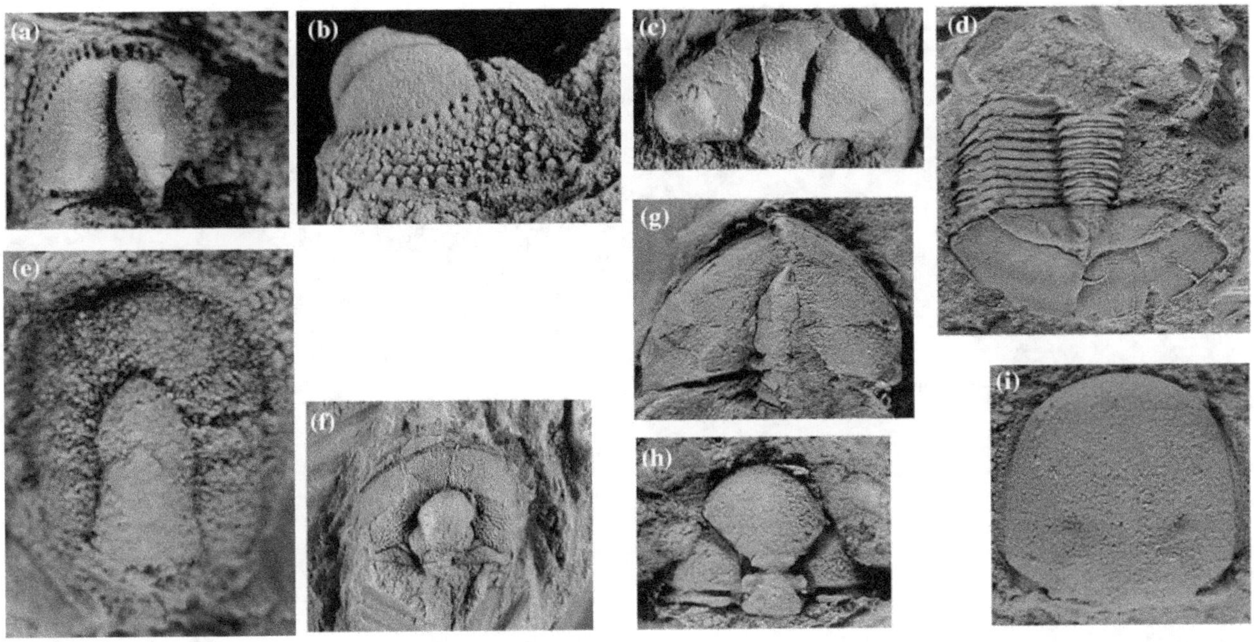

Fig. 2.51 Trilobite fossils in *Xiushuilithus* zone in lower part of Huangnigang Formation (O_3h) in Xinqiao profile (PM001) A.B-*Xiushuilithus* sp. (head, dorsal view); C-*Alceste sinensis* (Sheng) (head, dorsal view); D-*Alceste sinensis* (Sheng) thorax + tail (dorsal view); E-*Corrugatagnostus* sp. (head); F-*Dionidella subquadrata*, Han and Ju; (head, dorsal view); G-Saukia cf. superioris Zhou (head, dorsal view); H-*Ovalocephalus tetrasulcatus* (Kielan) (head cover, dorsal view); I-*Cyclopyge recurva* Lu (head cover)

K-bentonite interbeds developed at the highest degree in Huangnigang Formation, where a total of 42 K-bentonite interbeds were found and are divided into five cycles (K_1–K_5). In this Project, the samples were systematically collected followed by a series analysis including rock mineralogy, X-ray diffraction, silicates, rare earth, trace elements, rock spectra, and zircon SHRIMP U-Pb dating. The mineral composition and the characteristics of geochemistry and isotopic geochronology of the K-bentonite were studied. Furthermore, the genesis, tectonic environment, and event stratigraphic significance of the K-bentonite were discussed.

1. **Lithological Characteristics**

The bentonite: grayish (celadon), tuffaceous structure, blocky tectonics, weak bedding structure, composition: volcanic ash and a small amount of fine-grained subangular feldspar (5–10%), minerals mostly turning into kaolin owing to compression and hydrolysis, with weak waxy luster and soap-like slippery feeling, distinctly different from the upper and lower gray silicon-bearing mudstone and silty mudstone (Fig. 2.52). The K-bentonite in the Huangnigang Formation of the Area features a stable distribution horizon. It is visible in the outcrop of the Huangnigang Formation in Baishiwu Village, Songkengwu Village, and Baofu Town. Furthermore, stable K-bentonite interbeds are also visible in the Huangnigang Formation in the boreholes of Songkengwu Village.

X-ray diffraction analysis was made on six samples of K-bentonite collected from K_2 to K_7. The samples were tested in the State Key Laboratory of Endogenous Metal Deposit Mineralization, Nanjing University. Rigaku Dmax III-a $\theta/2\theta$-type powder diffractometer with fixed Cu target was adopted for test, and the settings are as follows: operating voltage: 40 kV; operating current: 25 mA; slit system: 1° scatter silt/antiscatter silt, accepting 0.3 mm slit; monochromator system: graphite curved crystal monochromator; receiving device: Ta-activated sodium iodide scintillation counting device; scanning method: step scan, 0.02°/step; preset time: 0.3 s/step.

The main mineral components of sample D006X1: quartz and illite; the main mineral components of samples D006X2, D006X4, D006X5, D006X6: illite, illite–montmorillonite mixed-layer clay minerals, a small amount of quartz, an extremely small amount of albite; the main mineral components of sample D006X3: montmorillonite, illite, kaolinite, and quartz (Fig. 2.53). There are different amounts of amorphous material (siliceous) in all the six samples. Furthermore, the mineral association above features the characteristics of the minerals as a result of weathering and

Fig. 2.52 Characteristics of K-bentonite in Layers 19 and 38, Xinqiao section

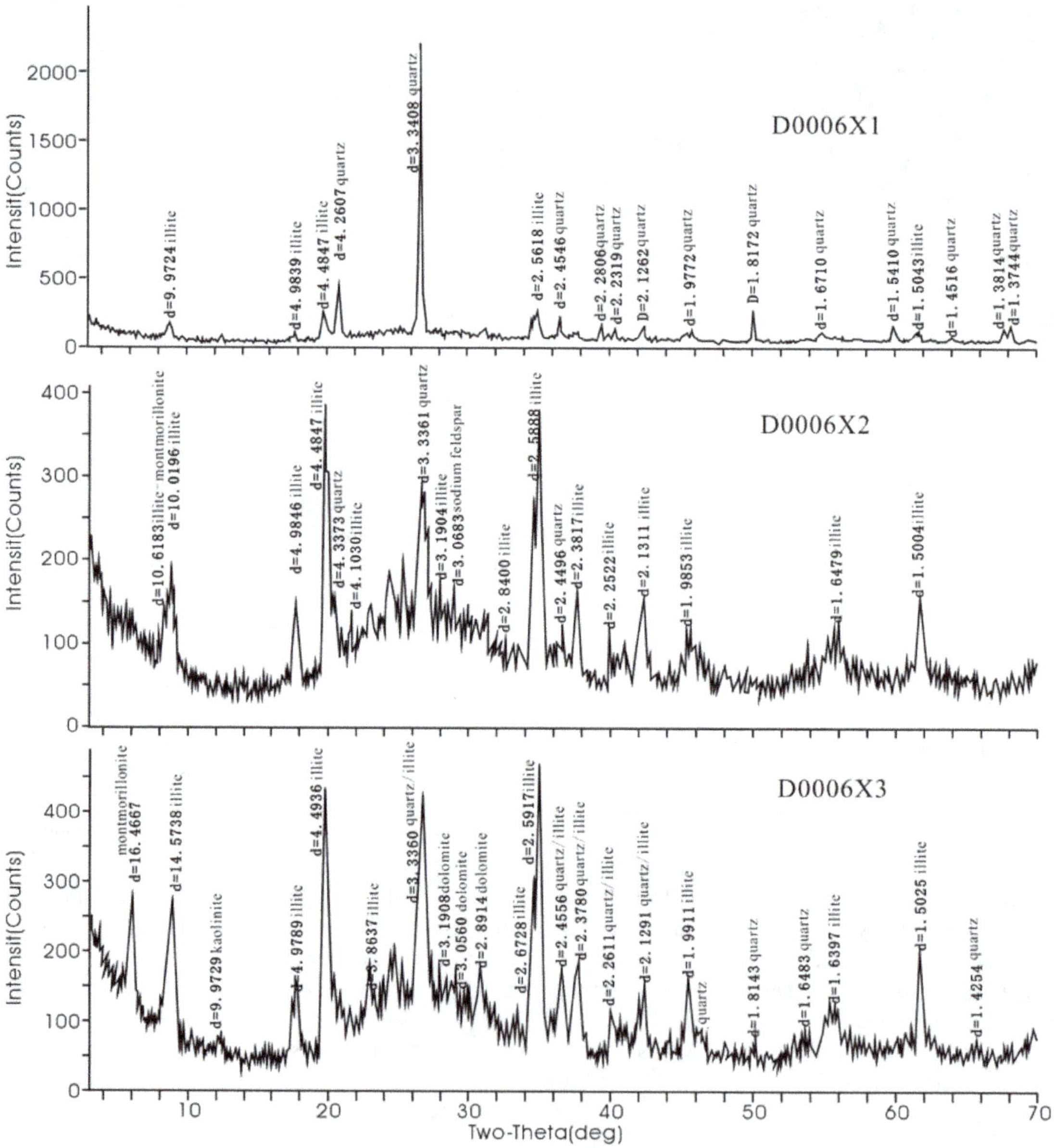

Fig. 2.53 X-ray diffraction pattern of K-bentonite

alteration of volcanic rocks, similar to the K-bentonite in Late Ordovician Wufeng Formation and Guanyinqiao Layer in Wangjiawan and Huanghuachang profiles in Yichang City, Hubei Province. However, the lithology varies among the six samples, suggesting that there may be two or more kinds of volcanic rocks.

2. **Major elements**

The respective content of the major elements in the K-bentonite is as follows. The SiO_2 content: 49.62–58.59%, indicating that the K-bentonite belongs to intermediate-basic rocks; σ (Rittman index): 3.57–12.40, indicating that the K-bentonite belongs to alkaline rocks–peralkaline rocks; K_2O content: generally high, 7.24–9.11%; ω (K_2O/Na_2O): >30, much higher than that of globally known Cambrian and Ordovician K-bentonite; Na_2O content: The maximum is only 0.22%, lower than that of globally known Cambrian and Ordovician K-bentonite; compared with the clastic rocks in Changwu Formation in the Area, the contents of Al_2O_3, MgO, K_2O, and TiO_2 are significantly higher; the consolidation index (SI) of the K-bentonite is 17.22–19.99, indicating a high differential degree. Therefore, the K-bentonite in Huangnigang Formation generally features high potassium (K_2O), high aluminum (Al_2O_3), high water (H_2O+), and low sodium (Na_2O), greatly different from normal sedimentary rocks.

In K_2O-SiO_2 illustration (Fig. 2.54), K-bentonite is all located in the shoshonite series. In Zr/TiO_2-Nb/Y illustration (Fig. 2.55), K-bentonite is mainly located in the areas of trachyte and rhyodacite, indicating that the original rocks contain rhyolitic and alkaline components.

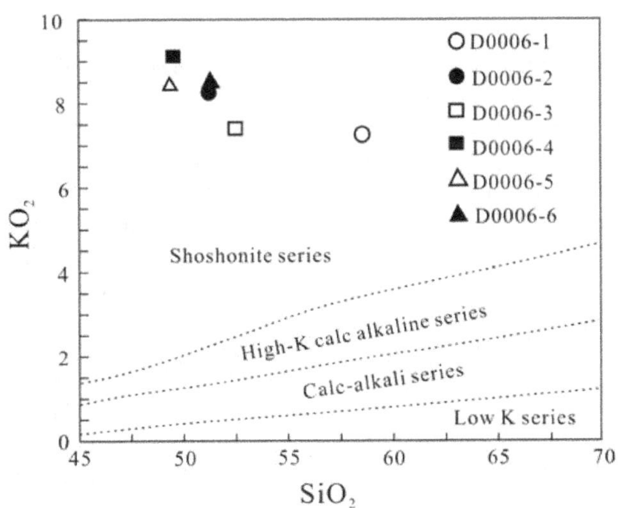

Fig. 2.54 K_2O-SiO_2 illustration of K-bentonite

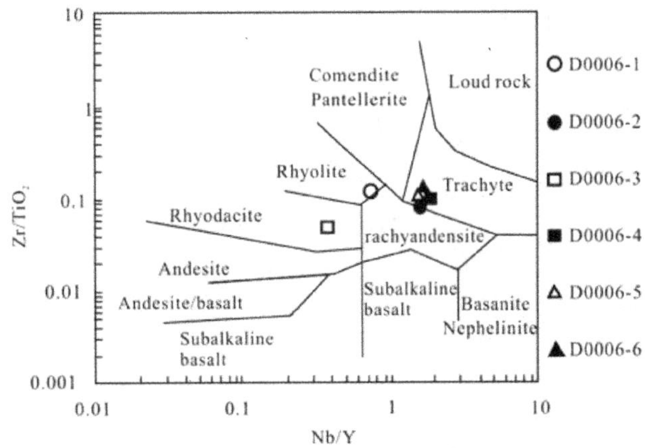

Fig. 2.55 Zr/TiO_2-Nb/Y bentonite of K-bentonite

3. **Rare-earth elements**

The total amount of rare-earth elements of K-bentonite (ΣREE) in the samples varies significantly with the range of (291.15–1452.51) × 10^{-6}, mainly in (1146.75–1452.51) × 10^{-6}. LREE is obviously enriched, and HREE is poor. The LREE/HREE ratio is 28.04–93.92, mainly in 87.60–93.92. The La_N/Yb_N ratio is 6.27–26.31, mainly in 16.45–26.31. δEu value is 0.53–0.68, indicating medium deficit. Although all samples feature quite different total amounts of REE, they enjoy relatively consistent LREE/HREE ratios, La_N/Yb_N values and δEu values. The diagenetic materials of K-bentonite in the Area should be consistent, given that the differences among the samples and experimental errors are excluded. The high contents of potassium, aluminum, and zircon are consistent with those of the upper crust, indicating that some materials may come from the part of the melting source areas of the upper crust. Low silicon is the characteristic of argillaceous and calcareous sedimentary rocks or lower crust rocks. Therefore, it can be inferred that the K-bentonite mainly comes from magmatic rocks and part of them are sedimentary rocks. The distribution curves of the rare-earth elements in K-bentonite are rightward generally (Fig. 2.56), showing similar distribution patterns. The distribution curves of rare-earth elements of the K-bentonite are slightly steeper than those of the K-bentonite of Late Ordovician Wufeng Formation and Guanyinqiao Bed in Wangjiawan and Huanghuachang profiles in Yichang City, Hubei Province.

4. **Trace elements**

The trace elements generally feature enriched large-ion lithophile elements such as Rb, Th, U, K, and Pb and poor high field strength elements such as Nb, P, and Ti (Fig. 2.57). In addition, the large-ion lipophilic elements Ba and Sr show strong negative anomaly, while the high field strong elements

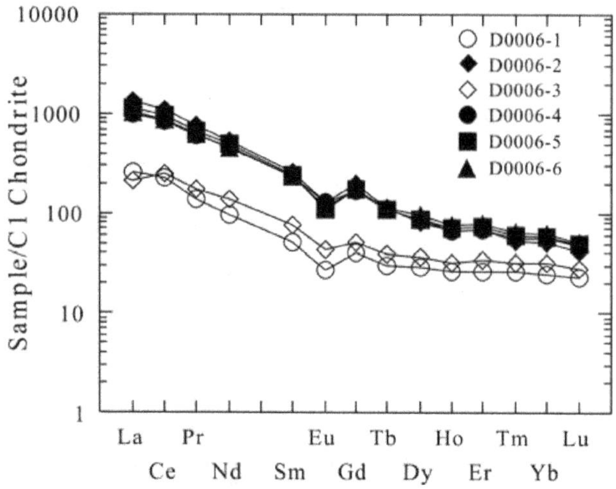

Fig. 2.56 Normalized pattern of rare-earth element chondrite of K-bentonite (Sun and McDonoungh 1989)

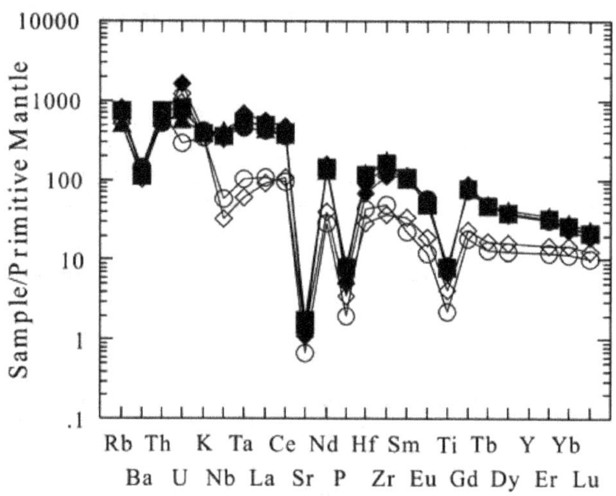

Fig. 2.57 Primitive mantle-normalized spidergram of trace elements of K-bentonite (Sun and McDonoungh 1989)

such as Zr and Hf show some positive anomaly. The Rb/Ba ratio is 0.44–0.77, far higher than that of the primitive mantle that is 0.088 (Hofmann 1988). The K/Rb ratio was 158.5–242, lower than that of normal magmatic rocks (242). In the primitive mantle-normalized spidergram of trace elements, Nb and Ti elements tend to show negative anomaly and Pb tends to show positive anomaly. The contents of Zr, Hf, Nb, Ba, Rb, Pb, Cu, Ag, and F in most K-bentonite in the Area are obviously 1–2 times higher than those in general clastic sedimentary rocks and claystone (Zhang et al. 1997).

5. **Zircon isotope chronology**

The zircon used to conduct zircon U-Pb dating was selected in the laboratory of Hebei Institute of Regional Geological and Mineral Resources Survey. In detail, the zircon was selected with binocular from 5 kg sample by crushing, heavy concentrate panning, and electromagnetic separation. The selected zircon was sent to Beijing Zircon Year Navigation Techonology Co., Ltd., where it was cemented into a sample target by using epoxy resin containing curing agent and then polished to reveal the center of zircon particles. Furthermore, cathode luminescence (CL) images and backscattering electron (BSE) imaging of zircon samples were analyzed, in order to study the internal structural characteristics. Zircon SHRIMP U-Pb isotope dating was done with the SHRIMP-II ion microprobe mass spectrometer of Institute of Geology, Chinese Academy of Sciences. The beam spot size is 30 μm, and the specific test conditions and workflow are described in the literature (Liu et al. 2006).

The zircon in K-bentonite is mostly irregularly columnar, with a length of about 80–120 μm and length/width ratio of about 1.5:1–1:1. Zonal structures developed in zircon (Fig. 2.58). The content of U in the zircon is (22–1643) $\times 10^{-6}$, the content of Th is (35–789) $\times 10^{-6}$, and the Th/U ratio is 0.28–4.17, indicating typical features of magmatic zircons (Hoskin and Schaltegger 2003). Most of the sample points were projected on or near the concordant curve (Fig. 2.59). Among the six samples of K-bentonite, the single-point age of $^{206}Pb/^{238}U$ is 482.1 ± 5–401.9 ± 8.2 Ma, except sample point D0006-2-1, of which the age of $^{206}Pb/^{238}U$ is 748.6 Ma, indicating possible captured zircon. The weighted mean ages of the six samples were 453 ± 4 Ma (MSWD = 1.7), 444.7 ± 3.3 Ma (MSWD = 1.6), 445.3 ± 2.6 Ma (MSWD = 0.48), 446 ± 4 Ma (MSWD = 1.5), 448 ± 8 Ma (MSWD = 3.2), and 449 ± 3 Ma (MSWD = 0.42), which were consistent within the error range. Therefore, the crystallization age of K-bentonite should be 453–444. 7 Ma.

6. **Comparison, genesis, and tectonic significance of regional distribution**

During the early Sandbian, a stratum interface with significant lithological difference developed between the Hule Formation and Yanwashan Formation in the Area. The part below the interface is the relatively stable abyssal deposition of graptolite-bearing siliceous argillaceous shale of the Early–Middle Ordovician Ningguo Formation and Hule Formation. The part above the interface is the carbonate deposition of neritic shelf of the Late Ordovician Yanwashan Formation. After stable abyssal subcompensational basin stage of the Early and Middle Ordovician, the crust uplifted rapidly. As a result, the Sandbian neritic shelf environment formed and sedimentary facies was changed from graptolite shale facies to shellfish facies. From Late Sandbian to Early Katian, the sedimentation was characterized by flat knotlike

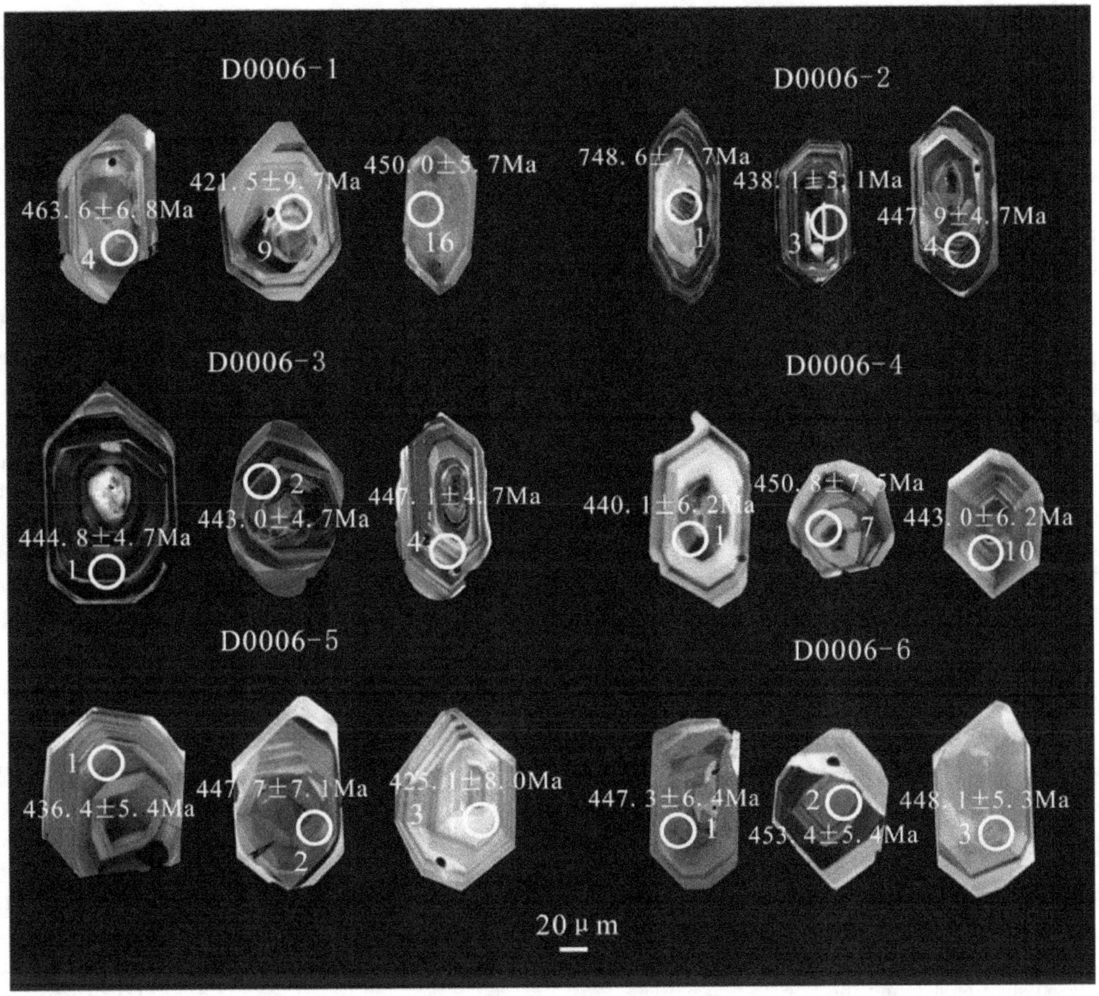

Fig. 2.58 Some CL images of zircon and analytical locations of K-bentonite samples

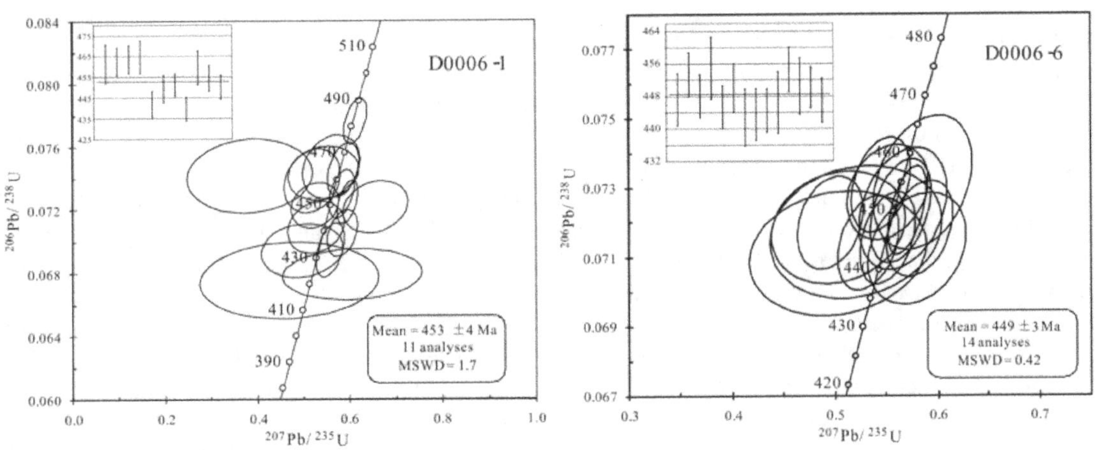

Fig. 2.59 SHRIMP U-Pb harmonic curve of zircon in K-bentonite

and banded limestone and marl of Yanwashan Formation, and lithofacies transition occurred in the siliceous calcareous ooze with the basin facies on the margin of the shallow shelf–bathyal basin facies of Huangnigang Formation. About 453–444.7 Ma, owing to the most intense crustal movement, the Yangtze landmass experienced internal or marginal tension. Consequently, dozens of layers of K-bentonite formed in the middle and upper parts of the Huangnigang Formation. In later Katian, with the rise of relative sea level, the deposition space increased rapidly; extremely thick flysch facies rocks of Changwu Formation were deposited in the boundary of Zhejiang Province and Anhui Province; the ancient organisms dominated by benthic shellfish changed to the ones mainly consisting of planktonic organisms such as pentolite; the sedimentary environment was changed to peripheral foreland basin from passive continent-edge basin.

The special characteristics of the event stratigraphy of Huangnigang Formation are as follows. In the Area, K-bentonite is visible in many outcrops in places such as Baofu Town in Hanggai map sheet. Besides, in the Huangnigang Formation in Liujia profile of Hecun Town, Tonglu County, which is about 83 km to the southwest of the Area, there are also 25 interbeds of K-bentonite. The Huangnigang Formation in Liujia profile is 34.2 m thick, and the lithologic association is similar to that of the Hanggai area. It can also be divided into two rhythm beds. The K-bentonite interbeds in the rhythm beds are pale gray, and the thickness of single layer is 1–4 mm, indicating the regional distribution of K-bentonite in Zhejiang Province. In the South Putangkou, Changshan County, the Jiangshan–Changshan–Yushan area along the border of Zhejiang Province and Jiangxi Province, the middle part of Huangnigang Formation is interbedded with a layer of dark gray blocky marl with thickness of 5.14 m, bears large scale of cross-bedding, slump structure, and slump breccia, and contains the fossils of brachiopoda and cephalopods. Upward from the middle part of the formation, there is a layer with the thickness of 18.53 m consisting of the interbeds of gray thin laminated calcareous mudstone and thin laminated mudstone-bearing pie-strip shaped and knotlike micritic limestone. Further upward, there is argillaceous limestone of the Sanqushan Formation (O_3s), with a thickness of more than 86 m and a large slump-fold structure developing. In the first member of the Sanqushan Formation (689 m thick) in the north of Yushan County and Changshan County of Jiangxi Province, 10 layers of slump-fold limestone and breccia developed in the argillaceous limestone and marl. It is generally believed that the origin of these slump structures is that the mud that is unconsolidated into rocks on the platform slid down and rolled down along the slope subjected to the seismic action. The contemporaneous multiple layers of K-bentonite in Hanggai and Liujia areas and seismic events in Jiangshan–Changshan–Yushan area indicate that northwest Zhejiang Province experienced regional tectonic movement and intense volcanic eruption in Early Katian.

Ordovician and Silurian K-bentonites are distributed globally. Since the 1970s, many scholars at home and abroad have conducted detailed studies on K-bentonite in sedimentary rocks (Huff et al. 1992, 1998; Huff et al. 2003; Huff 2008; Hu et al. 2008; 2009; Yang 2011; Xie et al. 2012a, b). In the areas such as Yichang City of Hubei Province, Tongzi County of Guizhou Province, and Haoping Town, Taoyuan County, Hunan Province, K-bentonite interbeds developed in Late Ordovician Linxiang–Wufeng Formation (corresponding to the Wenchang Formation in the Area). Along Hanggai of Zhejiang (453-444.7 Ma) → Wangjiawan–Huanghuachang of Hubei (445-442 Ma) (Hu et al., 2008) → Haoping, Taoyuan of Hunan (442 Ma) (Xie et al. 2012a, b) → Nanbazi of Guizhou (441 Ma), the age of K-bentonite tends to decrease and the total thickness of distribution also tends to reduce. In detail, the layer number and thickness are as follows. Hanggai of Zhejiang: 47 layers in total, total thickness: 60.7 cm, the maximum thickness of a single layer: 9 cm; Wangjiawan–Huanghuachang of Hubei: 11 layers in total, the maximum thickness of a single layer: 4.5 cm; Haoping, Taoyuan of Hunan: total thickness: 20 cm; Nanbazi of Guizhou: 12 layers in total, the maximum thickness of a single layer: 4 cm. In addition, foreign scholars also reported the research results of K-bentonite in North America, Britain, Sweden, and Estonia. Among these research results, 38 are about isotopic age dating by the K-Ar, U-Pb, Ar-Ar, and Sm-Nd methods of K-bentonite in Kentucky, Missouri, Alabama, and Tennessee, and the age is 462 ± 15–444 ± 20 Ma (Huff 2008). This can further confirm that the K-bentonite of the Sandbian-Hirnantian period is characterized by global distribution, and therefore, they may represent the global tectonic movement and volcanic activity events.

The distribution of the six K-bentonites in the Area in tectonic geochemical discrimination diagrams is as follows. In Rittmann-Gottiry diagram, they are all located in the alkaline and meta-alkaline rocks derived from the orogenic and anorogenic zones (Fig. 2.60). In Th-Hf-Ta diagram, they all fall in the alkaline intraplate basalt area and volcanic arc basalt area (Fig. 2.61). In Rb-Nb + Y diagram, most of them are located in the intraplate granite area (Fig. 2.62). In Th/Zr-Nb/Zr bilogarithmic diagram, they mainly fall in the tholeiite area of intracontinental rift and epicontinental rift types and in the basalt area in the continental collision zone (Fig. 2.63). The results show that the volcanic eruption is related to the intraplate or marginal tensile tectonic movement. The K-bentonite is mainly derived from the intraplate tectonic environment, and this is consistent with the previous conclusions of the early Paleozoic intraplate folding orogeny according to the lack of ophiolite and the presence of

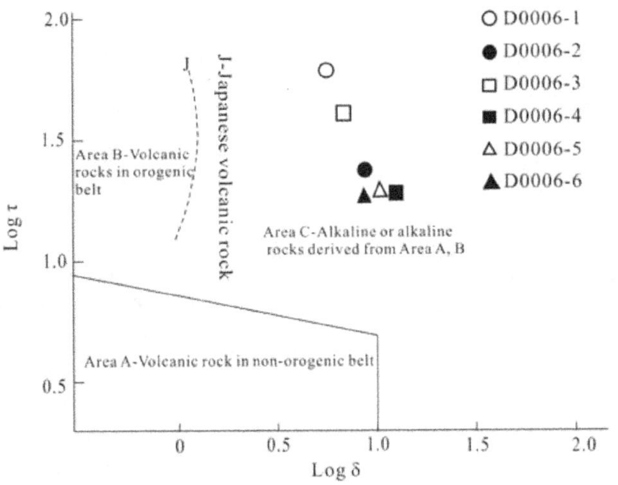

Fig. 2.60 Rittmann-Gottiry diagram of K-bentonite

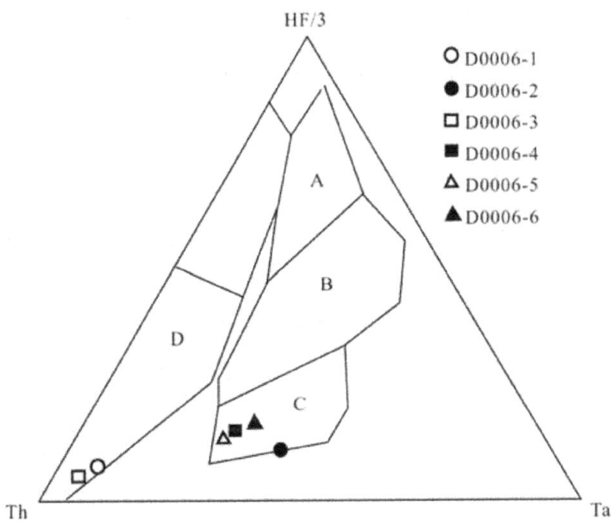

Fig. 2.61 Th-Hf-Ta tectonic setting discrimination diagram of K-bentonite (Pearce et al. 1984)

high-pressure metamorphic belt in South China in Caledonian period (Shu 2006, 2012).

Area A: N-type MORB; Area B: E-type MORB, and intracontinental tholeiites.
Area C: Alkaline intraplate basalt; Area D: volcanic arc basalt.

At present, there are two main viewpoints about the sources of the materials consisting of Late Ordovician bentonite. Most scholars believe that they are the production of volcanic ash deposition during submarine volcanic eruption, while a few scholars believe that it may be formed by lifting, denudation, transportation, sedimentation, and metamorphism of the

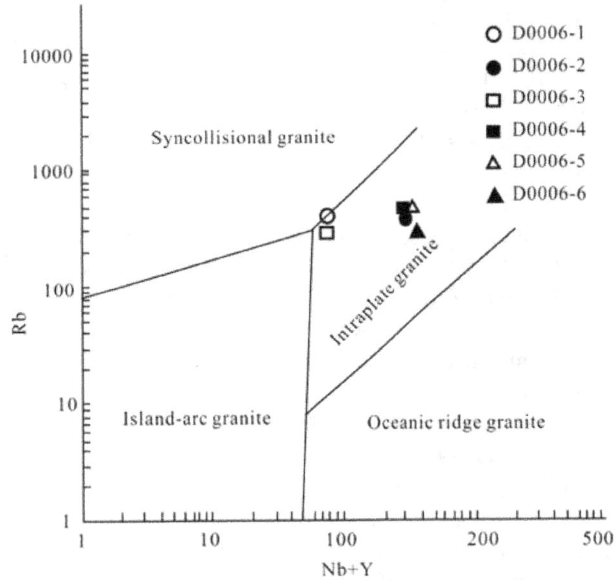

Fig. 2.62 Nb + Y-Rb tectonic setting discrimination diagram (Pearce et al. 1984)

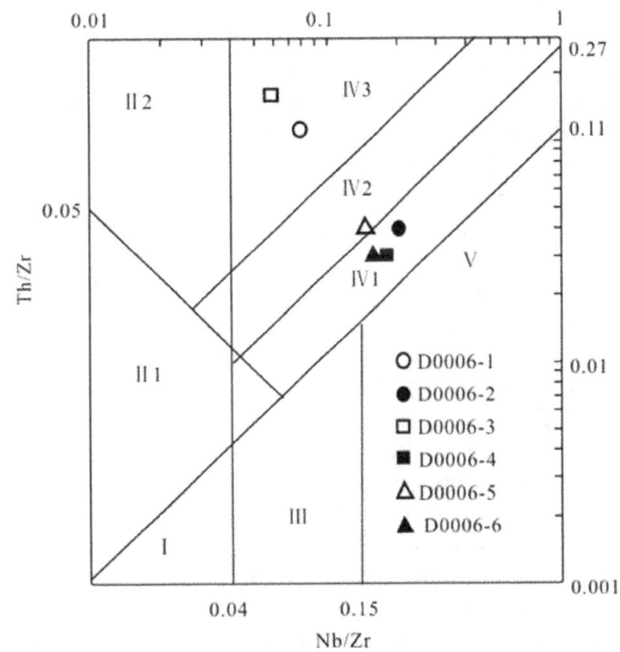

Fig. 2.63 Th/Zr-Nb/Zr bilogarithmic discrimination diagram (Sun et al. 2007) I. N-MORB area on the margin of divergent oceanic plate; II. convergent margin of plate (II_1. oceanic island arc basalts area; II_2. epicontinental island arc and continental volcanic basalt area); III. oceanic intraplate (oceanic island, seamount basalt area, T–MORB area, E–MORB area); IV. continental intraplate (IV_1. tholeiite area of intracontinental rift and epicontinental rift types; IV_2. basalt area in tension zone or (initial rift); IV_3. basaltic area in the continental collision zone); V. mantle plume basalt area

Caledonian intrusion. For the first point of view, on conducting the research of bentonite near the Ordovician–Silurian boundary in Hubei and Guizhou of the upper Yangtze landmass, the scholars believed that bentonite was widely distributed with small thicknesses. However, volcanic ash can drift in the air for a long distance and there are no enough corresponding magmatic activities in the interior of the Yangtze landmass, indicating that the materials consisting of these bentonite can only come from the periphery of the Yangtze landmass. This may be related to the volcanic eruption caused by plate subduction during the closing of Early Paleozoic Qinling palaeoocean in the north margin of the Yangtze landmass (Hu et al. 2009; Yang 2011; Xie et al. 2012a, b). However, some scholars believe that it is related to the converging of Cathaysia block toward Yangtze landmass (Su et al. 2009; Xie et al. 2012a, b). It may also be rated to the subducting of an ancient oceanic crust in the outside of the southeastern margin of the Early Paleozoic South China plate to the South China plate (Hu et al. 2009). According to the research results of *Jiangshan–Shaoxing Joint Belt Geological Structure Research Report* (2015) of Zhejiang Province, with the ultimate closing and extinction of the ancient South China Ocean in the end of the Early Paleozoic in the south of the Area (450 Ma), Wuyi block, respectively, converged and collided with the Yangtze landmass to the northwest and the southeast landmass to the southeast, and the tectonic evolution of the Jiangshan–Shaoxing joint zone entered the stage of continental collision (450–390 Ma). As a result, the Jiangshan–Shaoxing jointing zone formed owing to the northwest margin of Wuyi block converging and colliding with the southeastern margin of the Yangtze landmass, forming (450–420 Ma, Chencai subduction–accretionary complex), and Lishui–Yuyao jointing zone formed owing to the southeast margin of the Wuyi block converging and colliding with the northwest margin of the southeast landmass (Longyou tectonic melange).

According to relevant research, Caledonian magmatic activities are mainly distributed in two areas around the Yangtze landmass. One is the Qinling–Dabie area in the north margin of the Yangtze landmass, where intrusion and volcanic rocks are widely distributed. The other area is near the jointing zone on the southern margin of the Yangtze landmass, which is dominated by intermediate-acid intrusion and mainly distributed in Longyou, Zhejiang, from the southern section of the Jiangshan–Shaoxing jointing zone to Fujian (450–390 Ma) (Zhejiang Institute of Geological Survey 2015). At present, no magmatic activity in the northern part of the Jiangshan–Shaoxing jointing zone during that period has been reported. The possible reason is as follows. After the peak collision, the orogenic zone collapsed and stretched rapidly, and deep metamorphic rocks were uplifted out of the surface owing to strong post-orogeny. Among the metamorphic rocks, the garnet amphibolite (below 30 km) in the Longyou tectonic mélange area also turned back and emerged from the surface. All these indicate that a large number of magmatic activity products in the upper part may have been ablated and a large amount of sediments was provided for the later peripheral foreland basin deposition. After Yangtze landmass converged and collided with Wuyi block, the southeastern margin of the Yangtze landmass was transformed from a passive continental margin basin to peripheral foreland basin from the Late Ordovician and remained as peripheral foreland basin until the Early Devonian. As a result, the sedimentary environment and material deposition rate in the southeastern margin of the Yangtze landmass experienced great change. During this period, the terrigenous clasts in sedimentary strata increased significantly and thickened, and the sedimentary rate suddenly accelerated (the sedimentary rate of the Changwu Formation was even dozens of times higher than that of the Huangnigang Formation). Meanwhile, typical slump structure and reverse graded structure formed in the strata such as Huangnigang Formation and the Sanqushan Formation, reflecting the impact of collision orogeny on foreland basin.

In this Project, it is believed that just like the bentonite near the Ordovician–Silurian boundary of the upper Yangtze landmass, the Late Ordovician bentonite in the Area is also the products of volcanic ash deposition during the Caledonian collision and volcanic eruption, but not caused by the lifting, denudation, transportation, sedimentation, and metamorphism of the Caledonian intrusion. The main evidence is as follows:

(1) Material composition: The bentonite is mainly composed by montmorillonite, illite, kaolinite, illite–montmorillonite mixed-layer clay mineral, and quartz followed by a very small amount of albite. Compared with surrounding clastic rocks, bentonite is characterized by abrupt changes in color, composition, and structure. Furthermore, there is no obvious alteration in the clastic rocks. All these indicate that the bentonite and clastic rocks are sourced from different areas, and bentonite is not the products of metamorphism of clastic sediments.
(2) Geochemical evidence: The content of SiO_2, Na_2O, CaO, Fe_2O_3, FeO in bentonite is significantly lower than that in clastic rocks of the same period. The contents of Al_2O_3 and K_2O, as well as the K_2O/Na_2O ratio, are significantly higher than those in clastic rocks of the same period. The content of F (1549×10^{-6}) is also significantly higher than that in clastic rocks ($(596–1215) \times 10^{-6}$). All these indicate that the bentonite and clastic rocks are sourced from deposition of different materials.
(3) Zircon characteristics: According to CL image features of zircon, the zonal structure of zircon developed in the bentonite. Besides, the bentonite features obvious angular structure, indicating that the bentonite did not experience long-distance transportation.

(4) Chronological evidence: A total of 87 points were tested for the six zircon samples collected in this Project. Except for one point whose age is 748.6 Ma, the age of all the other points is 482.1–401.9 Ma and mainly 455–440 Ma. This indicates that the zircon in bentonite was from relatively uniform sources but was not transported from sedimentary rocks in different areas.

Therefore, the K-bentonite in the Area formed owing to large-scale magmatic intrusion and the disposition of volcanic ash (caused by volcanic eruption) in the peripheral foreland basin in the northwest during the collision and converging (450–420 Ma) between the Caledonian Yangtze landmass and the Cathaysian block. As a result, bentonite (sedimentary tuff) interbeds formed in the Late Ordovician Huangnigang Formation and Changwu Formation, and the interbeds in Huangnigang Formation were the most developed.

2.4.5.5 Analysis of Sedimentary Environment

The Huangnigang Formation is mainly composed of mudstone and knotlike limestone. In the mud, ancient organisms lack, and micro-fine horizontal bedding developed, indicating the anoxic quiet environment of deep shelf. The knotlike limestone is dominated by microcrystalline calcite and contains the fossils of benthic organisms, representing the oxygen-rich environment of shallow shelf.

According to two rhythm beds and the ancient organisms contained in Huangnigang Formation, it can be decided that the Huangnigang Formation experienced two similar rise and fall processes of the sea level. In the lower parts of the rhythm beds, the sediments are dominated by siliceous argillaceous matter interbedded with knotlike limestone layers. From bottom to top, the siliceous argillaceous matter gradually increases and the knotlike limestone interbeds gradually decrease, indicating a sedimentary environment with alternate vibration of abyssal and neritic and gradual rise of sea level. The upper parts of the rhythm beds only consist of siliceous argillaceous sediment, indicating that the Area was in the deep shelf environment near the stable redox zone after the maximum flooding. Within the Huangnigang Formation, there are a total of 47 layers of K-bentonite interspersed from the top of the first rhythm bed where siliceous mudstone-bearing knotlike limestone is distributed to the top of the second member of the Changwu Formation. Among these layers, 42 layers are interspersed in the Huangnigang Formation and five layers interspersed between the first and second members of the Changwu Formation. All these verify that frequent crustal movements and volcanic activities occurred in the middle and late sedimentary period of Huangnigang Formation, which affected the transformation of Middle and Late Ordovician sedimentary environment from closed basin to open basin. According to Zircon SHRIMP U-Pb isotope dating, the age of the K-bentonite is 453–445 Ma, basically consistent with the time of Guangxi movement (Chen et al. 2010, 2012, 2014). After experiencing strong Guangxi movement, the sedimentary environment in the Area gradually shifted from the shallow shelf to (bathyal) abyssal basin. Accordingly, the carbonate facies–siliceous-argillaceous facies of the shallow shelf–deep shelf in the Huangnigang Formation ended, and flysch sedimentation of abyssal slope began in Changwu Formation.

Regionally, Huangnigang Formation features slight lithofacies transition and mainly contains benthic ancient organisms, indicating that Zhejiang Province was in a stable and open shallow shelf environment during this period.

2.4.5.6 Trace Elements in Strata (Ore-Bearing)

In this Project, ten rock spectra were collected from the Huangnigang Formation of Xinqiao profile (PM001) and the statistics of the arithmetic mean values and concentration coefficients of the analysis results were made. According to the statistics, Huangnigang Formation is enriched in Bi, Pb, and W, among which Bi is extremely rich.

2.4.6 Changwu Formation (O_3c)

The name Changwu Formation (O_3c) was created by Lu et al. (1955) for Nanshan of Chengbei–Dongshan of Changwu area in Jiangshan City, Zhejiang Province. In 1937, Jie Xu discovered the Late Ordovician strata in Tashan Village, Yuqian Town, Lin'an City, and created the name "Yuqian System" for them. Zhejiang Regional Geological Survey Team (1965) plotted a complete Late Ordovician stratigraphic profile during regional geological survey of Jiande map sheet on a scale of 1:200,000. It divided the profile into three parts, i.e., the lower, middle, and upper parts. The former two parts were collectively called Changwu Formation, and the upper part was called Wenchang Formation. Zhejiang Regional Geological Survey Team (1967) divided the sedimentary strata into a southeast area and a northwest area during regional geological survey of Lin'an map sheet on a scale of 1:200,000. In the southeast area, Changwu Formation and Wenchang Formation were still adopted, while in the northwest area, the Yuqian System was divided into Yuqian Formation and Zhangcunwu Formation. In *Lithostratigraphy of Zhejiang Province* (1995), Yuqian Formation was renamed Changwu Formation and Zhangcunwu Formation was renamed Wenchang Formation. In this Project, the name Changwu Formation (O_3c) was still adopted given the lithological characteristics of Xinqiao profile (PM001) and Shangangshang map (PM002) in

Hanggai map sheet, Xiajia profile (PM030) and Maotan profile (PM041) in Xianxia map sheet, and the geological observation traverse.

2.4.6.1 Lithostratigraphy

Changwu Formation (O_3c) in the Area belongs to typical Ordovician flysch sedimentation in the Lower Paleozoic. It features basically the same outcrop range as Huangnigang Formation. That is, it mainly outcrops in the eastern part of the Hanggai map sheet and the south-central part of Xianxia map sheet. In addition, it is sporadically exposed in the southeast and northwest corners of the Chuancun map sheet. The total outcrop area is about 83.56 km^2, accounting for 6.57% of the bedrock area.

According to lithology and lithologic association, Changwu Formation (O_3c) is divided into the first member of Changwu Formation (O_3c^1), the second member of Changwu Formation (O_3c^2), and the third member of Changwu Formation (O_3c^3). They are in conformable contact with the underlying Huangnigang Formation and the overlying Wenchang Formation (Fig. 2.64).

The first member of Changwu Formation (O_3c^1): rhythm interbeds consisting of gray thick–medium laminated silt-fine sandstone, thin–medium laminated siltstone, and silicon-bearing mudstone; occasionally interspersed thick laminated K-bentonite; Bouma sequences such as ABE, ACE, ADE, AE, BE, and CE developing; from bottom to top, the sandstone gradually decreasing and the mudstone gradually increasing; the thickness of a single layer of silty-fine sandstone: 20–50 cm generally; the thickness of a single layer of silicon-bearing mudstone: 3–20 cm, varying greatly. The thickness of the member: 118.37–155.00 m.

The second member of Changwu Formation (O_3c^2): rhythm interbeds consisting of gray thin–medium laminated silt-fine sandstone, thin–medium laminated siltstone or argillaceous siltstone, thin–medium laminated silicon-bearing mudstone, and black micro-laminated carbonaceous mudstone; occasionally interbedded with thick laminated K-bentonite in the upper part; Bouma sequences such as ABE, ACE, ADE, AE, BE, and CE developing, dominated by CE followed by BCE, black micro-laminated graptolite-bearing carbonaceous mudstone visible in each sequence generally (the basis used to distinguish Bouma sequences of the member from the ones of the first and third member); from bottom to top, the sandstone gradually reducing, siltstone and mudstone increasing. The respective thickness of a single layer of siltstone and fine sandstone, siltstone or argillaceous siltstone, silicon-bearing mudstone, and black carbonaceous mudstone: 4–25 cm generally, 2–15 cm, 1–16 cm generally, and 0.2–3 cm generally. The thickness of the member: 95.70–201.48 m.

The third member of Changwu Formation (O_3c^3): the lower part: rhythm interbeds of gray thin–medium laminated feldspathic quartz silt-fine sandstone, siltstone or silty mudstone; the middle-lower part: rhythm interbeds consisting of thin–medium laminated argillaceous siltstone and mudstone; the middle-upper part: mainly rhythm interbeds consisting of feldspathic quartz silty-fine sandstone or sandstone and silty mudstone, interbedded with thick laminated mudstone. Generally, feldspathic quartz silty-fine sandstone constitutes the main body of the member, and the sandstone gradually increases upward. Graded bedding with thickness decreasing upward develops in small amounts of sandstones in the lower part of the layer; load casts develop at the bottom. The respective thickness of the single layer of silt-fine sandstone and siltstone: 10–30 cm generally and 5–15 cm. The thickness of the member: 161.39–167.45 m.

1. **Stratigraphic section**

The lithology of Changwu Formation is described by taking the example of Ordovician Changwu Formation (O_3c) profile in Xinqiao (Fig. 2.65) Village, Hanggai Town, Anji County, Zhejiang Province. The details are as follows:

Wenchang Formation — Total thickness: >3.93 m

59. Thick-bedded sandstone, no bedding developing, the thickness of a single layer: 50–60 cm. 3.93 m

———— Conformable contact ————

The third member of Changwu Formation (O_3c^3) — Total thickness: 167.45 m

58. Interbeds consisting of khaki thin–medium laminated fine-silty sandstone and thick laminated silty mudstone after weathering; the fine-silty sandstone: cross-bedding developing, the thickness of a single layer: 8–25 cm; the thickness of a single layer of the mudstone: 5–10 cm. 36.85 m

57. Khaki thin–medium laminated fine-silty sandstone after weathering. 42.07 m

56. Rhythmic interbeds consisting of khaki medium laminated fine-silty sandstone and thick laminated silty mudstone after weathering. The respective thickness of a single layer of the fine-silty sandstone and the mudstone:

10–20 cm and 2–6 c 46.65 m

55. Khaki medium–thin laminated fine-silty sandstone interbedded with thick laminated mudstone after weathering. The fine-silty sandstone: cross-bedding developing, the thickness of a single layer: 5–12 cm; the thickness of a single layer of the mudstone: 3–6 cm. 5.33 m

54. Khaki thick laminated silty mudstone after weathering, occasionally interbedded with medium laminated fine-silty sandstone. 18.88 m

53. Gray medium laminated silt-fine sandstone interbedded with thick laminated silty mudstone. The silty-fine sandstone: mainly composed of feldspar and quartz, normal graded bedding developing. The rocks are usually khaki owing to weathering. 17.67 m

——————————— Conformable contact ———————————

The second member of Changwu Formation (O_3c^2) Total thickness: 201.48 m

52. Interbeds consisting of gray thick laminated sandstone, micro–thin laminated argillaceous siltstone, mudstonc, and black carbonaceous mudstone. Bouma sequences BCE and CE developing, dominated by CE, BE visible occasionally. The sandstone (B): the thickness of a single layer: 4 cm; the argillaceous siltstone (C): cross-bedding developing, the thickness of a single layer: 2–9 cm; the mudstone (E): the thickness of single bed: 1–11 cm; The thickness of single layer of carbonaceous mudstone: 0.5–0.7 cm. Graptolite visible in the mudstone 25 cm away from the top; a layer of K-bentonite with a width of 2 cm respectively visible and 12 cm and 20 cm away from the top. 32.92 m

51. Interbeds consisting of gray thin–medium laminated sandstone, micro–thin laminated siltstone, mudstone, and black micro laminated carbonaceous mudstone, Bouma sequence BCE developing, gradual contact between the sandstone (B) and the siltstone (C) while the mudstone (E) in sharp contact with its underlying and overlying strata. The sandstone (B): parallel bedding developing, the thickness of a single layer: 3–12 cm; the siltstone (C): cross-bedding developing, the thickness of single bed: 1–4 cm; the mudstone (E) single-layer thickness 2–5 cm; the carbonaceous mudstone: the thickness of a single layer: 0.8–1 cm, containing graptolite, only appearing in the upper part. Producing Chitinozoa, *Belonechitina americana*(Taugourdeau, 1965), *Conochitina* aff. *dolosa* Laufeld, 1967, *Conochitina* sp. *Conochitina* sp. *Eisenackitina songtaoensis* Chen *et al.*, 2009, and *Rhabdochitina gallica* Taugourdeau 1961. 12.46 m

50. Interbeds consisting of gray medium laminated fine-silty sandstone and thick laminated silty mudstone, Bouma sequences AE developing. The fine-silty sandstone (A): graded bedding developing, the thickness of a single layer: 17–25 cm; the silty mudstone (E): the thickness of single bed: 3–5 cm. 1.14 m

49. Interbeds consisting of gray thick laminated siltstone, thin–medium laminated silty mudstone, and black micro laminated carbonaceous shale. Bouma sequence CE developing. The siltstone (C): cross-bedding developing, the thickness of a single layer: 3–5 cm; the siltstone (E) the thickness of a single layer: 4–22 cm. Black carbonaceous mudstone, single layer 0.5–1 cm, containing graptolite. 15.48 m

48. Interbeds consisting of gray medium laminated sandstone, thick laminated siltstone, thick laminated mudstone, and black micro laminated carbonaceous mudstone, Bouma sequence ACE developing. The sandstone (A): the thickness gradually decreasing upwards, the thickness of a single layer: 10–25 cm generally and even 43cm (maximum); the siltstone (C): cross-bedding developing, the thickness of a single layer: 3–9 cm; gradual transition existing from A to C. The respective thickness of a single layer of the mudstone (E) and the black carbonaceous mudstone: 2–10 cm and 0.1–1 cm. 2.96 m

47. Interbeds consisting of gray thin–medium laminated sandstone, thick laminated siltstone, silty mudstone, and black micro laminated graptolite-bearing carbonaceous shale. The sandstone (B): parallel bedding developing, the grain size gradually decreasing upwards, the thickness of a single layer: 2–22 cm; the siltstone (C): small cross-bedding developing, the thickness of a single layer: 2–6 cm; the silty mudstone (E): the thickness of a single layer: 1–16 cm, massive bedding (homogeneous bedding) developing; the carbonaceous shale: horizontal bedding developing, graptolite visible, the thickness of the single bed: 0.2–3 cm. The sandstone (C): groove casts visible in the bottom with trending of 120°–300°, steep in the southeast and gentle in the northwest, indicating southeast water source. Chitinozoa, *Eisenackitina songtaoensis* Chen *et al.*, 2009 produced. 7.74 m

46. Interbeds consisting of gray thin–medium laminated siltstone, mudstone, and micro laminated black shale, Bouma sequence CE developing. The siltstone (C): wavy or cross-bedding developing, the thickness of a single layer: 3–14 cm; the mudstone (E): massive bedding (homogeneous bedding), the thickness of a single layer: 3–16 cm; the black carbonaceous mudstone: micro-grained horizontal bedding developing, the thickness of a single layer: 0.3–1 cm. 17.79 m

45. Interbeds consisting of gray thick laminated siltstone, medium–thin laminated silicon-bearing mudstone, and micro laminated black carbonaceous mudstone, interbedded with medium laminated fine sandstone, Bouma sequence CE developing. The siltstone (B) in the upper part: sandstone with graded bedding visible locally, the thickness of a single layer: 10–20 cm; the siltstone (C): cross-bedding developing, the thickness of a single layer: 1–8 cm; the silicon-bearing mudstone (E): the thickness of a single layer: 4–11 cm. This layer is interbedded with graptolite-bearing micro laminated carbonaceous mudstone. 29.85 m

44. Interbeds consisting of gray thick laminated siltstone, silicon-bearing mudstone, and micro laminated black shale, Bouma sequence CE developing. The siltstone (C): cross-bedding developing, the thickness of a single layer: 2–8 cm; the silicon-bearing mudstone (E): the thickness of a single layer: 4–8 cm, massive bedding (homogeneous bedding) developing, the thickness of a single layer of carbonaceous mudstone: 1–4 cm. 21.37 m

43. Polycyclic interbeds consisting of gray thin–medium laminated fine sandstone, thick laminated siltstone, and thin–medium laminated mudstone, interbedded with black and micro laminated carbonaceous mudstone, Bouma sequence ACE developing. The fine sandstone (A): gray, scouring surface as the bottom, normal graded bedding developing; less obvious parallel bedding developing on the top, the thickness of a single layer: 5–20 cm; the siltstone (C): cross-bedding developing, the thickness of a single layer: 3–9 cm; the mudstone (E): the thickness of a single layer: 5–22 cm, massive bedding (homogeneous bedding) developing; the carbonaceous mudstone: the thickness of a single layer: 1–2cm, a small amount distributed in the bottom and stable in the upper part. 7.57 m

42. Rhythmic interbeds consisting of gray thin–medium laminated siltstone and mudstone, interbedded with micro laminated black shale, Bouma sequence CE developing. The siltstone (C): cross-bedding developing, uneven bottom, the thickness of a single layer: 3–11 cm; the mudstone (E): the thickness of a single layer: 7–12 cm, massive bedding (homogeneous bedding) developing. Black carbonaceous mudstone interbeds with a thickness of 1cm visible in the upper part. 10.06 m

41. Interbeds consisting of gray medium laminated fine-silty sandstone, thick laminated siltstone, medium–thin laminated mudstone, and black carbonaceous mudstone, Bouma sequence BCE developing. The gray fine-silt sandstone (B): normal graded bedding developing, the thickness of a single layer: 15–21 cm; the siltstone (C): cross-bedding developing, the thickness of a single layer: 2–4 cm; the respective thickness of a

single layer of the mudstone (E) and the carbonaceous mudstone: 6–15 cm and 1–3 cm. 10.1 m

40. Gray medium laminated siliceous mudstone, no bedding developing, the thickness of a single layer: 10–27 cm. 3.06 m

39. Interbeds consisting of gray medium laminated fine-silty sandstone and silty mudstone, interbedded with black thick laminated carbonaceous mudstone, Bouma sequences AE developing. The fine-silty sandstone (A): graded bedding developing, uneven bottom, the thickness of a single layer: 27–29 cm; the silty mudstone (E): no bedding developing, the thickness of a single layer: 14–25 cm; the black carbonaceous mudstone: horizontal bedding developing, the thickness of a single layer: 1–5 cm. 1 m

38. Interbeds consisting of gray thick laminated fine-siltstone and silicon-bearing mudstone, interbedded with black carbonaceous mudstone, two beds of fine sandstone with a thickness of 20cm of each layer interspersing in the middle and lower parts, one layer of iron oxide with a thickness of 8cm interspersing in the middle part, Bouma sequence CE developing. The fine-siltstone (C): cross-bedding developing, the thickness of a single layer: 3–3.5 cm; the respective thickness of single layer of the silicon-bearing mudstone (E) and the black carbonaceous mudstone: 3.5–4 cm and 1–2 cm. 6.1 m

37. Interbeds consisting of gray medium laminated fine sandstone, siltstone, and medium–thin laminated mudstone, Bouma sequences ABE developing. The fine sandstone (A): graded bedding developing, uneven bottom, the thickness of a single layer: 13–18 cm; the fine-sand siltstone (B): weak parallel bedding developing, the grain size gradually decreasing upwards, the thickness of a single layer: 7–16 cm; the mudstone (E): the thickness of a single layer: 7–15 cm. 8.21 m

36. Gray thin–medium laminated mudstone, interbedded with a layer of fine-silt sandstone with a thickness of 10cm in the middle part. The mudstone: silt visible occasionally, the thickness of a single layer: 10–15 m. 5.24 m

35. Interbeds consisting of grayish thick laminated siltstone and mudstone, interbedded with dark ash-black graptolite-bearing carbonaceous mudstone, Bouma sequences CE developing. The siltstone (C): cross-bedding developing, the thickness of a single layer: 3–7 cm; the mudstone (E): no bedding developing, the thickness of a single layer: 6–10 cm; the carbonaceous mudstone: graptolite visible, the thickness of a single layer: 0.5–3 cm. A layer of 0.5 cm thick K-bentonite visible in the middle-lower part of this bed. 6.22 m

——— Conformable contact ———

The first member of Changwu Formation (O_3c^1) Total thickness: 154.91 m

34. Interbeds of gray-caesious medium- to thick laminated fine-siltstone and dark gray thick laminated silicon-bearing mudstone, Bouma sequences AE and ABE developing. The fine-siltstone (A): graded bedding developing, the thickness of a single layer: 30–58 cm; the fine-silty sandstone (B): less obvious parallel bedding developing, the thickness of single bed: 40 cm; the siliceous mudstone (E): in sharp contact with its underlying and overlying strata, the thickness of a single layer: 4–7 cm. The top of this bed consisting of 12 cm thick K-bentonite. 1.67 m

33. Interbeds consisting of gray thin–medium laminated fine-silty sandstone and gray micro–thin laminated mudstone, Bouma sequence BE developing. The fine-silty sandstone (B): scour surface developing in the bottom, less obvious parallel bedding developing in the beds, the thickness of a single layer: 6–16 cm; the thickness of single layer of mudstone (E): 1–7 cm. 2.61 m

32. Interbeds consisting of gray thin–medium laminated fine sandstone and dark gray thin–medium

laminated silty mudstone, Bouma sequence BE developing. The fine sandstone (B): less obvious parallel bedding developing, grain size of the clastics decreasing gradually upwards, scour surface developing in the bottom, the thickness of a single layer: 6–23 cm; the silty mudstone (E): the thickness of a single layer: 4–13 cm, in sharp contact with B. The top of this bed: 30 cm thick siliceous mudstone, followed by 3 cm thick K-bentonite, bioturbation structures visible occasionally. 1.88 m

31. Interbeds consisting of gray medium laminated fine-siltstone and thin–medium laminated silicon-bearing mudstone, Bouma sequence AE developing. The fine-siltstone (A): gradually turning into silty sand from fine sand, from bottom to top (graded bedding developing), flute casts developing in the bottom, the thickness of a single layer: 15–47 cm. The thickness of a single layer of silicon-bearing mudstone (E): 8–26 cm. 15.17 m

30. Interbeds consisting of gray thin–medium laminated fine-silty sandstone, thick laminated siltstone, and silicon-bearing mudstone, Bouma sequences BCE and DCE developing dominated by BCE, C and E. The fine-silty sandstone (B): scour surface developing at the bottom, normal parallel bedding developing, the thickness of a single layer: 6–35 cm; the siltstone (C): fine sand with cross-bedding developing, in gradual contact with B, the thickness of a single layer: 2–5 cm; the siltstone (D): horizontal bedding developing, the thickness of a single layer: 3–8 cm; the silicon-bearing mudstone (E): bioturbation structures mostly developing in the upper part. 8.52 m

29. Interbeds consisting of gray medium- to thick laminated fine-silty sandstone, thin–medium laminated siltstone, and medium–thin laminated silicon-bearing mudstone, Bouma sequence AE developing in the lower part and Bouma sequences CE and BE developing upwards. The respective thickness of a single layer of the fine-silt sandstone (A), fine-silt sandstone (B), siltstone (C), and silicon-bearing mudstone (E): 23–110 cm,15–30 cm, 3–6 cm, and 4–15 cm. 6.27 m

28. Interbeds consisting of gray medium laminated fine-silty sandstone, gray – dark gray thin–medium laminated argillaceous siltstone, Bouma sequence BE developing. The fine-silty sandstone (B): parallel bedding developing, the grain size of the clastics gradually decreasing upwards, the thickness of a single layer: 13–18 cm. The argillaceous siltstone (E): no obvious bedding developing, the thickness of a single layer: 9–15 cm, 3–4 cm width (diameter: 7–8 cm) biological crawling trace developing near the top. 12.38 m

27. Interbeds consisting of gray medium- to thick laminated boulder silty-fine sandstone, fine-siltstone, dark gray medium–thin laminated silty mudstone, and siliceous mudstone; incomplete Bouma sequence ACDE developing, ACE distributed in the lower part and ADE mostly upwards. The sandstone (A): boulder clay visible in the bottom, the thickness of a single layer: 10–123 cm; the siltstone (C): small cross-bedding developing, the thickness of a single layer: 2–6 cm; the silty mudstone (D): parallel bedding developing, the thickness of a single layer: 3–15 cm; the siliceous mudstone (E): no bedding developing, the thickness of a single layer: 5–33 cm. 2.87 m

26. Interbeds consisting of medium laminated fine-silty sandstone and dark gray medium–thin laminated mudstone, Bouma sequence AE developing. The fine-silty sandstone (A): uneven scour surface developing in the bottom, the layer thickness gradually decreasing upwards, parallel bedding visible, the thickness of a single layer: 10–36 cm; the siliceous mudstone (E): horizontal bedding developing, the thickness of a single layer: 7–22 cm; bioturbation structures developing near the top. 11.59 m

25. Interbeds consisting of gray medium laminated boulder fine-silty sandstone, fine-silty sandstone, and dark gray thick laminated silty mudstone. Bouma sequence AE developing in the lower part, Bouma sequence

ADE developing in the upper part. The boulder fine-silty sandstone (A): grain size decreasing upwards, containing 5%–10% syngenetic boulder clay. The fine-silty sandstone(D): grain size decreasing upwards, parallel bedding developing. The silty mudstone (E): biological crawling trace visible. The respective thickness of a single of A in the upper part and the lower part: 7–14 cm and 10–100 cm; the thickness of a single layer of B in the upper part: 15–28 cm; the respective thickness of a single layer of E in the upper part: 4–13 cm and 3–6 cm. 7.36 m

24. Interbeds consisting of thin–medium laminated fine-silty sandstone, silty mudstone, and mudstone, Bouma sequence developing. The fine-siltstone (A): uneven scour surface developing in the bottom, the grain size gradually decreasing upwards, the thickness of a single layer: 7–35 cm; the siliceous mudstone (E): horizontal bedding developing, the thickness of a single layer: 7–22 cm; bioturbation structures visible near the top of this bed. 15.13 m

23. Rhythmic interbeds consisting of caesious medium- to thick laminated fine sandstone and gray thin–medium laminated siliceous mudstone, Bouma sequence AE developing. The fine sandstone (A): irregular syngenetic boulder clay visible occasionally, grain size: 1–5 cm and gradually decreasing from bottom to top, the thickness of a single layer: 50–80 cm generally and 20–30 cm locally; the siliceous mudstone (E): no bedding developing, the thickness of single bed: 5–30 cm. The top: 8 cm thick K-bentonite and 2 cm thick tan iron oxide layer downwards. 58.92 m

22. Interbeds consisting of caesious middle bedded fine sandstone, thin–medium laminated silty mudstone, and medium laminated mudstone, Bouma sequence ADE developing. The fine sandstone (A): gradually changing from fine sand to silt from bottom to top, the thickness of single bed: 10–30 cm; the silty mudstone (D): horizontal bedding developing, the thickness of a single layer: 5–10 cm; the grey mudstone (E): no bedding developing, the thickness of a single layer: 10–20 cm. 9.96 m

——— Conformable contact ———

Huangnigang Formation (O_3h) Total thickness: >2.14 m

21. Caesious medium laminated silicon-bearing mudstone, no bedding developing, the thickness of single bed: 30–50 cm generally. 2.14 m

The original stratigraphic lithologic association of the third member of Changwu Formation (O_3c^3) cannot be fully reflected from the above profile owing to strong weathering. The specific lithological characteristics of the third member can be shown in the third member of Changwu Formation (O_3c^3)–the middle and lower parts of Wenchang Formation (O_3w) profile (PM002) of Hanggai map sheet in Shangangshang, Hanggai Town, Anji County, Zhejiang Province.

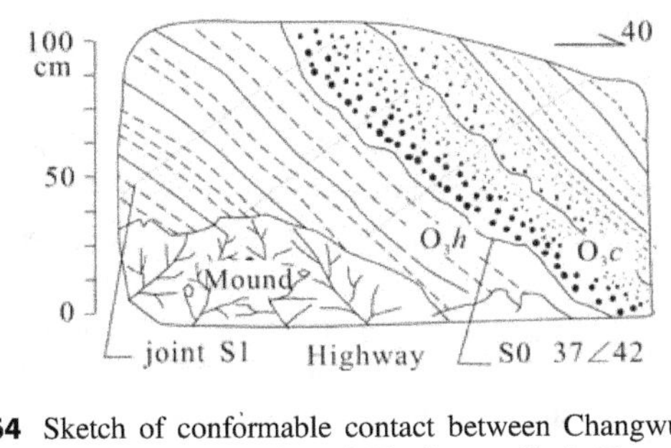

Fig. 2.64 Sketch of conformable contact between Changwu Formation and Huangnigang Formation of Xinqiao profile in Hanggai Town

2. **Lithological characteristics**

The lithology of the first member (O_3c^1) and the second member (O_3c^2) of Changwu Formation is mainly characterized by gray silt-fine sandstone, siltstone, silty mudstone or silicon-bearing mudstone, carbonaceous mudstone. Besides, there is a small amount of K-bentonite in the two members.

(1) The silty-fine sandstone: gray to caesious; composition: feldspar and quartz (80–85%), argillaceous material (15–20%), a small amount of detritus, and occasionally visible syngenetic boulder clay; the bottom: uneven

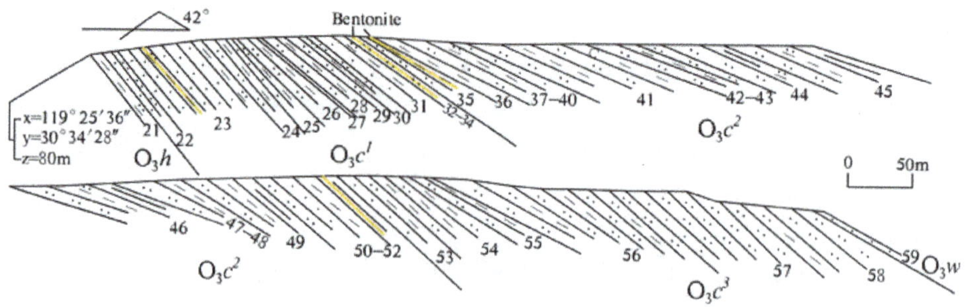

Fig. 2.65 Ordovician Changwu Formation (O_3c) profile in Xinqiao Village, Hanggai Town, Anji County, Zhejiang Province

sour surface; the thickness of a single layer: 15–50 cm generally; the siltstone: constituting A of Bouma sequences in case of normal graded bedding developing or constituting B of Bouma sequences in case of parallel bedding developing.

(2) The siltstone: gray; composition: feldspar and quartz (75–80%) and a small amount of argillaceous matter; constituting C of Bouma sequences in case of small cross-bedding, wavy bedding, or occasionally convolute bedding developing, or constituting D of Bouma sequences in case of parallel bedding developing; the thickness of single bed: 2–15 cm.

(3) The silicon-bearing mudstone: gray; composition: argillaceous matter (75–80%), siliceous matter (15–25%), and occasionally visible silt; no bedding developing generally; constituting C of Bouma sequences. However, biological disturbance relicts such as boring trace and crawling trace often visible in the upper part.

(4) The carbonaceous mudstone: dark gray–grayish black; composition: argillaceous matter and a small amount of silt and carbonaceous matter; horizontal bedding developing generally; graptolite fragments visible.

5) The K-bentonite: shallow gray yellow (green); tuff structure, blocky structure, and weak micro-bedding structure developing; composition: volcanic ash (90–95%) and a small amount of fine-grained angular feldspar (5–10%); the minerals mostly turning into kaolin owing to compression and hydrolysis, with weak waxy luster and soap-like slippery feeling, distinctly different from the upper and lower gray silicon-bearing mudstone and silty mudstone.

The lithology of the third member of Changwu Formation (O_3c^3) is mainly characterized by gray feldspathic quartz sandstone, feldspathic quartz silt-fine sandstone, silty mudstone, and mudstone.

(1) The feldspathic quartz sandstone: gray; composition: quartz and feldspar (5–80%) and silty argillaceous matter (20–25%); the feldspar and quartz: subangular to sub-well rounded; grain size: 0.05–0.4 mm, unevenly distributed, much distributed locally, mostly sub-well rounded.

(2) The feldspathic quartz silt-fine sandstone: gray; main component: quartz and a small amount of feldspar (75–80%), sub-well rounded; grain size: 0.05–0.1 mm; argillaceous matter: 20–25%.

(3) The silty mudstone: gray—dark gray, composition: argillaceous matter (50–90%) and a small amount of silt (10–50%), main components of the silt: quartz and feldspar.

(4) The mudstone: gray–dark gray; composition: argillaceous matter (80–90%) and occasionally visible silt and fine sand; part of the mudstone containing graptolite.

3. **Basic sequences**

In the first member (O_3c^1) of Changwu Formation, the basic sequences (Fig. 2.66) are consistent with the Bouma sequences. There are five types of basic sequences developing in this member, i.e., Bouma sequences ADE, AE, BE, CE, and ABE. The lower part of this member mainly consists of ADE and AE; the middle part is mainly composed of AE and BE followed by the small amount of CE; the upper part mainly includes AE, ABE, and BE. The thickness of the lower beds is greater than that of the upper bed. Therefore, these sequences belong to cyclic basic sequence with layer thickness generally decreasing upward.

(1) Bouma sequence ABE: composed of ① gray–caesious medium to thick laminated fine-silty sandstone, ② middle-bedded fine-silty sandstone, and ③ dark gray thick laminated silicon-bearing mudstone. The fine sandstone (A): graded bedding developing, the thickness of a single layer: 30–58 cm. The fine-silty sandstone (B): less obvious parallel bedding developing, in gradual contact with A, the thickness of a single layer: 40 cm. The silicon-bearing mudstone (E): no bedding developing; in sharp contact with its underlying and overlying strata, the thickness of a single layer: 4–7 cm.

(2) Bouma sequence ADE: composed of the interbeds of ① caesious middle-bedded sandstone, ② thick laminated silty mudstone, and ③gray middle-bedded mudstone. The fine sandstone (A): grayish–caesious, gradually changing from fine sand into silt from bottom to top, the

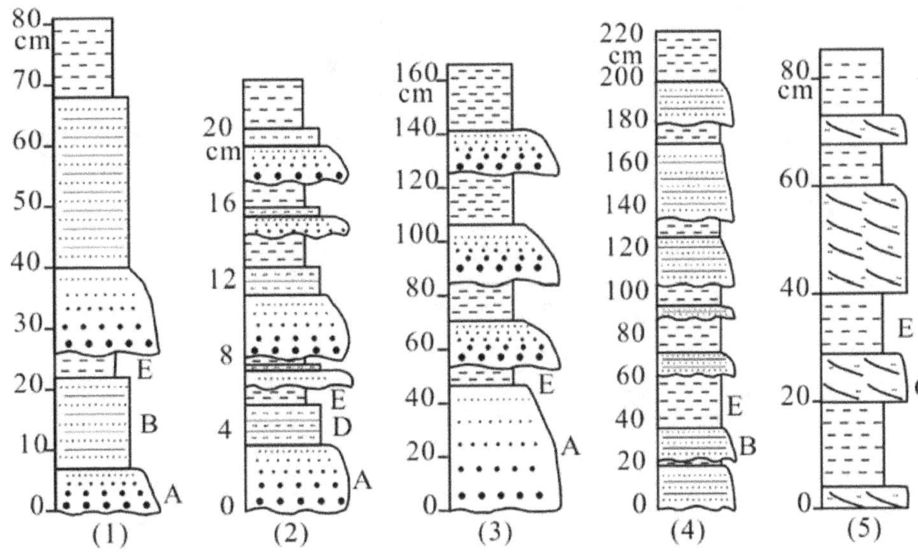

Fig. 2.66 Basic sequences of the first member of Changwu Formation (O_3c^1)

thickness of a single layer: 10–30 cm. The silty mudstone (D): horizontal bedding developing, the thickness of a single layer: 5–10 cm. The mudstone (E): no bedding developing, the thickness of a single layer: 10–20 cm.

(3) Bouma sequence AE: composed of rhythm interbeds of ① caesious medium to thick laminated fine sandstone and ② thin–medium laminated gray siliceous mudstone. The sandstone (A): 1–5 cm irregular syngenetic boulder clay visible occasionally, the grain size of the sand grain gradually decreasing from bottom to top, the thickness of a single layer: 20–80 cm. The siliceous mudstone (E): no bedding developing, the thickness of a single layer: 5–30 cm.

(4) Bouma sequence BE: consisting of ① gray thin–medium laminated fine sandstone and ② dark gray thin–medium laminated silty mudstone. The fine sandstone (B): parallel bedding developing, the grain size of the clastics gradually decreasing, scour surface developing in the bottom, the thickness of a single layer: 6–30 cm. The silty mudstone (E): the thickness of a single layer: 4–15 cm, in sharp contact with B, biological disturbance structures visible occasionally.

(5) Bouma sequence CE: consisting of ① gray thick laminated fine-silty sandstone and ② medium–thin laminated silicon-bearing mudstone. The fine sand siltstone (C): cross-bedding developing, convolute bedding developing locally, the thickness of a single layer: 3–6 cm. The siliceous mudstone (E): a small number of biological disturbance structures or crawling traces visible, the thickness of a single layer: 4–15 cm. Sharp contact between the two components.

The sediments in this member are mainly silt and fine sand, followed by siliceous and argillaceous matter. Crawling trace is visible in the upper part of the mudstone. The rock beds are thick, and scour structure generally developed in the bottom. All these reflect the sedimentary environment near oxidation–reduction zone with abundant terrigenous clastics sources and scarce ancient organisms. Therefore, this member belongs to turbidite facies of inner margin of bathyal basin slope.

In the second member of Changwu Formation (O_3c^2), the basic sequences (Fig. 2.67) are consistent with the Bouma sequences. There are five types of basic sequences developing in this member, i.e., Bouma sequences CE, ABE, AE, BCE, and ACE, dominated by CE. The lower part of this member mainly consists of CE, ABE, and AE, the middle part is mainly composed of CE, BCE, and ACE, and the upper part mainly includes CE, AE, and BCE. All these sequences belong to cyclic basic sequence with layer thickness and grain size decreasing from bottom to top.

(1) Bouma sequence ABE: consisting of ① gray medium laminated fine sandstone, ② thin–medium laminated fine-silty sandstone, ③ medium–thin laminated mudstone, and ④ black micro-laminated carbonaceous mudstone. The fine sandstone (A): normal graded bedding developing, uneven scour structure developing in the bottom, the thickness of a single layer: 13–25 cm. The fine-silty sandstone (B): weak parallel bedding developing, the grain size of clastics gradually decreasing upward, the thickness of a single layer: 3–20 cm. The mudstone (E): no bedding developing, the

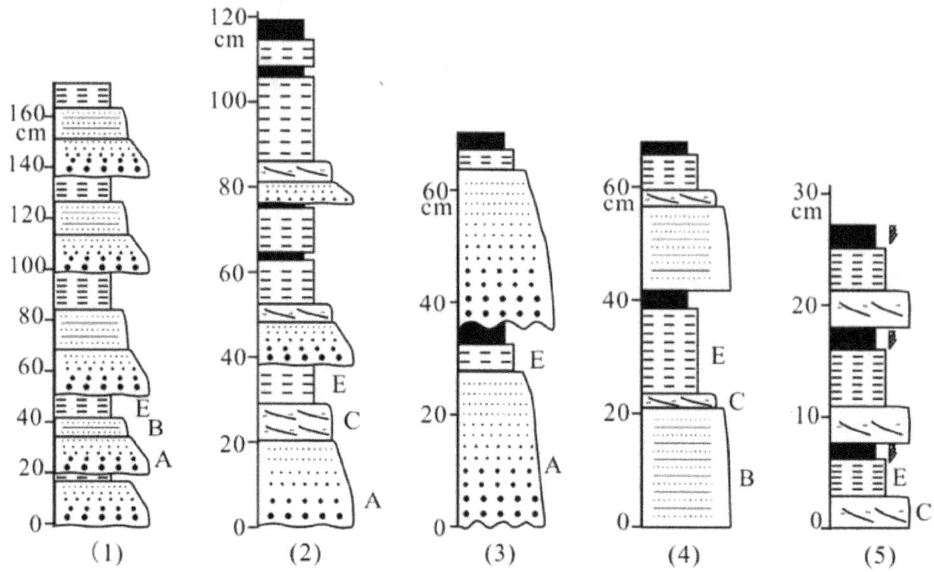

Fig. 2.67 Basic sequences of the second member of Changwu Formation (O_3c^2)

thickness of a single layer: 5–15 cm. The black micro-laminated carbonaceous graptolite-bearing mudstone: distributed in the upper part of this sequence, weak horizontal bedding visible occasionally.

(2) Bouma sequence ACE: consisting of ① gray thin–medium laminated fine sandstone, ② thick laminated siltstone, ③ thin–medium laminated mudstone, and ④ black micro-laminated carbonaceous mudstone. The fine sandstone (A): scour surface developing in the bottom, normal graded bedding developing, the thickness of a single layer: 5–20 cm. The siltstone (C): cross-bedding developing, the thickness of a single layer: 3–9 cm. The mudstone (E): the thickness of a single layer: 5–22 cm. The black carbonaceous mudstone: the thickness of a single layer: 1–2 cm, distributed in the upper part of the sequence.

(3) Bouma sequence AE: consisting of ① gray middle-bedded fine-silty sandstone, ② middle-bedded silty mudstone, and ③ black micro-laminated carbonaceous mudstone. The fine-silty sandstone (A): graded bedding developing, scour structure developing in the bottom, the thickness of a single layer: 27–29 cm. The silty mudstone (E): no bedding developing, the thickness of a single layer: 14–25 cm. The black carbonaceous mudstone: horizontal bedding developing, the thickness of a single layer: 1–5 cm. Sharp contact among the three components.

(4) Bouma sequence BCE: consisting of ① gray thin–medium laminated sandstone, ② thick laminated siltstone, ③ micro-thin laminated silty mudstone, and ④ black micro-laminated carbonaceous mudstone. The sandstone (B): parallel bedding developing, the thickness of a single layer: 3–12 cm. The siltstone (C): cross-bedding developing, the thickness of a single layer: 1–4 cm. The silty mudstone (E): the thickness of a single layer: 2–5 cm. Among them, sandstone (B) is in gradual contact with siltstone (C), while silty mudstone (E) is in sharp contact with its underlying and overlying strata. The black carbonaceous mudstone: the thickness of a single layer: 0.8–1 cm, containing graptolite, only distributed in the upper part.

(5) Bouma sequence CE: It consists of ① gray thick laminated siltstone and ② mudstone, often interbedded with ③ dark gray–black thick laminated graptolite-bearing carbonaceous mudstone. The lower siltstone (C): cross-bedding developing, the thickness of a single layer: 3–7 cm. The mudstone (E): no bedding developing, the thickness of a single layer: 6–10 cm. The black carbonaceous mudstone: graptolite commonly visible, the thickness of a single layer: 0.5–3 cm, mostly distributed above the mudstone; sharp contact among the three components.

The sediments in this member are mainly siliceous and argillaceous matters followed by silt and fine sand. Graptolites are common in black carbonaceous mudstone. Therefore, the second member belongs to the turbidite facies of outer margin of bathyal basin slope–abyssal basin.

There are three types of basic sequences (Fig. 2.68) developing in the third member of Changwu Formation (O_3c^3).

(1) Basic sequences in the lower part of the third member. Consisting of rhythm interbeds of ① thin–medium laminated feldspathic quartz silt-fine sandstone and ② silty mudstone. The silty-fine sandstone: mainly

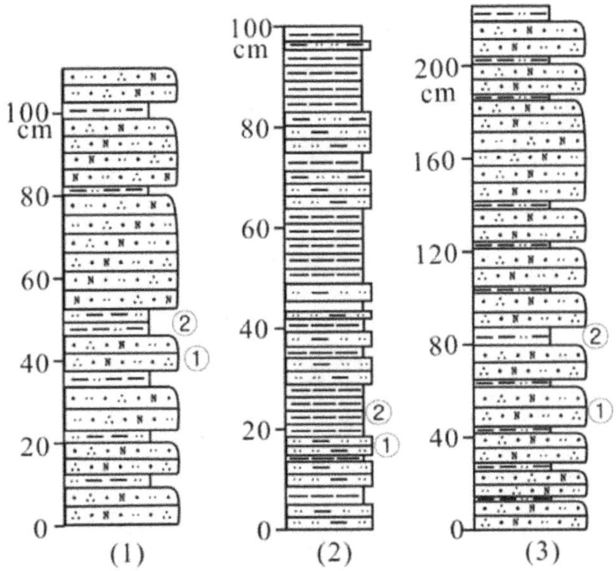

Fig. 2.68 Basic sequences of the third member of Changwu Formation (O_3c^3)

composed of quartz followed by feldspar, subangular–sub-well rounded, top and bottom mostly even, uneven scour surface developing in the bottom visible occasionally, normal graded bedding developing in the beds, the thickness of a single layer: 5–20 cm generally or even up to 40 cm locally. The silty mudstone: mainly composed of quartz and feldspar, sedimentary structure not obvious, horizontal bedding developing in a small amount of the silty mudstone, graptolite obtained in partial mudstone, the thickness of a single layer: 5–15 cm generally and 25–30 cm locally. Sandstone is approximately 1–2 times silty mudstone.

(2) Basic sequences in the middle part of the third member. Composed of rhythm interbeds of ① thin–medium laminated argillaceous siltstone and ② mudstone. The argillaceous siltstone: mainly consisting of feldspar and quartz with low maturity, the thickness of a single layer: 5–15 cm generally, the sedimentary structure not obvious. The mudstone: the thickness of a single layer: generally 5–25 cm, accounting for similar proportion to that of the argillaceous siltstone, distributed in the middle and lower parts.

(3) Basic sequences in the upper part of the third member. Consisting of rhythm interbeds of ① medium laminated feldspathic quartz silt-fine sandstone and ② thin–medium laminated silty mudstone. The silty-fine sandstone: mainly composed of quartz and feldspar, subangular–sub-well rounded, even top and bottom, print casts developing in the bottom, the thickness of a single layer: 10–25 cm generally and up to 40 cm locally. The silty mudstone: mainly composed of quartz and feldspar, sedimentary structure not obvious, horizontal bedding developing locally, the thickness of a single layer: 2–15 cm generally and 25–30 cm locally. The sandstone is about 3–10 times of silty mudstone.

Compared with the lower two lithological members, there is no Bouma sequence developing in the third member (O_3c^3) of Changwu Formation generally, and only the sequences similar to Bouma sequences developed locally. Furthermore, there is only a small amount of grapholite-bearing mudstone. Therefore, the third member is of flyschoid facies of inner margin of shallow shelf slope, and the clastic rock association belongs to non-cyclic basic sequence with width and thickness of beds increasing upward generally.

2.4.6.2 Sequence Stratigraphy

According to the characteristics of the lithology, paleontology, lithofacies association, and sequence boundary Xinqiao profile in Hanggai map sheet, there is one third-order sequence Sq23 in Changwu Formation (O_3c), which belongs to the second-order sequence SS8 (Fig. 2.69).

Third-order sequence Sq23

It consists of the transgressive systems tract (TST), starved section (CS), and highstand systems tract (HST). The bottom is the interface between the sandstone of the first member of Changwu Formation and the siliceous mudstone of Huangnigang Formation. Although the interface is uneven owing to sour of the turbidity current, there is only a short sedimentation interval between them. Therefore, they are still in conformable contact with each other. The top is the

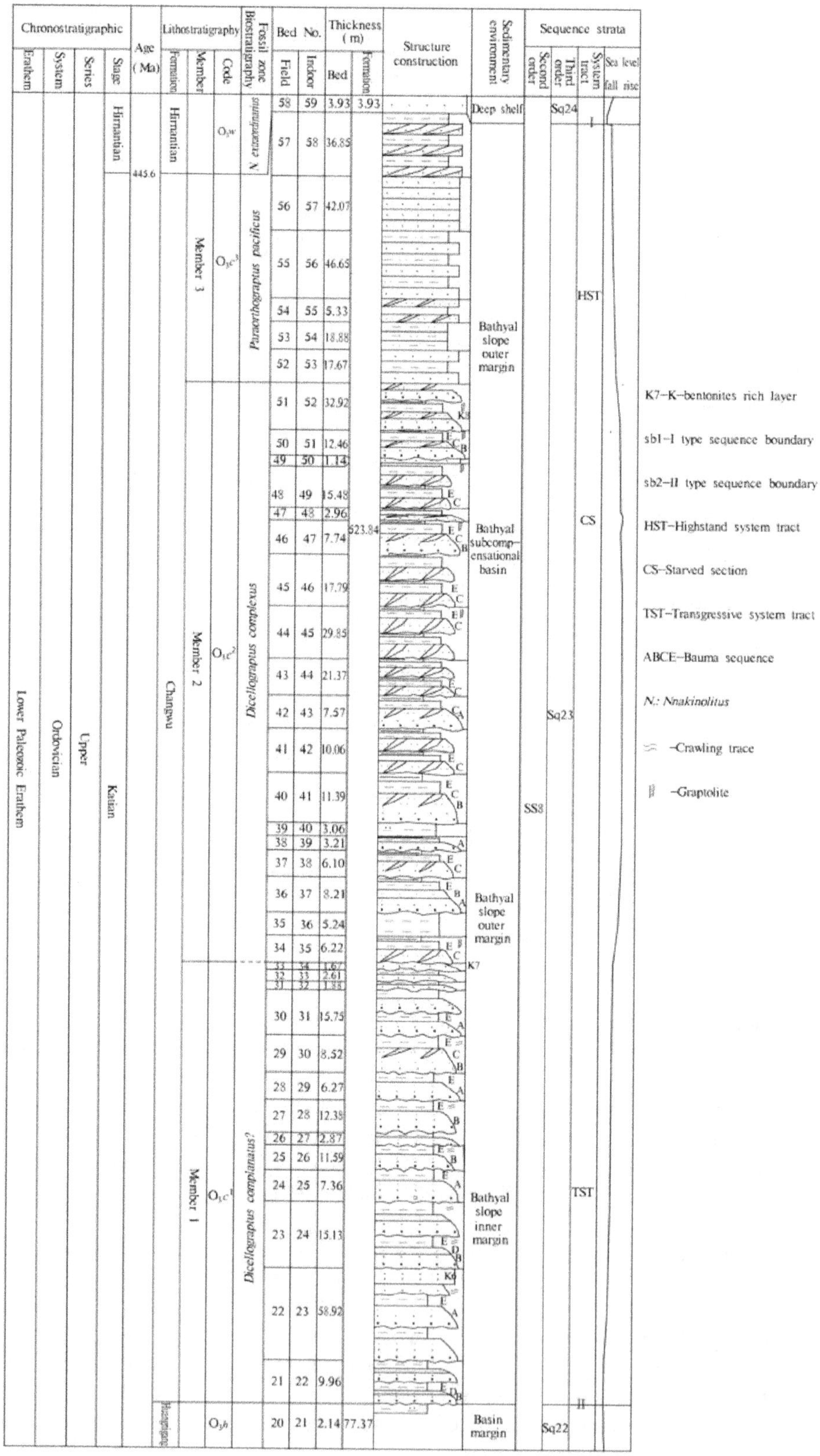

Fig. 2.69 Sequence of stratigraphic framework of Ordovician Changwu Formation in the area

conformity interface between the medium laminated silty mudstone of the third member of Changwu Formation and the thick laminated sandstone of the Wenchang Formation. Both the bottom and the top are of type-II sequence boundary.

Transgressive systems tract (TST) is located in the first member and the lower part of the second member of Changwu Formation. The sediments in TST are mainly terrigenous feldspathic quartz sand grains, silt, and argillaceous matter, followed by carbonaceous argillaceous matter. Typical Bouma sequences AE, ABE, ADE, ACE, BE, BCE, and CE developed from bottom to top. The sandstone (constituting A and B) is medium laminated (30–45 cm) mainly in the lower part and changes into thin–medium laminated shape upward, and the layer thickness and the grain size of the clastics decrease upward. However, the thick laminated siltstone (constituting D and C) and mudstone (constituting E) increases upward. All these indicate that the sea level of the sedimentary basin in the Area rose gradually, and consequently, the supplies of the terrigenous clastics gradually decreased and the hydrodynamic force gradually weakened.

In the middle and upper parts of the second member of Changwu Formation, the transgression reached the maximum, i.e., the maximum flooding surface. Owing to that, these parts are located in the outer margin of abyssal basin slope far away from the continent. The sediments are scarce and mainly include fine sand and silt, argillaceous matter, and carbonaceous argillaceous matter. Typical Bouma sequence CE developed in these parts, accompanied by graptolite-bearing carbonaceous shale. Therefore, these parts constitute the starvation section (CS) with the most deficient sediments.

The highstand systems tract (HST) is located from the upper part of the second member to the third member of Changwu Formation. The sediments in the second member mainly include terrigenous feldspathic quartz sand, silt, argillaceous matter, and carbonaceous argillaceous matter. Typical Bouma sequences CE, ACE, BE, and BCE developed in this member, accompanied by graptolite-bearing carbonaceous shale. The sediments in the third member are mainly terrigenous feldspathic quartz sand grains, silt, and argillaceous matter. No Bouma sequences developed in this member. All these indicate that the sea level of the sedimentary basin in the Area gradually fell, and accordingly, the supplies of the terrigenous clastics gradually increased and the hydrodynamic force increased.

To sum up, the third-order sequence (Sq23) experienced a relatively complete transgression–regression process and is a sequence of retrogradation–progradation type.

2.4.6.3 Biostratigraphy and Chronostratigraphy

After experiencing the Late Ordovician tectonic–volcanic events, the Area entered the abyssal slope environment. The ancient organisms in the Area mainly included planktonic graptolites, ancient microorganisms, and benthic mollusca, and the fossils were only preserved in mudstone or carbonaceous shale. At present, the biostratigraphic division of the Changwu Formation in China is still mainly based on graptolite zones and the study on chitinozoan is in its initial stage. There are three graptolite zones in Changwu Formation, which are *Dicellograptus complanatus* zone, *Dicellograptus Complexus* zone, and *Paraothograptus pacificus* zone, respectively. According to the *Stratigraphic Chart of China* (2014), all these three zones in Changwu Formation belong to the Late Ordovician Katian Stage.

In this Project, only a small number of graptolite fragments and biological crawling traces were found in the first member of Changwu Formation in the Area, and *Dicellograptus Complexus* zone and *Paraothograptus pacificus* zone were identified in the second and third members, respectively. The details are as follows:

1. ***Dicellograptus complexus* zone**

Dicellograptus complexus zone is located in the second member of Changwu Formation (O_3c^2). Abundant graptolites are obtained in the micro-laminated carbonaceous shale in the third layer of the second member of Changwu Formation in Shangangshang profile (PM002), including *Amplexograptus disjunctus yangtzensis* Mu et Lin, *Amplexograptus suni* (Mu) *Appendispinograptus supernus* (Elles and Wood), *Appendispinograptus venustus* (Hsu), *Climacograptus hastatus* (Hall 1902), *Yinograptus disjunctus* (Yin et Mu), *Paraplegmatograptus uniformis* Mu, and *Leptograptus extremus* Mu et Zhang. No standard fossil of *Dicellograptus Complexus* zone was obtained. However, the above fossils are the main molecules of *Dicellograptus complexus* zone, indicating the *Dicellograptus complexus* zone. The above fossils may be corresponding to the upper part of the zone (Figs. 2.70 and 2.71m).

2. ***Paraothograptus pacificus* zone**

Paraothograptus pacificus zone is located in the third member of Changwu Formation (O_3c^3). A few genera of graptolites were found in Layer 6 and Layer 7 of the third member of Changwu Formation (O_3c^3) in Shangangshang profile (PM002) including: *Paraothograptus pacificus* (Ruedemann), *Paraothograptus angustus* Mu et Lee 1977, *Amplexograptus disjunctus yangtzensis* Mu et Lin,

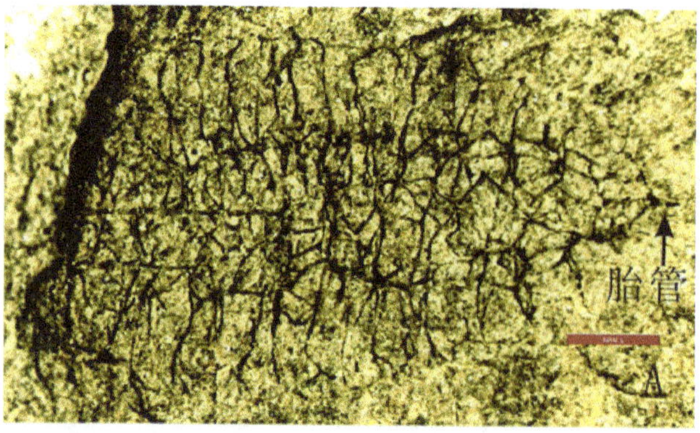

Fig. 2.70 *Paraplegmatograptus uniformis Mu(A)*

Climacograptus chiai Mu 1949 (= Amplexograptus hubeiensis Mu et al.), *Amplexograptus* sp. indet., *Appendispinograptus supernus* (Elles & Wood), and amplexograptid gen. et sp. indet. (Fig. 2.72).

In addition, 16 chitinozoan samples were collected from the first and the second members of Changwu Formation of Xinqiao profile (PM001) and four chitinozoan samples were taken from the third member of Changwu Formation of Shangangshang profile (PM002). Only three of these samples contain chitinozoan fossils. The following chitinite fossils were obtained from the samples collected from the Layer 47 and Layer 50 of the second member of Changwu Formation in Xinqiao profile: *Belonechitina americana* (Taugourdeau 1965), *Conochitina aff. dolosa* (Laufeld 1967), *Conochitina* sp. 1, *Conochitina* sp. 2, *Eisenackitina songtaoensis* (Chen et al. 2009), and *Rhabdochitina gallica* (Taugourdeau 1961). The chitinozoan fossil *Eisenackitina songtaoensis* Chen et al. 2009 was obtained from Layer 20 in the third member of Changwu Formation in the Shangangshang profile. All these chitinozoan fossils obtained belong to E. *Songtaoensis* zone (Fig. 2.73), which is corresponding to *Didymograptus complanatus* zone–*Nemagraptus extraordinarius* zone. The time of these chitinozoan fossils is the Late Katian–Late Ordovician.

2.4.6.4 Event Stratigraphy

There are five interbeds of K-bentonite developing in the first and the second members of Changwu Formation of Hule Formation ($O_{2\text{-}3}h$)–Changwu Formation (O_3c)

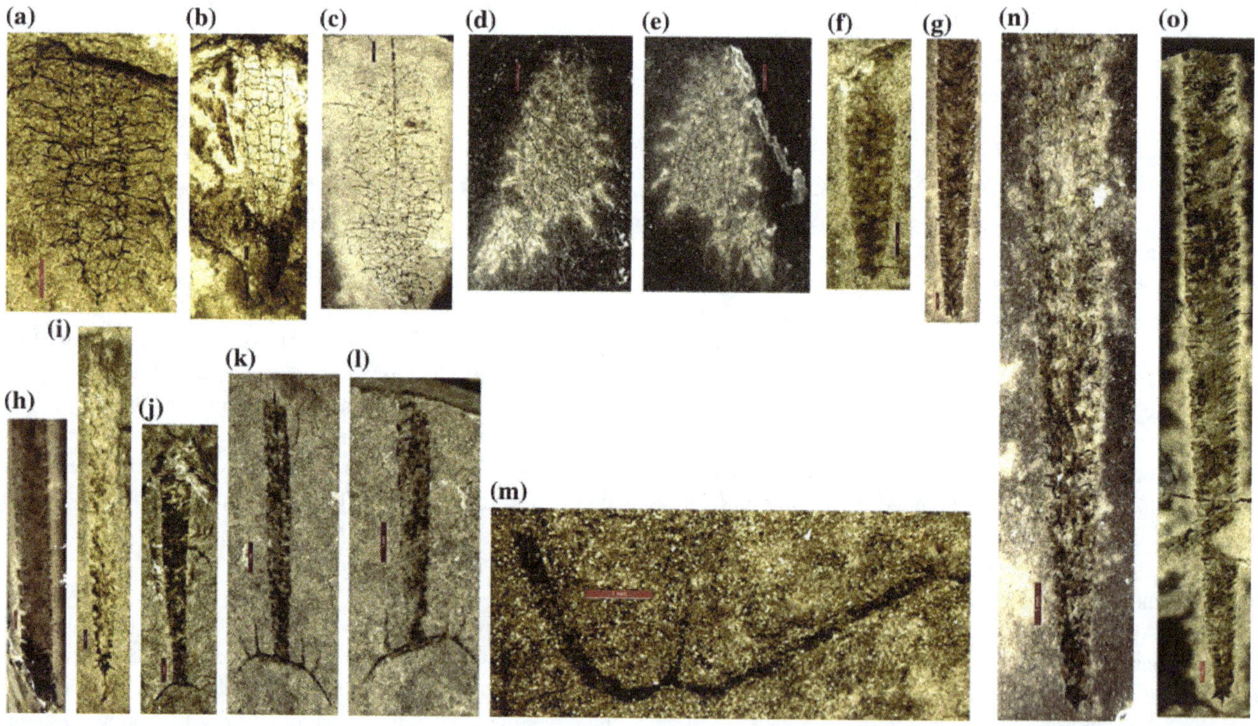

Fig. 2.71 Graptolite fossils in Dicellograptus complexus zone of the second member of Changwu Formation (O3c2): **a** *Paraplegmatograptus uniformis* Mu (Pm002-38a-21-1); **b** *Paraplegmatograptus uniformis* Mu (Pm002-38a-41); **c** *Yinograptus brevispinus* Mu(Pm002-38a-55a); **d** *Reteograptus geinitzianus* Hall (Pm002-38a-12-1); **e** *Reteograptus geinitzianus* Hall (Pm002-38a-11-2); **f** *Climacograptus hastatus* (Hall 1902) (Pm002-38a-10); **g** *Amplexograptus disjunctus yangtzensis* Mu et Lin, Pm002-Hb38b-6; **h** *Amplexograptus suni* (Mu) (Pm002-Hb38-6a); **i** *Diplograptus palaris* Lin (Pm002-38a-13-1); **j** *Appendispinograptus supernus* (Elles and Wood) (Pm002-38a-24-1); k-*Appendispinograptus venustus* (Hsu) (Pm002-38a-17); **l** *Appendispinograptus venustus* (Hsu) (Pm002-38a-18); **m** Leptograptus *extremus* Mu et Zhang (Pm002-38a-43-1); **n** *Amplexograptus suni* (Mu) (Pm002-Hb38a-18a); **o** *Amplexograptus suni* (Mu) (Pm002-38a-48-2)

Fig. 2.72 Graptolite fossils in *Paraothograptus pacificus* zone of the third member of Changwu Formation (O_3c^3): **a** *Paraothograptus pacificus* (Ruedemann) (Pm002-Hb6-2a); **b** *Paraothograptus pacificus* (Ruedemann) (Pm002-7-1-2-2); **c** *Paraothograptus pacificus* (Ruedemann) (Pm002-7-1-2-3); **d** *Paraothograptus pacificus* (Ruedemann) (Pm002-7-1-19); **e** *Paraothograptus pacificus* (Ruedemann) (Pm002-7-1-23-3); **f** *Paraothograptus angustus* (Pm002-7-1-3-3); **g** *Amplexograptus disjunctus yangtzensis* Mu et Lin (Pm002-7-1-7); **h** *Amplexograptus disjunctus yangtzensis* Mu et Lin (Pm002-7-1-8); **i** *Amplexograptus disjunctus yangtzensis* Mu et Lin (Pm002-7-1-40a); **j** *Amplexograptus disjunctus yangtzensis* Mu et Lin (Pm002-7-1-41); **k** *Amplexograptus hubeiensis* Mu et al. (Pm002-7-1-23-1); **l** *Amplexograptus disjunctus yangtzensis* Mu et Lin (Pm002-Hb6-1a)

Fig. 2.73 Chitinozoa fossil zone of Changwu Formation in Late Ordovician 1, 6, 7, 10. *Belonechitina americana* (Taugourdeau 1965), sample No. : 12ZAX40(PM001WH50); 2-5. *Eisenackitina songtaoensis* (Chen et al. 2009), sample No. of (2)(4) : 12ZAX38 (PM001WH46), sample No.of (3)(5):12ZAX40 (PM001WH50); 8. *Conochitina* sp. 1, sample No. : 12ZAX40 (PM001WH50); 9. Conochitina sp. 2, sample No. : 12ZAX40 (PM001WH50); 11. *Rhabdochitina gallica* (Taugourdeau 1961), sample No. : 12ZAX40 (PM001WH50); 12-14. *Conochitina aff. dolosa* (Laufeld 1967), sample No. : 12ZAX40 (PM001WH50). The white scale representing 100 μm

stratigraphic profile (PM001) in Xinqiao Village, Hanggai Town. All these interbeds concentrate on three K-bentonite enrichment cycles (K_6–K_8). A detailed study of the event stratigraphy is described in "2.4.5.5 Huangnigang Formation (O_3h)–Late Ordovician Volcanic Event Stratum."

2.4.6.5 Characteristics of Stable Isotopes

A total of seven samples were collected from the gray–dark gray mudstone of the Changwu Formation of Shangangshang profile (PM002) to conduct whole-rock stable isotope test and analysis of $\delta^{13}C$ and $\delta^{34}S$. The $\delta^{13}C$ value is from −30.99 to −29.45‰, with a small variation range and an average value of −30.07‰. This is consistent with the content range $\delta^{13}C$ (−35 to −30‰) of organic carbon in normal shale, indicating the normal reduction deep shelf sedimentary environment. Owing to the relatively low sulfur content in the samples, there is no $\delta^{34}S$ detected in four samples and the range of $\delta^{34}S$ value in the three effective samples was 4.70‰–20.60‰, with an average value of 12.33‰, indicating slightly positive excursion compared with the $\delta^{34}S$ content range (−10 to 10‰) in normal shale. This may be caused by partial isotope fractionation caused by sulfur-reducing bacteria.

2.4.6.6 Analysis of Sedimentary Environment

After Guangxi movement, the sedimentary environment in the Area changed from carbonate lithofacies–siliceous-argillaceous facies of shallow shelf–deep shelf of Huangnigang Formation period to the flysch-flyschoid facies of deepwater slope–deep shelf of Changwu Formation period.

In the first member of Changwu Formation, the Bouma sequences mainly include AE, ABE, ADE, and BE, followed by a small amount of CE. The thickness of the sandstone beds gradually decreases from 35 to 45 cm to less than 20 cm. The mudstone accounts for a small proportion while the crawling trace of mollusca can only be visible in the mudstone. All these indicate the sedimentation of the inner margin of abyssal slope near the continent. Owing to abundant sources of terrigenous clastics, frequent turbidity current is unfavorable to the survival of organisms and the preservation of fossils. Therefore, there are very few paleontological fossils in this member.

In the second member of Changwu Formation, the Bouma sequences mainly include CE followed by AE, ABE, BCE, and ACE. Furthermore, this member is often interbedded with black micro-laminated graptolite-bearing carbonaceous shale, which is mostly thick laminated mostly. All these indicate the sediments are deposited in the outer margin of abyssal slope far away from the continent with relative scarce sources of terrigenous clastics.

In the third member of Changwu Formation, the sediments mainly include thin–medium laminated sand and silt, followed by argillaceous matter. No graded bedding developed in the sandstone, and no diagonal bedding and convolute bedding developed in the siltstone. A small amount of gray mudstone contains planktonic graptolites. All these belong to a flyschoid formation and indicate a hypoxic deep shelf sedimentary environment.

According to the identification and analysis of the fine-silty sandstone samples collected from the first member and the second member of Changwu Formation, the main clastics in the sandstone is sub-well rounded feldspar followed by quartz, with a total content of 80%. Besides, the clastics feature good sorting and argillaceous–siliceous cementation. Therefore, the clastics belong to the graywacke with medium quartz content. The content of SiO_2 in the clastics is 65.03–68.97%, with a small change range. Besides, the content of other components is as follows: $Fe_2O_3 > FeO$, $K_2O > Na_2O$. All these indicate that the clastics come from relatively mature and stable platform. It is inferred that the main body of this member is granite and high-grade metamorphic rocks, followed by part of the product coming from the denudation area of low-grade metamorphic rock and sedimentary rocks.

To sum up, the sedimentary environment of the Area is the abyssal slope and its adjacent areas near the discontinuous zone of the geosphere at the boundary between active continental crust and oceanic crust. Regionally, the sedimentary environment of Changwu Formation varies greatly. In Xianlin area, Yuhang, carbonaceous shale is missing in the second member of Changwu Formation, indicating that the water body became shallower. In Jiangshan–Changshan–Yushan area, the upper and middle parts of Changwu Formation feature lithofacies transition and consist of mud and patch of Sanqushan Formation, belonging to the sedimentary environment of inner margin of the slope.

2.4.6.7 Trace Elements in Strata (Ore-Bearing)

Thirty-seven rock spectra were systematically collected from Changwu Formation of Xinqiao profile (PM001) including 17 collected from the first member and 20 from the second member. Besides, 19 rock spectra were collected from the third member of Changwu Formation of Shangangshang profile (PM002). Analysis was made on 14 trace elements, and statistics were made on the arithmetic mean values and

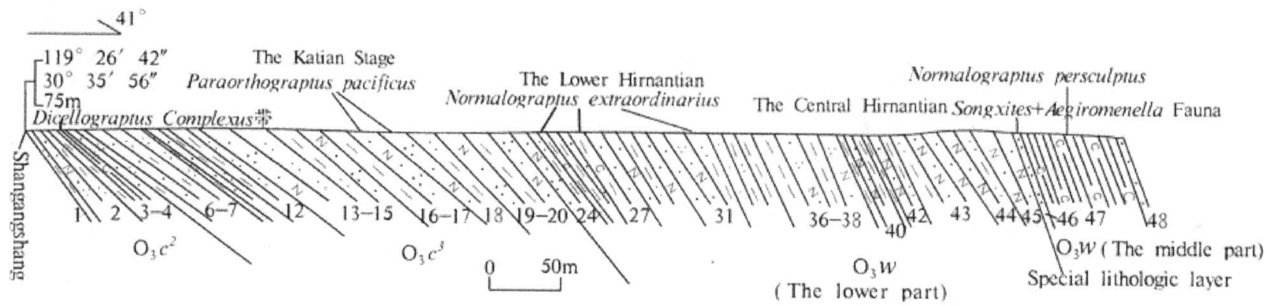

Fig. 2.74 Third member of Changwu Formation (O_3c^3)–middle and lower parts of Wenchang Formation (O_3w) profile in Shangangshang, Hanggai Town, Anji County, Zhejiang Province

concentration coefficients. The enriched elements of the three members of Changwu Formation vary slightly. The enriched trace elements in the first member are Bi and Pb, while those in the second member are Bi, Pb, and Zn. These two members share basically similar content of common enriched elements and other elements. The enriched elements in the third member are Bi, Sb, and Sn, and the content of other elements in the third member varies at a certain degree, indicating the influence of sedimentary environment on the distribution of trace elements in sediments.

2.4.7 Wenchang Formation (O_3w)–Standard Section of Lower Yangtze Region in the Upper Ordovician (Hirnantian)

The name Wenchang Formation (O_3w) was created by Zhejiang Regional Geological Survey Team (1965) in the south of Tantou Village, Wenchang Town, Chun'an County, Zhejiang Province. It was used by the preparation group of regional stratigraphic table of Zhejiang Province (1979). Wenchang Formation used by *Regional Geology of Zhejiang Province* (1989) referred to the former Wenchang Formation and the Yankou Formation collectively (including *Diplograptus bohemicus* and *Dalmanitina* cf. *mucronata* horizons and pebbled sandstone). Yankou Formation (including Zhangcunwu Formation and Yankou Formation) was adopted in the north Zhejiang. Late Ordovician Wenchang Formation was used in *Lithostratigraphy of Zhejiang Province* (1995) and northwest Zhejiang. In this Project, the name Wenchang Formation (O_3w) was still adopted according to the lithological characteristics of Shangangshang profile (PM002, Saoshe profile (PM008), and Zhuwukou profile (PM013) in Hanggai map sheet, and the geological observation traverse.

2.4.7.1 Lithostratigraphy

As the topmost Ordovician strata in the Area, Wenchang Formation (O_3w) is mainly distributed in the eastern part of Hanggai map sheet and the south-central part of Xianxia map sheet. In addition, it is sporadically exposed in the southeast and northwest corners of the Chuancun map sheet. The outcrop area is about 50.97 km^2, accounting for 4.01% of the bedrock area.

The characteristics of the lithology and lithologic association of Wenchang Formation (O_3w) are as follows. The lithology and lithologic association of the lower part: medium–thick laminated feldspathic quartz silty-fine sandstone interbedded with thin–medium laminated silty mudstone or their interbeds, and a small amount of grayish black micro-thin laminated graptolite-bearing silty mudstone. The lithology and lithologic association of the middle part: from bottom to top, 3-m-thick grayish siliceous nodule siltstone-bearing benthos fossils, and 8–10-m-thick black carbonaceous shale-bearing graptolites and sponge fossils. The lithology and lithologic association of the upper part: gray medium–thick laminated (blocky locally) medium-fine feldspathic quartz sandstone interbedded with gray–dark gray thin laminated graptolite-bearing silty mudstone. From bottom to top, the grain size: Rock clasts change from medium → fine → medium. The thickness of this formation is 363.04 m.

1. **Stratigraphic section**

The lithology of the lower and middle parts of Wenchang Formation (O_3w) is described by taking the example of the third member of Changwu Formation (O_3c^3)–the middle and lower parts of Wenchang Formation (O_3w) profile (Fig. 2.74) of Hanggai map sheet in Shangangshang, Hanggai Town, Anji County, Zhejiang Province. The details are as follows:

The upper part of Wenchang Formation Total thickness: >2.00 m

48. Khaki medium–thick laminated fine-silty sandstone owing to weathering. 2.00 m

——— Conformable contact ———

Special lithologic layer in the middle part of Wenchang Formation Total thickness: 11.76 m

47. Grayish black thin laminated carbonaceous shale, the thickness of a single layer: 2–6 cm generally, very thin horizontal bedding developing, 7–8 kinds of sponges produced, aboundant graptolite fossils obtained including: *Normalograptus persculptus* (Elles and Wood, 1907), *Normalograptus mirneyensis* (Obut and Sobolevskaya, 1967),*Normalograptus aff. indivisus* (Davies, 1929),*Normalograptus rhizinus* (Li and Yang, 1983), *Normalograptus madernii* (Koren and Mikhaylova, 1980), *Neodiplograptus* sp. nov. (= *Normalograptus* aff. *tamariscus* Nicholson, 1868), and *Normalograptus laciniosus* (Churkin and Carter, 1970). 8.77 m

46. Light grayish-yellow siltstone, interbedded with one siliceous mudstone nodule layer in the middle part with nodule size of 7–81 to 5–22 cm, rich in fossils such as brachiopodas, crinoid stems, trilobites and gastropods as well as *Aegiromena planissima* (Reed,1915) and *Mucronaspis* (Songxites) *wuning-ensis* (Lin). 3.09 m

——— Conformable contact ———

The lower part of Wenchang Formation Total thickness: 226.15 m

45. Grayish-yellow thick bedded–blocky quartz arkose, boulder clay visible locally, interbedded with a layer of silt with the thickness of about 1m in the middle-lower part. 30.88 m

44. Interbeds consisting of grayish yellow medium laminated quartz arkose and thin–medium laminated silty mudstone. The arkose: spherical weathering developing, the thickness of a single layer: 10–50 cm generally, argillaceous bands visible locally. The thickness of a single layer of the mudstone: 4–17 cm generally. 13.82 m

43. Grayish yellow thick laminated blocky lithic quartz sandstone, interbedded with 2 layers of silty mudstone with the thickness of a single layer of about 30 cm at the bottom. 32.67 m

42. Interbeds consisting of grayish yellow thin–medium laminated quartz arkose and silty mudstone. The arkose: globular weathering of φ =5–10cm developing, the thickness of a single layer: 8–23cm generally and 90cm locally. The silty mudstone: horizontal bedding developing, the thickness of a single layer: 1–30cm.

10.63 m

41. Weathering is brownish yellow mudstone with horizontal bedding, the thickness of a single layer: 2–5 cm thick and thin laminated siltstone interbedded with on top. 1.87 m

40. Interbeds consisting of brownish yellow thin–medium laminated quartz arkose and mudstone owing to weathering. The thickness of a single layer of the arkose: 8–25 cm. The mudstone: horizontal bedding developing, the thickness of a single layer: 2–12 cm. 6.95 m

39. Brownish yellow silty mudstone owing to weathering, interbedded with two layers of fine-silty sandstone upwards with respective thickness of 10 and 20 cm. The silty mudstone: horizontal bedding developing. 2.78 m

38. Two cycles constituting of gray thin–medium laminated, fine- to medium-grained, fine quartz arkose interbedded with mudstone. The respective thickness of a single layer of the arkose and the mudstone: 6–20 cm and 2–15cm. The ratio of the arkose to mudstone: ≈ 20:1. 19.27 m

37. Rhythmic interbeds consisting of caesious medium laminated quartz feldspathic silty-fine sandstone and thin–medium laminated mudstone. The sandstone: the thickness of a single layer: 12–40 cm, less obvious argillaceous micro-thin bands visible only on the top of local sandstone. The thickness of a single layer of the mudstone: 2–35 cm thick. 7.39 m

36. Interbeds consisting of ①caesious medium–thick laminated fine- to medium- grained quartz arkose and ②gray medium laminated fine sandstone and thin-to medium laminated mudstone, containing 8 cycles. The arkose: the thickness of a single layer: 10–115 cm. The fine sandstone; cross-bedding developing locally. The mudstone: the thickness of a single layer: 6–17 cm, horizontal bedding developing. 15.4 m

35. Gray thin–medium laminated quartz feldspathic fine sandstone, interbedded with silty mudstone. The fine sandstone: the thickness of a single layer: 10–17 cm generally and 70cm locally. The mudstone: the thickness of a single layer: 5–16 cm, tidal bedding and cross bedding developing. 9.77 m

34. Interbeds consisting of gray thin–medium laminated quartz arkose with thin laminated mudstone. The arkose: the thickness of a single layer: 7–18cm. The mudstone: containing a small amount of silt, no obvious bedding developing, the thickness of a single layer: 2–6 cm. 2.16 m

33. Caesious blocky quartz arkose. 2.54 m

32. Interbeds consisting of silty sandstone and silty mudstone are gray thin–medium laminated quartz feldspar silty sand. Silty-fine sandstone, the thickness of a single layer: 4–11cm thick, a small amount being 40cm. The silt in silty mudstone is fine banding with the thickness of a single layer: 2–20 cm. 2.22 m

31. Rhythmic interbeds consisting of caesious thick blocky–laminated quartz arkose, medium–thin laminated sandstone, and thin laminated silty mudstone, with a blocky–laminated sandstone as the upper part. The sandstone: the thickness of a single layer: 8–10cm generally and 150cm locally. The silty mudstone: the thickness of a single layer: 3–5cm. 8 rhythms totally. Fossils of chitinozoan *Belonechitina americana* (Taugourdeau, 1965) and the fossils of the flowing graptolite obtained: *Normalograptus extraordinarius* (Sobolevskaya 1974), *Normalograptus ojsuensis* (Koren et Mikhailova) (graptolite), and a small amount of *Normalograptus mirnyensis* (Obut et Sobolevskaya) and *Normalograptus normalis* (Lapworth). 27.36 m

30. Interbeds consisting of gray thin–medium laminated quartz arkose and silty mudstone. The arkose: the thickness of a single layer: 5–48 cm; The mudstone: containing many silty strips, which constitute flaser bedding, the thickness of a single layer: 1–8 cm generally and 12–20 cm locally. 2.58 m

29. Gray thick laminated quartz arkose, the thickness of a single layer: greater than 50 cm. 1.46 m

28. Gray medium–thick laminated quartz arkose intercalated with thin laminated silty mudstone. The arkose: medium–thick laminated in the lower part and thin–medium laminated in the upper part, the thickness of a single layer: 6–130 cm thick. The silty mudstone: the thickness of a single layer: 1–9 cm, the amount increasing upwards, fine-banded in the upper part. 8.18 m

27. Consisting of black carbonaceous mudstone, feldspathic quartz sandstone and silty mudstone, three cycles in total. The carbonaceous mudstone: graptolite visible. The lithic sandstone: containing syngenetic boulder clay locally, distributed in flat manner along the layer, length × width: (1–2) cm × (0.1–0.2) cm. The silty mudstone: fine strips of silt developing. Graptolite visible about 3m above the bottom of this layer. Chitinozoa *Belonechitina Americana* (Taugourdeau, 1965) and graptolite *Normalograptus extraordinarius* (Sobolevskaya, 1974) obtained. 7.77 m

26. Gray thick laminated quartz arkose, no sedimentary structure developing. 1.75 m

25. Interbeds consisting of thin–medium laminated sandstone and thin laminated mudstone, interbedded with a small amount of black thin laminated carbonaceous mudstone. The respective thickness of a single layer of the sandstone, the mudstone, and the carbonaceous mudstone: 6–22 cm, 1–7 cm, and 0.5–1 cm. 5.48 m

24. Interbeds consisting of dark gray thin–medium laminated carbonaceous mudstone and thin laminated quartz feldspathic fine sandstone and silty mudstone. The respective thickness of a single layer of the carbonaceous mudstone, the fine sandstone, and the silty mudstone: 2–19cm, 4–8cm, and 4–10cm. The upper part: interbeds consisting of caesious thin laminated silty mudstone and fine-silty sandstone, interbedded with carbonaceous mudstone, the respective thickness of a single layer of the sandstone and the mudstone: 2–7 cm and 2–6 cm. A large number of graptolite fossils visible in the black shale in the lower part of this layer including *Normalograptus extraordinarius* (Sobolevskaya, 1974), *Normalograptus ojsuensis* (Koren et Mikhailova), diplograptids gen. et sp. indet. *Neodiplograptus charis* (Mu et Ni 1983), and *Amplexograptus* sp. Indet. Chitinozoan fossil obtained such as *Belonechitina americana* (Taugourdeau, 1965)and *Eisenackitina songtaoensis* Chen et al. 2009. 2.91 m

--- Bottom of Hirnantian Stage

23. Gray thick laminated fine- to medium-grained quartz arkose, the thickness of a single layer: greater than 50 cm. 0.87 m

22. Rhythmic interbeds consisting of gray thin–medium laminated fine- to medium-grained quartz arkose and thin laminated mudstone. The arkose: the thickness of a single layer: 8–28 cm generally and 90cm locally. The mudstone: horizontal bedding developing, the thickness of a single layer: 2–10 cm. The ratio of the sandstone to the mudstone: (2–4):1 approximately. 6.25 m

21. Brownish yellow thick laminated fine- to medium-grained quartz arkose, intercalated with a layer of 5cm thick mudstone in the middle; horizontal bedding developing. 3.19 m

———— Conformable contact ————

The third member of Changwu Formation Total thickness: 161.39m

20. Three cycles of interbeds consisting of gray medium–thin laminated fine- to medium-grained and fine-grained quartz arkose and mudstone, the respective thickness of single layer of the arkose and the mudstone: 5–13cm and 2–14cm. The fossils of chitinozoan *Eisenackitina songtaoensis* Chen *et al.*, 2009 obtained. 9.39 m

19. Brownish yellow thin laminated feldspathic quartz fine-grained sandstone owing to weathering, intercalatede with mudstone. The sandstone: diagonal bedding developing, the thickness of a single layer: 5–10 cm. The mudstone: no obvious bedding developing, the thickness of a single layer: 1–3 cm, flaky owing to weathering. The ratio of the sandstone to the mudstone: (3–10):1 approximately. The top: medium laminated sandstone intercalated thin laminated mudstone, the respective thickness of a single layer of the sandstone and the mudstone: 10–20 cm and 2–5 cm. 29.88 m

18. The upper and lower parts of this layer: interbeds consisting of gray medium laminated feldspathic

quartz fine sandstone and dark gray medium laminated silty mudstone. The fine sandstone: diagonal bedding visible locally, the thickness of a single layer: 11–18 cm. The mudstone: no obvious horizontal bedding developing, the thickness of a single layer: 10–34 cm. The middle part of this layer: interbeds consisting of thin laminated sandstone and mudstone, the respective thickness of a single layer of the sandstone and the mudstone: 5–9 cm and 2–7 cm. The 2–3 cm thick dark gray shale 2m above the bottom of this layer producing the following graptolites: the species possibly in the lower part of *Paraorthograptus pacificus* zone including *Paraorthograptus pacificus* (Ruedemann), *Paraorthograptus angustus* Mu et Lee 1977, *Amplexograptus disjuctus yangtzensis* Mu et Lin, and *Climacograptus chiai* Mu, 1949 (= *Amplexograptus hubeiensis* Mu et al.), as well as common species in *Dicellograptus complexus* zone including *amplexograptid gen.* et sp. indet. and *diplograptid gen.* et sp. Indet. No *Tangyagraptus* and *Diceratograptus* visible. 10.61 m

17. Dark gray medium–thin laminated mudstone intercalated with thin laminated feldspathic quartz fine sandstone. The fine sandstone: less obvious normal graded bedding developing upwards, cross bedding developing on the top locally, the thickness of a single layer: 3–10 cm. The mudstone: less obvious horizontal bedding developing, the thickness of a single layer: 4–10 cm generally and 22–40 cm locally, thin laminated carbonaceous mudstone constituting the top locally. The ratio of the mudstone to the sandstone: (2–4):1 approximately. Carbonaceous mudstone producing *Amplexograptus disjuctus yangtzensis* Mu et Lin and *Paraorthograptus pacificus* (Ruedemann), corresponding to the lower part of *Paraorthograptus pacificus* zone of the Wufeng Formtion in Yangtze Region. 5.71 m

16. Gray medium laminated feldspathic quartz silty-fine sandstone, intercalated with dark gray thin laminated mudstone. The silty-fine sandstone: dominating the lower part, the thickness of a single layer: 12–18 cm. The mudstone: the thickness of a single layer: 2–3 cm. The ratio of the sandstone to the mudstone: 20:1 approximately and 2:1 at the top. 16.61 m

15. Interbeds consisting of dark gray medium laminated silty mudstone and thin laminated feldspathic quartz sandstone. The silty mudstone: horizontal bedding developing locally, the thickness of a single layer: 11–22 cm. The sandstone: sedimentary structure uncertain, the thickness of a single layer: 5–10 cm. The ratio of the mudstone to the sandstone: (2–4):1 approximately. A layer of 20 cm thick sandstone intercalating in the middle part of the layer. 4.92 m

14. Rhythmic interbeds consisting of dark gray thin–medium laminated siltstone and gray feldspathic quartz fine sandstone. The siltstone and the mudstone mainly consisting the lower part and the upper part of a single rhythm respectively. The siltstone: the thickness of a single layer: 5–15 cm. The mudstone: horizontal bedding developing, the thickness of a single layer: 5–30 cm. The ratio of the sandstone to the mudstone: 1:1.5 approximately. 19.02 m

13. Interbeds consisting of gray medium laminated feldspathic quartz silty-fine sandstone and thin laminated silty mudstone. The silty-fine sandstone: graded bedding and less obvious parallel bedding developing, the thickness of a single layer: 16–32 cm. The silty mudstone: blocky, no bedding developing, the thickness of a single layer: 5–10 cm. The ratio of the sandstone to the mudstone: (3–5):1 approximately. 24.98 m

12. Interbeds consisting of yellow medium–thin laminated mudstone and argillaceous siltstone owing to weathering. The respective thickness of a single layer of the argillaceous siltstone and the mudstone: 5–25 cm and 5–12 cm. The ratio of the mudstone to the siltstone: 1.2:1 approximately. 13.45 m

11. Interbeds consisting of gray medium–thin laminated silty mudstone and feldspathic quartz silty-fine

sandstone. The silty-fine sandstone: no bedding developing, diagonal bedding visible occasionally, the thickness of a single layer: 6–20 cm. The silty mudstone: composed of argillaceous matter and a small amount of silt, the thickness of a single layer: 5–20 cm. The ratio of the silty-fine sandstone and the mudstone: 1:1 approximately. 0.75 m

10. Dark gray thin laminated silty mudstone intercalated with micro laminated feldspathic quartz sandstone. The silty mudstone: no obvious bedding developing. The sandstone: diagonal bedding developing, the thickness of a single layer: 1–2 cm. The ratio of the silty mudstone and the siltstone: 20:1 approximately. 1.13 m

9. Interbeds consisting of gray thin laminated silty mudstone and feldspathic quartz silty-fine sandstone. The sandstone bottom is in a soothing wave shape, with occasional oblique bedding and the single layer of the rock layer is 2–5 cm thick. The ratio is about 1. 5.73 m

8. Gray thin laminated silty mudstone intercalated with micro laminated siltstone. The silty mudstone: the thickness of a single layer: 3–5 cm, very thin horizontal bedding developing. The siltstone: the thickness of a single layer: 1–2 cm. The ratio of the mudstone to the siltstone: (30–50):1 approximately. 2.64 m

7. Interbeds consisting of gray thin–medium laminated feldspathic quartz fine sandstone and thin–medium laminated silty mudstone. The respective thickness of the fine sandstone and the silty mudstone: 5–24 cm and 2–17 cm. The ratio of the fine sandstone to the silty mudstone: 2:1. 16.07 m

6. Interbeds consisting of gray medium laminated feldspathic quartz silty-fine sandstone and thin laminated mudstone. The silty-fine sandstone: the thickness of a single layer: 23–40 cm. The mudstone: no bedding developing, the thickness of a single layer: 6–10 cm. The ratio of the sandstone and the mudstone: 2–3:1. Differential weathering and costate landform shown in the surface, the siltstone protruding from the surface. 2.02 m

5. Interbeds consisting of gray thin–medium laminated argillaceous siltstone and thin laminated mudstone. The argillaceous siltstone: the thickness of single bed: 7–14 cm, undulating uneven bottom, groove and gravity casts developing. The thickness of a single layer of the mudstone: 3–10 cm. The ratio of the argillaceous siltstone to the mudstone: (1–2):1. 4.43 m

4. Interbeds consisting of gray thick–medium laminated feldspathic quartz silty-fine sandstone and thick–medium laminated mudstone. The thickness of a single layer of the silty-fine sandstone: 16–63 cm. The mudstone: horizontal bedding visible locally, the thickness of a single layer: 5-18cm. The ratio of the silty-fine sandstone to the mudstone: (2–3):1. 4.16 m

——————— Conformable contact ———————

The second member of Changwu Formation Total thickness: > 15.73 m

3. Gray - dark gray thin–medium laminated mudstone intercalated with thin laminated feldspathic quartz fine sandstone and black micro laminated carbonaceous shale. The mudstone: very thin horizontal bedding developing. The fine sandstone: the thickness of a single layer: 2–5 cm, The ratio of the mudstone to the sandstone: 3–5:1. The carbonaceous shale: the thickness of a single layer: 0.2–8 cm, rich in abundant graptolites of *Dicellograptus complexus* zone including *Amplexograptus disjunctus yangtzensis* Mu et Lin, *Amplexograptus suni* (Mu), *Appendispinograptus supernus* (Elles and Wood), *Appendispinograptus venustus* (Hsu), *Climacograptus hastatus* (Hall, 1902), *Yinograptus disjunctus* (Yin et Mu), *Paraplegmatograptus uniformis* Mu, and *Leptograptus extremus* Mu et Zhang *et al*. 7.41 m

2. Interbeds consisting of grayish thin–medium laminated silty mudstone and thin–medium laminated

feldspathic quartz silty-fine sandstone. The silty-fine sandstone: the thickness of a single layer: 6-12 cm. The silty mudstone: the thickness of a single layer: 7–20 cm, very thin horizontal bedding visible locally. The ratio of the silty-fine sandstone to the silty mudstone is 1:(2–3). 6.52 m

1. Interbeds consisting of gray thin–medium laminated silty mudstone and thin laminated feldspathic quartz silty-fine sandstone. The respective thickness of a single layer of the silty-fine sandstone and the silty mudstone: 2–7 cm and 8–28 cm. The ratio of themis 1:(2–3). 2.80 m

The lithology of the special lithological layer in the middle part of Wenchang Formation (O_3w) is described by taking the example of the profile of the special lithological layer in the middle of the Late Ordovician Wenchang Formation (O_3w) of Zhuwukou, Hanggai Town, Anji County, Zhejiang Province (Fig. 2.75). The details are as follows:

The upper part of Wenchang Formation (O_3w) Total thickness: >5.42 m

6. Gray thick laminated blocky fine-grained quartz arkose intercalated with thin laminated siltstone. The thickness of a single layer: 100–200 cm. The thickness of a single layer of the siltstone: 1–2 cm. 5.42 m

Special lithological layer of Wenchang Formation (O_3w) Thickness: 10.88 m

The second special lithological bed Thickness: 10.43 m

5. Black thin laminated carbonaceous siliceous silty mudstone, taxitic after weathering with a small amount of black mudstone remaining, very thin horizontal bedding developing, rich in graptolite fossils fragments.2.13 m

4. Gray medium laminated fine-grained quartz arkose. 0.36 m

3. Interbeds consisting of gray medium laminated fine-grained quartz arkose and thin–medium laminated silicon-bearing silty mudstone. The thickness of a single layer of the arkose: generally 10–20 cm. The silicon-bearing silty mudstone: the thickness of a single layer: generally 2–15 cm, very thin horizontal bedding developing. The two components featuring similar proportion. 0.80 m

2. Black thin laminated carbonaceous siliceous silty mudstone, very thin horizontal bedding developing, rich in graptolite fossils and sponges followed by a small amount of chitosan. 7.14 m

Graptolites: *Nomalograptus angustus* (Pemer)

Sudburigraptus? Similaris Chen (*in press*)

Chitinozoan: *Eisenackitina rectangularis Zaslavskaya* 1983

The first special lithological bed Total thickness: 0.45m

1. Grayish cataclastic siltstone after diagenesis, brachiopoda fragments visible. > 0.45 m

——— Conformable contact ———

Lower part of Wenchang Formation (O_3w) Total thickness: >5.00 m

0. Gray medium–thick laminated medium- to fine-grained feldspar lithic sandstone intercalated with thin laminated silty mudstone. The respective thickness of a single layer of the sandstone and the silty mudstone: 38–150 cm and 1–3 cm. 5.00 m

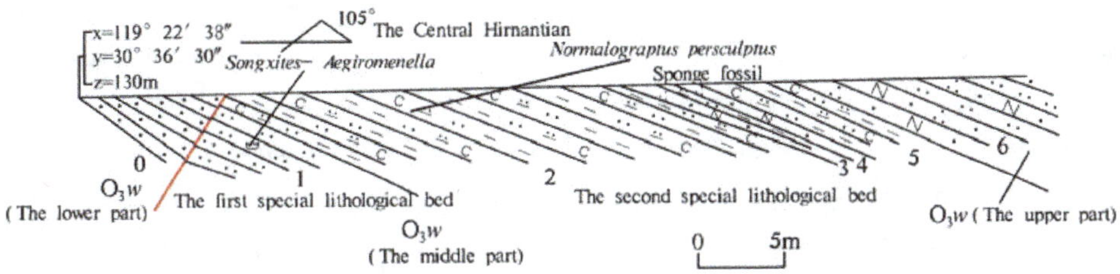

Fig. 2.75 Profile of the special lithological layer in the middle part of the Late Ordovician Wenchang Formation (O_3w) in Zhuwukou, Hanggai Town, Anji County, Zhejiang Province

The lithology of the upper part of Wenchang Formation (O_3w) is described by taking Ordovician Wenchang Formation (O_3w) profile (Fig. 2.76) of Hanggai map sheet in Saoshe Village, Hanggai Town, Anji County, Zhejiang Province. The details are as follows:

The first member of Xiaxiang Formation Total thickness: >64.20 m

65. Covering, black carbonaceous shale visible on the slopes about 200m to the southwest. 18.58 m

64. Gray medium–thick laminated feldspathic quartz fine sandstone. 32.99 m

63. Interbeds consisting of gray thin–medium laminated sand-bearing shale and black carbonaceous siliceous mudstone. The mudstone: the thickness of a single layer: 5–20 cm, very thin horizontal bedding developing. 6.69 m

62. Gray medium–thick laminated feldspathic quartz fine sandstone. The thickness of a single layer: 40–150 cm. One layer of medium laminated sandstone respectively intercalating in the top and bottom of the layer; 2–10 cm thick shale intercalating between the sandstone. 5.37 m

61. Interbeds consisting of black thin laminated carbonaceous silty mudstone and grayish shale. The carbonaceous silty mudstone: the thickness of a single layer: 1–3 cm, abundant fossils of graptolite obtained includes: *Akidograptus ascensus* Davies 1929, *Normalograptus laciniosus* (Churkin and Carter, 1970), *Normalograptus rhizinus* (Li and Yang, 1983), and *Nomalograptus ojsuensis*, *Normalograptus angustus* (Perner, 1895). 0.57 m

--- Ordovician - Silurian Boundary

———————— Conformable contact ————————

The upper part of Wenchang Formation Total thickness: 125.03m

60. Gray blocky–laminated feldspathic quartz sandstone intercalated with micro laminated shale. A single layer of sandstone: the thickness: >200cm, less obvious parallel bedding developing, intercalated with a layer of 18cm thick black shale in the middle part. The shale: very thin horizontal bedding developing. 16.19 m

59. Gray medium laminated feldspathic quartz fine sandstone intercalated with micro–thin laminated siltstone or silty mudstone. The thickness of a single layer of the fine sandstone: 10–40 cm. The siltstone or silty mudstone: very thin horizontal bedding developing, the thickness of a single layer: 3–30 mm. 9.14 m

58. Dark gray thin laminated silty mudstone containing a small amount of micro laminated carbonaceous

matter and pyrite, very thin horizontal bedding developing. The fossils of the graptolites of *Normalograptus persculptus* zone at the end of Hirnantian obtained includes: *Normalograptus persculptus* (Elles & Wood), *Normalograptus avitus* (Davies, 1929),*Normalograptus ojsuensis* (Koren and Mikhailova, 1980),*Normalograptus rhizinus* (Li and Yang, 1983),*Normalograptus angustus* (Perner, 1895,*Normalograptus* cf. *minor* Huang 1982, *Normalograptus laciniosus*, *Normalograptus* (Koren and Melchin, 2000), *Glyptograptus* aff. *tamariscus* (Nicholson, 1868) sensu Koren and Melchin 2000, *Normalograptus zhui* (Yang, 1964), *Sudburigraptus? angustifolius* Chen et Lin 1978, *Normalograptus normalis* Lapworth 1877 and *Normalograptus persculptus, Atavograptus* sp. nov. etc. No *Akidograptus ascensu* is visible owing to abundant N. *avitus*. The age is belonging to the final Hirnantian *Normalograptus persculptus* graptolite belt. 2.06 m

57. Gray–purple-gray thick blocky–laminated feldspathic quartz fine sandstone intercalated with medium–thin laminated siltstone or silty mudstone. The thickness of a single layer of the fine sandstone: 100–200 cm. The siltstone: the thickness of a single layer: 3–15 cm, very thin horizontal bedding developing. 12.05 m

56. Interbeds consisting of gray thin–medium laminated feldspathic quartz fine sandstone and black micro laminated carbonaceous shale. Parallel bedding developing in the fine sandstone. The shale: mostly 1–2 mm thick, containing carbon. 1.41 m

55. Gray blocky–laminated feldspathic quartz silty-fine sandstone, the thickness of a single layer: > 1 m. 4.66 m

54. Interbeds consisting of gray thin laminated fine-silty sandstone and micro laminated black carbonaceous siliceous mudstone. The fine-silty sandstone: very thin horizontal bedding developing, carbon and pyrite cast distributed along the bedding. The mudstone: containing carbonized graptolites, the following fossils obtained: *Normalograptus mirneyensis* Obut et Sobolevskaya and *Normalograptus laciniosus* Churkin et Carter, *Sudburigraptus angustifolius* Chen et Lin. 6.11 m

53. Gray thick laminated fine sandstone interbedded with silicon-bearing silty mudstone. Less obvious parallel bedding developing in the sandstone. 1.62 m

52. Gray thin laminated feldspathic quartz fine sandstone intercalated with micro laminated carbonaceous shale. The fine sandstone: the thickness of a single layer: 1–3 cm thick, very thin horizontal bedding developing, a small amount of pyrite casts distributed along the bedding surface. The carbonaceous shale: containing graptolite fossil fragments, the thickness of a single layer: 1–3 mm. 1.44 m

51. Gray blocky–laminated feldspathic quartz sandstone intercalated with micro–thin laminated mudstone. The sandstone: less obvious parallel bedding developing, the thickness of a single layer: > 2 m. 43.02 m

50. Gray medium laminated feldspathic quartz sandstone with the thickness of a single layer of 30–40 cm. A single layer of the sandstone: intercalated with thin strips and lenses of siltstone 10–20 cm above the bottom, then the sandstone becoming pure upwards. 3.00 m

49. Interbeds consisting of gray thin–medium laminated argillaceous siltstone, and black shale bearing thin laminated siliceous carbonaceous matter and gray siltstone. The argillaceous siltstone: the thickness of a single layer: 5–15 cm, horizontal bedding developing. The shale: the thickness of a single layer: 3–10 cm, rich in graptolites, horizontal bedding developing. *Monograptid* gen. & sp., *Normalograptus* sp. obtained. 2.06 m

48. Gray thick laminated feldspathic quartz fine-silty sandstone intercalated with gray - dark gray thin laminated argillaceous siltstone and shale. The fine-silty sandstone: the thickness of a single layer: 50–150 cm, smooth layer surface. The argillaceous siltstone: the thickness of a single layer: 0.5–5 cm, very thin horizontal

bedding developing. The fossils of the following graptolites obtained in the shale at the bottom: *Normalograptus avitus* (Davies), *Normalograptus* cf. *avitus* (Davies 1929, *Normalograptus* cf. *laciniosus* (Churkin and Carter), and *Normalograptus mirneyensis* (Obut and Sobolevskaya). The fossils of the following graptolites obtained in the shale on top: Normalograptus rhizinus (Li and Yang), Normalograptus avitus (Davies, 1929), Normalograptus cf. avitus (Davies), Normalograptus laciniosus (Churkin & Carter), and Normalograptus mirneyensis (Obut & Sobolevskaya), Neodiplograptus sp. 22.27 m

——— Conformable contact ———

Special lithological layer in the middle part of Wenchang Formation Total thickness: >7.28 m

47. Grayish-black thin laminated carbonaceous shale, in which micro-grain horizontal bedding is developed, rich in graptolites. 7.28 m

2. **Lithological characteristics**

The lithology of Wenchang Formation (O_3w) is mainly characterized by medium–thick laminated feldspathic quartz sandstone, feldspathic quartz fine-silty sandstone, followed by thin–medium laminated silty mudstone, thin laminated argillaceous siltstone, silty siltstone, and carbonaceous mudstone.

(1) The feldspathic quartz sandstone: gray, composition: quartz (45–55%), feldspar (10–25%), and detritus (10–25%), subangular to sub-well rounded, grain size: 0.1-0.5 mm, argillaceous cementation.
(2) The feldspathic quartz fine-silty sandstone: gray, the main components: quartz and feldspar (55–80%), subangular to sub-well rounded, grain size: 0.05–0.1 mm generally and 0.1–0.2 mm locally, argillaceous cementation, the thickness of a single layer: 10–40 cm and 50–150 cm, layer surface: even.
(3) The argillaceous siltstone: gray–dark gray, composition: quartz and feldspar (60–75%) and argillaceous matter (25–40%), thickness of a single layer: 0.5–8 cm, horizontal bedding developing generally, cross-bedding developing locally.
(4) The silty mudstone: grayish gray; composition: argillaceous matter (40–60%) and a small amount of silt (20–40%); composition of silt: feldspar and quartz; grain size: 0.01–0.05 mm, subangular to sub-well rounded; horizontal bedding developing.
(5) The carbonaceous shale: black, composition: argillaceous matter (75–80%), silt (10–15%), and carbonaceous matter (5–15%), the thickness of a single layer: 2–6 cm generally, very thin horizontal bedding developing, rich in graptolite fossils and sponge.

3. **Basic sequences**

There are five types of basic sequences developing in Wenchang Formation (O_3w) (Fig. 2.77).

(1) Composed of ① gray thick laminated feldspathic quartz sandstone and ② rhythmic interbeds of thin–medium laminated silty mudstone, intercalated with ③ thin–medium laminated fine-silty sandstone. The sandstone: mainly composed of quartz and feldspar, low maturity, subangular to subrounded, the thickness of a single layer: 50–80 cm generally. The fine-silty sandstone and silty mudstone: 4–12 cm thick generally and 15–20 cm thick locally per single bed, horizontal bedding generally or micro-fine-silty strips developing in silty mudstone. The thickness of the thick laminated sandstones is 2–5 times that of the rhythmic layers generally. Therefore, the basic sequences of this type are common basic sequences.
(2) Distributed in the middle-lower and upper parts in Wenchang Formation (O_3w), composed of the interbeds of ① gray thin–medium laminated silty-fine sandstone and ② thin laminated silty mudstone, intercalated with

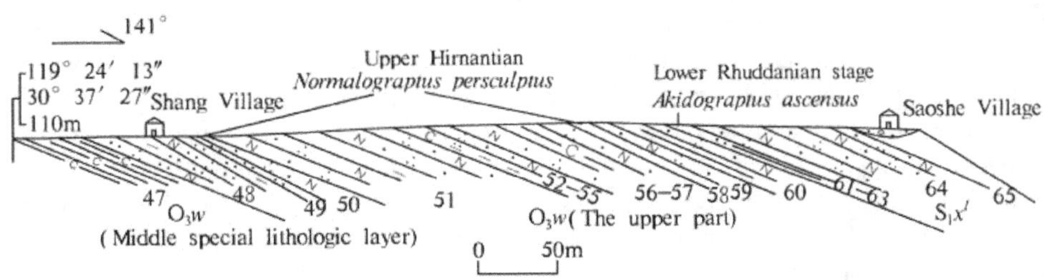

Fig. 2.76 Profile of the upper part of Ordovician Wenchang Formation (O_3w) in Saoshe Village, Hanggai Town, Anji County, Zhejiang Province

Fig. 2.77 Basic sequences of Wenchang Formation (O_3w)

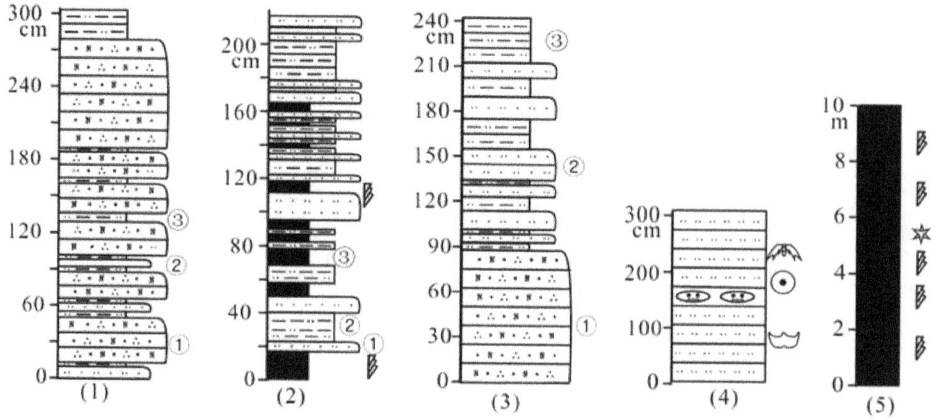

black thin laminated silicon-bearing carbonaceous shale. The silty-fine sandstone: mainly composed of feldspar and quartz (60–85%), and argillaceous matter (15–40%), the thickness of a single layer: 5–15 cm. The carbonaceous shale: composed of argillaceous matter (70–75%) and a small amount of carbonaceous matter (10–15%) and silt (10%), the thickness of a single layer: 3–10 cm, rich in graptolites, very thin horizontal bedding developing. The ratio of the silty-fine sandstone and the mudstone is 1–5:1 approximately.

(3) Developing in the lower and middle-upper parts of Wenchang Formation (O_3w), composed of ① gray thick blocky–laminated medium-grained feldspathic quartz sandstone, and rhythmite consisting of ② thin–medium laminated fine-silty sandstone and ③thin–medium laminated silty mudstone. The sandstone: mainly composed of quartz and feldspar, low maturity, subangular to subrounded. The thickness of single layer of medium-grained sandstone: 50–200 cm generally. The fine-silty sandstone and silty mudstone: the clasts mainly composed of feldspar and quartz, the thickness of single bed: 2–15 cm generally, horizontal bedding or less obvious bedding developing in silty mudstone. The thickness of the thick blocky–laminated sandstone is generally 3–10 times that of the rhythmic bed.

(4) Developing in the middle part of Wenchang Formation (O_3w), composed of ① grayish medium laminated siltstone and ② silicon-bearing mudstone nodule layers, the main components: feldspar, quartz silt followed by a small amount of argillaceous matter, rich in benthic organisms including trilobites, deepwater brachiopoda, pteropods, and gastropods. The nodules: (5–9) × (15–25) cm in size, circle structures developing. Therefore, the basic sequences of this type are of non-cyclic monotonous type.

(5) Developing in the middle part of Wenchang Formation (O_3w), composed of black silicon-bearing carbonaceous mudstone, rich in graptolites and sponge fossils, belonging to monotonous basic sequences.

The main sediments of Wenchang Formation (O_3w) are thick blocky–laminated feldspathic quartz silt and fine sand, followed by thin–medium laminated silt and fine sand, all of which are terrigenous clasts. Part of the silt is rich in benthos fossils. All these indicate clastic sedimentary rock facies of shallow shelf in weak oxidization environment. The graptolite-bearing carbonaceous shale in the lower and upper parts of Wenchang Formation (O_3w) features positive excursion of stable isotopes of organic carbon and monotonous graptolite species, indicating subcompensational basin facies of bathyal stagnant reduction environment in Hirnantian global glacial period. The 8–10-m-thick black carbonaceous graptolite-bearing shale in the middle part of Wenchang Formation (O_3w) features no positive excursion of the stable isotopes of organic carbon and many kinds of sponges and graptolites, indicating subcompensational basin facies of bathyal reduction environment in normal climate.

2.4.7.2 Sequence Stratigraphy

According to the characteristics of the lithology, paleontology, lithofacies association, and sequence boundary of Shangangshang profile (PM002) and Saoshe profile (PM008) in Hanggai map sheet, there is one third-order sequence Sq24 (Fig. 2.78) in Wenchang Formation (O_3w), which belongs to the second-order sequence SS8.

Third-order sequence Sq24

It includes transgressive systems tract (TST), starved section (CS), and highstand systems tract (HST). The bottom of the sequence is the conformity interface between thick laminated sandstone of Wenchang Formation and the silty mudstone of the third member of Changwu Formation. The top of the sequence is the conformity interface between thick laminated sandstone of Wenchang Formation and carbonaceous silty mudstone of Xiaxiang Formation. The bottom and top are both type-II boundaries.

TST is located in the lower part of Wenchang Formation, with the thick blocky–laminated feldspathic quartz

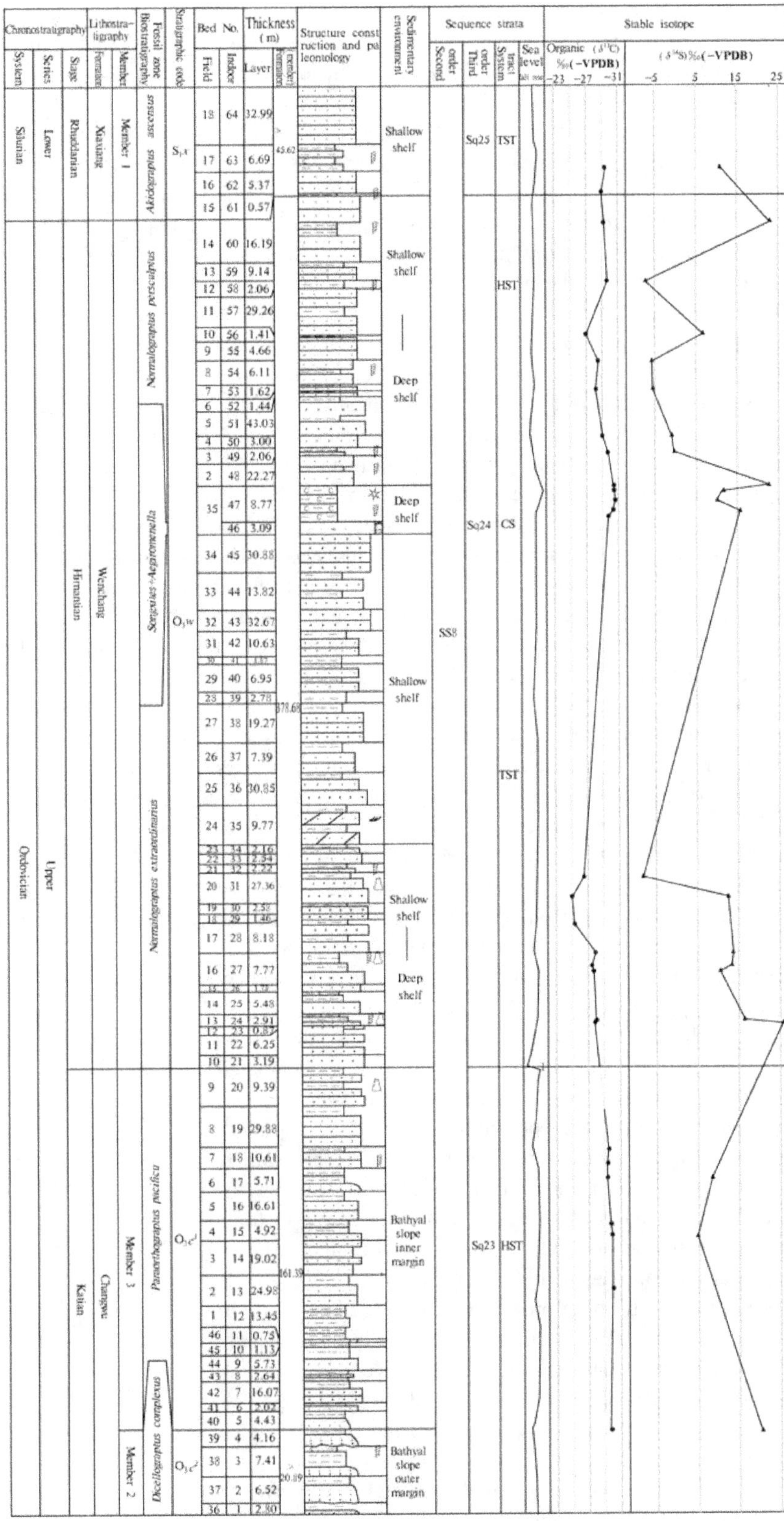

Fig. 2.78 Sequence of stratigraphic framework of Ordovician Wenchang Formation in the area

sandstone as its main component and the medium laminated fine-silty sandstone and carboniferous mudstone-bearing graptolite and chitinozoan, which are distributed alternatively as its minor components. The main component is of the clastic sedimentary rock facies of shallow shelf with abundant sources of terrigenous clasts. The minor components are of deepwater mudstone facies. All these indicate the two kinds of sedimentary environment. The positive drift of $\delta^{13}C$‰ in the carbonaceous mudstone indicates that the Area was affected by the global glacial events in the Late Ordovician. The monotonous genera and species of the graptolites obtained in the multilayers of the shale also indicate that glacier climate is unfavorable for the evolution of the genera and species of organisms. TST of Sq24 features frequent oscillation of the sea level owing to the glaciation.

CS is located in the middle part of Wenchang Formation and consists of siltstone and carbonaceous mudstone, belonging to the shallow shelf–deep shelf facies of siliceous-argillaceous matter-bearing terrigenous clasts. The lower part of CS is composed of grayish siltstone intercalated with a layer of siliceous mudstone nodules. It is rich in the fossils of brachiopoda, trilobites, cephalopods, and crinoid but features monotonous genera and species and small figure of the organisms, indicating the sedimentation above the oxidation zone of shallow shelf. The upper part of the CS is composed of carbonaceous silty mudstone and contains abundant graptolites and well-preserved sponges of 7–8 genera and species in the middle part. Furthermore, the carbonaceous shale in the upper part of CS features no positive excursion of $\delta^{13}C$‰. All these indicate the mudstone facies between the oxidation and reduction zones of deep shelf. Therefore, the CS indicates that there was a rapid sea level rise and the maximum transgression reached during the interglacial period, when the low-velocity sedimentary strata were strongly scarce.

HST is located in the upper part of Wenchang Formation and mainly composed of medium–blocky–laminated fine-silty sandstone, thin–medium laminated siltstone, and carbonaceous mudstone. The mudstone contains graptolite, with horizontal bedding developing. HST features similar rock association with the lower part of Wenchang Formation. Besides, the carbonaceous mudstone also features positive excursion of $\delta^{13}C$‰. Therefore, HST is similar to the lower part of Wenchang Formation.

To sum up, the third-order sequence (Sq24) formed in shallow shelf–the deep shelf. It was controlled by the Hirnantian glacier event and the crustal movement, and the product is subjected to both climate and crustal stress field. Furthermore, it is a sequence of aggradation type.

2.4.7.3 Biostratigraphy and Chronostratigraphy

Wenchang Formation was deposited during the Late Ordovician Hirnantian global glacial period and experienced two major glacial periods and one interglacial period. As a result, a different biocenosis formed. In this Project, two fossil zones and one shellfish fauna were identified, i.e., *Normalograptus extraordinarius* graptolite zone, *Songxites–Aegiromenella* fauna, and *Normalograptus persculptus* graptolite zone from bottom to top. Among them, chitinozoan fossils were found in *Normalograptus extraordinarius* graptolite zone, no graptolite was discovered in *Songxites–Aegiromenella* fauna, and sponge fossils were found in *Normalograptus persculptus* graptolite zone.

1. ***Normalograptus extraordinarius* zone**

Normalograptus extraordinarius graptolite zone is located in the lower part of Wenchang Formation. The large fossils and microfossils obtained are graptolite fossils and chitinozoan fossils, respectively, and they are produced in the thin laminated black carbonaceous shale. The graptolite fossils obtained include main fossils of *Normalograptus extraordinarius* (Sobolevskaya 1974) (Figs. 2.79 and 2.80) and associated graptolites including *Normalograptus ojsuensis* (Koren and Mikhailova, 1980), *Normalograptus mirnyensis* (Obut and Sobolevskaya, 1967), *Normalograptus normalis* (Lapworth), *Neodiplograptus charis* (Mu and Ni 1983), *Paraclimacograptus innotatus, and Normalograptus laciniosus* (Churkin and Carter 1970). Among them, *Normalograptus* cf. *persculptus* (Elles and Wood, 1907) is small in figure. The chitinozoan fossils obtained include

Fig. 2.79 *Normalograptus extraordinarius* graptolite (Sobolevskaya 1974)

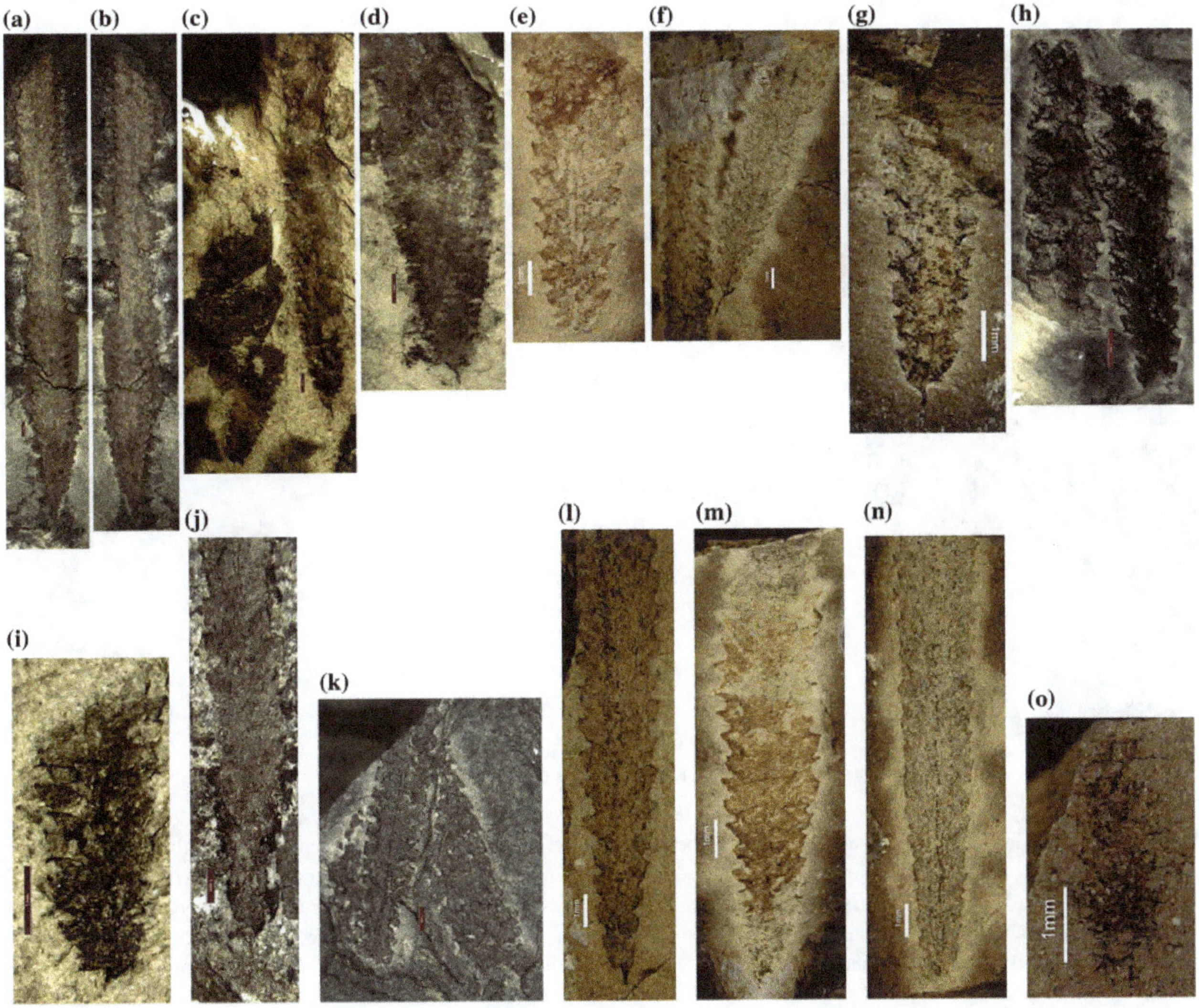

Fig. 2.80 *Normalograptus extraordinarius* graptolite zone in the lower part of Wenchang Formation (O_3w): **a** *Normalograptus extraordinarius* (Sobolevskaya 1974) (Pm002-20-5a); **b** *Normalograptus extraordinarius* (Sobolevskaya 1974) (Pm002-20-6-1a); **c** *Normalograptus extraordinarius* (Sobolevskaya 1974) (Pm002-20-13-3); **d** *Normalograptus extraordinarius* (Sobolevskaya, 1974) (Pm002-20-18-3); **e** *Normalograptus extraordinarius* (Sobolevskaya, 1974) (Pm002-wh 13-11-1a); **f** *Normalograptus extraordinarius* (Sobolevskaya 1974) (Pm002-wh13-1); **g** *Normalograptus ojsuensis* (Koren et Mikhailova) (Pm002-wh13-15); **h** *Normalograptus ojsuensis* (Koren et Mikhailova) (Pm002-20-8-4a); **i** *Normalograptus ojsuensis* (Koren et Mikhailova) (Pm002-20-23); **j** *Normalograptus mirnyensis* (Obut et Sobolevskaya) (Pm002-20-8-2); **k** *Normalograptus normalis* (Lapworth) (Pm002-20-26-2); **l** *Neodiplograptus charis* (Mu et Ni, 1983) (Pm002-wh13-45); **m** *Neodiplograptus charis* (Mu et Ni,1983) (Pm002-wh13-71); **n** *Neodiplograptus charis* (Mu et Ni, 1983) (Pm002- wh13-60); **o** *Paraothograptus pacificus* (Ruedemann) (Pm002-wh13-50)

Belonechitina americana (Taugourdeau 1965) and *Eisenackitina songtaoensis* Chen et al., which belongs to *Eisenackitina songtaoensis* chitinozoan zone. Since the chitinozoan zone features a time span, it can only be used for reference in chronostratigraphy.

2. ***Songxites–Aegiromenella* fauna**

Songxites–Aegiromenella fauna is located in the first special lithological bed in the middle part of Wenchang Formation, and the large fossils obtained are all the fossils of benthic shellfish including *Mucronaspis* (*Songxites*) *wuningensis* (Lin) (Fig. 2.81d, e) and *Aegiromena* (*Aegiromenella*) *planissima* (Reed 1915) (Fig. 2.81a, b, c), as well as gastropod and crinoid stem. The age of this fauna is between the ages of *Normalograptus extraordinarius* zone and *Normalograptus persculptus* graptolite zone. This fauna is called Hernantabean faunain in the Upper Yangtze Region and formed in the age of the lower part of *Normalograptus persculptus* graptolite zone. This fauna corresponds to the

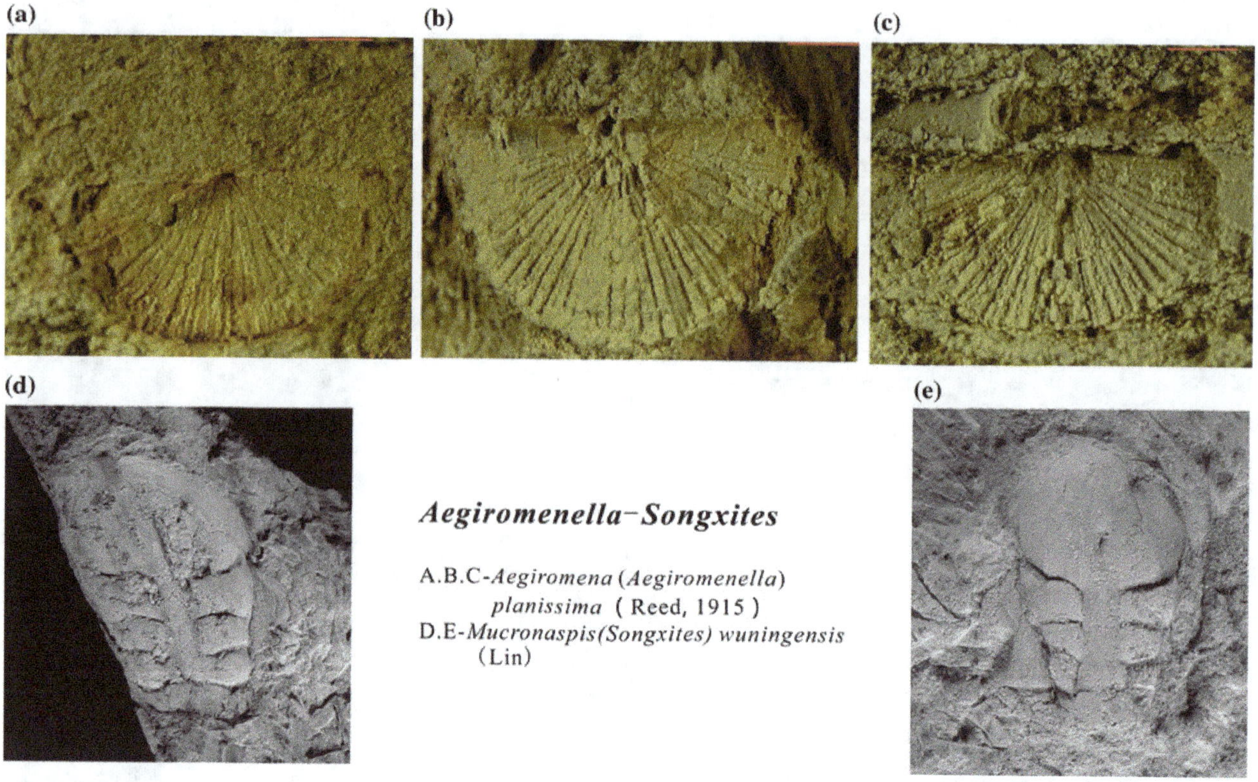

Fig. 2.81 *Songxites–Aegiromenella* fauna in the first special lithological bed in the middle part of Wenchang Formation (O_3w)

Guanjianqiao bed in the Wangjiawan profile of Hubei Province and is the product of the life explosion during the sea level rise period in Hirnantian interglacial period.

3. ***Normalograptus persculptus*** **graptolite zone**

Normalograptus persculptus graptolite zone is located in the upper part and the second special lithological bed in the middle part of Wenchang Formation, and the large fossils obtained are mainly graptolite fossils, which are all produced in the micro-thin laminated black carbonaceous shale. The 10-m-thick black shale in the second special lithological bed in the middle part of Wenchang Formation features the most abundant fossils including the main graptolite fossils of *Normalograptus persculptus* (Elles and Wood E M 1907, Fig. 2.82a) and associated graptolites such as *Normalograptus parvulus* (Lapworth), *Normalograptus laciniosus* (Churkin and Carter 1970), *Normalograptus angustus* (Perner 1895), *Neodiplograptus charis* (Mu et Ni 1983), *Normalograptus rhizinus* (Li and Yang 1983), *Neodiplograptus shanchongensis* (Li 1984), *Normalograptus aff. indivisus* (Davies 1929), *Normalograptus mirneyensis* (Obut and Sobolevskaya 1967), *Normalograptus zhui* (Yang 1964), *Normalograptus* aff. *tamariscus Nicholson*, 1868, *Normalograptus acceptus* (Koren and Mikhaylova 1980), "*Glyptograptus*" *jerini* Koren and Melchin 2000, and *Normalograptus madernii* (Koren and Mikhaylova 1980) (Figs. 2.82 and 2.83). The genera and species of the fossils decrease upward. Especially in carbonaceous shale with a positive excursion of $\delta^{13}C$‰, the graptolite fossils feature monotonous genera and species and increased individual figure. This may be affected by glaciation.

4. **First-ever global discovery of fossil sponges of Late Ordovician**

The fossils of the sponge of multiple genera and species were first discovered (Fig. 2.84) in the Late Ordovician Wenchang Formation in the Area. This is of great significance in the following three aspects. (1) They are the ones firstly discovered in this period. (2) They are of the Burgess Shale type, and the sponges of this type were previously believed to have already been extinct during the first main extinction event in the early stage of the Late Ordovician Hirnantian. The sponge in the Area developed together with the *Normalograptus persculptus* graptolite zone at the end of the Hirnantian Stage, indicating that the sponge extended to the late stage of the Hirnantian Stage. (3) The sponge fossils should have developed in an oxygen-rich environment, while the sponge in the Area developed in an anoxic

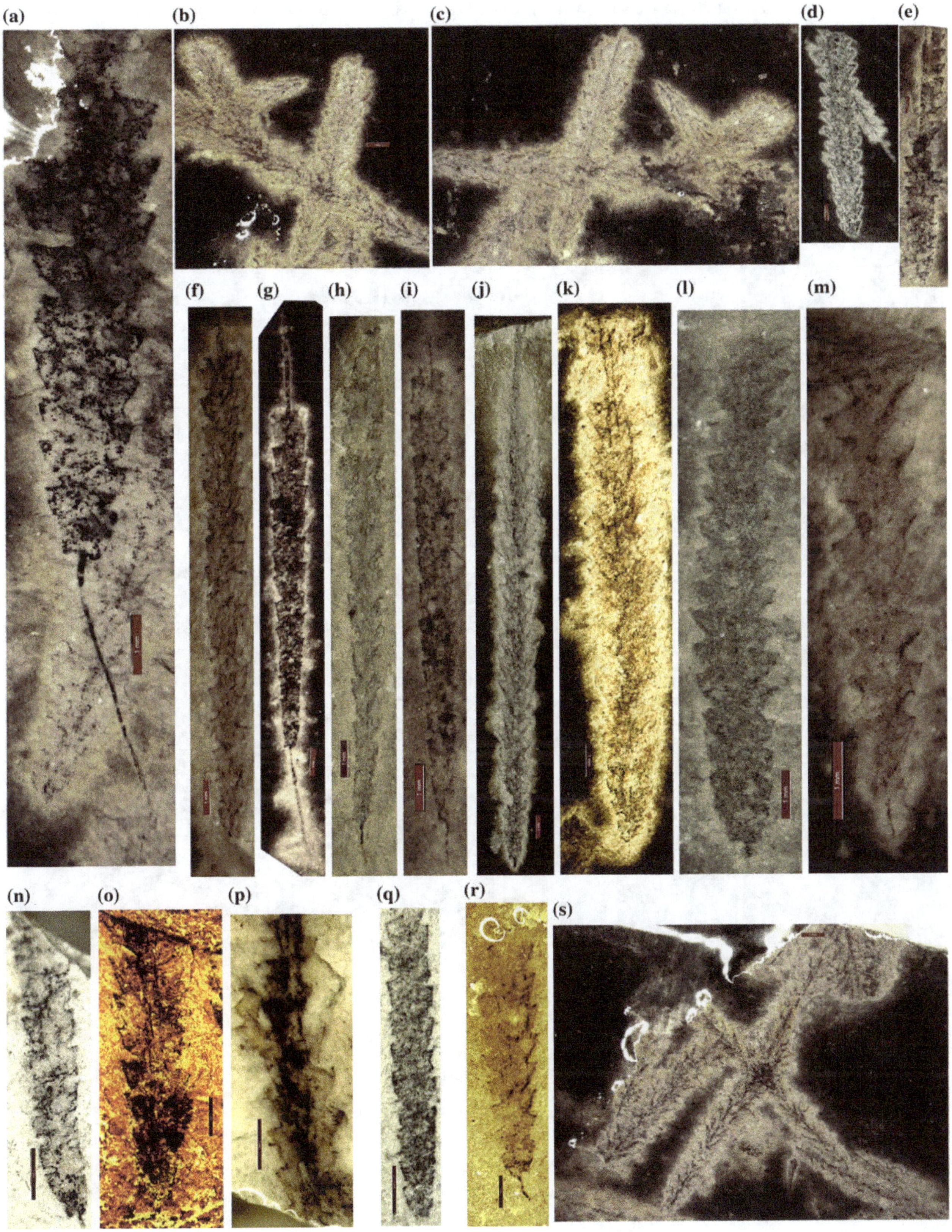
(a)
(b)
(c)
(d)
(e)
(f)
(g)
(h)
(i)
(j)
(k)
(l)
(m)
(n)
(o)
(p)
(q)
(r)
(s)

Fig. 2.82 *Normalograptus persculptus* graptolite zone in the middle and upper parts of Wenchang Formation (O_3w): **a, b, c** *Normalograptus persculptus* (Elles and Wood 1907) (Pm002-Hb35-4-2b, 2d, 3b); **d, e** *Normalograptus laciniosus* (Churkin and Carter 1970) (Pm002-35-4-4, -4-3a); **f, g** *Normalograptus rhizinus* (Li et Yang 1983) (Pm002-35-4-9-1, -4-4a); **h, i** *Normalograptus angustus* (Perner 1895) (Pm002-35-4-5, -4-13-1); **j**, P-*Normalograptus* aff. *tamariscus Nicholson* (1868 Pm002-35-4-7, -15a); **k** *Neodiplograptus charis* (Mu et Ni 1983) (Pm002-35-4-8); **l** *Neodiplograptus shanchongensis* (Li 1984) (Pm002-35-4-11-1); **m** *Normalograptus* aff. *indivisus* (Davies 1929) (Pm002-35-4-12); **n** *Normalograptus mirneyensis* (Obut and Sobolevskaya 1967) (Pm002-35x-2a); **o** *Normalograptus zhui* (Yang 964) (Pm002-35x-10a); **p** *Normalograptus acceptus* (Koren and Mikhaylova 1980) (Pm002-35x-16a); **q** "*Glyptograptus*" jerini Koren et Melchin (2000) (Pm002-35x-21a); **r** *Normalograptus parvulus* (Lapworth) (Pm002-Hb35-4-2a1); scale length: 1 mm

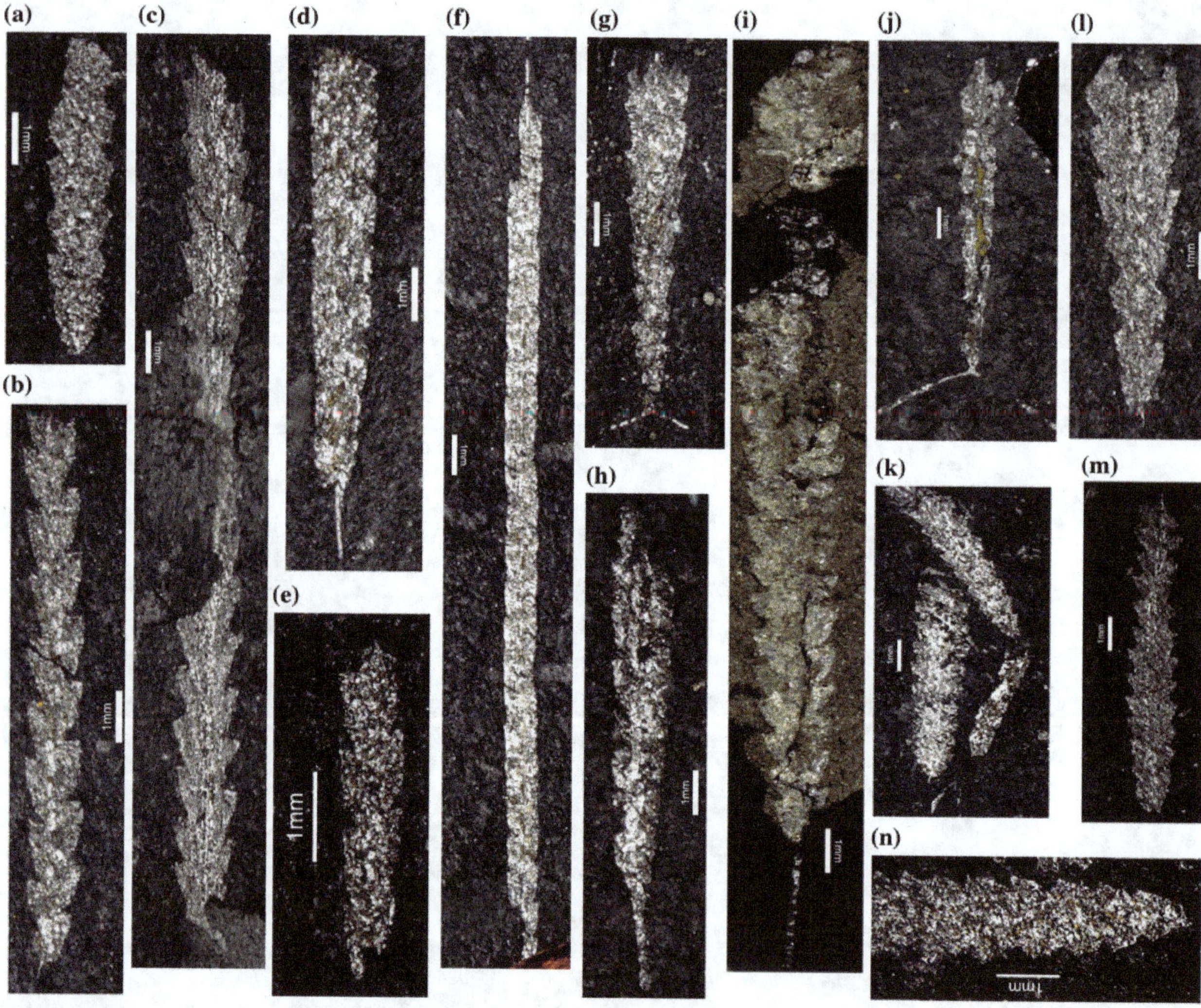

Fig. 2.83 *Normalograptus persculptus graptolite* zone in the upper part of Wenchang Formation (O_3w): **a** *Normalograptus* (Koren and Melchin 2000) (Pm008-Hb12-9); **b, c** *Glyptograptus* aff. *tamariscus* (Nicholson 1868) sensu Koren and Melchin 2000 (Pm008-Hb12-15, 12a); **d, e, f** *Normalograptus angustus* (Perner 1895) (Pm008-Hb12-13, 19, 14a); **g** *Normalograptus* cf. *minor* (Huang 1982) (Pm008-Hb12-11a); **h** *Normalograptus rhizinus* (Li and Yang 1983) (Pm008-Hb12-7); **i** *Normalograptus ojsuensis* (Koren and Mikhaylova in Apollonov et al. 1980) (Pm008-Hb12-32a); **j** *Normalograptus zhui* (Yang 1964) (Pm008-Hb12-23); **k, l** *Normalograptus avitus* (Davies 1929) (Pm008-Hb12-35a Volume 1, 26); *Normalograptus normalis* (Lapworth 1877) (Pm008-Hb12-35a Volume 2); **m** *Normalograptus laciniosus* (Pm008-Hb12-38b); **n** *Normalograptus persculptus* (Pm008-Hb12-30b)

environment such as carbonaceous shale, indicating that the Area was in a very special paleogeographic environment and thus significant in the research of paleoecology, biological paleogeography, and evolutionary paleontology.

To sum up, the age range of Wenchang Formation is basically consistent with that of Hirnantian Stage. The Hirnantian global glaciation exposed huge impact on organisms and caused two significant events of biological extinction,

Fig. 2.84 Sponge fossils of *Normalograptus persculptus* zone in the middle part of Wenchang Formation (O_3w)

i.e., the extinction of graptolite fauna of Katian *Paraothograptus pacificus* zone and the extinction of the vast majority of graptolites of *Normalograptus extraordinarius* zone. *Songxites–Aegiromenella* fauna was also extinct.

2.4.7.4 Characteristics of Stable Isotope

A total of 29 mudstone (shale) samples were collected from Shangangshang profile (PM002), the Saoshe profile (PM008), and Zhuwukou profile (PM013) to conduct whole-rock stable isotope analysis of $\delta^{13}C$ and $\delta^{34}S$.

In the lower part of the Wenchang Formation, the $\delta^{13}C$ value is from −28.09 to −24.96‰ with an average of −26.99‰, and the $\delta^{34}S$ value is −2.1‰–25.7‰ with an average of 12.83‰. Compared with the $\delta^{13}C$ value (−35 to 30‰) and $\delta^{34}S$ value (−10 to 10‰) of normal argillaceous shale, they feature weak positive excursion, indicating shallow shelf sedimentary environment. In addition, the $\delta^{13}C$ value and the $\delta^{34}S$ value of the samples PM002C17 and PM002C20-1 feature the relatively obvious positive excursion, proving that the Area may experience a short glacial age in the Early Hirnantian. The reason is that a large number of organisms (including plants, animals, and microorganisms) died and were buried and deposited in the Area in the short cooled glacial environment, leading to the enrichment of $\delta^{13}C$ in the sediments and highlighting the positive anomaly. This anomaly is completely comparable with the Wangjiawan profile in Yichang City, Hubei Province, of the same period.

In the middle part of Wenchang Formation of Zhuwukou profile, the $\delta^{13}C$ value is from −31.00‰ to 30.76‰ with an average of −30.89‰, and the $\delta^{34}S$ value is 5.5‰–13.7‰ with an average of 10.36‰. In the middle part of the Wenchang Formation of Shangangshang profile and Saoshe profile, the $\delta^{13}C$ value is from −30.64 to −29.71‰ with an average of −30.22‰, and the $\delta^{34}S$ value is from 9.3 to −22.7‰ with an average of 14.92‰. Compared with the values of normal argillaceous shale, they feature weak negative excursion. The possible reason is that after the early Wenchang period, the climate in the Area gradually warmed up, the Area entered into a short interglacial period, and the sedimentary environment evolved into the deep shelf sedimentation.

In the upper part of Wenchang Formation, the $\delta^{13}C$ value is from −29.73 to −26.96‰ with an average of −28.91‰, and the $\delta^{34}S$ value is from −7.0 to 23.3‰ with an average of 1.54‰, showing a weak positive anomaly in general. The values of PM008C10 sample feature relatively obvious positive excursion. All these may indicate that during the Late Hirnantian, the Area experienced glacial period again after the short interglacial period, and the sedimentary environment evolved into shallow shelf facies. Therefore, the Area experienced two short glacial periods and one interglacial period during the whole Wenchang Formation period.

2.4.7.5 Analysis of Sedimentary Environment

The sediments in the lower and upper parts of Wenchang Formation are mainly thick blocky–laminated feldspathic quartz sand grains followed by medium laminated fine-silty sandstone, indicating the sedimentary environment of shallow shelf with abundant sources of terrigenous clasts. The interbedded graptolite and chitinozoan fossils-bearing carbonaceous argillaceous layer within Wenchang Formation represent the quiet deep shelf sedimentary. All these indicate that two kinds of sedimentary environment frequently alternate, suggesting varied sedimentary environment and frequent crust movements. The positive excursion of $\delta^{13}C$ in the carbonaceous mudstone shows that the sedimentary environment is related to the Late Ordovician global glacial events, and this is basically consistent with the characteristics of Wangjiawan stratigraphic profile in Yichang City, Hubei Province.

There are two different sedimentary strata in the middle part of Wenchang Formation. The lower stratum is composed of grayish siltstone intercalated with a layer of siliceous mudstone nodules. It is rich in the fossils of brachiopoda, trilobites, cephalopods, and crinoid but features monotonous genera and species and small figure of the organisms, indicating the sedimentation above the oxidation zone of shallow shelf. In the upper stratum, the carbonaceous mudstone with the thickness of about 10 m is rich in graptolites, representing the quiet reduction environment of deep shelf. However, there are well-preserved sponge fossils of 7–8 genera and species associated with the graptolites. Since sponges are benthic sessile organisms and lived in seawater containing enough oxygen and there is no excursion of $\delta^{13}C$‰ in the carbonaceous shale, it is inferred that there was an interglacial period during Hirnantian period.

Regionally, the lithology of Wenchang Formation and the sedimentary environment reflected change greatly. In Jiangshan–Changshan area, amaranthine sandstone appears in the middle and upper parts of Wenchang Formation, suggesting a locally hot and dry shallow shelf environment. In Wenchang, Chun'an County–Tonglu–Fuyang–Huangshi Village, Anji County area, tidal bedding developed in the sandstone of Wenchang Formation, and the strata located between *Songxites–Aegiromenella* fauna and *Normalograptus persculptus* graptolite zone are intercalated with the gravels, with quartz as main component and good rounding, indicating a sedimentary environment of fluvial facies during glacial period.

2.4.7.6 Trace Elements in Strata (Ore-Bearing)

A total of 28 rock spectra were collected from the Wenchang Formation of Shangangshang profile (PM002), including 27 systematically collected in the lower part and one collected from about 10–m-thick carbonaceous shale in the middle part. Furthermore, 16 rock spectra were collected from the Wenchang Formation of Saoshe profile (PM008), including one collected from about 10-m-thick carbonaceous shale in the middle part and 15 systematically collected in the upper part. Statistics were made on the arithmetic mean values and concentration coefficients of 14 main trace elements. According to spectral analysis and statistics, the lower part of Wenchang Formation is enriched in Be, Sb, and Sn; the middle part is enriched in Be, Sb, Pb, W, Mo, and Ag, especially Sb; the upper part is enriched in Sb, Mo, and Sn, indicating weak glacial chemical weathering resulting in less enriched elements. During the interglacial period, the quantity and quality of deposited elements from terrigenous supplies increased sharply with the increase of oxidation, reflecting that the climate exposed great influence on the distribution of trace elements in sediments.

2.4.7.7 Standard Section of Lower Yangtze Region in the Upper Ordovician Hirnantian

The Hirnantian Stage refers to an informal but widely recognized stratum with an age range of less than 2 Ma at the end of the Ordovician (Cooper and Sadler 2004; Gradstein et al. 2004). It is of great significance in spite of short age range since it records the second largest biological extinction event in geological history. This extinction event is related to the cooling of climate and the rise of global sea level caused by the expansion of Antarctic ice sheet, resulting in the extinction of about 85% of the species (Sheehan 2001). Furthermore, these events left unique records on lithostratigraphy, biostratigraphy, sedimentary rocks, and chemostratigraphy in the Area. All China Commission of Stratigraphy organized competent experts to conduct a field investigation and demonstration for the "Upper Ordovician Wenchang Formation Profile of Hanggai Town, Anji County, Zhejiang Province." All the experts unanimously agreed to rank the profile as the "Standard Section of Lower Yangtze Region in the Upper Ordovician Hirnantian" (Fig. 2.85) and believed that the profile obtained the following innovative achievements:

1. Hirnantian Stage and Ordovician—Silurian boundary are exposed completely in Hanggai profile in Anji County, Zhejiang Province. Besides, about 360-m-thick cyclic sedimentation consisting of graptolite-bearing shale and clastic rocks developed in the profile. According to chemostratigraphical studies of carbon isotopes, no structural damage was observed near the major boundary, and the geological events and biological sequences were well recorded. Therefore, Hanggai profile is qualified to be established as the Standard Section of Lower Yangtze Region in the Upper Ordovician Hirnantian.
2. Well-preserved fossils have been discovered in the profile. Among them, six graptolite zones and one shellfish fauna were determined in successive strata in Upper Ordovician Katie Stage–Hirnantian Stage and Lower Silurian Rhuddanian Stage, i.e., *Dicellograptus complexus*, *Paraothograptus pacificus*, *Normalograptus extraordinarius*, *Normalograptus persculptus*, *Akidograptus ascensus*, *Parakidograptus acuminatus* graptolite zones and *Songxites–Aegiromenella* shellfish fauna. In addition, the fossils of diverse graptolites,

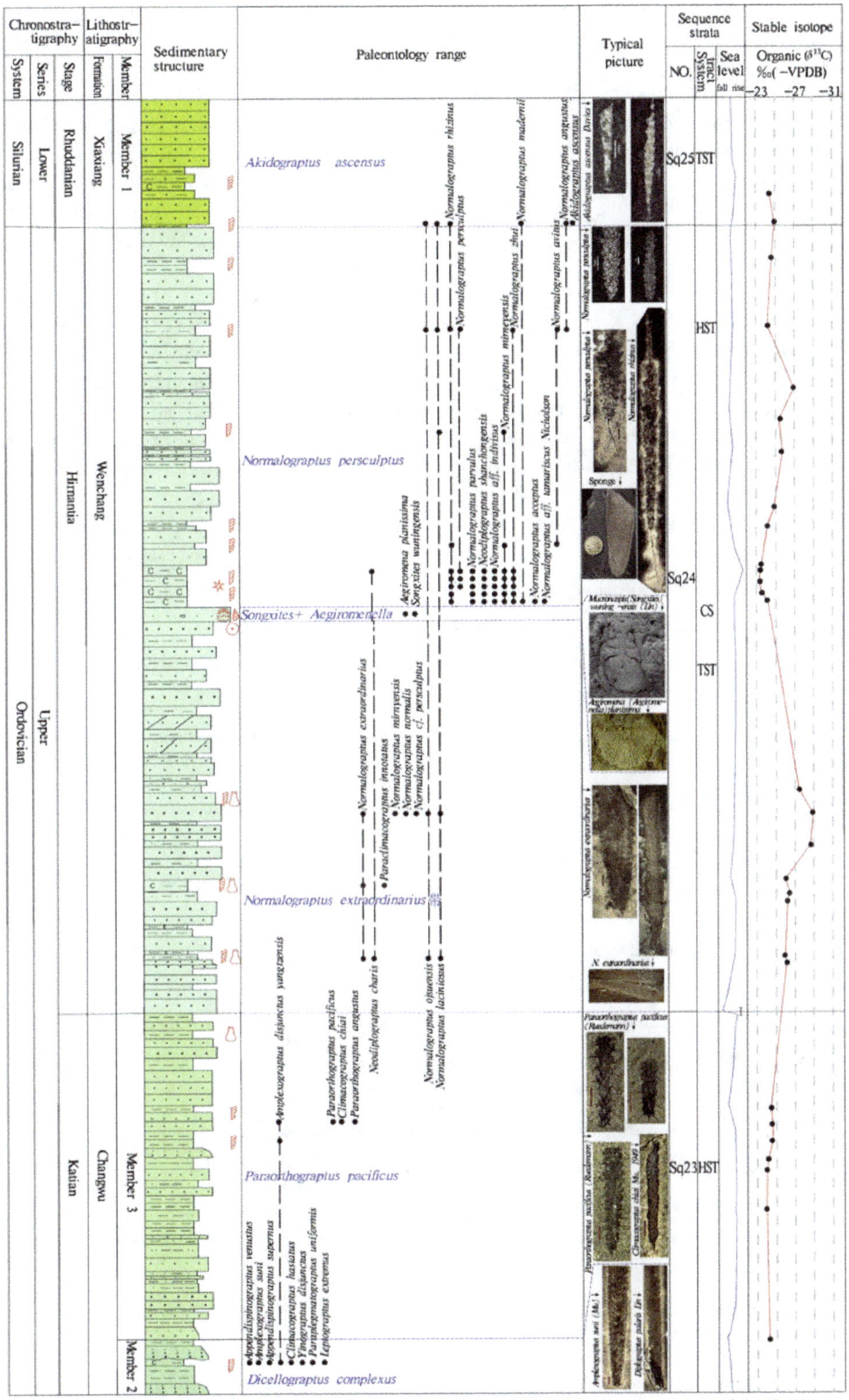

Note: the section was approved as "Standard Section of Lower Yangtze Region of Upper Ordovician Hirnantian Stage" by All China Commission of Stratigraphy

Fig. 2.85 Histogram and paleontology range of the standard section of Lower Yangtze Region in the Upper Ordovician (Hirnantian) in the area chitinozoans, sponges, trilobites, gastropods, brachiopoda, cephalopods, etc., have been identified in the successive strata.

3. Abundant sponge fauna (involving more than 10 genera and species) was first discovered in the Hirnantian strata, opening an important window to the further understanding of the global biosphere of the Hirnantian period and the sponge evolution after the Cambrian life explosion.

2.4.8 Xiaxiang Formation (S_1x)

As the only exposed Silurian stratum in the Area, Xiaxiang Formation is mainly distributed in the Fushi Reservoir in the eastern part of Hanggai map sheet and sporadically exposed in Ma'anshan area in the southern part of the Xianxia map sheet. The distribution area is about 21.98 km^2, accounting for 1.73% of the bedrock area.

The name Xiaxiang Formation (S_1x) was created by the 317 Geological Team of Bureau of Geology and Mineral Exploration of Anhui Provincial (1965) on the side of the railway near Xiaxiang Village, which lies about 5 km to the north of Hulesi Town, Ningguo County, Anhui Province. In Zhejiang Province, Xiaxiang Formation was originally known as Anji Formation, which was created by Regional Geological Survey Team of Geological Exploration Bureau of Zhejiang Province (1967) in Xiaofeng Town, Anji County, Zhejiang Province. When the name was created, the bottom of conglomerate and pebbled sandstone was taken as the bottom of Anji Formation. The preparation group of regional stratigraphic table of Zhejiang Province (1979) divided the conglomerate, pebbled sandstone, and

Xiaxiang Formation Total thickness: > 493.18 m

The second member of Xiaxiang Formation Total thickness: 414.91 m

Top: undiscovered

31. Gray blocky–laminated quart-feldspar silty-fine sandstone, the thickness of a single layer: greater than 1 m. 53.99 m

30. Gray thick laminated silty mudstone, micro-very thin horizontal bedding developing, the thickness of a single layer: 0.6–1 m. 1.87 m

29. Gray thick bedded-blocky silty-fine sandstone, main components: sub-well-rounded quartz and feldspar, argillaceous cement, the thickness of a single layer: 0.5–1.5 m, micro horizontal bedding developing. 16.72 m

28. Gray thick laminated silty mudstone, the thickness of a single layer: 0.5–1 m, no bedding developing. 0.89 m

27. Gray thick laminated argillaceous silty-fine sandstone, main components: quartz and feldspar, argillaceous cement, the thickness of a single layer: 0.5–1 m. 16.01 m

26. Khaki thick laminated argillaceous silty-fine sandstone owing to weathering, main components: quartz and feldspar, argillaceous cementation, the thickness of a single layer: 0.5–1 m, very thin horizontal bedding developing. 2.92 m

25. Dark grayish–grayish-yellow thick laminated argillaceous siltstone, mainly composed of quartz-feldspar sand, argillaceous cement, pyrite and mica visible occasionally, the thickness of a single layer: 0.5–1 m, very thin horizontal bedding developing. 26.98 m

24. Gray argillaceous siltstone, main components: quartz and feldspar followed by a small amount of argillaceous matter, very thin horizontal bedding developing, containing a small amount of pyrite grains 4.57 m

23. Gray medium laminated argillaceous siltstone, mainly composed of quartz-feldspar sand, thickness of a single layer: 0.2–0.4 m, horizontal bedding developing. 3.27 m

22. Gray thick laminated argillaceous silty-fine sandstone, main components: quartz-feldspar sand followed by a small amount of argillaceous matter, the thickness of a single layer: 0.8–1.5 m, parallel bedding developing. 4.68 m

21. Gray thick laminated silty-fine sandstone, main components: quartz and feldspar followed by a small

amount of argillaceous matter, clear horizontal bedding developing, the bottom: a layer of mudstone with the thickness of about 50 cm. 11.73 m

20. Gray thick bedded-blocky quartz arkose, main components: quartz and feldspar followed by a small amount of argillaceous matter, the thickness of a single layer: 1–1.2 m, parallel bedding developing. 21.40 m

19. Gray medium laminated fine siltstone; main components: quartz and feldspar followed by argillaceous matter; the fine sandstone: locally concentrating, micro-fine banded; the thickness of a single layer: 30–40 cm; horizontal bedding developing. 15.26 m

18. Gray thick laminated argillaceous siltstone, with a few fine sand bands, developed horizontal bedding, local cross-bedding. 10.58 m

17. Covering. 25.40 m

16. Gray thick laminated argillaceous siltstone, intercalated with micro laminated silty mudstone; pyrite grains visible occasionally, horizontal bedding developing, cross-bedding developing locally. 113.29 m

15. Gray-caesious argillaceous fine sandstone, parallel bedding developing, spheroidal weathering. 34.15 m

14. Gray-caesious silty-fine sandstone, pyrite visible occasionally, parallel bedding developing. 33.06 m

13. Gray argillaceous siltstone; very thin horizontal bedding developing; the silt: micro-fine banded (1–2 mm), spheroidal weathering locally. 18.13 m

12. Gray silty mudstone intercalated with thin laminated of argillaceous siltstone. The thickness of a single layer of the argillaceous siltstone: 2–3 cm, spherical weathering, horizontal bedding developing. 8.78 m

——————— Conformable contact ———————

The first member of Xiaxiang Formation Total thickness is 78.27 m

11. Gray silty mudstone, very thin horizontal bedding developing. 34.88 m

10. Covering. 8.10 m

9. Gray silty mudstone, graptolite visible. 10.43 m

8. Gray thick laminated quartz feldspar fine sandstone, main mineral components: quartz and feldspar clasts followed by a small amount of argillaceous matter, no bedding developing. 2.09 m

7. Gray silty mudstone, major component: argillaceous matter, minor components: clasts of quartz and feldspar. 2.41 m

6. Gray thick laminated siltstone interbedded with mudstone. 3.76 m

5. Gray silty mudstone, main mineral components: argillaceous mater followed by clasts of quartzy silt. 1.32 m

4. Gray thick laminated argillaceous silty-fine sandstone, main mineral components: clasts of quartz and feldspar followed by a small amount of argillaceous. 1.74 m

3. Gray thin laminated siliceous carbonaceous shale, main mineral component: argillaceous matter, graptolites visible. 2.10 m

2. Interbeds consisting of gray thick–medium laminated argillaceous siltstone and micro laminated carbonaceous shale; the argillaceous siltstone: composed of clasts of quartz and feldspar followed by argillaceous matter, the thickness of a single layer: 25–55 cm; the carbonaceous shale: the thickness of a single layer: 0.5–1 cm, graptolites visible. 2.67 m

——————— Conformable contact ———————

Wenchang Formation Total thickness: >14.79 m

1. Khaki thick laminated silty-fine sandstone. 10.39 m

0. Grayish yellow thick sandstone interbedded with argillaceous siltstone. 4.40 m

the horizon producing *Dalmanitina* cf. *mucronata Mucronaspis* (*Songxites*) *wuningensis* (Lin) from Anji Formation and collectively named them Yankou Formation, and accordingly the Anji Formation only referred to the sandstone and mudstone-bearing *Normalograptus* (*Glytograptus*) *persculptus* and the strata upward in Anji Formation. This kind of division was adopted in *Regional Geology of Zhejiang Province* (1989). In *Lithostratigraphy of Zhejiang Province* (1995), Anji Formation was renamed Xiaxiang Formation. In this Project, according to the characteristics of the lithologic association, sedimentary structures, fossils, and geological observation traverse of Fushi Reservoir profile (PM009) of Hanggai map sheet in the Area, the strata dominated by thick blocky–laminated sandstone and intercalated with shale containing *Normalograptus* (*Glytograptus*) *persculptus* in the profile were classified under Wenchang Formation (O_3w), while the strata consisting of dark gray–black thin laminated carbonaceous shale containing *Akidograptus ascensus* and medium laminated fine-silty sandstone in the profile were classified as Xiaxiang Formation (S_1x).

2.4.8.1 Lithostratigraphy

According to the lithology and lithologic association, Xiaxiang Formation is divided into the first member (S_1x^1) and the second member (S_1x^2), which are in conformable contact with the underlying Ordovician Wenchang Formation.

The first member of Xiaxiang Formation (S_1x^1): grayish–grayish black thin laminated graptolite-bearing siliceous carbonaceous shale intercalated with gray medium laminated feldspathic quartz silty-fine sandstone or rhythmic interbeds constituting of them. The siliceous carbonaceous shale: micro-texture horizontal bedding developing, bearing micro-fine-grained pyrite, and rich in graptolites.

The second member of Xiaxiang Formation (S_1x^2): interbeds consisting of gray thick blocky–laminated silty-fine sandstone and gray–caesious medium laminated argillaceous siltstone, interbedded with a small amount of silty mudstone and mudstone. The silty-fine sandstone gradually increases, and the rock layers gradually thicken upward. The top of this member is not discovered in the Area.

1. **Stratigraphic section**

The lithology of Xiaxiang Formation is described by taking the example of Xiaxiang Formation (S_1x) profile (Fig. 2.86) of Hanggai map sheet in Fushi Reservoir, Anji County, Zhejiang Province. The details are as follows:

2. **Lithological characteristics**

The lithology of the first member of Xiaxiang Formation (S_1x^1) is mainly characterized by thin laminated silicon-bearing carbonaceous shale and medium–thick laminated feldspathic quartz silty-fine sandstone.

(1) Silicon-bearing carbonaceous shale: dark gray–grayish black; composition: argillaceous matter (65–70%), silt (10–20%), cryptocrystalline siliceous matter (10–15%), and carbonaceous matter (5–10%), micro-fine-grained pyrite visible occasionally along the bedding (0–5%); very thin horizontal bedding and many carbonized graptolite fossils developing.
(2) Feldspathic quartz silty-fine sandstone: gray; main component: feldspar and quartz (80–85%); grain size: 0.1–0.2 mm generally and 0.05–0.1 mm for a small amount, sub-well rounded; argillaceous cementation; argillaceous content: 15–20%.

The lithology of the second member of Xiaxiang Formation (S_1x^2) is mainly argillaceous siltstone, silty-fine sandstone, and silty mudstone.

(1) Silty-fine sandstone: gray–caesious; silty-fine sandstone structure; medium–thick laminated structure dominated by thick laminated structure; main component: quartz–feldspar sand (80–85%), sub-well rounded; grain size: 0.01–0.1 mm; minor component: a small amount of argillaceous matter (15–20%).
(2) Argillaceous siltstone: gray; silty structure; medium–thick laminated structure dominated by the medium laminated structure; main component: quartz–feldspar sand (60–80%), sub-well rounded; grain size: 0.01–0.05 mm; minor component: argillaceous matter (20–40%); pyrite grains visible occasionally; clear horizontal bedding developing.
(3) Silty mudstone: gray; thin laminated structure; main component: argillaceous matter (65–70%); minor component: quartz–feldspar sand (30–35%); grain size: 0.01–0.06 mm, sub-well rounded.

3. **Basic sequences**

There are two types of basic sequences developing in the first member of Xiaxiang Formation (S_1x^1) (Fig. 2.87).

(1) Mainly distributed in the upper part of Xiaxiang Formation, mainly composed of medium laminated silty mudstone, main component of the silt: quartz, horizontal bedding developing, belonging to monotonic basic sequences.

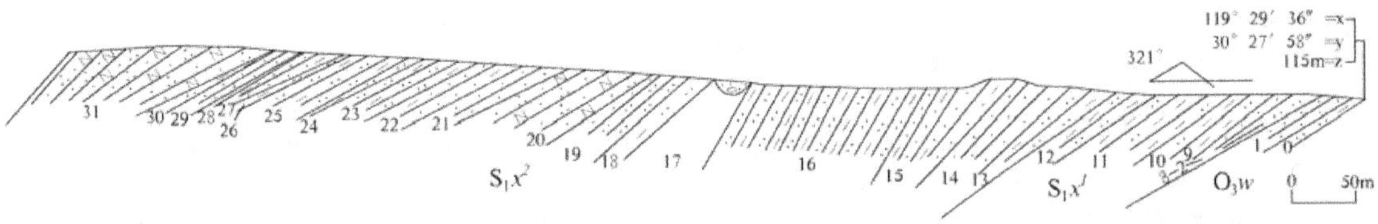

Fig. 2.86 Xiaxiang Formation (S_1x) profile in Fushi Reservoir, Anji County, Zhejiang Province

(2) Mainly distributed in the lower part of Xiaxiang Formation, composed of interbeds consisting of ① gray thin–medium laminated silty-fine sandstone and ② black carbonaceous silicon-bearing mudstone; the silty-fine sandstone: thin–medium laminated in the lower part, medium laminated mainly upward; the carbonaceous silicon-bearing mudstone: thin laminated mostly, micro-fine horizontal bedding developing; the ratio of silty-fine sandstone to the carbonaceous silicon-bearing mudstone: (3–5):1 approximately; the mudstone gradually increasing upward; belonging to the non-cyclic basic sequence with terrigenous clasts and thickness increasing upward.

The silty-fine sandstone in this member is mainly composed of feldspar and quartz. The clasts feature poor maturity and rounding and good sorting. They are the products of weathering, transportation, and deposition of far terrigenous granitic rocks and belong to the silt–mudstone facies of the outer margin of shallow shelf. The shale with high content of silicon and carbon and abundant graptolite is the sediments in bathyal reduction environment, belonging to the siliceous-argillaceous facies of shallow shelf–bathyal basin.

There is one type of basic sequences in the second member (S_1x^2) of Xiaxiang Formation.

(3) Consisting of ① thin–medium laminated silty mudstone, ② thin–medium laminated argillaceous siltstone, and ③ gray medium–thick laminated fine-silty sandstone; the fine-silty sandstone: mostly thick laminated, blocky–laminated occasionally, no obvious bedding developing; the argillaceous siltstone: medium laminated mostly, thick laminated for a small amount, parallel bedding developing; the mudstone: mostly thin laminated, thick laminated occasionally, pyrite visible locally, micro-fine horizontal bedding developing; the ratio of sandstone: siltstone: mudstone is (8–10):(3–4):1 approximately; mudstone often absent in the upper and lower parts of the second member; sandstone increasing and layer thickness increasing upward; belonging to the cyclic basic sequence with grain size decreasing and layer thickness increasing upward.

The strata in this member are mainly composed of feldspar–quartz silt and fine sand. The clasts feature poor maturity and rounding but finer grain size compared with Wenchang Formation. They are the products of weathering, transportation, and deposition of far terrigenous granitic rocks and belong to the far terrigenous clastic rock facies of the outer margin of shallow shelf.

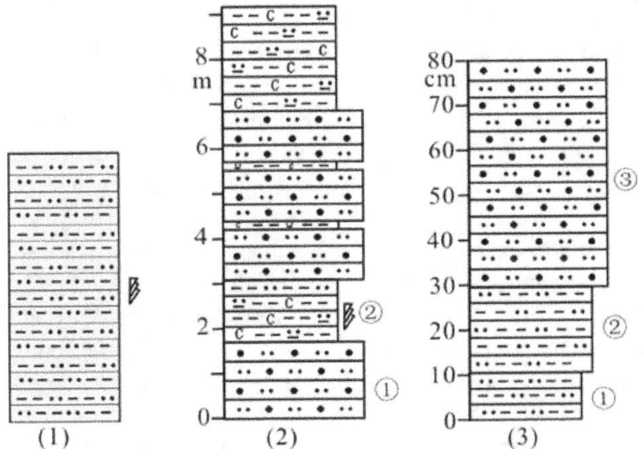

Fig. 2.87 Xiaxiang Formation (S_1x)—basic sequence chart

2.4.8.2 Sequence Stratigraphy

According to the characteristics of the lithology, paleontology, lithofacies association, and sequence boundary of the Fushi Reservoir profile of Hanggai map sheet, there is one third-order sequence Sq25 in Xiaxiang Formation (S_1x), which belongs to the second-order sequence SS8. It is briefly described as follows (Fig. 2.88).

Third-order sequence Sq25: consisting of the transgressive systems tract (TST) and highstand systems tract (HST). The bottom of Sq25 is the conformable interface between the black carbonaceous mudstone of the first member of Xiaxiang Formation and the thick laminated sandstone of Wenchang Formation, belonging to the type-II sequence boundary. The top of Sq25 is not discovered. TST is located in the first member and the bottom of the second member of Xiaxiang Formation. The sediments of TST mainly include carbonaceous argillaceous matter followed by silt and fine sand. The carbonaceous mudstone contains graptolite fossils. All these indicate that TST belongs to the sedimentation in oxidation–reduction environment below wave base of deep shelf. HST is located in the middle and upper parts of the second member of Xiaxiang Formation. The sediments of HST mainly contain fine sand, silt, and a small amount of argillaceous matter, indicating the clastic rock facies of shallow shelf during the gradual rise of sea level after the maximum flooding. The coarse clasts increase from bottom to top in this sequence, reflecting the gradual falling of the sea level. Therefore, this sequence is of aggradation–progradation type.

2.4.8.3 Biostratigraphy and Chronostratigraphy

Xiaxiang Formation (S_1x) in the Area inherited the lithofacies and paleogeographic conditions of Wenchang Formation. With the improvement of climatic conditions, planktonic graptolites feature new thriving but the monotonous genera and species. In this Project, a large number of graptolite fossils were obtained in the first member of Xiaxiang Formation (S_1x^1) around Fushi Reservoir, and two graptolite zones, i.e., *Akidograptus ascensus* zone and *Parakidograptus acuminatus* zone, were identified from bottom to top.

1. ***Akidograptus ascensus* zone**

The graptolites discovered in the carbonaceous silty mudstone in the lower part of the first member of Xiaxiang Formation in Saoshe profile include the zone fossil *Akidograptus ascensus* (Davies 1929) and the symbiotic molecules including *Normalograptus laciniosus* (Churkin and Carter 1970), *Normalograptus rhizinus* (Li and Yang, 1983), *Normalograptus ojsuensis*, and *Normalograptus angustus* (Perner 1895). Some graptolites suffered the metasomatism of pyrite (Fig. 2.89).

2. ***Parakidograptus acuminatus* zone**

Parakidograptus acuminatus zone is located in the silty mudstone in the upper part of the first member of Xiaxiang Formation. In the silty mudstone, which is khaki owing to weathering and about 10–12 m in the upper part of the first member of Xiaxiang Formation of Fushi Reservoir, a large number of graptolites were obtained, including the zone fossil *Parakidograptus acuminatus* Nicholson 1867 and other symbiotic molecules such as *Parakidograptus* sp. indet., diplograptid gen. & sp. indet., "*Climacograptus*" sp. indet., *Normalograptus* sp. indet., and *Normalograptus rhizinus* (Fig. 2.90).

According to the *Stratigraphic Chart of China* (*2014*), the *Akidograptus ascensus* and *Parakidograptus acuminatus* zones belong to Rhuddanian Stage of Early Silurian, and Xiaxiang Formation is a stratigraphic unit of the Lower Silurian.

2.4.8.4 Analysis of Sedimentary Environment

With the global glacier melting in the Late Hirnantian, a rapid sea level rise occurs, and the Area experienced a new round of abyssal sedimentation accordingly. In the first member of Xiaxiang Formation, the sediments mainly include terrigenous clasts consisting of fine sand silt followed by carbonaceous argillaceous matter. The terrigenous clasts feature relatively good sorting and mainly include quartz and feldspar. Therefore, they were brought by long-distance transportation. The carbonaceous argillaceous sediments are pelagic sediments, and the sulfide reflects the reduction condition. Therefore, the first member of Xiaxiang Formation should feature standing-water sedimentary environment of the deep shelf reduction zone. In the second member of Xiaxiang Formation, the sediments are mainly composed of gray and caesious medium–thick laminated silt and fine sand, interbedded with a small amount of silt and argillaceous matter. There is no biological fossil discovered. All these indicate the terrigenous clasts are abundant and the water body is anoxic and thus unfavorable for the survival of organisms. It can be inferred that there was a sea level rise, and the Area was in the weak reduction sedimentary environment in the transition zone between shallow shelf and deep shelf.

Regionally, the Xiaxiang formations in all areas all feature a set of argillaceous clastic rock formation of deep shelf–shallow shelf, but greatly a different stratum thickness. Northwestward to Ningguo City, Anhui Province, the thickness increases, the grain size of the clasts decreases, and silt becomes the main component. Southeastward to Lin'an,

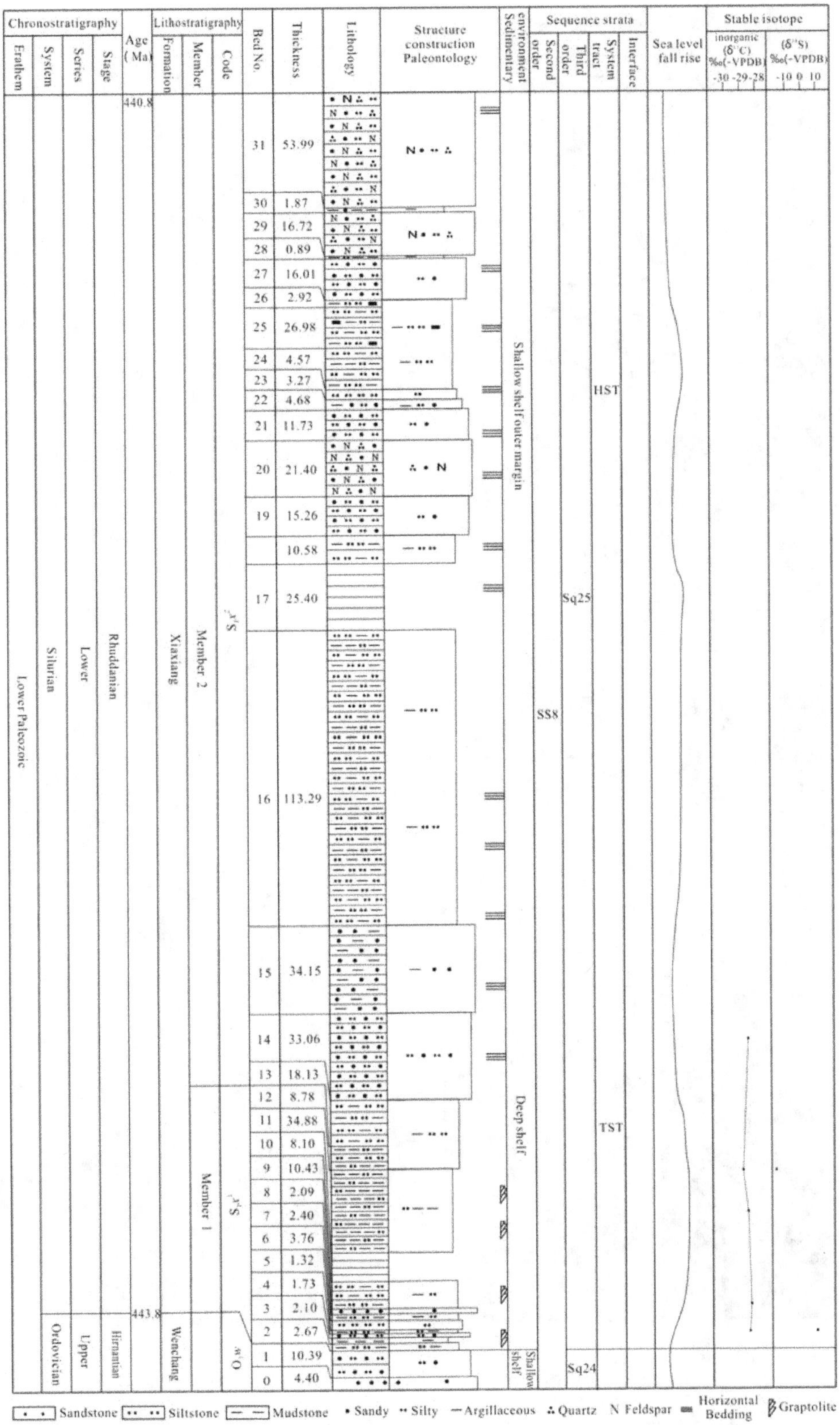

Fig. 2.88 Sequence of stratigraphic framework of Early Silurian Xiaxiang Formation in the area

Fig. 2.89 Graptolite in *Akidograptus ascensus* zone in the first member of Xiaxiang Formation (S_1x^1): **a**, **b**, **c**, **d** *Akidograptus ascensus* Davies 1929 (Pm008-Hb15-3, -1e, -16a, -6e); **e**, **f** *Normalograptus laciniosus* (Churkin and Carter 1970) (Pm008-Hb15-13, -20); **g** *Normalograptus angustus* (Perner 1895) (Pm008-Hb15-21); **h** *Normalograptus laciniosus* (Pm008-Hb15-2); **i** *Normalograptus madernii* (Pm008-Hb15-5); **j**, **k**, **l** *Normalograptus rhizinus* (Li and Yang 1983) (Pm008-Hb15-12a, -18, -4). The length of the white scale is 1 mm

Fig. 2.90 Graptolites in *Parakidograptus acuminatus* zone in the first member of Xiaxiang Formation (S_1x^1): **a** *Parakidograptus acuminatus* Nicholson, 1867 (Pm009-Hb4-3); **b** *Parakidograptus* sp. (Pm009-Hb1-1); **c** *Normalograptus* sp. (Pm009-Hb3-1); **d** *Diplograptid gen.* & sp. (Pm009-Hb3-2); **e**, **f** *Normalograptus rhizinus* (Pm009-Hb3-1, Hb4-4); **g** *Climacograptus sp*. (Pm009-Hb10-1); the length of the white scale: 1 mm

the thickness evidently decreases, fine and medium sands become the main components, and tidal bedding developed locally. All these indicate that paleogeographic relief was high in the north and low in the south.

2.4.8.5 Characteristics of Stable Isotope

A total of seven carbonaceous mudstone samples were collected from Xiaxiang Formation of Saoshe profile (PM008) and Xiaofeng profile (PM009) to conduct whole-rock stable isotope analysis of $\delta^{13}C$ and $\delta^{34}S$. According to the analysis, the $\delta^{13}C$ value of Xiaxiang Formation is from −29.48 to −28.91‰ with a small variation range and an average of −29.24‰; the $\delta^{34}S$ value of Xiaxiang Formation is from −9.2 to 11.1‰ with the average 3.57‰. These values are basically consistent with the $\delta^{13}C$ value (−35 to −30‰) and $\delta^{34}S$ value (−10 to 10‰) of normal shale, indicating deep shelf sedimentation.

2.4.8.6 Trace Elements in Strata (Ore-Bearing)

Thirty rock spectra were systematically collected from Xiaxiang Formation (S_1x) of Fushi Reservoir profile (PM009). Among these spectra, 10 were collected from the first member (S_1x^1) and 20 from the second member (S_1x^2). Analysis was made on 14 trace elements, and then statistics were made on the arithmetic mean values and concentration coefficients of the analysis results. According to the statistics, the first member is enriched in Be, Sb, and F, and the second member is enriched in Be, Sb, W, Sn, F, and S. The possible reason is that after long-distance transportation, the clasts were worn and finer, and the heavy minerals containing W and Sn deposited and concentrated near the shore accordingly.

2.4.9 Comparison of Regional Stratigraphy

In northwest Zhejiang Province, the topographic pattern in the Ordovician was inherited from the Sinian–Cambrian. However, with the change in paleogeographic conditions and paleoclimate, the sediments and ancient organisms of the Ordovician are different from those of the Sinian and Cambrian. Except Yuhang–Fuyang area where the sediments were still composed of carbonate in the Early and Middle Ordovician, the sediments in other areas were dominated by argillaceous, sandy, and silty terrigenous clastics (Fig. 2.91) in the Ordovician. The regional stratigraphy was compared by taking the samples of the profile in the Area in Hanggai area, and Banqiao (Lin'an) profile and Sanxikou profile in Yuhang–Fuyang area from northwest to southeast.

The regional stratigraphy in the early and middle period of the Early Ordovician is as follows. In the northwest Hanggai area, obvious transgression took place twice, and the carbonate-bearing argillaceous matter–silicon-bearing argillaceous matter of deep shelf in the third member of Xiyangshan Formation and the first member of Yinzhubu Formation was deposited, with respective deposition depth of 255 m and 76 m. In Yuhang–Fuyang area, the sediments were composed of the mud and knotlike carbonate of platform facies in Lunshan Formation and the knotlike and reticulate carbonate of platform–shallow shelf facies in Honghuayuan Formation from bottom to top, with respective thickness of 329 m and 92 m, respectively, representing high-energy turbulent environment and open and de-saline neritic environment. The sedimentary rocks in the two sedimentary zones generally reflect the trend of crust subsidence and sea level rise.

The regional stratigraphy in the late period of the Early Ordovician is as follows. In Hanggai area, the sediments consisted of black graptolite-bearing siliceous mudstone of bathyal facies, with a thickness of less than 100 m. In Yuhang–Fuyang area, the sediments were composed of reticulate carbonate with the platform facies in the lower part of Guniutan Formation and a thickness of less than 20 m.

The regional stratigraphy in the early period of the Middle Ordovician is as follows. In Hanggai area, the sediments consisted of the graphite-bearing carbonaceous shale of sub-deepwater subcompensational basin facies in the Ningguo Formation, with a thickness of 70 m. To the late period of the Middle Ordovician, the maximum flooding surface appeared, and the silicalite of deepwater basin facies was deposited in Hanggai area. Southeastward to Yuhang area, the sediments changed from argillaceous carbonate–knotlike carbonate of the platform and shallow shelf facies in the first member of Guniutan Formation, to micro- to argillaceous–crystalline carbonate from bottom to top. In the late period of the Middle Ordovician, a small amount of siliceous of deep shelf was deposited in Yuhang area. Based on the absence of transitional environmental sedimentation, it is inferred that a strong crustal movement event occurred during that period.

The regional stratigraphy in the early period of the Late Ordovician is as follows. In the northwest Zhejiang Province, the argillaceous carbonate and knotlike carbonate of the Yanwashan Formation were deposited owing to the shallow shelf sedimentary environment, with a thickness of 10–20 m. Then, a new round of uneven crustal subsidence and transgression started. The siliceous mudstone of Huangnigang Formation was initially deposited. The bottom and middle of the Huangnigang Formation were intercalated with sparse knotlike limestone layers, and the thickness was reduced from 80 m to 35 m from north to south.

In the middle and late period of the Late Ordovician, the water body deepened from the southeast to the northwest, with the further enhancement of crust subsidence. In the inner margin of the slope in Yuhang–Fuyang area and the outer margin of the slope in Hanggai area far away from the sediment sources, the flysch clastic rocks of sedimentary slope facies were deposited at the same time, with the thickness of sedimentary rocks of 480–500 m, and the grain size of the clasts was relatively small. In Yuhang–Fuyang area close to the sediment sources, the thickness of the sedimentary strata was up to more than 1500–2000 m and the grains of the clasts were relatively coarse.

The Late Ordovician was a global glacial period, during which the sea level fell rapidly. In Hanggai area, thick blocky–laminated sandstone intercalated with thin laminated siltstone and graptolite-bearing mudstone of shallow shelf facies in Wenchang Formation was deposited. While in Yuhang–Fuyang–Wenchang area, Wenchang Formation was composed of clastic rocks of littoral neritic facies with tidal bedding developing, and a layer of conglomerate intercalated in the middle part of the formation.

At the beginning of the Silurian, the sea level rose in northwest Zhejiang Province with the melting of glaciers, and a new round of deep shelf terrigenous clast deposition began.

2.5 Cretaceous System

Cretaceous strata in the Area mainly include the volcanic rocks in Huangjian Formation. They are distributed in most areas of the southeast of Xianxia map sheet and the southeast of Chuancun map sheet. Besides, they are locally exposed in the northwest corner of Chuancun map sheet. The total outcrop area is up to about 519.72 km^2, accounting for 40.88% of the total bedrock area.

The name Huangjian Formation (K_1h) was created by Zou (1964) in Huangjian Mount, which lies in the south of Shouchang Town, Jiande City, Zhejiang Province. The original lithology of the Huangjian Formation was characterized into two parts, i.e., the upper part and the lower part. The upper part consisted of caesious rhyolitic tuff lava, vitric tuff, and tuff breccia interspersed with amaranthine rhyolite, with tuff interspersed with amaranthine siltstone constituting its top. The lower part was composed of purple and gray rhyolite and rhyolite porphyry. The name has been used up to now by Lin et al. (1989), Ding et al. (1999), Chen (2000), etc., as well as in *Lithostratigraphy of Zhejiang Province* (1996). In this Project, the name Huangjian Formation was still adopted based on the lithological

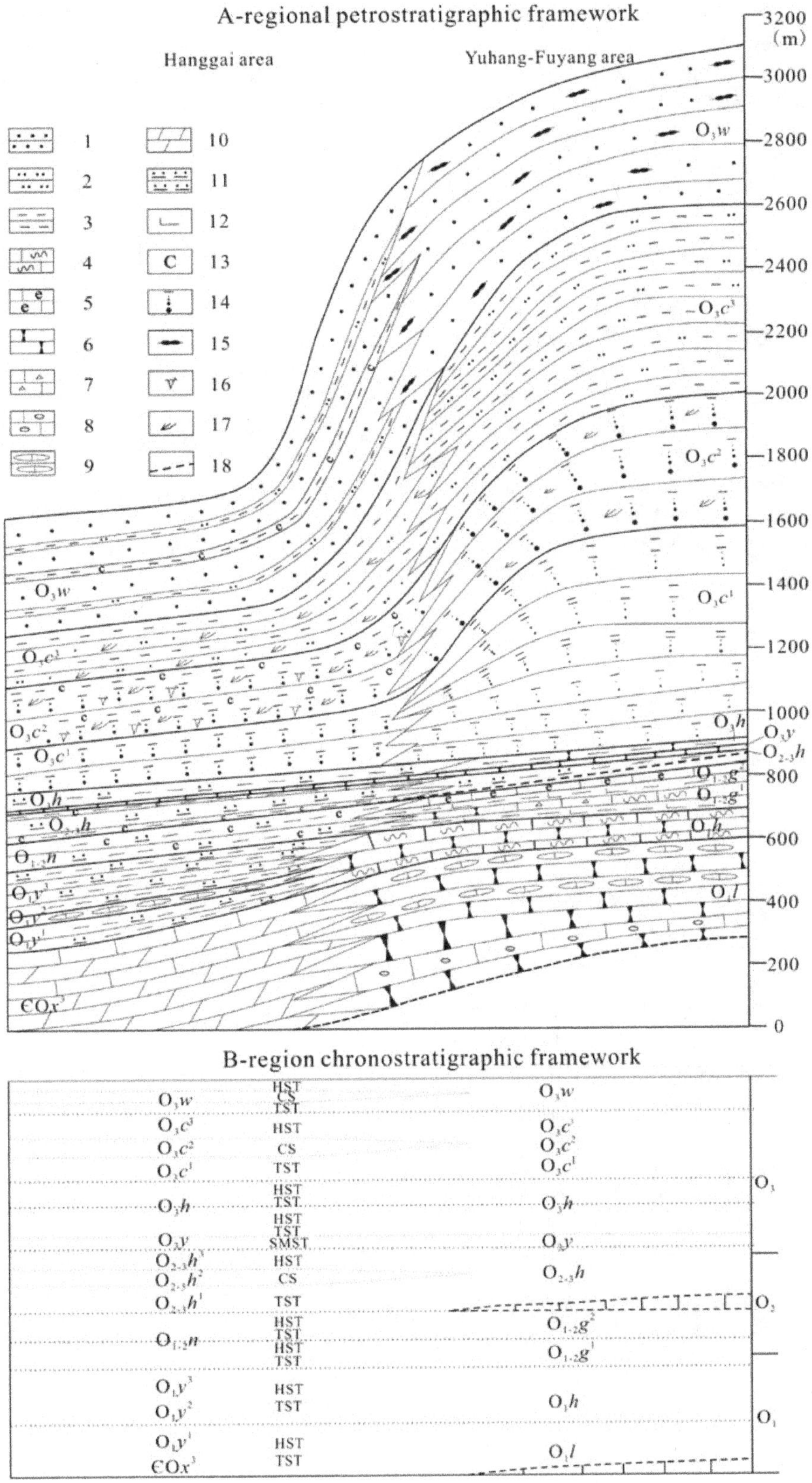

Fig. 2.91 Regional stratigraphic framework of Hanggai area and Yuhang–Fuyang Area in the Ordovician–Early Silurian: 1. sandstone; 2. siltstone; 3. mudstone; 4. reticulate limestone; 5. biogenic limestone; 6. knotlike limestone; 7. breccia limestone; 8. calcirudite; 9. pie-shaped limestone; 10. marl; 11. silicalite; 12. calcareous matter; 13. carbonaceous matter; 14. Bouma sequence; 15. veined structure; 16. groove casts; 17. cross-bedding; 18. disconformity; O_1y^1, O_1y^2, O_1y^3—the first, second, and third members of the Yinzhubu Formation; $O_{1\text{-}2}n$—Ningguo Formation; $O_{2\text{-}3}h$—Hule Formation; O_3y—Yanwashan Formation; $\text{€}Ox^3$—the third member of Xiangshan Formation; O_3h—Huangnigang Formation; O_3c—Changwu Formation; O_3w—Wenchang Formation; O_1l—Lunshan Formation; O_1h—Honghuayuan Formation; $O_{1\text{-}2}g$—Guniutan Formation; SB1—type-I unconformity interface; SMST—shelf marginal systems tract; TST—transgressive systems tract; CS—starved section; HST—Highstand systems tract

characteristics of the volcanic rocks in the Shanxi great valley profile (PM023) in Xianxia map sheet, the profiles of Chuancun map sheet including Dongtianmu profile (PM020) in Lin'an City, Dongkencun profile (PM026) in Lin'an City, and Gaohong–Tianhuangping profile (PM060) as well as along the geological observation traverse.

2.5.1 Lithostratigraphy

On the basis of the characteristics of the lithological association, lithofacies, and spatial distribution of the volcanic rocks, Huangjian Formation in the Area is divided into four lithological members, i.e., the first member (K_1h^1), the second member (K_1h^2), the third member (K_1h^3), and the fourth member (K_1h^4), which are in angular unconformable, intrusive, or fault contact with the underlying Paleozoic strata. The first member (K_1h^1) is mainly distributed in Zhangcun and Yunti area in the west of the volcanic rock area as well as Shanxi great valley and Yangshuling forest farm area in the north of the volcanic rock area. The second member (K_1h^2) is mainly distributed in Yangtianping and Baishujian scenic area in the central and eastern part of the volcanic rock area. It is the main lithological member in the volcanic rock area. The third member (K_1h^3) is mainly distributed in Tianhuangping in the northern part of the volcanic area and locally on the top of Dongtianmu Mount in the south of the volcanic area. The fourth member (K_1h^4) is mainly distributed in Nantianmu area in the north of the volcanic rock area as well as Xitianmu and Longwang Mount in the southwest of the volcanic area.

1. **Stratigraphic section**

The lithology of the first member and the second member of Huangjian Formation is described by taking the example of the first member (K_1h^1)–the second member (K_1h^2) profile (Fig. 2.92) of Early Cretaceous Huangjian Formation of Xianxia map sheet in Shanxi great valley, Baofu Town, Anji County, Zhejiang Province. The details are as follows:

36. Light flesh red - gray quartz-monzonite porphyry; grain size of the matrix and the content and size of phenocryst decreasing at a certain degree in the contact zone with felsophyric rhyolite porphyry.

———————— Intrusive contact ————————

The second memberof Huangjian Formation — Total thickness: 4466.0 m

35. Gray felsophyric rhyolite porphyry. 502.0 m

34. Celadon andesitic dike.

33.Gray felsophyric rhyolite porphyry. 1657 m

32. Ggreenish–grayish-yellow andesitic porphyrite dike.

31. Gray felsophyric rhyolite porphyry, the content of light flesh red K-feldspar phenocrysts: 5–10%, cementation of gray felsophyric lava. 157.9 m

30. Celadon basaltic porphyrite dike.

29. Grayish felsophyric rhyolite porphyry. 3.4 m

28. Celadon basaltic porphyrite dike.

27. Grayish felsophyric rhyolite porphyry. 1.7 m

26. Celadon basaltic porphyrite dike.

25. Grayish felsophyric rhyolite porphyry. 151.4 m

24. Dark flesh red orthophyre dike.

23. Grayish blocky nevadite, the content of K-feldspar phenocrysts: 10%−15%, matrix:cementation of grayish felsophyric lava, K-feldspar porphyroclast indistinctly visible in directional arrangement in NE60° trending on the weathered surface. 656.0 m

22. Dark gray felsophyric rhyolite porphyry; generally disintegrated; the content of K-feldspar phenocrysts: 3–5%, decreased at a certain degree comparatively; cementation of dark gray felsophyric lava. 137.5 m

21. Dark gray blocky nevadite; the content of phenocrysts: about 10%−15%,increased comparatively; the

matrix: dark gray felsophyric lava. 69.6 m

20. Off-white rhyolitic vitric tuff, bearing a small amount of quartz crystal pyroclast (3–5%) and of light off-white cryptocrystalline siliceous matter and crystalvitric tuff breccia (2%–3%). 17.2 m

19. Dark gray felsophyric rhyolite porphyry, the matrix: dark gray felsophyric lava. X-shaped joints developing locally. 369.6 m

18. Dark gray-celadon dacitic tuff breccia. 85.2 m

17. Dark gray breccia-bearing vitreous rhyolite porphyry; the breccia: early erupted Amaranthine rhyolitic crystalvitric ignimbrite, grain size: about 1–6 cm; platy joints developing with the thickness of 15–40 cm. 273.1 m

16. Gray dacitic breccia-bearing crystalvitric ignimbrite, interpenetrated with small apophysis of dark gray porphyric vitreous K-feldspar rhyolite porphyry; the apophysis: irregular composite twigs, general width: 5–25 cm, general strike: 330; recrystallization occurring locally and the grain size and content of the K-feldspar pyroclasts increasing with the apophysis intruding the border. 290.4 m

15. Dark gray breccia-bearing vitreous rhyolite porphyry, containing phenocryst (10%–15%), the matrix: dark gray vitreous lava. 15.6 m

14. Gray dacitic crystalvitric ignimbrite, apparently rhyolitic bands visible on weathered surface, attitude: 100°∠77°. Medium laminated bedding developing in the bottom and gradually becoming blocky upwards.

52.8 m

———————— Volcanic-eruption unconformable contact ————————

The first member of Huangjian Formation Total thickness: 855.1 m

13. Amaranthine medium–thin laminated tuffaceous sandstone; from bottom to top, the respective thickness of conglomerate-bearing gritstone, gritstone, fine sandstone, and dark-purple silty mudstone: 10–20 cm, 10–30 cm, 5–10 cm, and 5–8 cm; the lithology of the top: Grayish black carbonaceous argillaceous siltstone. 47.7 m

12. Amaranthine rhyolitic breccia-bearing crystalvitric tuff, locally interbedded with purple mudstone; the thickness of a layer: 10–50 cm, breccia, and crystal pyroclast reducing gradually but also increasing locally upwards. 125.2 m

11. Amaranthine tuffaceous breccia-bearing sandstone; the layers containing two rhythms consisting of fine sand–silt and gritstone; the respective thickness of fine sand–silt and the gritstone: about 20 cm, and 10–20 cm.

0.5 m

10. Darcitic breccia-bearing crystalvitric ignimbrite, overlying on dacitic conglomerate-breccia-bearing crystalvitric ignimbrite in zygomorphic way. 13.9 m

9. Darcitic conglomerate-breccia-bearing crystalvitric ignimbrite: the conglomerates are elongate and elliptical with a size of about (10–40) cm × (5–10) cm and a content in the ignimbrite of 5%; the breccia is subangular–sub-rounded with a grain size of 5–60 mm. 16.2 m

8. Dark gray dacitic crystalvitric ignimbrite. 27.6 m

7. Amaranthine dacitic breccia-bearing crystalvitric ignimbrite: the breccia with a small amount (5–10%) consist of components of light flesh red rhyolite, celadon crystalvitric ignimbrite, etc. 29.5 m

6. Amaranthine dacitic magma-fragment-bearing crystalvitric ignimbrite, obvious welded bands developing with length of 1–5 cm and magma fragments as the main component; locally interpenetrated with irregular fissure

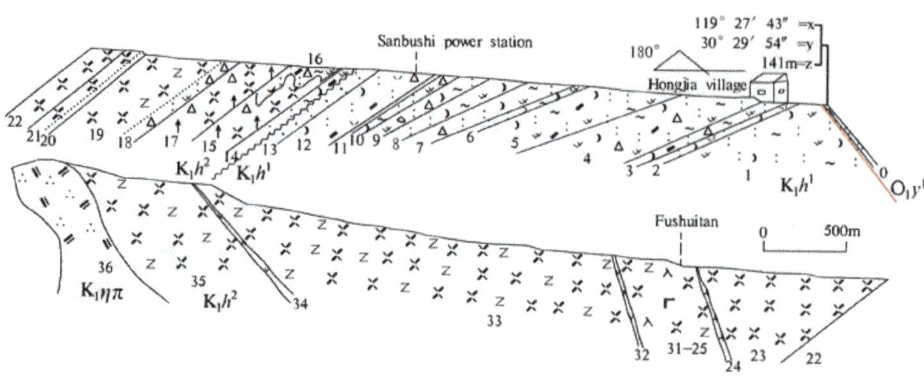

Fig. 2.92 First member (K_1h^1)–second member (K_1h^2) profile of the Early Cretaceous Huangjian Formation of Xianxia map sheet in Shanxi great valley, Baofu Town, Anji County, Zhejiang Province subvolcanic rock ($K_1\eta\pi$)

The lithology of the first member of Huangjian Formation is described by taking the example of the second member of Huangjian Formation(K_1h^2) profile (Fig. 2.93) of Chuancun map sheet in East Tianmu Mount scenic area, Lin'an City, Zhejiang Province. The details are as follows:

rhyolitic porphyry dike, which are 1–3 cm wide. 7.1 m

5. Grayish black dacitic crystalvitric ignimbrite, the crystal pyroclast of K-feldspar decreasing and the crystal pyroclast of plagioclase increasing in this bed, also containing a small amount of breccia. 280.6 m

4. Dark gray dacitic breccia-bearing crystalvitric ignimbrite. 211.3 m

3. Grayish purple dacitic breccia-bearing crystalvitric (strong) ignimbrite; welded bands developing very well; the bands: gray, distributed discontinuously, components: magma fragments and K-feldspar grains. 37.9 m

2. Amaranthine rhyolitic crystalvitric tuff, interbedded with unstable amaranthine thin laminated tuffaceous conglomerate-bearing sandstone in the lower part, a purple tuffaceous conglomerate-bearing sandstone agglomerate with the size of 2 m × 2 m visible in the upper. 18.9 m

1. Grayish purple dacitic vitric ignimbrite, welded bands (magma fragments) developing; the length and the width of the band: 1–3 cm and 0.2–1 cm. 38.7 m

———————— Fault contact ————————

The first member of Yinzhubu Formation of the Early Ordovician — Total thickness: >30.8 m

0. Dark gray thin laminated broken hornfelsic argillaceous siltstone, interbedded with gray–off-white dolomitic limestone. 30.8 m

The middle part of the second member of Huangjian Formation — Total thickness: > 303.1 m

25. Dark gray blocky rhyolite porphyry mainly composed of broken phenocrysts and matrix; the broken phenocrysts: cracks developing, fragments distributed in or close to its original place. 75.8 m

24. Light flesh red felsite vein.

23. Light flesh red rhyolite; the feldspar phenocryst: cracks developing generally; the broken fragments: distributed in or close to its original place; containing a small amount of quartz and dark minerals. 52.9 m

22. Purple-gray porphyritic rhyolite, rhyolitic structure obviously visible, rhyolitic bands developing and bypassing the phenocrysts. 111.7 m

21. Amaranthine blocky rhyolite porphyry, containing a small amount of breccia (10%) on the edge; intruded by a felsophyre vein with the width of about 8m inside this bed. 54.9 m

———————— Volcanic-eruption unconformable contact ————————

The lower part of the second member of Huangjian Formation — Total thickness: 889.2 m

20. Caesious rhyolitic crystalvitric ignimbrite, apparently rhyolitic bands developing and bypassing crystal pyroclast; the magma fragments: celadon, arranging directionally in a depressed and elongated manner, length: 3 cm, width: 0.1–0.2 cm. 57.5 m

19. Amaranthine rhyolitic crystalvitric tuff. 25.0 m

18. Celadon altered olivine-bearing basaltic porphyrite dike.

17. Amaranthine rhyolitic crystalvitric tuff; the crystal pyroclast near the vein: content and grain size increasing, about 2–5 mm; a conglomerate layer with the thickness of about 2 m exposing locally in this bed, the breccia in the layer: content is 30% and size is 2–5 cm; quartz vein locally developing with the width of about 5–8 cm.

172.3 m

16. Celadon altered olivine-bearing basaltic porphyrite vein.

15. Amaranthine rhyolitic crystalvitric tuff. 38.1 m

14. Amaranthine rhyolitic crystal pyroclast tuff. 63.3 m

13. Celadon altered olivine-bearing basaltic porphyrite vein.

12. Amaranthine rhyolitic crystalvitric tuff; the crystal pyroclast: quantity increasing and gain size increasing to 2–4 mm upwards: the gain size of the feldspar contained: large locally, up to 1 cm × 4 cm. 155.3 m

11. Celadon rhyolitic crystalvitric tuff, celadon in the lower part and becoming amaranthine upwards. 13.8 m

10. Gray rhyolitic breccia-bearing crystalvitric tuff, with less breccia in the lower part. 6.5 m

3. Generally amaranthine rhyolitic conglomerate-breccia-bearing crystalvitric tuff, and locally rhyolitic crystal pyroclast tuff. The thickness of a single bed: 1–5 m; interbedded with amaranthine siltstone (10cm) and celadon sedimentary tuff (1m) locally. The conglomerate breccia: a small amount locally, crystal pyroclast increasing and gradually becoming rhyolitic crystal tuff. 94.2 m

2. Amaranthine rhyolitic crystalvitric tuff, apparently rhyolitic structure locally developin 40.5 m

1. Gray – celadon cataclasite, components: broken altered tufaceous conglomerate-bearing sandstone, vitric tuff, etc.; the broken zone: about 12m wide, discontinously arranged lens visible; the lens: 20–40 cm long and 6–10 cm wide. 8.1 m

——————————— Fault contact ———————————

The third member of Late Cambrian Xiyangshan Formation Total thickness >11.7 m

0. Gray medium–thick laminated calcareous mudstone. 11.7 m

Volcanic intrusion

24. Flesh red fine-grained granite porphyry.

25. Dark-celadon crushed fine-grained quartz diorite vein.

——————————— Intrusive contact ———————————

The third member of Huangjian Formation

26. Celadon dacitic breccia-bearing crystalvitric tuff. Owing to strong silicified alteration, local tuff becoming dense, blocky dacitic vitric ignimbrite; wide inserted and quartz felsite vein with a width of about 5m intruding and interpenetrating locally.

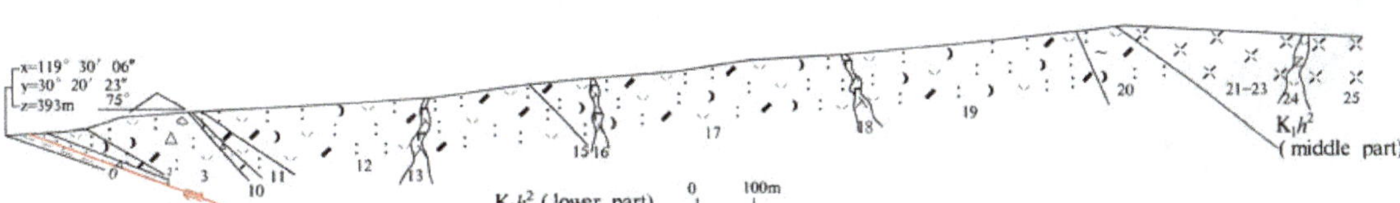

Fig. 2.93 Second member of Huangjian Formation (K_1h^2) profile in Tianmu Mount Scenic Area, Lin'an City, Zhejiang Province

The lithology of the third member of Huangjian Formation is described by taking the profile of the second member of Early Cretaceous Huangjian Formation (K_1h^2) (Fig. 2.94) of Chuancun map sheet in Dongken Village, Lin'an City, Zhejiang Province. The details are as follows:

The lithology of the third member of Huangjian Formation is described by taking the profile of the third member of Huangjian Formation (K_1h^3) (Fig. 2.95) of Chuancun map sheet in Dashancun–Tianhuangping area, Gaohong Town, Lin'an City, Zhejiang Province. The details are as follows:

27. Light flesh red felsite vein.

28. Gray rhyolitic crystalvitric tuff.

29. Light flesh red felsite vein.

30. Dark gray dacitic breccia-bearing crystalvitric ignimbrit e.

31. Light off-white thin–medium laminated tufaceous arkose, the thickness of a single layer: 5–30 cm, horizontal bedding extremely developing, diagonal bedding locally developing; components: fine sand generally, interbedded with 2–20 mm thick medium sand in the middle part.

32. Gray dacitic crystalvitric ignimbrite, apparently rhyolitic structure developing, containing an extremely small amount of breccia locally. The bioclastic pyroclast has been devitrified and becomes felsic aggregate, with the shape of plastic vitreous pyroclasts generally.

——————— Volcanic-eruption unconformable contact ———————

The second member of Huangjian Formation

33. Purple-gray–gray-black rhyolitic ignimbrite.

Volcanic intrusion

19. Grayish - light flesh red porphyritic mid-fine grained quartz monzonite.

——————— Intrusive contact ———————

The middle part of the second member of Huangjian Formation — Total thickness: 2317.9 m

18. Light flesh red–flesh red fine–micro grained rhyolitic tuff lava. — 149.9 m

17. Light flesh red –flesh red fine-grained granite porphyry.

16. Light flesh red–flesh red fine–micro grained rhyolitic tuff lava. — 555.8 m

15. Celadon andesite dikes.

14. Light flesh red–flesh red fine–micro grained rhyolitic tuff lava. — 82.5 m

13. Light flesh red medium–micro grained rhyolitic tuff lava, off-white locally, the content of plagioclase increasing. — 623.0 m

12. Light flesh red fine–micro grained rhyolitic tuff lava, the thickness of the matrix decreasing obviously compared with the beds above. — 88.9 m

11. Light flesh red–light ,off-white medium–fine grained rhyolitic tuff lava, the grain size of the minerals slightly decreasing compared with the beds above. 101.9 m

10. Celadon andesitic dike.

9. Grayish–gray, medium-fine grained rhyolitic tuff lava, two sets of columnar joints developing; the respective intervals between horizontal joints and between vertical joints: about 150 cm and 10–90 cm; light flesh red fine-grained orthophyre vein with a width of about 8m visible locally. 261.3 m

8. Celadon andesite and light flesh red orthophyre dikes.

7. Grayish–gray medium-fine grained rhyolitic tuff lava, a set of joints with the attitude of 10°∠80° developing; the interval between the joints: about 20–100 cm. 77.4 m

———————— Fault contact ————————

6. Light flesh red crushed medium-fine grained rhyolitic tuff lava; pretty broken in general, secondary fracture surface developing internally; the width of the broken zone: about 10m as a whole; cataclasite and broken mud with a thickness of a about 10–40cm developing near the main fracture surface; the main fracture surface and the secondary fracture surface constituting a lentoid broken zone with a size of about 6 m × 2 m, belonging to tensile-torsional normal fault. 20.6 m

5. Gray-yellow andesitic dike.

4. Light flesh red fine-grained porphyritic syenite vein.

———————— Intrusive contact ————————

The lower of the second member of Huangjian Formation Total thickness: >174.0 m

3. Gray-purple rhyolitic crystalvitric tuff, silicified alteration developing. 41.4m

2. Celadon andesitic dike.

1. Gray rhyolitic crystalvitric tuff; strong silicified alteration developing; vitric pyroclast: recrystallized and micro-fine grained, with strongly coarse fracture surface. The rocks in this bed become dark gray-purple and denser gradually frontwards. The crystal pyroclasts of K-feldspar and quartz feature clear contour, and the crystal pyroclasts of K-feldspar increases locally. 132.6 m

Subvolcanic rock

22. Celadon—tawny rhyolite porphyry.

———————— Intrusive contact ————————

The third member of Huangjian Formation Total thickness: 2040.2 m

21. Gray–grayish rhyolitic conglomerate-breccia-bearing crystal ignimbrite; the agglomerate: (6–10cm) × (2–6 cm) in size; the breccia: 2–6 mm in size, mainly composed of amaranthine and celadon ignimbrite and tuff. A small amount of celadon magma fragment bands developing; the bands: arranging directionally and discountinously, size: (5–30) mm (length) × (1–5) mm (width). 257.0 m

20. Gray–dark gray dacitic crystal ignimbrite; light flesh red syenite and dark gray dacitic breccia visible individually; the size of the breccia: 10 cm × 5 cm–3 cm × 1.5 cm. 313.2 m

19. Gray ivernite.

18. Grayish–gray rhyolitic dacitic breccia-bearing crystalvitric ignimbrite, apparently rhyolitic structures visible locally. 571.2 m

17. Dark gray dacitic crystalvitric ignimbrite generally and vitric ignimbrite locally. Tufaceous cement feasures obviously higher density and high welding degree compared with the beds above. 7.0 m

16. Light flesh red fine-grained syenite vein.

15. Gray–dark gray dacitic crystal-bearing crystalvitric ignimbrite. 19.3 m

14. Gray ivernite.

13. Gray–dark gray dacitic breccia-bearing crystalvitric ignimbrite; the welding degree of the rocks in this bed is obviously higher compared with the beds above, and local rocks become gray-purple. 8.2 m

12. Celadon andesitic dike.

11. Grayish rhyolitic dacitic breccia-bearing crystalvitric tuff with weak welding. 96.8 m

10. Off-white felsite vein.

9. Gray-purple dacitic crystal vitric ignimbrite, nearly upright cleavages developing in the rocks on the edge of the contact zone. 60.6 m

8. Light flesh red fine-grained granite porphyry.

7. Gray dacitic crystalvitric ignimbrite, the content of crystal pyroclasts obviously decreases locally. 260.7 m

6. Gray dacitic crystalvitric ignimbrite, apparently rhyolitic structure developing. 47.9 m

5. Gray-purple rhyolitic dacitic breccia-bearing crystalvitric ignimbrite, apparently rhyolitic structure developing, agglomerate invisible, the content of breccia obviously decreases. 36.2 m

4. Off-white rhyolitic dacitic conglomerate-breccia-bearing crystalvitric ignimbrite; the respective size of the agglomerate and the breccia: 6–10 cm and 2–3 cm; both the agglomerate and the breccia mainly composed of (crystal) vitric ignimbrite and cryptocrystalline siliceous matter, etc.; apparently rhyolitic structure developing. 65.2 m

3. Grayish–dark gray rhyolitic dacitic crystalvitric ignimbrite, containing a small amount of dark gray–grayish breccia with a size of 2–8 mm. 113.8 m

2. Light flesh red rhyolitic tuff lava. 27.8 m

1. Gray rhyolitic dacitic breccia-bearing crystal ignimbrite; compared with the lava in the beds above, the outcrops feature obviously coarser weathering surface and no developed joint structure. Intrusion of andesitic dike with a width of about 10 cm is visible locally. 65.3 m

Volcanic-eruption unconformable contact

The second member of Huangjian Formation Total thickness > 12.4 m

0. Light flesh red - dark gray rhyolitic tuff lava. 12.4 m

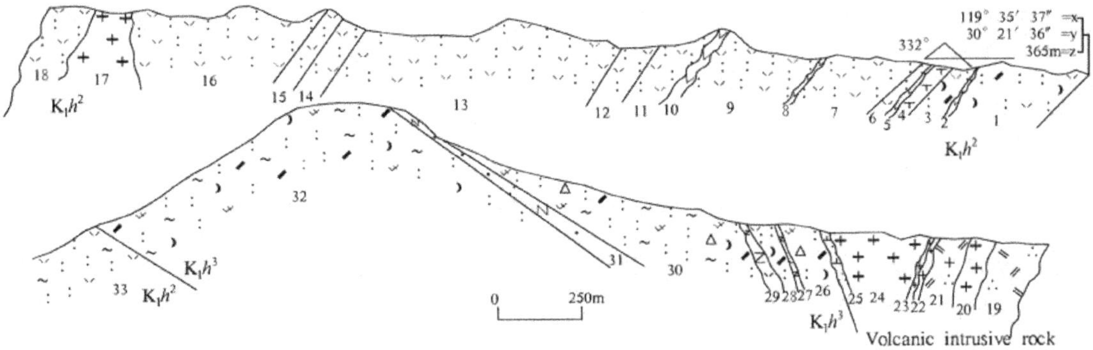

Fig. 2.94 Second member (K_1h^2)–third member (K_1h^3) tectonic–stratigraphic profile of the Early Cretaceous Huangjian Formation in Dongken Village, Lin'an City, Zhejiang Province

2. **Lithological characteristics**

The lithology of the first member of Huangjian Formation (K_1h^1) is mainly characterized by dacitic conglomerate-bearing (magma-fragment-bearing) crystalvitric (strong) ignimbrite, rhyolitic dacitic crystalvitric tuff, and the interbeds consisting of tufaceous conglomerate-bearing sandstone, tufaceous sandstone, and siltstone.

(1) The dacitic conglomerate-bearing (magma-fragment-bearing) crystalvitric (strong) ignimbrite: light purple and gray, interbedded with amaranthine rhyolitic conglomerate-bearing crystalvitric tuff locally, interbedded with unstable thin laminated amaranthine tuffaceous conglomerate-bearing sandstone in its lower part. The thickness: greater than 682.2 m.
(2) The rhyolitic dacitic crystalvitric ignimbrite: light gray, thick–medium laminated generally, locally interbedded with 10–50-cm-thickness purple mudstone. The thickness: 125.2 m.
(3) The interbed consisting of tufaceous conglomerate-bearing sandstone, tufaceous sandstone, and siltstone: light purple-gray, medium–thin laminated generally, featuring internal dipping. The thickness: about 47.7 m.

The lithology of the second member of Huangjian Formation (K_1h^2) is mainly characterized by rhyolitic crystalvitric ignimbrite, vitric blocky rhyolite porphyry, felsitic blocky rhyolite porphyry, felsitic blocky nevadite, porphyritic rhyolite, micro-(fine)-grained rhyolitic tuff lava, etc.

(1) The rhyolitic crystalvitric ignimbrite: grayish–gray; thick laminated–blocky rhyolitic agglomerate breccia, rhyolitic crystalvitric tuff, and tufaceous conglomerate-bearing sandstone of volcanic sedimentation facies developing locally at the bottom. The thickness: 445.6–2373.6 m.
(2) The vitric blocky rhyolite porphyry, felsitic blocky rhyolite porphyry, and felsitic blocky nevadite: gray–dark gray, no rhyolitic structure developing. The phenocrysts feature broken cloudy shape that can be spliced together locally. The matrix features vitric structure on the edge and felsitic structure in the middle part. The thickness: about 4466 m.
(3) The porphyritic rhyolite: gray, grayish and gray purple, rhyolitic structure and a small amount of bubble structure developing, the attitude of rhyolitic structure

Fig. 2.95 Profile of the third member of Huangjian Formation (K_1h^3) (Fig. 2.95) in Dashancun–Tianhuangping, Gaohong Town, Lin'an City, Zhejiang Province

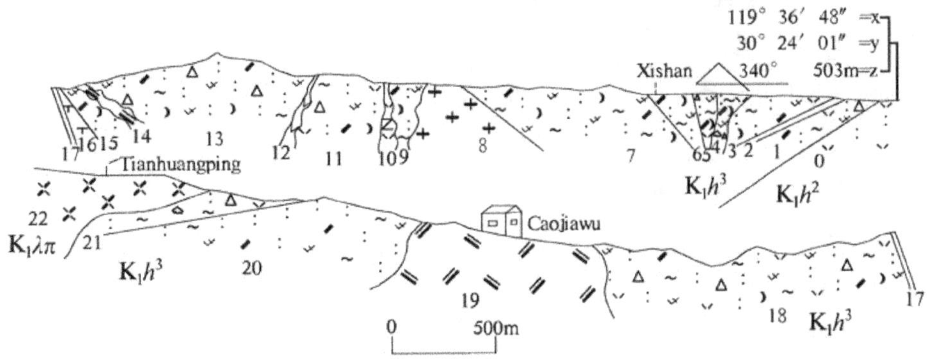

dipping northward or southward in general. The thickness: greater than 303.1 m.

(4) The micro-(fine)-grained rhyolitic tuff lava: grayish, no rhyolitic structure developing, composed of single component, the component of the phenocrysts: feldspar (20–40%). The thickness: greater than 2060 m.

The lithology of the third member of Huangjian Formation (K_1h^3) is mainly characterized by amaranthine and gray rhyolitic dacitic ignimbrite, and apparently rhyolitic structure band developed. The thickness of this member is about 341–2,701.8 m.

The lithology of the fourth member of Huangjian Formation (K_1h^4) is mainly characterized by light purple-gray–grayish bubble rhyolite or porphyritic rhyolite. The thickness of this member is about 560.1 m.

For specific lithological characteristics, refer to the sections about the features and types of volcanic rocks in Sect. 3.2.

2.5.2 Biostratigraphy and Chronostratigraphy

In this Project, no fossils were discovered in Huangjian Formation. According to previous information such as the regional geological survey on a scale of 1:200,000 of Lin'an map sheet in the Area, the biological fossils were mainly discovered locally in the first member (the former Laocun Formation) of Huangjian Formation all through in Huangjian Formation. The outcropped fossils mainly include the ones of *Ephemeropsis trisetalis* insects, *Zamites sp.* (new united), *Solenites sp.* (*approximate murrayana L. et H*), *Yong'an Sagenopteris, Podozamites lanceolatus, Coniopteris burejensis, Brachyphyllum obesum, midsized brachygrapta, Chinese Bairdestheria, Yumen bairdestheria,* and *Zhejiang bairdestheria*. Among these organisms, *Ephemeropsis trisetalis insects* are the most typical.

According to zircon U-Pb dating of the volcanic rocks and subvolcanic rocks in Huangjian Formation in the area, the age range of the volcanic rocks is 135.2–125.4 Ma (refer to the sections about the age and rhythm of volcanic eruption in Sect. 3.2 for specific chronological features). It is basically consistent with the age (134–115 Ma) obtained in the area where Jiande Zhou family is located by Li et al. (2011), geologically belonging to the Early Cretaceous.

2.6 Quaternary System

The Quaternary in the Area mainly consists of Yinjiangqiao Formation (Qh*y*). Yinjiangqiao Formation is about 61.25 km^2, accounting for about 4.6% of the total area. It is distributed in Hanggai Town and Baofu Town in southeast of Hanggai map sheet and Zhangcun Town and Xianxia Town in the northwest of Xianxia. Besides, it is exposed locally in Shanchuan Town and Baishuijian scenic area of Chuancun map sheet. The exposed strata are single, and only Holocene Yinjiangqiao Formation is visible. In terms of the origin types of formation, alluviation is the main type, and eluvial and proluvial rank the second.

Yingjiangqiao Formation (Qh*y*) is mainly distributed in the riverbeds of open and wide river valleys, first terrace, washland terrace as well as narrow and long gulley, such as the basin of Fushi Reservoir and Laoshikan Reservoir. A small amount of the formation is distributed in the foot-slope area in front of mountains. The formation is mainly formed by proluvial–alluviation and generally consists of banded landforms such as valley plains, river terrace, and proluvial fan. The lithology of this formation is mainly characterized by loose conglomerate stratum, sandy loam, and loam formed owing to alluviation and pluvial–alluvial, which constitute the dual structure of fluvial facies. In addition, there are small amounts of crushed rock layers and crushed-rock-bearing clay formed owing to eluvial and proluvial. The sand and conglomerate feature good sorting and roundness, and contain much cohesive soil locally. The formation is grayish yellow and gray generally, loosely structured, and 1–5 m thick. It overlies strata and migmatite of the pre-Cenozoic in the manner of angular unconformity.

References

Anbar AD, Knoll AH (2002) Proterozoic ocean chemistry and evolution: a bioinorganic bridge? Science 297:1137–1142

Blackwelder E (1907) Paleozoic In: Willis B, Blackw elder E, Sargent H (eds) Research in China, vol 1 Carnegie Institution of Washington, pp 136–152

Chen PJ (2000) Comments on the Classification and Correlation of Non-marine Jurassic and Cretaceous of China[J]. J Stratigr (2):114−119. (in Chinese with English abstract)

Chen X, Zhang YD, Fan JX, Cheng JF, Li QJ (2010) Ordovician graptolite-bearing strata in southern Jiangxi with a special reference to the Kwangsian Orogeny. Sci China Earth Sci 53(11):1602–1610. https://doi.org/10.1007/s11430-010-4117-6

Chen X, Zhang YD, Fan JX, Tang L, Sun MQ (2012) Onset of the Kwangsian Orogeny as evidenced by biofacies and lithofacies. Sci China Earth Sci 42(11):1617–1626 (in Chinese with English abstract)

Chen X, Fan JX,Chen Q, Tang L, Hou XD (2014) Toward a stepwise Kwangsian Orogeny. Sci China: Earth Sci 44(5):842–850. (in Chinese with English abstract)

Chen XH, Wang CS, Zhang M (2009) Ordovician Chitinozoan from South China. Geological Publishing House, Beijing pp 1 − 260. (in Chinese)

Churkin MJR, Carter C (1970) Early Silurian graptolites from southeastern Alaska and their correlation with graptolitic sequences in North America and the Arctic. US Geol Surv Prof Pap 653:1–51

Claypool GE, Holser WT, Kaplan IR et al (1980) The age curves of sulfur and oxygen isotopes in marine sulfate and their mutual interpretation. Chem Geol 28:199–260

Cooper RA, Sadler PM (2004) Ordovician System [M]. In: Gradstein FM, Ogg JG, Smith AG et al (eds) A geologic time scale 2004. Cambridge University Press, London, pp 165–187

Davies KA (1929) Notes on the graptolite faunas in the Upper Ordovician and Lower Silurian. Geol Mag 66:1–27

Ding Y (1935) Geological bulletin of Tongling and Xiuning, Anhui. Mine Test Newsletter

Ding BL, Li YX, Wang YP, Feng NS, Zhang Y, Yan YK (1999) The detailed study and advance of cretaceous in Zhejiang province[J]. Volcanol Miner Resour (4):241–286. (in Chinese)

Dong L, Song WM, Xiao SH, Yuan XL, Chen Z, Zhou CM (2012) Micro-and macrofossils from the piyuancun formation and their implications for the ediacaran-cambrian boundary in southern Anhui. J Stratigr 36(3):600–610 (in Chinese with English abstract)

Elles GL, Wood EMR (1907) A monograph of British graptolites. Part 6. Monograph of the Palaeontographical Society, pp 217–272

Gao LZ, Zhang H, Ding XZ, Liu YX, Zhang CH, Huang ZZ, Xu XM, Zhou ZY (2014) SHRIMP zircon U-Pb dating of the Jiangshan-Shaoxing faulted zone in Zhejiang and Jiangxi. Geol Bull China 33(6):763–775 (in Chinese with English abstract)

Gradstein FM, Ogg JG, Smith AG et al (2004) A new geologic time scale, with special reference to Precambrian and Neogene. Episodes 27(2):83–100

Hall TS (1902) Reports on graptolites. Rec Geol Surv Victoria 1:104–118

Hofmann AW (1988) Chemical differentiation of the Earth: the relationship between mantle, continental crust, and oceanic crust. Earth Planet Sci Lett 90(3):297–314

Hoskin PWO, Schaltegger U (2003) The composition of zircon and igneous and metamorphic petrogenesis. Rev Mineral Geochem 53:27–62

Hu YH, Sun WD, Ding X, Wang FY, Ling MX, Liu J (2009) Volcanic event at the Ordovician-Silurian boundary: the message from K-bentonite of Yangtze Block. Acta Petrol Sinica 25(12):3298–3308. (in Chinese with English abstract)

Hu YH, Zou JB, Song B, Li W, Sun WD (2008) SHRIMP zircon U-Pb dating from K-bentonite in the top of Ordovician of Wangjiawan Section, Yichang, Hubei, China. Sci China (Series D: Earth Sci) 38 (1):72–77. (in Chinese with English abstract)

Huang ZG (1982) Late Ordovician and early Silurian graptolite assemblages of shenzha area, Tibet and Ordovician—Silurian boundary[C]. Contrib Geol Qinghai-Tibet Plateau (5):27–52, 171–173. (in Chinese)

Huff WD, Bergström SM, Kolata DR et al (1998) Ordovician K–bentonites in the Argentina Precordillera: relations to Gondwana margin evolution. In: Pankhurst RJ, Rapela CW (eds) The Proto–Andean Margin of Gondwana. Geological Society of London Special Publications, 142, 107–126

Huff WD, Bergström SM, Kolata DR et al (2003) Ordovician K–bentonites in the Argentine Precordillera and their relation to Laurentian volcanism. In: Albanesi GL, Beresi MS, Peralta SH (eds) Ordovician from the Andes. Universidad Nacional de Tucumin, INSUGEO, Serie Correlacih Geolbgica, 17, 197–202

Huff WD, Bergström SM, Kolata DR (1992) Gigantic Ordovician volcanic ash fall in North America and Europe: biological, tectonomagmatic, and event–stratigraphic significance. Geology 20:875–878

Huff WD (2008) Ordovician K–bentonites: Issues in interpreting and correlating ancient tephras. Quatern Int 178(1):276–287

Jaanusson V (1960) Graptoloids from the Ontikan and Viruan (Ordov.) limestones of Estonia and Sweden. Bull Geol Inst Uppsala 38 (4):289–365

Ju TY (1983) Early Cambrian Trilobites From the Hotang and Dachenling Formations of Zhejiang. Acta Palaeontol Sinica 22 (6):628–639 (in Chinese with English abstract)

Koren TN, Melchin MJ (2000) Lowermost Silurian graptolites from the Kurama Range, eastern Uzbekistan. J Paleontol 74(6):1093–1113

Koren TN, Mikhaylova NF (1980) Class Graptolithina. In: Nikitkin IF (ed) The Ordovician–Silurian Boundary. Nauka Kazakhstan SSR Publishing House, Alma–Ata, pp 1–300. (in Russian)

Lapworth C (1877) The graptolites of County Down. Belfast Nat Field Club 125–144

Laufeld S (1967) Caradocian chitinozoa from Dalama, Sweden. Geol Foren Stockholm Forh 89:275–349

Li J, Li YY (1930) Stratigraphic comparison map of the lower Yangtze River in China. Former Central Academia 19th Annual Report (in Chinese)

Li JJ, Yang XC (1983) Palaeontological Atlas of East China, Part 1, Early Palaeozoic. In: Nanjing Institute of Geology and Mineral Resources (ed). Geological Publishing House, Beijing, pp 1–657. (In Chinese)

Li JJ (1984) Graptolites across the Ordovician–Silurian boundary from Jingxian, Anhui, 309–370. In: Nanjing Institute of Geology and Paleontoloy, Chinese Academy of Sciences (ed) Stratigraphy and palaeontology of systemic boundaries in China, Ordovician—Silurian boundary. Anhui Science and Technology Publishing House, Hefei

Li SG, Zhao YZ (1924) Xiadong Geology and the history of the Yangtze River. Chinese Geol Soc 3(3–4). (in Chinese)

Li WN, Yu CL (1965) Discovery of *Arthricocephalus* in the western Zhejiang. Geol Rev 23(6):510–511 (in Chinese)

Li XH, Chen SD, Luo JH, Wang Y, Cao K, Liu L (2011) LA-ICP-MS U-Pb isotope chronology of the single zircons from early Cretaceous Jiande Group in Western Zhejiang, SE China: significances to Stratigraphy.Geol Rev 57(6):825–836. (in Chinese with English abstract)

Lin YR, Jiang WS, Xu KD, Zhen JS (1989) Zhejiang Cretaceous. In: Selected Papers of the Cretaceous Conference of Southern China. Nanjing University Press, Nanjing, pp 63–68. (in Chinese)

Liu DY, Jian P, Kröner A et al (2006) Dating of prograde metamorphic events deciphered from episodic zircon growth in rocks of the Dabie-Sulu UHP Complex, China. Earth Planet Sci Lett 250:650–666

Lu YH, Lin HL (1983) Zonation and correlation of Cambrian Faunas in western Zhejiang [J]. Acta Geological Sinica 4:317–340 (in Chinese with English abstract)

Lu YH, Mu EZ, Hou YT, Zhang RD, Liu DY (1955) New insights into the Paleozoic strata in western Zhejiang. Geol Knowledge 2:1–6 (in Chinese)

Marshall CR (2006) Explaining the Cambrian "explosion" of animals. Annu Rev Earth Planet Sci 34:355–384

Mu EZ, Ni YN (1983) Uppermost Ordovician and Lowermost Silurian graptolites from the Xainza area of Xizang (Tibet) with discussion on the Ordovician-Silurian boundary[J]. Palaeontologia Cathyana, (1):155−180 (in Chinese)

Nicholson H (1868) On the Graptolites of the Coniston Flags; with Notes on the British Species of the Genus Graptolites[J]. Q J Geol Soc 24:521−545. https://doi.org/10.1144/GSL.JGS.1868.024.01-02.67

Obut AM, Sobolevskaya PF (1967) In: Obut AM, Sobolevskaya RF, Nikolaev, AA (eds) Graptolites and stratigraphy of the lower silurian along the margins of the Kolyma Massif. Akademiya Nauk SSR, Sibirskoe otdelenie, Institut geologii i geofiziki, Ministerstvo geologii SSSR, Nauchno–issledovatel'sky institut geologii Arktiky, pp 1–164. (in Russian)

Pearce JA, Harris NBW, Tindle AG (1984) Trace element discrimination diagrams for the tectonic interpretation of granitic rocks. J Petrol 25:952–983

Peng SC (2011) Jiangshanian Stage (Cambrian, Furongian) and the GSSP for the base of the Stage established foramally. J Stratiraphy 35(4):393–396 (in Chinese with English abstract)

Perner J (1895) Studie o českých graptolitech II. 52 pp. Česká akademie císaře Františka Josefa pro vědy, slovesnost a umění, Praha

Qi JY (2006) Modern analytical testing techonology. Tongji University Press, Shanghai, pp 1–476. (in Chinese)

Qian YY, Li JJ, Li WN, Jiang NY, Bi ZG, Gao YX (1964) New Understanding of Sinian and Lower Paleozoic in Southern Anhui[J]. J Nanjing Inst Geol Palaeontol, Chinese Academy of Sciences, Strata Corpus (1):21–66 (in Chinese)

Reed FRC (1915) Supplementary memoir on new Ordovician and Silurian fossils form the Northern Shan States. Palaeontol Indica (New Ser) 6:1–98

Sheehan PM (2001) The Late Ordovician mass extinction. Annu Rev Earth Planet Sci 29:331–364

Sheng XF (1951) Zhejiang strata. Geol Zhejiang 2:1–18 (in Chinese)

Shu LS (2006) Predevonian tectonic evolution of South China: from Cathaysian block to Caledonian period folded orogenic belt. Geol J China Univ 12(4):418–431 (in Chinese with English abstract)

Shu LS (2012) An analysis of principal features of tectonic evolution in South China Block.Geol Bull China 31(7):1035–1053. (in Chinese with English abstract)

Sobolevskaya RF (1974) New Ashgill graptolites in the middle flow basin of the Kolyma–river. In: Obut AM (ed) Graptolites of the USSR. Nauka Siberian Branch, Novosibirsk, pp 63–71 (in Russian)

Su WB, Huff WD, Ettensohn FR et al (2009) K–bentonite, black–shale and flysch successions at the Ordovician-Silurian transition, South China: possible sedimentary responses to the accretion of Cathaysia to the Yangtze Block and its implications for the evolution of Gondwana. Gondwana Res 15:111–130

Sun SQ, Zhang CJ, Zhao SJ (2007) Identification of the tectonic settings for continental intraplate by trace elements. Geotectonica Et Metallogenia 31(112):104–109. (in Chinese with English abstract)

Sun SS, McDonough WF (1989) Chemical and isotopic systematics of oceanic basalts: Implication for mantle composition and processes. In: Saunder AD, Norry MJ (eds) Magmatism in the ocean basins. Geol Soc Spec Publ 42:313–345

Tang F, Yin CY, Gao LZ (1997) A New Dea Of Metaphyte Fossils From The Late Sinian Doushantuo Stage At Xiuning, Anhui Province[J]. Acta Geologica Sinica 71:289–296 (in Chinese with English abstract)

Taugourdeau P (1961) Chitinozoaires du Silurien d'Aquitaine. Revue Micropaleontologie 4(3):135–154

Taugourdeau P (1965) Chitinozoaires de l'Ordovicien des U.S.A.; comparaison avec les faunes de l'Ancien Monde. Rev Inst Fr Pet Ann Combust Liq. 20:463–485

Wang HZ, Shi XY (1998) Hierarchy of depositional sequences and eustatic cycles a discussion on the mechanism of sedimentary cycles. Geoscience 12(1):1–16 (in Chinese with English abstract)

Wang ZJ, Wang J, Jiang XS, Sun HQ, Gao TS, Chen JS, Qiu YS, Du QD, Deng Q, Yang F (2015) New progress for the stratigraphic division and correlation of neoproterozoic in Yangtze Block, South China. Geol Rev 61(1):1–22 (in Chinese with English abstract)

Xie JC, Chen S, Rong W, Li QZ, Yang XY, Sun WD (2012a) Geochronology, geochemistry and tectonic significance of Guniujiang A-type granite in Anhui Province. Acta Petrologica Sinica 28 (12):4007–4020 (in Chinese with English abstract)

Xie SK, Wang ZJ, Wang J, Zhuo JW (2012b) LA-ICP-MS zircon U-Pb dating of the bentonites from the uppermost part of the Ordovician Wufeng Formation in the Haoping section, Taoyuan, Hunan [J]. Sedimentary Geology and Tethyan Geology 32(4):65–69 (in Chinese with English abstract)

Xing YS, Gao ZJ, Liu GZ et al (1989) Stratigraphy of China No.3: the Upper Precambrian of China. Geological Publishing House, Beijing, pp 1–150. (in Chinese)

Yan YK, Jiang CR, Zhang SE et al (1992) Research on the Sinian System in the south of Zhejiang-Jiangxi-Anhui [J]. J Nanjing Institute of Geol Mineral Res Chinese Acad Geol Sci 20(supp.):44–49. (in Chinese)

Yang DQ (1964) Some lower silurian graptolites From Anji, Northwestern Zhejiang(Chekiang). Acta Palaeontologica Sinica 12(4). (in Chinese with English abstract)

Yang JD, Tao XC, Xue YS, Sun WG, Wang ZZ, Zhou CM (1997) Origin and age of Mn deposits of the Nantuo Formation in the Sinian System. Chin Sci Bull 42(14):1538–1541 (in Chinese with English abstract)

Yang Y (2011) Zircon U-Pb Geochronology and Genesis of K-bentonite at the Paleozoic-Mesozoic Key Stratigraphic Boundaries of South China. Master's degree thesis of China University of Geosciences. (in Chinese with English abstract)

Yin CY, Liu YQ, Gao LZ et al (2007) Phosphate fauna in the Early Sinian (Idikala): the characteristics of the Chun'an biota and its environmental evolution. Geological Publishing House, Beijing, pp 1–132. (in Chinese)

Yu GH (1996) Lithostratigraphy in Zhejiang Province. China University of Geosciences Press, Wuhan (in Chinese)

Yuan XL, Chen Z, Xiao SH, Wan B, Guan CG, Wang W, Zhou CM, Hua H (2012) The Lantian biota: a new window onto the origin and early evolution of multicellular organisms. Chin Sci Bull 58 (7):701–707. https://doi.org/10.1007/s11434-012-5483-6

Zhang LG (1989) Petrogenetic and minerogenetic theories and prospecting—Stable isotopic geochemistry of mian type ore deposits and granitoids of china[M]. Press of Beijing University of Technology, Beijing, pp 1−200. (in Chinese)

Zhang JM, Li GX, Zhou CM (1997) Geochemistry of light colour clayrock layers from the early Cambrian Meishucun Stage in Eastern Yunnan and their geological significance. Acta Petrologica Sinica 13(1):100–110 (in Chinese with English abstract)

Zheng YF, Chen JF (2000) Stable isotope geochemistry. Geological Publishing House, Beijing, pp 143–217

Zhejiang Bureau of Geology (1965) The reginal geological and minerals survey of Jiande map sheet on a scale of 1:200,000 (in Chinese). National Geological Archives of China [distributor], 2018. http://ngac.org.cn/Data/FileList.aspx?MetaId=E928A0F473A47A73E0430100007F3D67%26Mdidnt=x00079952

Zhejiang Bureau of Geology (1967) The reginal geological and minerals survey of Lin'an map sheet on a scale of 1:200,000 (in Chinese). National Geological Archives of China [distributor], 2018. http://ngac.org.cn/Data/FileList.aspx?MetaId=E928A11F195D7A73E0430100007F3D67%26Mdidnt=x00098215

Zhejiang Institute of Geological Survey (2007) The evaluation report of the Cambrian limestone resources survey in Zhejiang. (in Chinese)

Zhejiang Institute of Geological Survey (2015) Scientific report on the geology and structure of Jiangshan-Shaoxing Collision Belt (in Chinese). National Geological Archives of China [distributor], 2018. http://ngac.org.cn/Data/FileList.aspx?MetaId=7B2A429A8EC71AF0E05341015A0A70BD%26Mdidnt=x00144980

3 Intrusive, Volcanic and Vein Rocks

3.1 Intrusives

3.1.1 Overview

Intrusives in the survey area are mainly formed in the Mesozoic acid to intermediate-acid magmatism, which belongs to the Shunxi–Huzhou tectono-magmatic subbelt, north Zhejiang Province. There are 39 plutons of different sizes in the Anji–Chun'an area and some span the provinces of Zhejiang and Anhui, which are mainly stock and apophysis, followed by bosse and a few batholith. Distribution of the intrusives is closely related to structures, and regional faults and folds axial zone provided space for magmatic emplacement. In addition, the contact zone along the Mesozoic volcanic rock and the Paleozoic sedimentary rock is also a favorable emplacement location. The formation time of intrusives can be roughly divided into the Late Jurassic and Early Cretaceous, the former a total of 13 places accounting for one-third of the total number of plutons, the latter about 26 places accounting for two-thirds of the total number. However, based on the study of modern isotope geochronology, there is no obvious temporal interval between both periods, often in a transitional relationship, indicating that there were frequent magma activities from Late Jurassic to Early Cretaceous in northwestern Zhejiang. Intrusives are mainly distributed in the northern part of the survey area (the Anji–Lin'an belt). The outcroppings are less in the southern part of the survey area (the west Chun'an belt), only a small number of granodiorite plutons, but geophysical exploration data indicate that there may be concealed plutons under the overlying strata. The intrusives in the survey area are mainly granodiorite, monzonitic granite, and syenogranite, etc., and secondly quartz (porphyritic)-syenite and quartz monzonite, etc. Metallogenesis is mainly closely related to granodiorite, monzonitic granite, and syenogranite. The contact zone between plutons and wall rocks often developed alterations such as hornfelsic, silicification, pyritization, skarnization, and marbleization, while alterations such as albitization, potassic, greisenization, and sericitization were usually visible on plutons.

(1) Distribution Characteristics

In the survey area, the intrusives are widely distributed in large scale, with total outcropping about 254.38 km^2, and the area of plutons varies greatly ranging from 0.7 to 84.18 km^2 with stock or apophysis occurrence. From north to south and from west to east, there are mainly seven composite plutons: Ma'anshan, Tangshe, Tonglizhuang, Xianxia, Zhinanshan, Dongkeng, and Wushanguan. Among them, Ma'anshan, Tangshe, and Xianxia plutons mainly intruded along the northeast fracture zone or the core of the Nanhua–Cambrian anticlinorium. Tonglizhuang and Wushanguan plutons intruded mainly along the contact zone between the Paleozoic strata and the Mesozoic volcanic rocks, which were under the joint control of nearly EW, NE and NW fractures. Zhinanshan–Dongkeng pluton mainly intruded into the volcanic rocks. All these plutons were affected and damaged by the later NE and NW fracture structures to different degrees. The intrusive distribution, contact relationship, lithology, and rock-formation age of these intrusives are detailed in Table 3.1.

(2) Rock Classification

During the geological survey on intrusive rocks, the map expression mode of "lithology + grain size structure + age + stage" is used to analyzing and dividing lithological compositions and intrusive periods. First classify lithological categories and then structures, and finally, determine facies-change transitional contact or different periods of intrusive contact according to contact relationship. The results show that in the survey area, the lithology of intrusive rocks is mainly monzonitic granite and syenogranite, followed by syenite, quartz syenite, and quartz diorite. Monzonitic granite mainly has textures such as megacrystic-pegmatitic porphyritic and medium-grained

Table 3.1 Geological characteristics of intrusives in the survey area

Name	Geographical location	Area (km^2)	Geological features	Lithological combination		Zircon U-Pb age (Ma)	Lithologic symbol
				Lithology	Mineral assemblage		
Ma'anshan	Northeast: 119°27′30″ 30°40′00″ Southwest: 119°25′40″ 30°34′52″	83.86	In the strike of northeast, distributed in Shangshanling–Tongkengcun–Lingxi, its southeast side contacts by the NE50°–60° faults with the Paleozoic strata, the fault occurrence is 300°–310°∠60°–80°; its east side from Longxianshan to Xiayangcun is fine-grained syenogranite and megacrystic porphyritic monzonitic granite, in intrusive contact with the Ordovician strata; the border in the northeast side is not seen in the survey area, and based on previous study, it is in intrusive contact with the Silurian strata, and the contact surface occurrence is 300°–340°∠70°–80°, locally in fault contact, and the contact surface between plutons and strata has hornfelsic alteration	Pegmatitic porphyritic monzonitic granite	Quartz 35%, K-feldspar 50%, plagioclase 15%, a few biotite, and hornblende, etc., apatite and zircon occasionally seen; minerals grain size was generally <2 mm; quartz is in the anhedral granular shape, feldspar in the shape of block and strip, quartz and feldspar are distributed alternately in equal size	132.2 ± 1.6 SHRIMP	$\eta\gamma^{(W)}K_1^2$
				Phenocryst-porphyritic monzonitic granite	Phenocryst: plagioclase 5–8%, K-feldspar 2–7%, quartz 2%, with grain size of 1–3 cm; matrix: quartz 25%, K-feldspar 20%, plagioclase 20%, biotite 10–20%, a handful of hornblende and metallic minerals, with grain size of 2–5 mm in general, and >5 mm for a handful	127.7 ± 1.2 SHRIMP	$\eta\gamma^{(G)}K_1^2$
				Fine-grained syenogranite	Phenocryst: plagioclase 5–10%, K-feldspar 2–4%, and a handful of quartz 3–5%, generally the grain size >3 cm; matrix: quartz 25%, K-feldspar 20%, plagioclase 20%, biotite 10–20%, a handful of hornblende 1–2%, and the grain size is 3–8 mm	128.3 ± 1.1 SHRIMP	$\gamma^{(x)}K_1^2$
Tangshe	North: 119° 19′20″ 30°32′09″ East: 119°21′ 10″ 30°31′02″ Southwest: 119°15′57″ 30°29′34″	18.90	In the strike of northeast, intruded in the anticline core. The north part of the medium-coarse-grained syenogranite in Bijia Mount–Tali Mount is in intrusive contact with the Nanhua–Sinian strata, the contact zone has developed strong skarnization and hornfelsic, with the dip angle of 60°–80°; medium-grained monzonitic granite in Nanshan–Tangshe, in its south is in intrusive contact with the Nanhua–Cambrian strata, the contact zone has developed hornfelsic, or locally strong	Medium-grained monzonitic granite	Quartz 20%, K-Na-feldspar 30%, plagioclase 35%, biotite 5–10%, hornblende 5%, a handful of pyrite, and magnetite, etc., zircon and apatite visible occasionally; minerals have smaller grain sizes, generally 2–3 mm or few 0.5–2 mm	140.9 ± 3.4 LA-ICP-MS	$\eta\gamma^{(z)}K_1^1$
				Medium-coarse-grained syenogranite	K-feldspar 40%, plagioclase 15%, quartz 30%, biotite 5–10%, hornblende 3%, a handful of metallic minerals, and zircon occasionally seen; minerals have a bigger grain size, generally 3–9 mm, up to 1–2 cm for some K-feldspar	132.2 ± 1.6 LA-ICP-MS	$\gamma^{(c)}K_1^2$

(continued)

Table 3.1 (continued)

Name	Geographical location	Area (km^2)	Geological features	Lithological combination		Zircon U-Pb age (Ma)	Lithologic symbol
				Lithology	Mineral assemblage		
			skarnization; in the southeast side of main pluton, there are developed small stocks such as medium-grained syenogranite, fine-grained syenogranite, and fine-grained monzonitic granite		and >3 cm for a handful of them		
				Fine-grained syenogranite	Quartz 30%, K–Na-feldspar 45%, plagioclase 20%, biotite 5%, a handful of metallic minerals, zircon, and apatite occasionally seen; general grain size is 0.5–2 mm, and a handful of grain size up to 2–3 mm	125.0 ± 2.0 LA-ICP-MS	$\gamma^{(x)}K_1^2$
Tonglizhuang	North: 119° 29′44″ 30°30′40″ South: 119° 28′56″ 30°29′57″	1.54	It partially outcrops near Baofu Town in the survey area. Fine-grained syenogranite at the north side of the pluton is in intrusive contact with the Cambrian–Ordovician strata, the contact surface occurrence is 300°–330°∠60°–80° and it has developed hornfelsic alteration. In the south side of the pluton at Shimendong, medium-coarse-grained porphyritic quartz syenite is in intrusive contact with the first member of the Huangjian Formation volcanic strata. The medium-grained monzonitic granite in the southeast side also intrude into the first member of the Early Cretaceous Huangjian Formation, with the intrusive contact surface occurrence of 320°∠70°	Coarse-medium-grained porphyritic quartz syenite	Phenocryst: K–Na-feldspar 25%, plagioclase 5%, and dark-colored minerals 2%, minerals have bigger grain sizes, generally over 2 mm, even up to 6–7 mm individually; a handful of plagioclase phenocryst has eroded into clay minerals. Matrix: feldspar 48% and quartz 12%, with grain size below 0.5 mm in general	126.0 ± 3.0 LA-ICP-MS	$\xi o^{(z)}K_1^3$
				Fine-grained syenogranite	K-feldspar 40–50%, plagioclase 20–25%, quartz 20–30%, and a handful of biotite 5%, with grain size of 0.2–1 mm; K-feldspar is mainly orthoclase, wide plate euhedral crystal, Carlsbad twin visible in some of them; plagioclase is in the shape of subhedral wide plate and plate-column, albite bicrystal visible in some of them	About 130	$\gamma^{(x)}K_1^2$
				Medium-grained monzonitic granite	Quartz 25%, K–Na-feldspar 30%, plagioclase 30%, biotite 5–10%, hornblende 5%, zircon and apatite occasionally seen; grain size is 2–3 mm and a handful of 1–2 mm	142.3 ± 1.8 LA-ICP-MS	$\eta\gamma^{(z)}K_1^1$
Zhinan Mount	119°34′22″ 30°22′10″	0.7	The NW-strike small apophysis intruded in the rhyolitic tuff lava of the second member of the Huangjian Formation, about 1 km long and 700 m wide	Medium-grained quartz diorite	Quartz 22%, K–Na-feldspar 5%, plagioclase 50%, hornblende 20%; minerals' grain size is generally 0.4–3 mm. Quartz is granular and distributed pretty evenly; plagioclase is plate-column, most weathered, based on the refractive index it is mainly andesine—oligoclase; hornblende is column	130.5 ± 1.7 LA-ICP-MS	$\delta o^{(z)}K_1^3$

(continued)

Table 3.1 (continued)

Name	Geographical location	Area (km^2)	Geological features	Lithological combination		Zircon U-Pb age (Ma)	Lithologic symbol
				Lithology	Mineral assemblage		
Xianxia	Northeast: 119°25′43″ 30°29′20″ Southwest: 119°15′05″ 30°20′18″	84.18	In the NE-strike, it spreads in Xianxia–Zhangcun, the southeast side of the pluton is mainly in NE30°–40° fault contact with the Paleozoic strata, with the occurrence of 320°–330°∠60°–80°, and locally intruded into the Cambrian–Ordovician strata; its northwest side is in intrusive contact with the Nanhua–Ordovician strata, with the contact surface occurrence of 310°–350°∠60°–90° and fault contact exists locally	Medium-coarse-grained quartz syenite	K-Na-feldspar 85%, plagioclase 10%, quartz 5%, biotite 5–10%; grain size is 3–10 mm, and a handful of minerals' grain size is up to 1 cm; quartz is granular; feldspar is mostly plate and short strip, locally is arranged directionally, and individual plagioclase 1–2 cm	About 130 Ma	$\xi o^{(c)} K_1^3$
				Fine-grained syenogranite	Quartz 30%, K-feldspar 55%, plagioclase 10%, biotite 5%; grain size 0.5–2 mm	132.4 ± 2.4 LA-ICP-MS	$\gamma^{(x)} K_1^2$
				Medium-coarse-grained and coarse-medium-grained syenogranite	Quartz 30%, K-feldspar 45%, plagioclase 20%, biotite (5%); grain size generally 5–10 mm for internal-facies minerals and up to 2 cm for a few, and locally it becomes medium-coarse-grained syenogranite, the grain size of margin minerals is 3–6 mm	132.9 ± 3.3 LA-ICP-MS	$\gamma^{(c)/(z)} K_1^2$
				Phenocryst and medium-coarse-grained porphyritic monzonitic granite	Phenocryst: feldspar and quartz, in the content of 10–15%, grain size up to 1–3 cm; Matrix: quartz 20%, K–Na-feldspar 20%, plagioclase 30%, biotite 5–10%, hornblende 5% the grain size is 2–5 mm, occasionally 5–8 mm	144.2 ± 1.0 SHRIMP	$\eta\gamma^{(G)/(c)} K_1^1$
				Medium- and fine-grained (porphyritic) monzonitic granite	Quartz 15%, K-feldspar 30%, plagioclase 40%, biotite 5–10%, hornblende 5%; minerals' grain size is generally 2–5 mm, a handful of plagioclase is bigger in grain size, up to about 1 cm and dark-colored minerals increased in the margin of the pluton	145.1 ± 1.2 SHRIMP	$\eta\gamma^{(z)} K_1^1$
Wushanguan	Northwest: 119°38′04″ 30°29′59″ Southwest: 119°36′38″ 30°25′03″ Southeast: 119°43′18″ 30°24′29″	64.00	It spreads in the NW-strike, its west side is mainly in fault contact with the Huangjian Formation volcanic strata, the fault's strike is NW 320°–330° and NE 40°, it dips toward southwest and northwest, its dip angle varies between 65°–80°; locally in intrusive contact with volcanic strata (or the Cambrian Yangliugang Formation). Its northeast	Coarse-medium-grained syenite	K–Na-feldspar 85%, plagioclase 10%, quartz 5%, biotite 5–10%; the grain size is 3–8 mm, but up to 1 cm for a handful of minerals; quartz is granular; feldspar is mostly plate and locally short strip, it is arranged directionally, and individual plagioclase 1–2 cm	About 130	$\xi^{(z)} K_1^3$
				(Porphyritic) fine-grained syenogranite	K-feldspar 60–65%, quartz 25–35%,	128.1 ± 0.82 SHRIMP	$\gamma^{(x)} K_1^2$

(continued)

Table 3.1 (continued)

Name	Geographical location	Area (km^2)	Geological features	Lithological combination		Zircon U-Pb age (Ma)	Lithologic symbol
				Lithology	Mineral assemblage		
			and south sides are in intrusive contact with the Huangjian Formation volcanic strata; its southeast side is in intrusive contact with the Cambrian strata, the intrusive contact surface is not constant in attitude and dip, with dip angle of 60°–90°, and hornfelsic and skarnization are often seen near the intrusive contact zone		plagioclase 5–10%; the grain size is mainly fine, about 0.5–1.5 mm		
				Phenocryst-porphyritic monzonitic granite	Phenocryst: feldspar and quartz, in the content of 10–15%, grain size up to 1–3 cm; matrix: quartz 20%, K-Na-feldspar 20%, plagioclase 30%, biotite 5–10%, hornblende 5%, the grain size is generally 2–5 mm, occasionally 5–8 mm	About 130	$\eta\gamma^{(G)} K_1^2$
				Medium-coarse- and medium-grained syenogranite	K-feldspar 45–55%, plagioclase 15–25%, quartz 25–35%, biotite 1%; the grain size is 2–7 mm with few 7–10 mm, and locally, K-feldspar is approximately porphyritic (1–2 cm × 1–1.5 cm); from southeast to northwest, it evolves from medium-coarse-grained to coarse-medium-grained	132.0 ± 3.0 LA-ICP-MS	$\gamma^{(z)} K_1^2$
				Medium-grained monzonitic granite	Quartz 20–30%, K-feldspar 15–30%, plagioclase 30–40%, biotite 2–6%, very few zircon and apatite; the grain size is 0.5–4 mm. Inside the lithology there are developed mafic micro-granular enclaves (MMEs) such as biotite or biotite, plagioclase, in the ellipse, long strip and irregular shapes, in the sized of 1–30 cm	136.3 ± 1.2 SHRIMP	$\eta\gamma^{(z)} K_1^1$
Dongkeng	119°35′34″ 30°24′02″	1.2	Ellipse-like small apophysis in NEE strike, intruded in the contact zone between the second–third member of the Huangjian Formation, from outside to inside they are coarse-medium-grained quartz monzonite, medium-coarse-grained syenogranite, and fine-grained syenogranite	Fine-grained (porphyritic) syenogranite	Phenocryst: K–Na-feldspar 10%, plagioclase 4%; the grain size is 0.1–1 mm and 0.5–2 mm, but 2–6 mm for very few of it; matrix: K–Na-feldspar 46%, plagioclase 8%, quartz 30%, a handful of metallic minerals, and the grain size is 0.1–1 mm	127.6 ± 1.2 LA-ICP-MS	$\gamma^{(x)} K_1^3$
				Medium-coarse-grained syenogranite	Quartz 40%, K–Na-feldspar 54%, plagioclase 6%, a handful of biotite; the grain size is 3–8 mm, and 8–12 mm for a handful	127.9 ± 1.3 LA-ICP-MS	$\gamma^{(c)} K_1^3$

(continued)

Table 3.1 (continued)

Name	Geographical location	Area (km^2)	Geological features	Lithological combination		Zircon U-Pb age (Ma)	Lithologic symbol
				Lithology	Mineral assemblage		
				Fine-medium-grained porphyritic quartz monzonite	Phenocryst: quartz (very few), plagioclase 20%, K–Na-feldspar 8%, hornblende 4%; the grain size is generally 1–3 mm, few 3–4 mm; matrix: quartz 12%, K–Na-feldspar 35%, plagioclase 20%, and the grain size is generally 0.2–0.5 mm	About 130	$\eta o^{(z)} K_1^3$

(porphyritic) and fine-grained; syenogranite mainly has textures such as coarse-grained, medium-coarse-grained, coarse-medium-grained, medium-grained, and fine-grained; (quartz) syenite is mainly medium-coarse-grained and coarse-medium-grained texture; and quartz diorite is mainly medium-grained texture.

3.1.2 Main Plutons

3.1.2.1 Ma'anshan Composite Pluton

1. **Geological Features**

Ma'anshan composite pluton is located in the northwest of the survey area, with 83.86 km^2 area, and its main body spreads along the NE-strike, distributed at the intersection of northern Zhejiang and southern Anhui, and intruded into the northeastern part of the Hanggai–Fushi NE-strike anticlinorium between the NE-strike Maotan-Luocun Fault and the Tongkengcun–Qiguancun Fault. The northwestern part of the pluton (outside the survey area, in Anhui Province) is in intrusive contact with the Silurian strata, and the dipstrike of the contact surface is 300°–330°∠45°–60°, the dip angle is up to 80° at very few places, and locally it is in NE-strike fault contact with the strata; the southwestern part in its southeastern side borders on NE-strike Tongkengcun–Qiguancun Fault, and is in fault contact with the Cambiran–Ordovician strata, and the overall dipstrike of the fault surface is 150°∠45°. Its northeastern part is in intrusive contact with the Silurian strata, dipping toward southeast in general (Fig. 3.1).

2. **Petrological Features**

The composite pluton is mainly composed of early (pegmatitic and megacrystic) monzonitic granite and late fine-grained syenogranite, and later was intruded by a series NE-strike granite (porphoritic) veins.

(Pegmatitic and megacrystic) porphyritic monzonitic granite

It is the main lithology of Ma'anshan composite pluton, pegmatitic or megacrystic porphyritic granitoid texture, massive structure, and both are in progressive transitional contact and there is no obvious borderline between the two in the area. Pegmatitic phenocrysts grain size is generally >3 cm (Fig. 3.2a), up to 5 cm, megacryst grain size is 1–3 cm and both mainly contains plagioclase (5–10%) and K-feldspar (2–4%); Matrix: grain size is generally 3–8 mm, mainly quartz (20–25%), K-feldspar (20–25%), plagioclase (20–25%), biotite (10–15%), and a handful of hornblende (1–2%), and accessory minerals are mainly zircon and apatite, etc. Quartz is anhedral granular, some irregular quartz is inlaid in K-feldspar to form graphic texture; biotite crystallized in the gap of feldspar and quartz (Fig. 3.2g) is present as foliated aggregates (Fig. 3.2e); K-feldspar is mostly anhedral granular, plagioclase is plate, and it is visible that plagioclase wrapping early-crystallized K-feldspar formed growth zoning (Fig. 3.2f). Porphyritic monzonitic granite has developed many dark fine-grained dioritic enclaves (MMEs), in ellipse or irregular shapes, are 1–20 cm in length, with clear border (Fig. 3.2b) and enclaves at some places contain very few plagioclase phenocryst in the size of 3–5 mm. Regionally, the intrusive contact between porphyritic monzonitic granite and fine-grained syenogranite or pegmatitic granite veins is clear, and in the contact zone with pegmatitic granite lens, it is usually seen that biotite concentrates in flow line in the shape of strip (Fig. 3.2c), and locally the occurrence is 145°∠25°, similar to the contact surface. Pegmatitic and megacrystic porphyritic monzonitic granite progressively become medium-fine-grained porphyritic monzonitic granite at the margin.

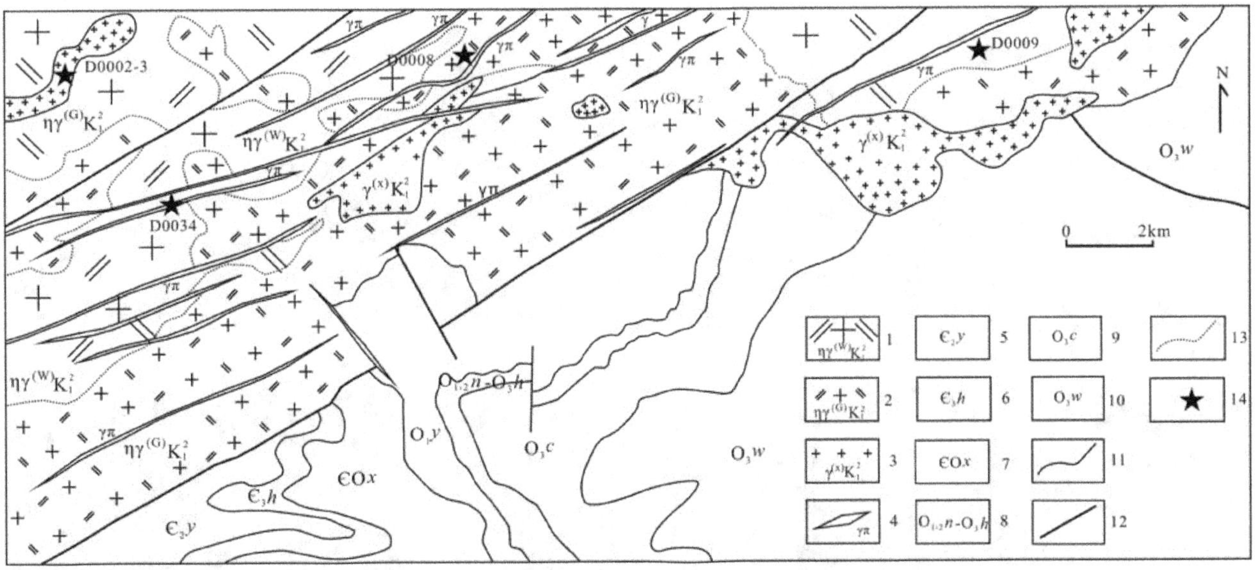

Fig. 3.1 Regional geological sketch map of Ma'anshan composite pluton. 1. Pegmatitic porphyritic monzonitic granite in second phase, Early Cretaceous; 2. Early Cretaceous Period-2 Phenocryst–porphyritic monzonitic granite in second phase, Early Cretaceous; 3. Early Cretaceous Period-2 fine-grained syenogranite in second phase, Early Cretaceous; 4. Granite porphyry vein; 5. Cambrian Yangliugang Formation; 6. Cambrian Huayansi Formation; 7. Cambrian–Ordovician Xiyangshan Formation; 8. Ordovician Ningguo Formation–Huangnigang Formation; 9. Ordovician Changwu Formation; 10. Ordovician Wenchang Formation; 11. Geological boundary; 12. Fault; 13. Lithofacies boundary; 14. Sampling location

Fine-grained syenogranite

It is mainly in the form of small apophysis, intruded into porphyritic monzonitic granite (Fig. 3.3) or near the contact zone between porphyritic monzonitic granite and wall rocks, irregular long strip, and its length and width vary between 100 and 1500 m. Rocks are of fine-grained granitoid texture and massive structure; minerals mainly are quartz (30–35%), K-feldspar (45–50%), plagioclase (15–20%), and a handful of dark-colored minerals such as biotite and hornblende, and accessory minerals are apatite and zircon; the grain size is generally <2 mm (Fig. 3.2d); quartz is in anhedral granular, feldspar is lath-shaped, quartz and feldspar are distributed alternately in equigranular texture, and crystals such as hornblende and biotite are developed at the contact margin (Fig. 3.2i). In fine-grained syenogranite, layer joints is often seen, and flow line and planar flow structure are seen sometimes; especially, fine-grained syenogranite in Guocun has developed parallel joints which are basically consistent with the platy joints in early pegmatitic porphyritic monzonitic granite, and weakly oblique joints of which structure are closely related to mineralization of W-Be-quartz veins.

3. Wall-rock Alteration

Regionally, pegmatitic and megacrystic porphyritic monzonitic granite is strongly weathered and has developed alteration. Plagioclase and K-feldspar frequently show albitazation, chloritization, and kaolinization alterations, etc., along cracks. The fracture zone inside the pluton has developed silicification alteration or filled with quartz vein. Greisenization is seen in margin of the pluton.

Fine-grained syenogranite is weakly weathered, a handful of feldspar probably developed epidotization along cracks, and the contact zone developed joints structure and silicified quartz vein. The Silurian sandstone strata in the southeast side of the pluton mainly have developed hornfelsic alteration. Since the contact surface's dip angle is gentle, the hornfelsic zone is generally 1–3 km wide, the inner zone is about 1.5 km wide and mainly contains dark purple gray hornfels, and the outer zone is about 1.2 km wide and mainly contains dark purple gray and hornfelsic siltstone. The Ordovician calcareous siltstone around the Yonghe Forest Farm in its southeast which is interbedded with micro-crystal limestone, and the strata there have developed silicification, poorly skarnization, and marbleization.

4. Geochemical Features

Lithologies in Ma'anshan composite pluton are high in SiO_2 content (67.09–76.74%), enrichment in alkali (Alk = K_2O + Na_2O, 8.02–8.74%), high in the K_2O/Na_2O ratio (1.24–1.78%); low MgO (0.18–1.26%), low P_2O_5 (0.01–0.2%), and low TiO_2 (0.06–0.56%). Both monzonitic granite and syenogranite have higher differentiation index (DI) (78.62–89.28, 93.73–95.01), similar to the highly differentiated I-type granite (82–94) in

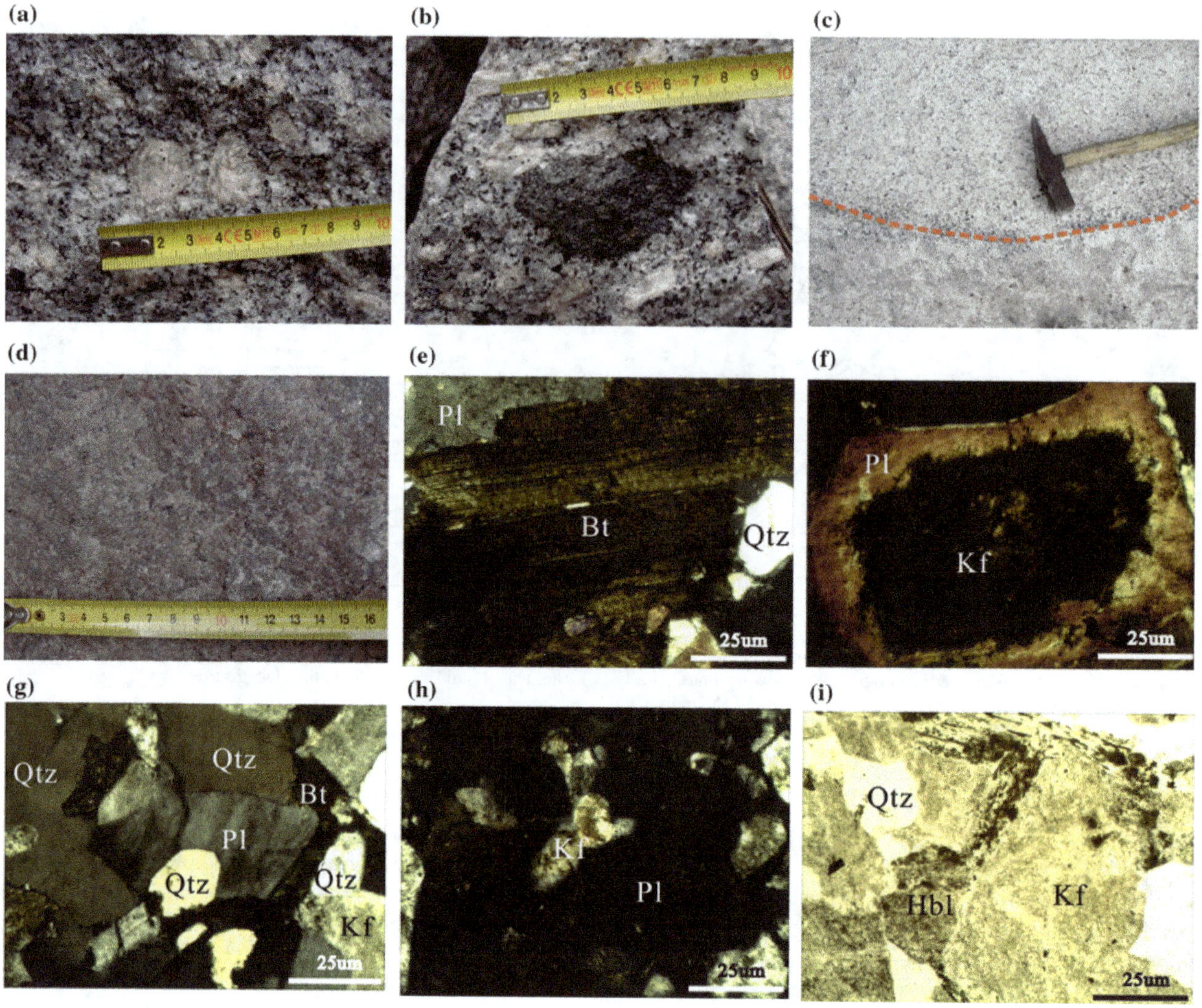

Fig. 3.2 Field pictures and petrographical microscopic pictures of Ma'anshan composite pluton. **a** K-feldspar phenocryst in pegmatitic porphyritic monzonitic granite; **b** dioritic enclaves developed in pegmatitic porphyritic monzonitic granite; **c** biotite concentrates in strips at the margin of phenocryst-porphyritic monzonitic granite and fine-grained syenogranite; **d** fine-grained syenogranite; **e** biotite (Bt), plagioclase (Pl) and quartz (Qtz) in phenocryst-porphyritic monzonitic granite; **f** plagioclase (Pl) containing K-feldspar (Kf) core grown in phenocryst-porphyritic monzonitic granite; **g** biotite (Bt) gets crystallized at the contact edge of quartz (Qtz), plagioclase (Pl) and K-feldspar (Kf) in pegmatitic porphyritic monzonitic granite; **h** plagioclase (Pl) wrapping K-feldspar (Hbl) in pegmatitic porphyritic monzonitic granite; **i** quartz (Qtz), K-feldspar (Kf), and hornblende (Hbl) in fine-grained syenogranite

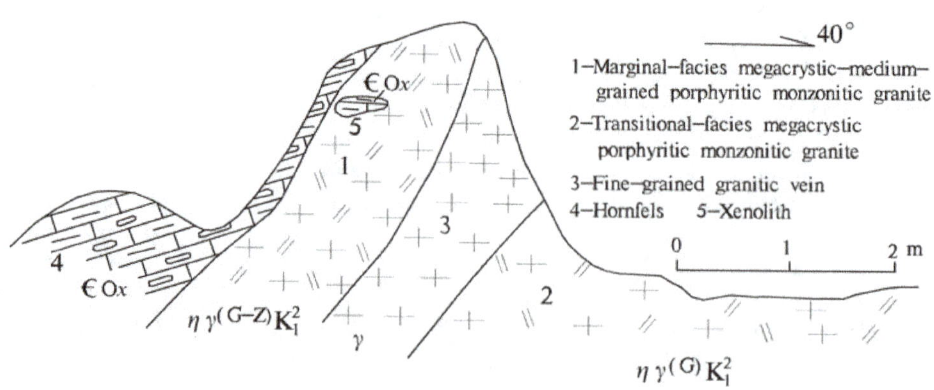

Fig. 3.3 Features of internal-facies zones of megacrystic porphyritic monzonitic granite in Ma'anshao composite pluton

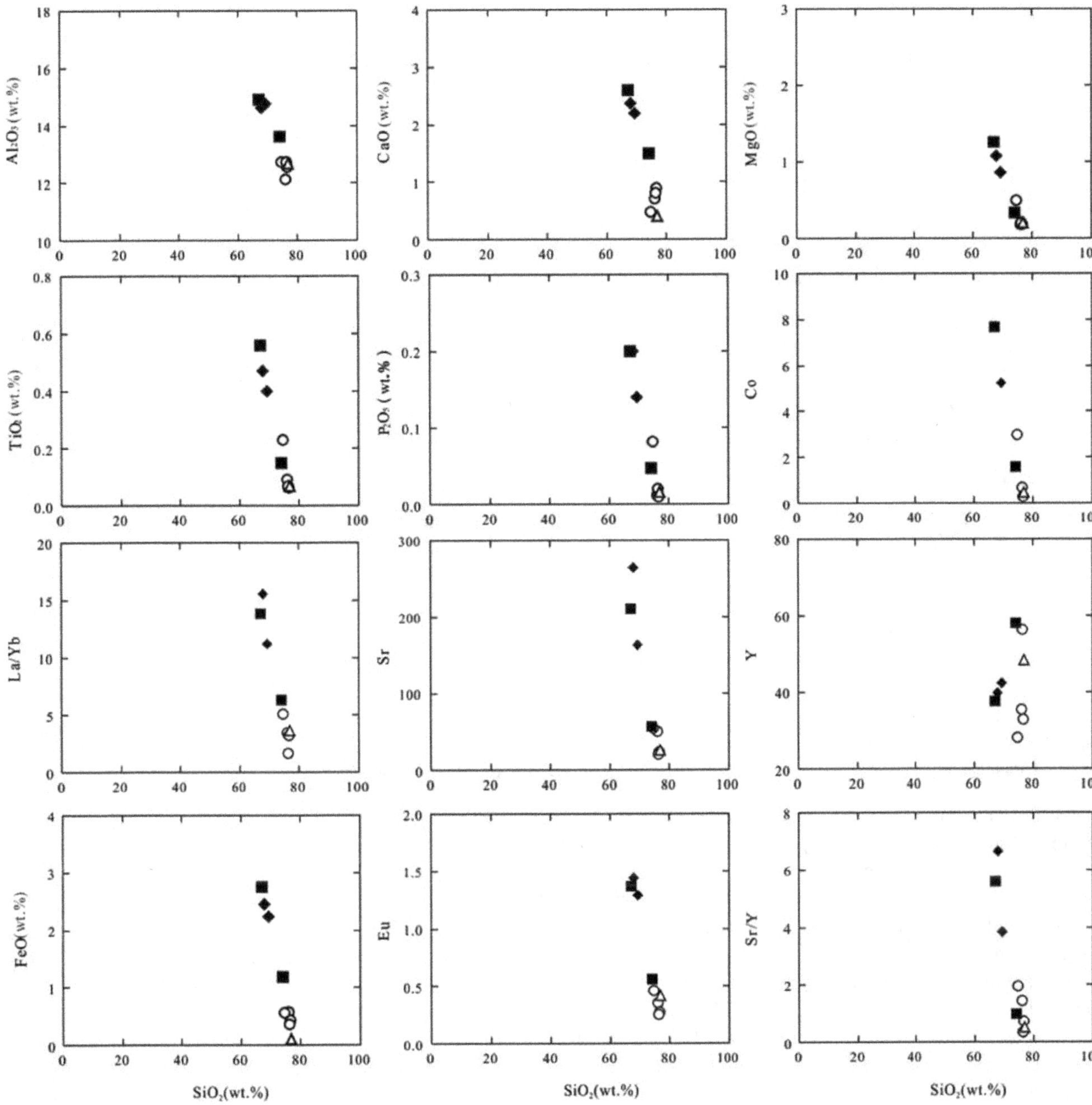

Fig. 3.4 Harker diagram of SiO_2 in Ma'anshan composite pluton (its legends are the same as those in Fig. 3.5)

Fogang, South China. In some samples of CIPW standard minerals, a handful of corundum (0.16–0.47%, a few 1.63%) and diopside (0.02–0.58%, a few 1.37%).

From pegmatitic and megacrystic porphyritic monzonitic granite to fine-grained syenogranite, the content of SiO_2 and K_2O, the K_2O/Na_2O ratio, and DI increase gradually, the contents of TiO_2, Al_2O_3, TFeO, MgO, CaO, and P_2O_5 decrease gradually, and the contents of MnO and Na_2O vary small. In Harker diagram (Fig. 3.4), Al_2O_3, CaO, MgO, FeO, TiO_2, P_2O_5, and Fe_2O_3 have a negative correlation with SiO_2, K_2O is in weakly positive correlation with SiO_2, and Na_2O has no clear relationship with SiO_2. All these results suggested that with differentiation evolution becoming more sufficient, lithologies evolved toward acidity and alkalinity did not change greatly. A/CNK is 0.95–1.14, increasing gradually, featuring evolution from quasi-aluminous to weakly peraluminous (Fig. 3.5). Rittmann Index (σ) is 1.92–2.77, featuring evolution from the shoshonite series to the high K–Ca alkaline series. TFeO/(TFeO + MgO) ratio is 0.75–0.81 (0.77) and 0.77–0.89 (0.81), respectively.

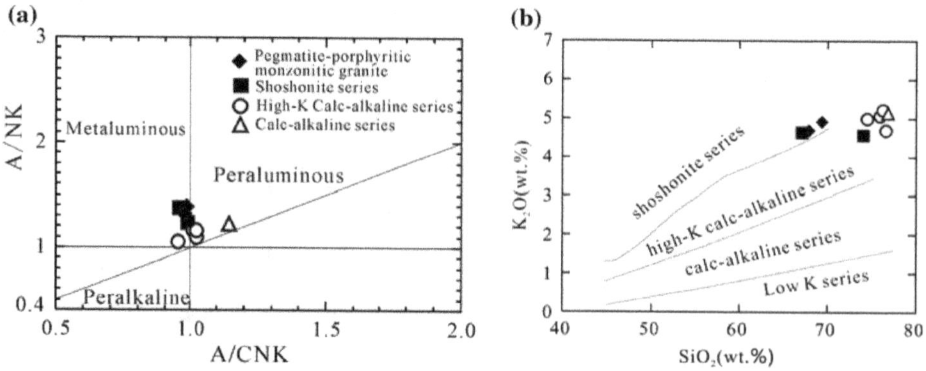

Fig. 3.5 A/CNK-A/NK diagram (**a**) and SiO_2–K_2O diagram (**b**) for Ma'anshan composite pluton

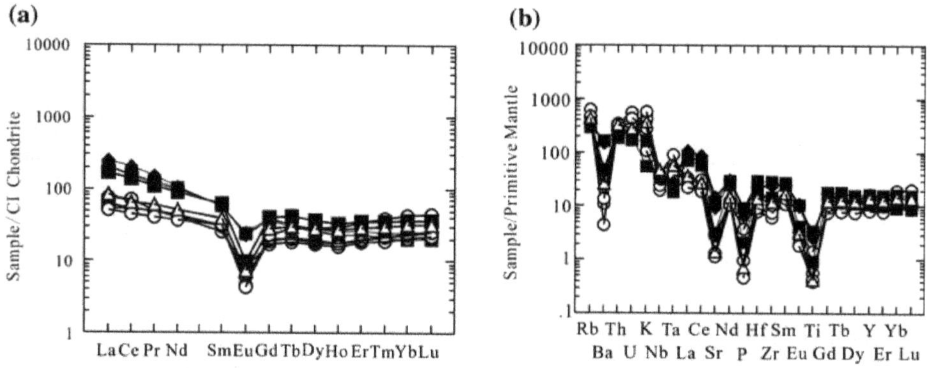

Fig. 3.6 Chondrite-normalized REE distribution mode (**a**) and primitive mantle-normalized trace element spider diagram (**b**) for Ma'anshan composite pluton (the normalized values of chondrite and primitive mantle come from Sun and McDonough 1989)

TFeO/MgO ratio is 3.04–4.37 and 3.36–7.77, respectively, and mean value is close to the I-type granite (2.27) but very different from the A-type granite (13.4).

Pegmatitic and megacrystic porphyritic monzonitic granite is high in content of $\sum$REE (218.89×10^{-6}–283.63×10^{-6}), and the chondrite-normalized REE patterns show a feature of weakly dipping toward right, light rare-earth elements (LREE), and heavy rare-earth elements (HREE) differentiate pretty obvious, La_N/Yb_N is 4.53–11.18 and δEu is 0.19–0.52, Eu showing stronger negative anomaly; rocks are enriched in K, Th, U, and Rb, weakly depleted in LILE such as Ba; weakly depleted in high field-strength elements (HFSE) such as Sr, P, Nb, Ta, and Ti. Fine-grained syenogranite is low in the content of $\sum$REE (94.28×10^{-6}–122.78×10^{-6}). The chondrite-normalized REE patterns show a feature of "V" shape, and LREE and HREE differentiate not so obviously. La_N/Yb_N is 1.17–3.64 and δEu is 0.15–0.38, Eu showing a strong anomaly; similarly, it is enriched in large-ion lithophile elements (LILE) such as K, Th, U, and Rb, but strongly depleted in elements such as Ba and Sr; strongly depleted in HFSE such as P, Nb, and Ti (Fig. 3.6).

Pegmatitic porphyritic monzonitic granite $(^{86}Sr/^{87}Sr)_i$ is 0.70825, εNd(t) is −6.77 and the two-stage model ages (T_{DM2}) is 1.47 Ga; fine-grained syenogranite $(^{86}Sr/^{87}Sr)_i$ is 0.69215, εNd(t) is −6.59, and T_{DM2} is 1.46 Ga; both have similar features, indicating their same magma evolution.

5. **Isotope geochronology**

In porphyritic monzonitic granite, zircon mostly is irregular long strip, 100–200 μm in length, the length-width ratio is about 2:1, and zircon has developed oscillatory zoning. In fine-grained syenogranite, zircon mostly is irregular short strips and granular, 50–100 μm in length, or occasional 100–200 μm, the length-width ratio is about 2:1–1:1 and also zircon has developed oscillatory zoning (Fig. 3.7). In pegmatitic porphyritic monzonitic granite, megacrystic porphyritic monzonitic granite and fine-grained syenogranite, the U content in zircon is $(242–2041) \times 10^{-6}$, $(100–2886) \times 10^{-6}$, and $(114–1863) \times 10^{-6}$, respectively, and the Th content is $(97–349) \times 10^{-6}$, $(38–499) \times 10^{-6}$, and $(52–497) \times 10^{-6}$, respectively, and the Th/U ratio is 0.11–0.41, 0.16–1.05, and 0.22–0.97, respectively, a typical feature of magmatic original zircon (Hoskin and Schaltegger 2003). Sample locations mostly are projected on or near the concordant curve, $^{206}Pb/^{238}U$ weighted mean age is 132.2 ± 1.6 Ma (MSWD = 1.9), 127.7 ± 1.2 Ma (MSWD = 1.3), and 128.3 ± 1.1 Ma (MSWD = 1.7), respectively (Fig. 3.8), representing the

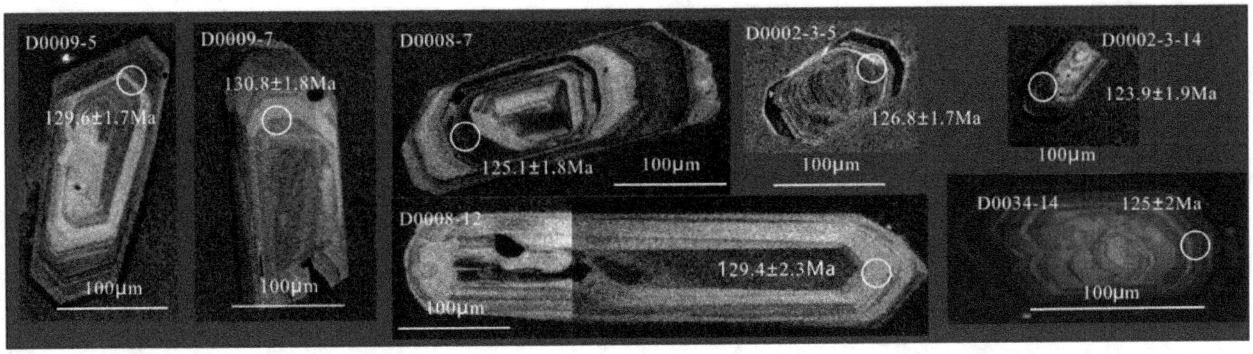

Fig. 3.7 Main zircon cathodoluminescence (CL) imaging and dating spots diagram of samples from Ma'anshan composite pluton

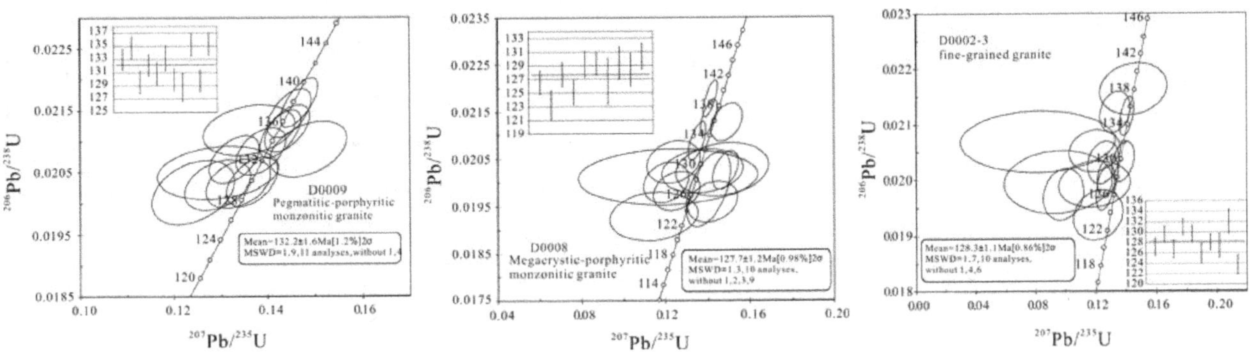

Fig. 3.8 Zircon U–Pb concordant diagrams for Ma'anshan composite pluton

crystallization age of Ma'anshan pluton. Of these, $^{206}Pb/^{238}U$ ages at the sample locations D0009-14 and D0002-3-8 are 414 Ma and 310.5 Ma, showing older ages and possibly representing inherited zircon.

Based on the contact relationships, geochronological and geochemical features, Ma'anshan composite pluton experienced two stages of magmatism, the early stage is pegmatitic and megacrystic porphyritic monzonitic granite (132.2 ± 1.6 Ma—127.7 ± 1.2 Ma) and the later stage is fine-grained syenogranite (128.3 ± 1.1 Ma).

3.1.2.2 Tangshe Composite Pluton

1. **Geological Features**

Tangshe composite pluton is situated in the midwest of the survey area, with outcrop area of about 18.9 km^2, NE-strike, distributed in Tangsheling which is the intersection of Zhejiang and Anhui Province. The pluton, in the shape of stock, intruded into the core of Wangjia-Tangshecun anticline (Fig. 3.9), of which the distribution direction is consistent with the anticline axis. It is in intrusive contact with wall rocks on both sides, and the dip angle of contact surface is generally steep with a range of 50°–80°, all dipping outwards; wall rocks are mainly the Neoproterozoic Nanhuaian to Cambrian strata. Many secondary faults in NE, NEE, and NW-strike have developed inside and surrounded the pluton.

2. **Petrological Features**

The composite pluton is mainly composed of early medium-grained monzonitic granite and late medium-coarse-grained (coarse-medium-grained) syenogranite, and fine-medium-grained syenogranite.

Medium-grained monzonitic granite

It is distributed in the southern part of the composite pluton, NEE strike, 10–12 km in length and 1–2 km in width, and its north side is in surging intrusive contact with medium-coarse-grained syenogranite. Rocks are offwhite, medium-grained texture, and massive structure, and mainly contain quartz (20%), K-Na-feldspar (30%), plagioclase (35%), biotite (5–10%), hornblende (5%), a handful of pyrite and magnetite, etc., and zircon and apatite occasionally seen; the grain size is generally 2–3 mm, or a few 0.5–2 mm; quartz is anhedral granular, not evenly distributed; plagioclase is platy, and mainly andesine-oligoclase (Fig. 3.10a, d).

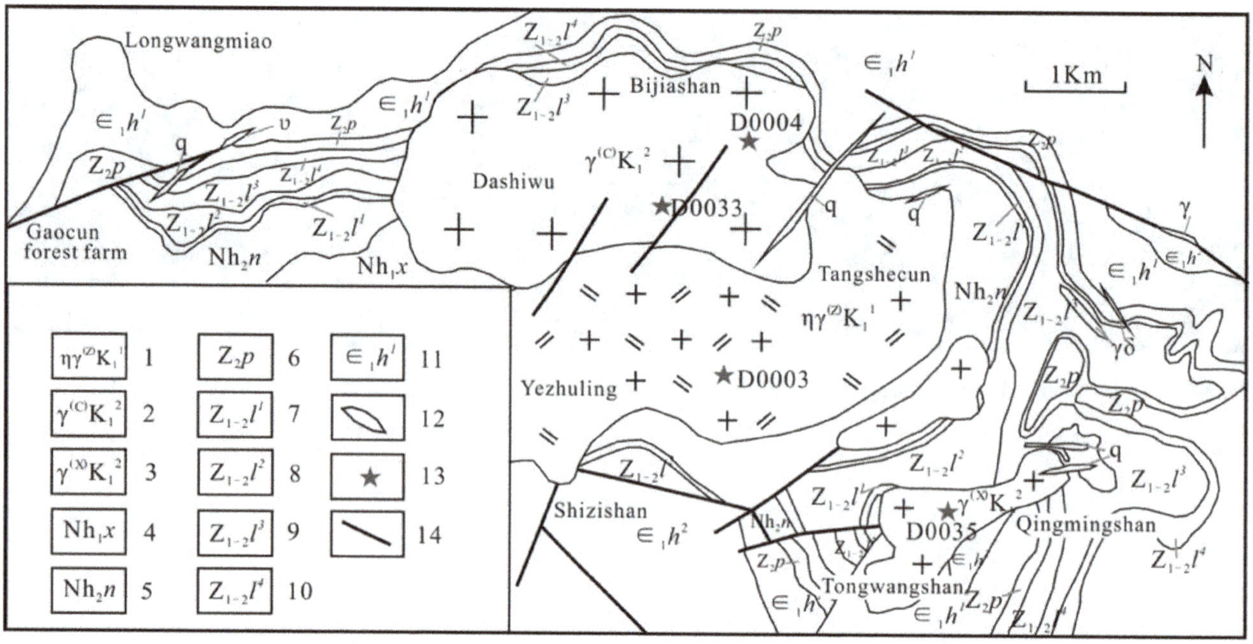

Fig. 3.9 Regional geological sketch map of Tangshe composite pluton. 1. medium-grained monzonitic granite; 2. medium-coarse-grained syenogranite; 3. medium-fine-grained syenogranite; 4. the Xiuning Formation; 5. the Nantuo Formation; 6. the Piyuancun Formation; 7. the first member of Lantian Formation; 8. the second member of Lantian Formation; 9. the third member of Lantian Formation; 10. the fourth member of Lantian Formation; 11. the first member of Hetong Formation; 12. veins; 13. sampling location; 14. faults

Fig. 3.10 Outcrop and microscopic petrographical pictures of Tangshe composite pluton. **a** medium-grained monzonitic granite; **b** medium-coarse-grained syenogranite; **c** fine-grained syenogranite; **d** plagioclase (Pl), biotite (Bt), and quartz (Qtz) in medium-grained monzonitic granite; **e** K-feldspar (Kf), plagioclase (Pl) showing polysynthetic twin, quartz (Qtz), and biotite (Bt) in coarse-grained syenogranite; **f** plagioclase (Pl) showing polysynthetic twin, K-feldspar (Kf), and quartz (Qtz) in fine-grained syenogranite

Medium-coarse-grained syenogranite

The rocks are distributed in the northern part of the composite pluton, EW strike, short strip, 3.5 km in length, and 1.5 km in width, it is transitioned in local parts of the contact zone with early stage medium-grained monzonitic granite to become coarse-medium-grained (porphyritic) syenogranite, and the margins of its east and west margins are progressively transitioned to become medium-fine-grained syenogranite. Rocks are pink, medium-coarse-grained (porphyritic) granitic texture and massive structure, and mainly contain K-feldspar (40%), plagioclase (15%), quartz (30%), biotite (5–10%), hornblende (3%), a handful of metallic minerals, and zircon occasionally seen; mineral's grain size is bigger, K-feldspar's grain size up to 1–2 cm, a few's grain size > 3 cm, K-feldspar mostly in the shape of square, highly idiomorphic; quartz is in idiomorphic texture, inlaid in K-feldspar, and forms graphic texture (Fig. 3.10b, e).

Fine-medium-grained monzonitic granite

The rocks are distributed in the southeastern of the composite pluton, NE-strike, about 2 km in length, and 0.5–1 km in width. Rocks are pink, medium-grained granitic texture and massive structure, and mainly contain quartz (30%), alkaline feldspar (45%), plagioclase (20%), biotite (5%), a handful of metallic minerals, zircon, and apatite occasionally seen; the mineral grain size is generally 2–6 mm, and a few up to 6–7 mm. Quartz is anhedral granular, with more content, not evenly distributed; feldspar is platy, mainly alkaline feldspar, less plagioclase; biotite is schistose, not evenly distributed, but locally concentrated (Fig. 3.10c, f).

3. **Wall-rock Alteration**

Tangshe composite pluton is weathered strongly, showing albitazation and greisenization. The interior pluton has developed secondary fractures, its both sides developed silicification and sericitization; its margins developed joints structure, and tungsten-bearing quartz vein in the northern margin of medium-grained monzonitic granite is mainly related to the NNW-strike joints. Alterations in wall-rock strata are mainly skarnization and hornfelsic, locally silicification and marbleization; the Nanhuaian siltstone and sandy conglomerate in the western and eastern sides of the composite pluton developed strong hornfelsic alteration, with 1–2 km in alteration width; the Sinian Lantian Formation strata in northern side developed strong skarnization and marbleization, the skarn zone is 5–100 m generally, minerals assemblage included garnet, pyroxene, epidote, minor idocrase, etc., layer-like Pb–Zn ore (mineralization) bodies can be seen, scheelite mineralization can be seen locally, the borderlines of the contact surface in metallogenic areas are often irregular, accompanied with tongue-shape protrusion, and the dip angle has direct impact on mineralization alteration zones with respect to width and strength.

4. **Geochemical Features**

Tangshe composite pluton is high in SiO_2 content (67.83–75.88%), enriched in alkali (Alk = K_2O + Na_2O, 7.18–8.84%), high in K_2O/Na_2O ratio (0.95–1.63%) (Fig. 3.11); low in MgO (0.06–0.98%), low in P_2O_5 (0.01–0.28%), and low in TiO_2 (0.09–0.52%). The DI indexes of medium-grained monzonitic granite, medium-coarse-grained and coarse-medium-grained syenogranite and medium-fine-grained syenogranite are higher (76.54–77.94, 90.23–92.95, 92.53), of which the latter two are similar to the highly differentiated I-type granite (82–94) in Fogang in South China.

From medium-grained monzonitic granite to medium-coarse-grained and coarse-medium-grained syenogranite to medium-fine-grained syenogranite, the content of SiO_2 and K_2O, the K_2O/Na_2O ratio, and DI increase gradually, the contents of TiO_2, Al_2O_3, TFeO, MgO, CaO, and P_2O_5 decrease gradually, and the contents of MnO and Na_2O vary fewly. The contents of Al_2O_3, CaO, MgO, FeO, TiO_2, P_2O_5, and Fe_2O_3 have a negative correlation with SiO_2, K_2O is in weakly positive correlation with SiO_2, and Na_2O has no clear relationship with SiO_2. All these features suggest that with differentiation evolution becoming more sufficient, lithologies evolve toward acidity and their alkalinity does not change greatly. A/CNK is 0.93–1.10, increasing gradually, featuring evolution from quasi-aluminous to weakly peraluminous (Fig. 3.12). Rittmann Index (σ) is 1.88–2.47, consistent with high K–Ca alkaline

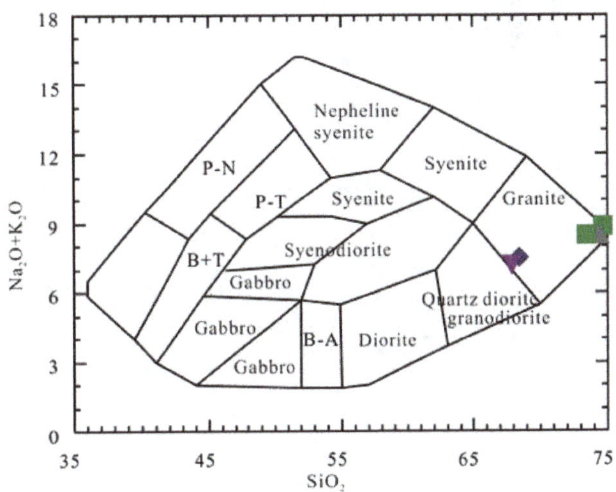

Fig. 3.11 SiO_2–Na_2O + K_2O diagram for Tangshe composite pluton (its legends are the same as those in Fig. 3.12)

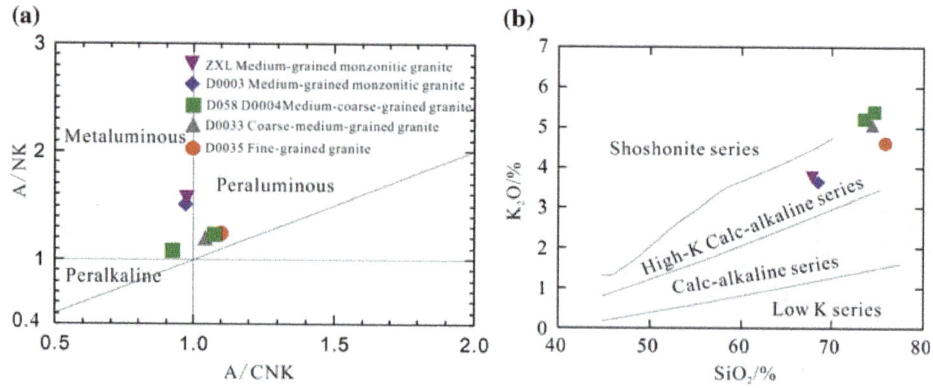

Fig. 3.12 A/CNK-A/NK diagram (**a**) and SiO_2-K_2O diagram (**b**) for Tangshe composite pluton

series. Their TFeO/(TFeO + MgO) ratio is 0.77–0.79 (0.78 in average), 0.81–0.86 (0.84 in average), and 0.96, respectively; their TFeO/MgO ratio is 3.42–3.70 (3.56 in average), 4.18–6.00 (5.08 in average), and 22.03, respectively.

Medium-grained monzonitic granite is low in content of $\sum$REE (136.16×10^{-6}–190.57×10^{-6}), and the chondrite-normalized REE patterns (Fig. 3.13) show a feature of weakly dipping toward right, LREE and HREE differentiate pretty obvious, La_N/Yb_N is 17.25–19.14 and δEu is 0.64–0.72, Eu showing weakly negative anomaly. Rocks are enriched in K, Th, U, and Rb, weakly depleted in LILE such as Ba and Sr; weakly depleted in HFSE such as P, Nb, Ta, and Ti.

Medium-coarse-grained and coarse-medium-grained monzonitic granite is high in content of $\sum$REE (179.39×10^{-6}–229.56×10^{-6}), and the chondrite-normalized REE patterns (Fig. 3.13) show a feature of weakly dipping toward right, LREE and HREE differentiate pretty obvious, La_N/Yb_N is 5.92–16.45 and δEu is 0.22–0.32, Eu showing strong negative anomaly. Similarly, rocks are enriched in K, Th, U, and Rb, weakly depleted in LILE such as Ba and Sr; weakly depleted in HFSE such as P, Nb, Ta, and Ti. $(^{86}Sr/^{87}Sr)_i$ value is 0.703,60, $\varepsilon Nd(t)$ value is −5.11 and T_{DM2} is 1.34 Ga.

Fine-grained syenogranite is high in content of $\sum$REE (214.70×10^{-6}), and the chondrite-normalized REE patterns (Fig. 3.13) show a feature of "V" shape, light, and heavy rare earth differentiates unobviously, La_N/Yb_N is 6.24 and δEu is 0.10, Eu showing strongly negative anomaly. Similarly, the rocks are also enriched in LILE such as K, Th, U, and Rb, strongly depleted in LILE such as Ba and Sr; in terms of HFSE, strongly depleted in P and Ti, weakly depleted in Nb and Ta.

5. **Isotope geochronology**

In medium-grained monzonitic granite (D0003), the zircons (Fig. 3.14) have a grain size of 100–250 μm, with well euhedral platy or column textures, showing obvious magmatic oscillatory zoning. The D0003-15 zircon is subrounded, without oscillatory zones, obviously brighter in the interior, like cloud and mist, indicating its typical metamorphic origin. There are 16 dating points for D0003, and the obtained $^{206}Pb/^{238}U$ ages can be divided into four groups. Th/U values of 8 dating points (1–5, 7, 9, 16) are 0.37–0.70, showing the magmatic origin feature, and $^{206}Pb/^{238}U$ weighted mean age is 140.4 ± 3.3 Ma (MSWD = 3.3) (Fig. 3.15), representing the crystallization

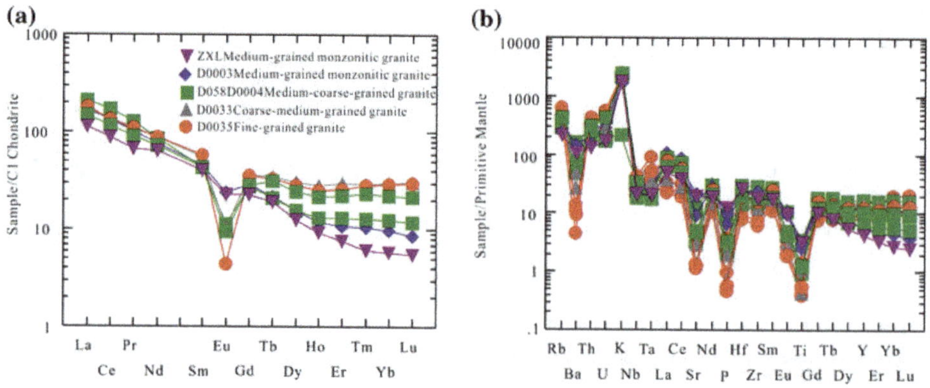

Fig. 3.13 Chondrite-normalized REE distribution mode (**a**) and primitive mantle-normalized trace-element spider diagram (**b**) for Tangshe composite pluton (the normalized values of chondrite and primitive mantle come from Sun and McDonough 1989)

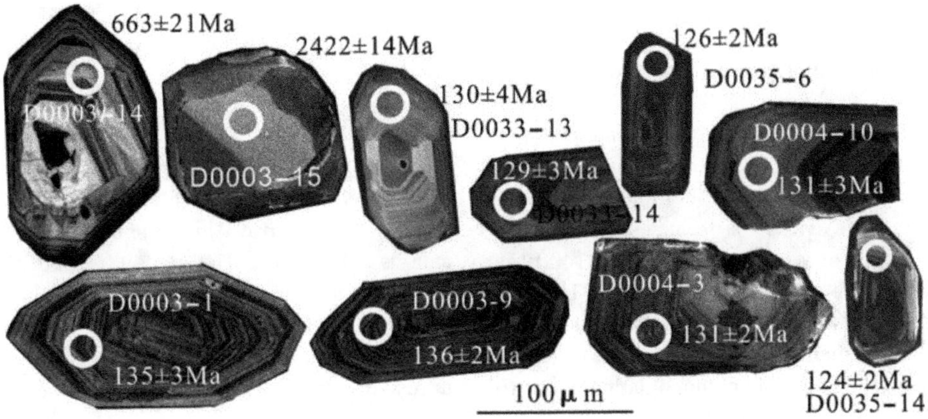

Fig. 3.14 CL images, analysis location ,and U–Pb age of some zircon points at Tangshe composite pluton

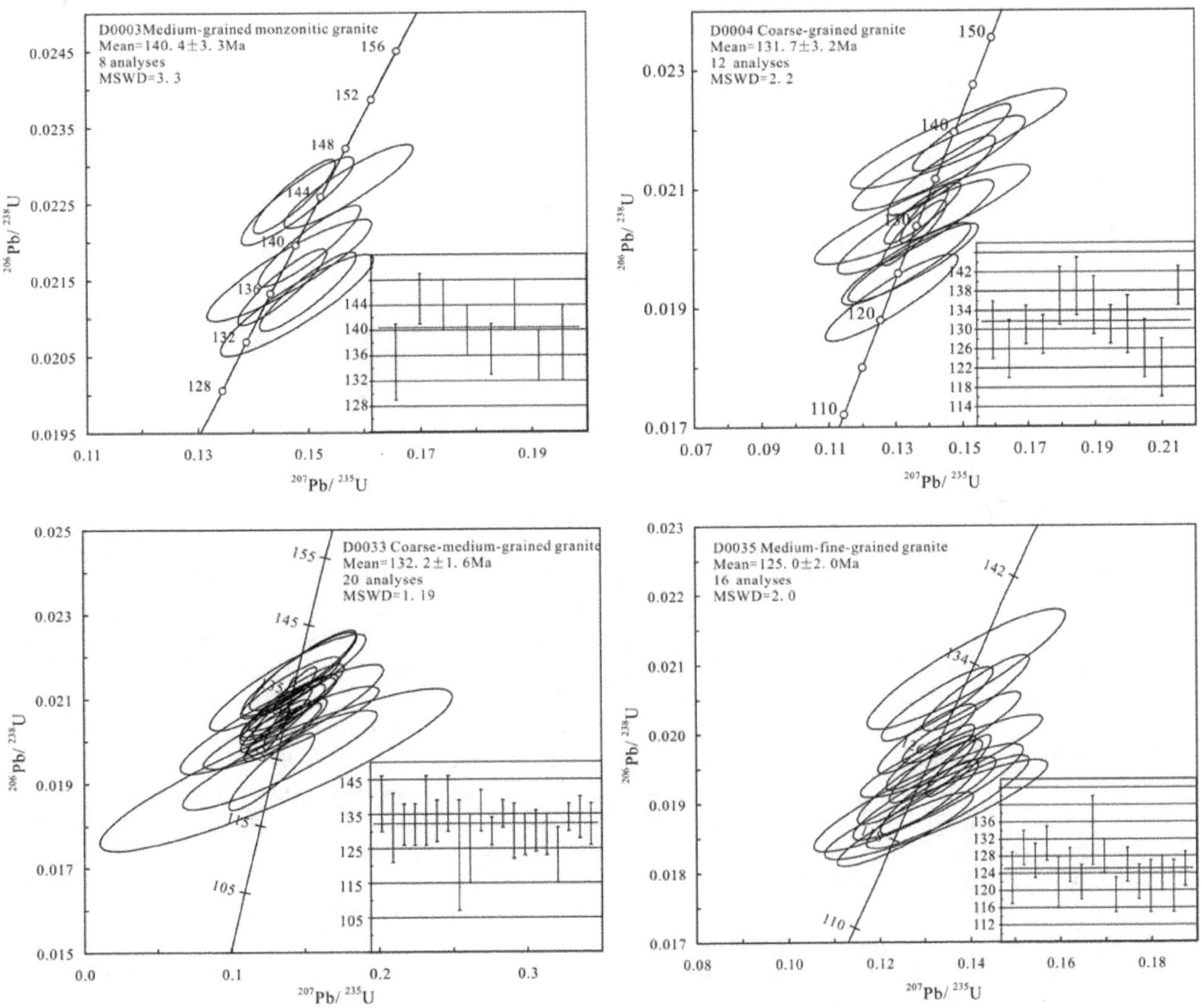

Fig. 3.15 Zircon U–Pb concordant diagrams for Tangshe composite pluton

age of medium-grained monzonitic granite. Vavra et al. (1999) argued that, for metamorphic zircon, Th/U values are mainly less than 0.1 for most of them and may be greater than 0.7 for a few of them. For zircon D0003-15, Th/U value is 1.72. By using the SHRIMP zircon U-Pb dating method, Gao et al. (2001) obtained that the age of trondhjemitic gneiss in Kongling high-graded metamorphic complex is 2947–2903 Ma, and argued the existence of the ancient basement of Yangtze block. For ancient zircon with age over 1400 Ma, due to Pb loss, $^{207}Pb/^{206}Pb$ ages are often choosed. Thus, the zircon age of D0003-15 is 2422 ± 14 Ma, which may indicate that the ancient basement of Jiangnan Terrane existed in the survey area. $^{206}Pb/^{238}U$ ages of another 6 dating points (6, 8, 11–14) are 675–588 Ma, and in combination with the feature of CL image, it may be inherited zircon.

Zircon in medium-coarse and coarse-medium-grained syenogranite (D0004 and D0033) is basically in consistent form, euhedral–subhedral platy, the grain size is 80–150 μm, its length-width ratio is 2:1–1:1, and almost all zircon have developed magmatic oscillatory zoning. There are totally 14 dating points for D0004. The ages from point 6 which is rather large and point 13 which is rather small are removed. The $^{206}Pb/^{238}U$ weighted mean age of the remaining 12 points is 131.7 ± 3.2 Ma (MSWD = 2.2), representing the crystallization age of medium-coarse-grained syenogranite. There are totally 20 dating points for D0033, and the $^{206}Pb/^{238}U$ weighted mean age is 132.2 ± 1.6 Ma (MSWD = 1.19), representing the crystallization age of medium-grained syenogranite. Th/U values of 12 zircons in sample D0004 and 20 zircon in sample D0033 are 0.37–0.76 and 0.30–0.83, respectively, featuring magmatic orgin.

In fine-grained syenogranite (D0035), the zircon grain size is 60–120 μm, its length-width ratio is 2:1–1:1, euhedral–subhedral, and zircon has well-developed magmatic oscillatory zoning, and the phenomenon of decrystallization may be seen in a few zircons. There are totally 18 dating points for D0035, of which the $^{206}Pb/^{238}U$ age of point 15 is 113 ± 3 Ma, greatly smaller than those of other dating points, which may represent a thermal event at the end of crystallization stage. The age data of point 17 are problematic, possibly resulting from a mistake during the experimental operation. After the ages of point 15 and 17 are removed, the $^{206}Pb/^{238}U$ weighted mean age of the remaining 16 points is 132.2 ± 2.0 Ma (MSWD = 2.0), representing the crystallization age of fine-grained syenogranite. Th/U value of 16 dating zircons in sample D0035 is 0.55, and in combination with the features of CL image, it is considered that the zircon in fine-grained granite is magmatic orgin.

In conclusion, Tangshe composite pluton has conspicuously experienced three stages of magmatism, the early stage is medium-grained monzonitic granite (140.4 ± 3.3 Ma), the middle stage is medium-coarse-grained and coarse-medium-grained syenogranite (132.2 ± 1.6 Ma–131.7 ± 3.2 Ma), and the later stage is fine-grained syenogranite (125.0 ± 2.0 Ma).

3.1.2.3 Xianxia Composite Pluton

1. **Geological Features**

Xianxia composite pluton is located in the southwest of the survey area; about 25 km in length and 1.5–7.5 km in width, extended in southwest into Anhui Province, and its outcrop area is 84.18 km^2. The pluton is under joint control of NE-strike Maotan-Luocun Fault and Shiling–Shangmei Forest Farm anticlinorium, propagated in NE-strike generally and narrows gradually from SW to NE; its northwestern side is in intrusive contact with the Neoproterozoic–Paleozoic strata, and its southeastern side is in intrusive contact with the Paleozoic strata or in NE-strike fault contact with the late Mesozoic volcanic rocks (Fig. 3.16). The composite pluton obviously has experienced intrusions in multiple stages. From early to late stages, magma intruded from southwest to northeast, and the rocks are sequentially (medium-grained, fine-grained, and megacrystic porphyritic) monzonitic granite → (medium-coarse, coarse-medium, and fine-grained) syenogranite → (medium-coarse-grained) quartz syenite, and later, it was intruded by plenty of fine-grained granite, aplite, quartz syenite, diorite porphyrite, and diabase veins, etc.

2. **Petrological Features**

The features of lithological compositions in Xianxia composite pluton are listed in Table 3.2. In the early stage, the center of megacrystic porphyritic monzonitic granite surged and intruded into medium-grained monzonitic granite, and the contact boundary is not clear for strong weathering, with the "abrupt change" of both seen within 0.5–1 m locally. Medium-grained monzonitic granite xenolith is developed in megacrystic porphyritic monzonitic granite of Zhongguling. Small megacrystic porphyritic monzonitic granite apophysis is intruded into medium-grained monzonitic granite stock, and it is visible on the contact zone that biotite is distributed as strip-like flow-line structure (Fig. 3.17). The rock gradually transitioned to coarse-medium-grained porphyritic monzonitic granite at the margin of NE-strike fault contact zone near the Shangyan Reservoir. The feldspar is generally long column-like, showing a certain directional arrangement, and the quartz minerals in the marginal rocks show cataclastic texture due to later fractures.

Medium-grained monzonitic granite

The rock is gray, mainly consisting of plagioclase (35–40%), K-feldspar (25–35%), quartz (20–25%), biotite (5–10%), as well as a handful of magnetite, apatite, and zircon, etc. The grain size is 1–3 mm, casually 3–6 mm. The biotite-bearing MMEs (Fig. 3.18a) and megacrystic plagioclase could be locally seen, and the plagioclase shows zoning texture (Fig. 3.18g).

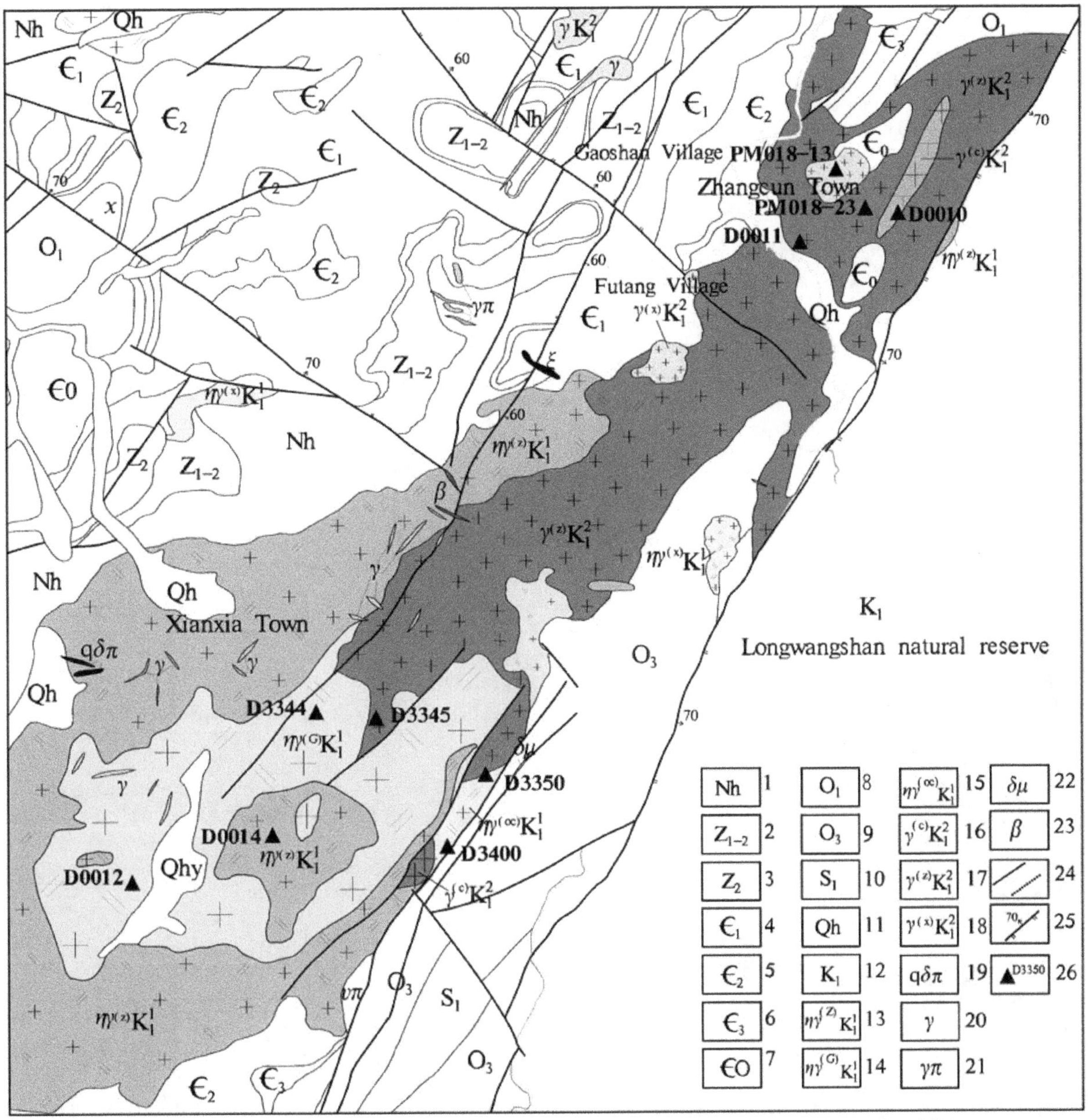

1–Nanhuaian 2–Sinian 3–Upper sinian 4–Early Cambrian 5–Middle Cambrian 6–Later Cambrian 7–Later Cambrian – early Ordovician 8– Early Ordovician 9–Later Ordovician 10–Lower silurian 11–Quaternary 12–Early Cretaceous 13–Medium–grained monzonitic granite 14–Megacrystic–porphyritic monzonitic granite 15–Coarse–grained porphyriticmonzonitic granite 16–Medium–coarse–grained syenogranite 17–Coarse–medium–grained syenogranite 18–Fine–grained syenogranite 19–Quartz diorite porphyry 20–Granite 21–Granite porphyry 22–Dioritic porphyrite 23–Diabase 24–Geological boundary/lithofacies boundary 25–Fault 26–Sampling location and number

Fig. 3.16 Regional geological sketch map of Xianxia composite pluton

Megacrystic porphyritic monzonitic granite

The rock is light gray-light pink. The phenocryst is mainly plagioclase (3–5%) and K-feldspar (5–7%), in the size of 1–2 cm (Fig. 3.18b, h), and matrix mainly is plagioclase (30–35%), K-feldspar (20–25%), quartz (25–30%), biotite (5–10%), and a handful of apatite and zircon, etc., 1–3 mm in size; the lithology at the margin of the NE-strike fault contact

Table 3.2 Lithological composition and geological features of the Xianxia composite pluton

Lithology	Contact relationship	Mineral assemblage	Fabric feature
Medium- grained monzonitic granite	The periphery of southwest part and the central top cap is in intrusive contact with wall-rock strata (345°∠20°), and locally fault contact.	Plagioclase (35–40%), K-feldspar (25–35%), quartz (20–25%), biotite (5–10%), a handful of magnetite, apatite, and zircon, etc.	Light gray, fine-medium-grained subhedral granular texture, generally 1–3 mm in size, occasionally 3–6 mm, locally seen biotite-bearing MMEs and megacrystic plagioclase, and plagioclase has developed zoning texture
Phenocryst-porphyritic monzonitic granite	The central part of its southwest section is in gushing intrusive contact with medium-grained monzonitic granite at the margin.	Phenocryst: plagioclase (3–5%), K-feldspar (5–7%); matrix: plagioclase (30–35%), K-feldspar (20–25%), quartz (25–30%), and biotite (5–10%)	Light gray–light pink, in facies-change zonation with medium-grained monzonitic granite, megacrystic porphyritic is subhedral granular, and phenocryst is about 1–2 cm in size; matrix is about 1–3 mm in size
Coarse-medium-grained porphyritic monzonitic granite	Mainly distributed in southwest section at the intersection of medium-grained granite and megacrystic porphyritic monzonitic granite, or distributed near the fault edge in southeast section	Plagioclase (35–40%), K-feldspar (25–35%), quartz (20–25%), a handful of apatite, and zircon, etc.	Light pink, coarse-medium-grained porphyritic granitoid texture; the grain size is 3–8 mm, feldspar is generally long column-like and had a certain directionally arrangement, and rocks at the margin minerals such as quartz have developed cataclastic texture due to later fractures
Coarse-medium-grained syenogranite	Mainly distributed in the middle and northeast sections of the pluton, and in intrusive contact with monzonitic granite in southwest section	Quartz (25–35%), K-feldspar (35–40%), plagioclase (15–20%), biotite (5–10%)	Light pink, coarse-medium-grained texture, with 1–7 mm in size, and from southwest to northeast, the grain size gradually decreased
Fine-grained syenogranite	Small apophysis or vein, intruded in coarse-medium-grained syenogranite in northeast section.	Quartz (30–35%), K-feldspar (35–40%), plagioclase (15–20%), biotite (3–5%)	Light pink, fine-grained texture, grain size 0.5–2 mm
Coarse-grained quartz syenite	Small apophysis, intruded in coarse-medium-grained syenogranite	Quartz (10–15%), K-feldspar (65–70%), plagioclase (5–10%), biotite (1–3%)	Coarse-grained granitic texture, grain size 5–8 mm

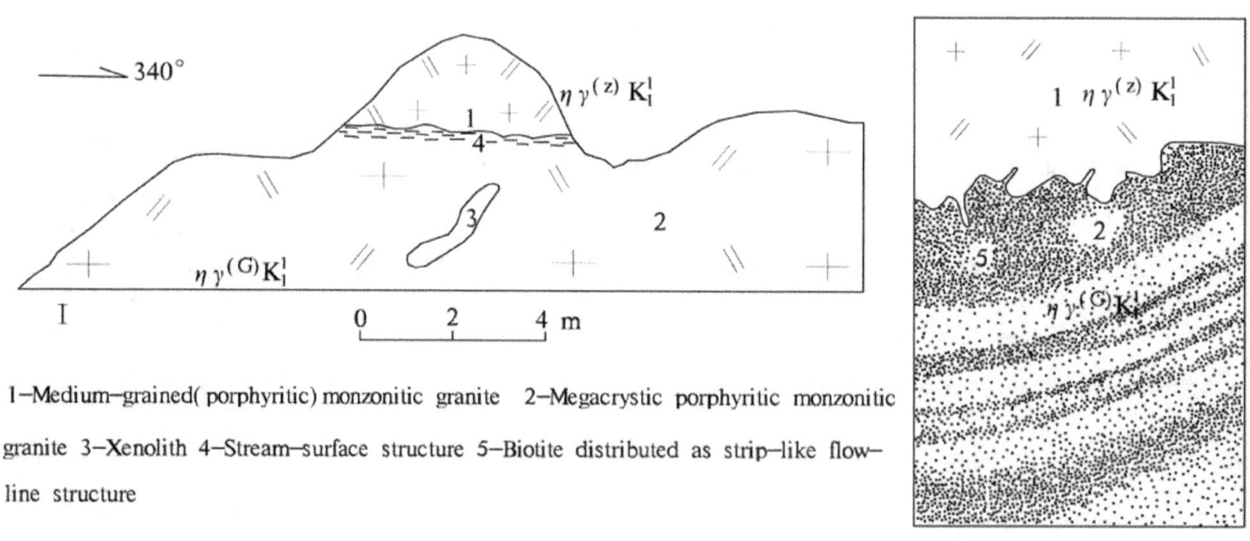

Fig. 3.17 Contact relationship between medium-grained monzonitic granite and megacrystic porphyritic monzonitic granite in Majiafan area

Fig. 3.18 Outcrops and micrographs images showing lithological features of the Xianxia composite pluton. **a** biotite-bearing MMEs in medium-grained monzonitic granite; **b** K-feldspar megacryst in megacrystic porphyritic monzonitic granite; **c** medium-coarse-grained porphyritic monzonitic granite long-strip shaped feldspar; **d** medium-coarse-grained syenogranite; **e** fine-grained syenogranite vein intruded in medium-grained monzonitic granite; **f** coarse-grained quartz syenite; **g** zoning texture plagioclase in medium-grained monzonitic granite; **h** feldspar phenocryst in megacrystic porphyritic monzonitic granite (K-feldspar wrapping plagioclase); **i** K-feldspar inside plagioclase in medium-coarse-grained porphyritic monzonitic granite, and quartz cataclastic phenocryst affected by later fracture

zone in the east side is transitioned to coarse-medium-grained porphyritic monzonitic granite, feldspar is generally long column-like and had a certain directionally arrangement (Fig. 3.18c), and the quartz in the margin rocks developed cataclastic texture due to later fractures (Fig. 3.18i).

Coarse-medium-grained granite

The area was intruded by later medium-coarse-grained syenogranite in the northeastern section, and from south-west to northeast the mineral grain size of the intrusion becomes smaller and the content of biotite decreases. The intrusion intruded as tree branch shape in the contact zone of monzonitic granite. Coarse-medium-grained granite is light pink, mainly consisting of quartz (25–35%), K-feldspar (35–40%), plagioclase (15–20%), and biotite (5–10%), in the size of 1–7 mm (Fig. 3.18d). Fine-grained syenogranite occurred as small apophysis intruding into medium-coarse-grained syenogranite, with each pluton generally about 0.5–1 km^2. It is light pink, and mainly composed of quartz (30–35%), K-feldspar (35–40%), plagioclase (15–20%), and biotite (3–5%), showing a grain size of 0.5–2 mm.

3. **Wall-rock Alteration**

The northwestern section of the pluton, generally as medium-grained monzonitic granite or coarse-medium-grained syenogranite, is in intrusive contact with the Neoproterozoic–Paleozoic strata. The boundary of the contact zone is irregularly, generally dipping outward trend, and the dipstrike of the contact surface is 310°–340°∠20°–60°, and locally in NE-strike fault contact. The southeastern section of the pluton is in intrusive contact with the Paleozoic strata, or in NE-strike fault contact with the Early Cretaceous volcanic rocks in the Huangjian Formation. It was controlled by the NE-strike Maotan-Luocun fault before and after magmatism. In the southeastern section of the pluton, argillaceous and siliceous rocks at the contact zones were strongly hornfelsic, and the secondary NW-strike fault was filled with silicification-fluorite veins. In the northeastern section, the strong skarnization alteration was developed in the Cambrian carbonate strata near the NW contact zone or carbonate xenolith in the pluton. The NE-strike fracture in the pluton was filled with quartz-fluorite veins. In the southeastern side, the wall-rock volcanic rock alteration was weak, with slightly silicification and argillization just locally.

4. **Geochemical Features**

The Xianxia composite pluton is high in SiO_2 content (67.6–76.11%), enriched in alkali (Alk = $K_2O + Na_2O$, 7.16–9.91%) (Fig. 3.19), and low in P_2O_5 (0.03–0.3%) and TiO_2 (0.17–0.55%). In general, compared with later medium-coarse-grained and fine-grained syenogranite, the early medium-grained and the megacrystic porphyritic monzonitic granite showed a trend of increase in SiO_2 and K_2O, and K_2O/Na_2O ratio, but a trend of contents decrease in TiO_2, TFeO, MgO, CaO, and Na_2O. A/CNK is mostly 0.95–1.03. A/CNK of medium-coarse-grained porphyritic monzonitic granite of D3350 is higher (1.18) possibly due to later fracturing and K-feldspathization locally. As a whole, early monzonitic granite is metaluminous, high K

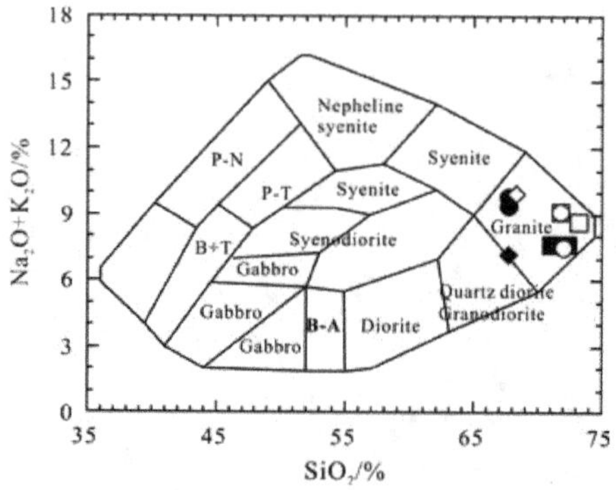

Fig. 3.19 SiO_2–K_2O diagram for Xianxia composite pluton (its legends are the same as those in Fig. 3.20)

calc-alkaline, while the late-stage syenogranite is metaluminous–peraluminous and shoshonitic (Fig. 3.20a, b), which are similar to late coarse-grained quartz syenite veins.

For the early medium-grained and megacrystic porphyritic monzonitic granite, the $\sum$REE content is $(112.66–193.55) \times 10^{-6}$, LREE/HREE ratio is 10.39–16.90, δEu is 0.52–0.75, LREE and HREE was strongly differentiated. The chondrite-normalized REE patterns are strongly dipping rightward, medium negative Eu anomaly, enriched in Rb, Th, U, and K, etc., depleted in elements such as Ba, Sr, Nb, P, and Ti (Fig. 3.21a, b). For the late medium-coarse-grained and fine-grained syenogranite in the northeastern section of the pluton, $\sum$REE content is $(155.87–244.51) \times 10^{-6}$, LREE/HREE ratio is 5.02–13.74, δEu is 0.22–0.67, LREE and HREE were weakly differentiated, the chondrite-normalized REE patterns were weakly dipping rightward, strongly negative Eu anomaly, similarly showing enrichment in Rb, Th, U, and K, strongly depleted in elements such as Sr, Nb, P, and Ti, possibly associated with crystal segregation of minerals such as plagioclase, apatite, ilmenite, rutile, and sphene during magmatism, and conforming to the gradually increasing differentiation. The REE distribution curve and trace element features in the late coarse-grained quartz syenite veins are greatly different from those of early monzonitic granite but similar to those of later granite, indicating their similar magma source.

In the Xianxia composite pluton, $(^{86}Sr/^{87}Sr)_i$ is 0.70988–0.70455 and $\varepsilon Nd(t)$ is −5.14 to −8.87, and from the early monzonitic granite to the late syenogranite both have a trend of decrease, and T_{DM2} is low (1.36–1.65 Ga); they are similar to $(^{86}Sr/^{87}Sr)_i$ (0.71030–0.70613) and $\varepsilon Nd(t)$ (−3.75 to −6.4) of the Mogan Mount granite pluton in the northern Zhejiang Province (Zhang et al. 2012) or slightly lower than $(^{86}Sr/^{87}Sr)_i$ (0.71010–0.70960) and $\varepsilon Nd(t)$ (−6.28 to −7.32) of Jinde granodiorite pluton in the southern Anhui Province (Zhou et al. 2014).

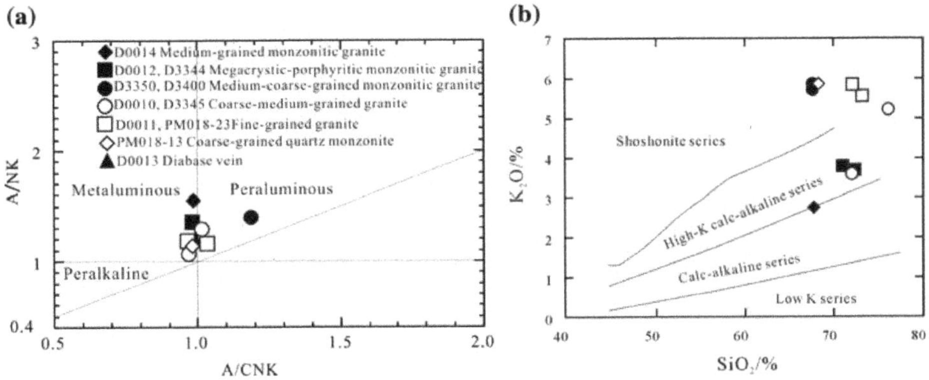

Fig. 3.20 SiO_2–Na_2O + K_2O and A/CNK-A/NK diagrams for Xianxia composite pluton

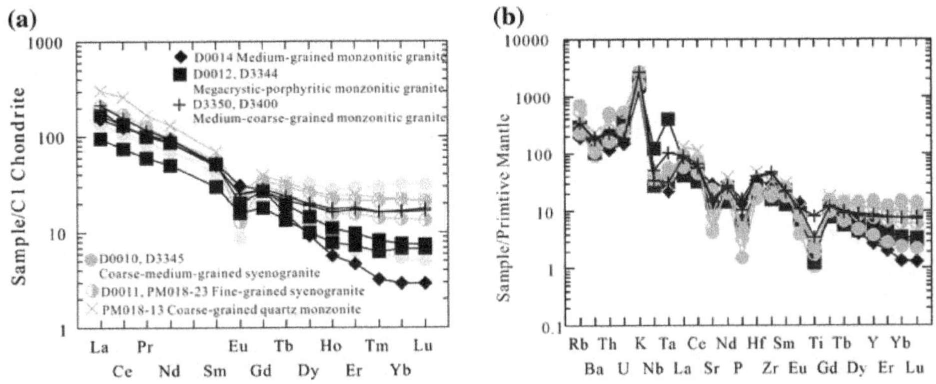

Fig. 3.21 Chondrite-normalized REE distribution mode (**a**) and primitive mantle-normalized trace element spider diagram (**b**) for the Xianxia composite pluton (the normalized values of chondrite and primitive mantle come from Sun and McDonough 1989)

5. **Isotope geochronology**

In the Xianxia pluton, zircon in main lithologies is mostly irregular long strip, 100–200 μm in length and with about 2:1 length-width ratio. Zircon developed magmatic oscillatory zoning and very few of zircons have developed inherited core (Fig. 3.22). The Th contents in zircons from the medium-grained monzonitic granite, the megacrystic porphyritic monzonitic granite, the medium-coarse-grained syenogranite, and the fine-grained syenogranit are (8–310) × 10^{-6}, (83–566) × 10^{-6}, (28–220) × 10^{-6}, and (72–466) × 10^{-6}, respectively; the U contents are (226–1041) × 10^{-6}, (175–1004) × 10^{-6}, (40–284) × 10^{-6}, and (80–433) × 10^{-6}, respectively; the Th/U ratios are 0.02–0.83, 0.19–0.91, 0.70–1.29, and 0.36–1.08, respectively, showing a typical magmatic zircon (Hoskin and Schaltegger 2003). Zircon dating points are mostly projected on or near the concordant curve, and $^{206}Pb/^{238}U$ weighted mean ages are 145.1 ± 1.2 Ma (MSWD = 1.8), 144.2 ± 0.97 Ma (MSWD = 0.94), 131.7 ± 1.6 Ma (MSWD = 2.3), and 130.8 ± 1.6 Ma (MSWD = 1.4), respectively (Fig. 3.23), indicating that the pluton was intruded in Early Cretaceous. The obtained ages can be divided into two stages (145.1–144.2 Ma and 131.7–130.8 Ma). Notably in the early monzonitic granite, the $^{206}Pb/^{238}U$ ages of some sample points concentrate in three stages: 1368–704 Ma (D0014-7, D0014-14, D0014-15, D0012-3), 497.8–268 Ma (D0014-9, D0014-2, D0012-4, D0012-5), and 200.1–163 Ma

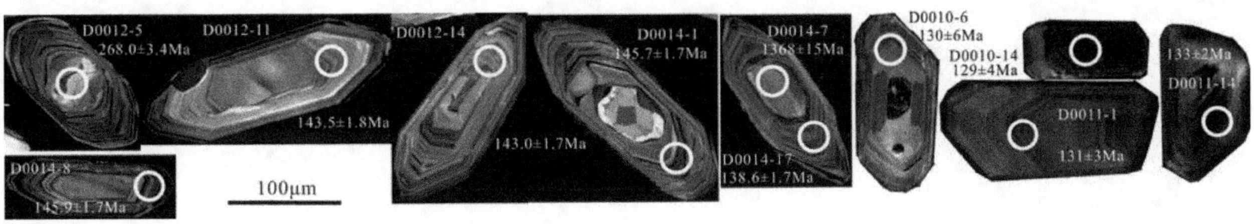

Fig. 3.22 Typical zircon CL images, dating points, and ages of the rocks in the Xianxia composite pluton

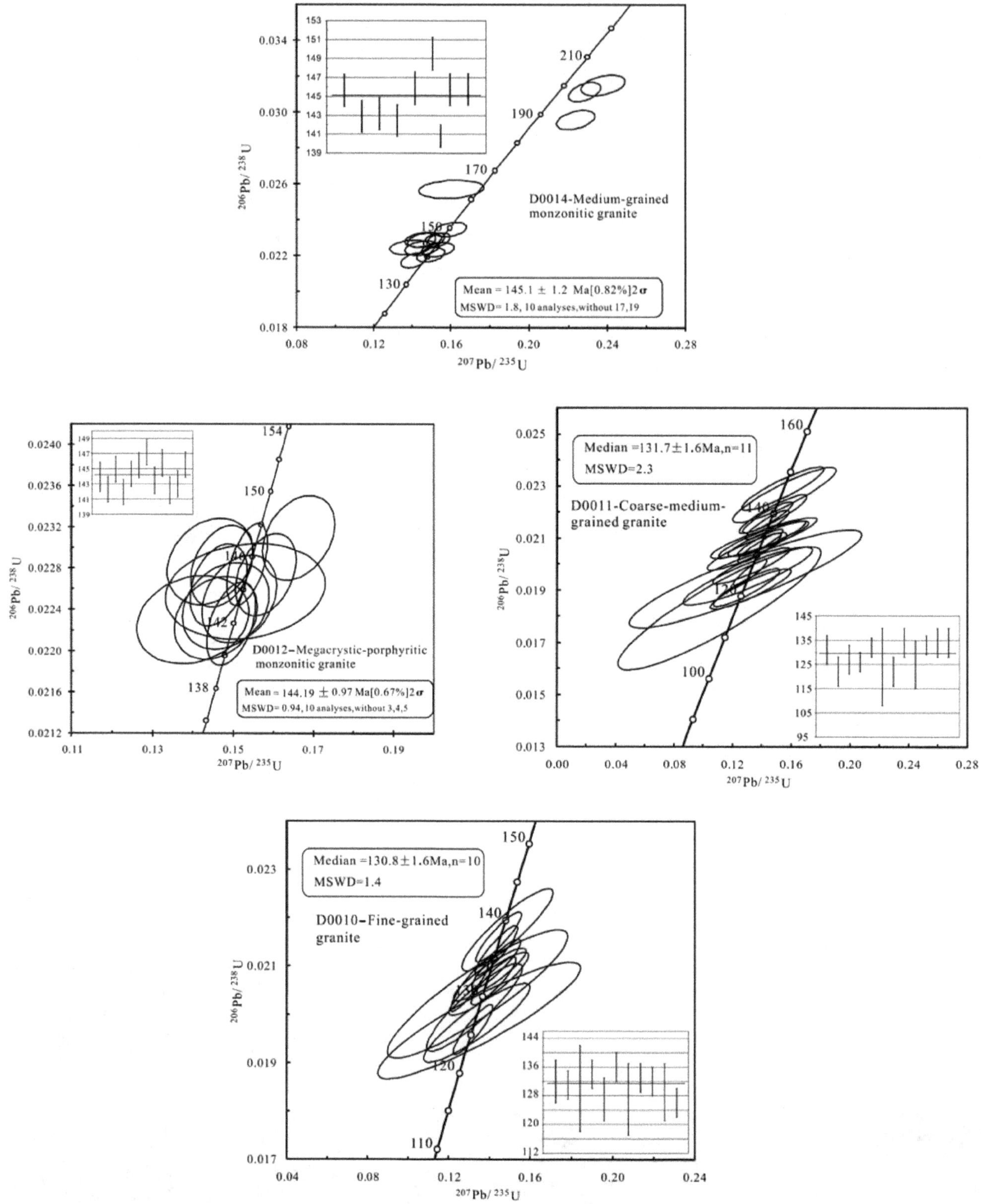

Fig. 3.23 Zircon U–Pb concordant diagrams of main rocks of the Xianxia composite pluton

(D0014-13, D0014-18, D0014-6, D0014-16), which possibly indicate that during the emplacement along the NE-strike fault, the upwell magma captured zircon or inherited core from the Middle Neoproterozoic to the Middle–Late Jurassic basement and wall rocks. For the later syenogranite, the $^{206}Pb/^{238}U$ ages in a few sample points involved four stages: 720 Ma (D0011-7), 154–145 Ma (D0011-18, D0011-15, D0011-19), 140–136 Ma (D0010-1, D0011-5, D0011-16, D0010-8, D0010-5, D0010-10), and 119–117 Ma (D0010-2, D0011-8). The first two stages indicated it also captured zircon or inherited core of the Neoproterozoic basement and the early monzonitic granite. The presence of the three-stage ages possibly meant that, the syenogranite emplacement crystallization last longer or inherited zircon growth of early monzonitic granite.

3.1.2.4 Wushanguan Composite Pluton

1. **Geological Features**

Wushanguan composite pluton is situated in the adjacent area of Yuhang District of Hangzhou and Shanchuan Town of Anji County, which is in the eastern part of the survey area. The pluton is generally NW-strike, about 10 km in length and 2–8 km in width, outcrop areas 64 km^2. Regionally, the pluton intruded into the northeastern margin of the volcanic low-lying land of Tianmu Mount was jointly controlled by the NE-strike Zaoxi-Mogan Mount Fault and the NW-strike fault.

2. **Petrological Features**

The Wushanguan pluton is mainly composed of the first stage medium-grained (biotite) monzonitic granite, the second stage medium-coarse-grained and medium-grained syenogranite, the third stage megacrystic porphyritic monzonitic granite (a handful of) and fine-grained syenogranite, and the fourth stage medium-grained syenite.

Medium-grained monzonitic granite

The first stage irregular and NE-strike medium-grained monzonitic granite are mainly distributed in the northwestern part of the pluton. The color of fresh rock is light pink–light offwhite, while the weathered rock is light gray yellow–light gray but with smooth and flat surface. The rocks are fine-medium-grained granitic texture and massive structure. The typical mineral assemblage is composed of plagioclase (30–35%), K-feldspar (25–30%), quartz (25–30%), biotite (10–15%), and hornblende (2–4%). The grain size is mainly 2–5 mm, and minor <2 mm. Plagioclase is subhedral plate or column and developed Na-feldspar bicrystal and zoning structure. K-feldspar is perthite, irregular plate, with plagioclase replacement and inclusions. Quartz is anhedral granular, unevenly distributed among feldspar. Biotite is dark brown, platy or laminated, a small portion alterated into chlorite along the joints. Hornblende is light green and long column-like. The MMEs are oval, long strip, or irregular, 1–30 cm with the size, composed of biotite or biotite-plagioclase in the pluton.

Medium-coarse-grained and medium-grained syenogranite

The second stage medium-coarse-grained and medium-grained syenogranite is located in southeastern part of the composite pluton, which is mainly distributed in the area of the Luniao Town, such as the Xianbaikeng Reservoir, Shangougou Village, and Taigongtang Village. Its outrcrop area is NEE, NE, and NW-strike, irregular shapes, about 20 km^2. The rocks are light gray. The weathered surface is coarse and full of sags and crests, mineral grains such as quartz show embossment. It is coarse-medium and medium-coarse-grained granitic texture, massive structure, and from southeast to northwest part of the pluton, the grain sizes gradually changed from medium-coarse to coarse-medium. The main mineral assemblages are 45–55% K-feldspar, 15–25% plagioclase, 25–35% quartz and 1% biotite, etc.; the grain size is generally 2–7 mm, and 7–10 mm for a handful, and locally K-feldspar looks porphyritic (1–2 cm × 1–1.5 cm). K-feldspar is mainly orthoclase, wide plate, Carlsbad bicrystals occurred. Plagioclase is hypidiomorphic wide plate, plate-column, and Na-feldspar bicrystal visible. Quartz is mostly anhedral granular, distributed rather evenly, and a handful of quartz is inlaid in a certain shape in K–Na-feldspar formed graphic texture. Rocks are rather weathered and broken locally, and plagioclase alterated into kaolinization and epidotization.

Megacrystic porphyritic monzonitic granite

In the third stage, a handful of lump (2–8 m × 2–8 m) megacrystic porphyritic monzonitic granite intruded into medium-coarse-grained syenogranite, locally wrapping ellipse small-lump-like xenolith of medium-coarse-grained syenogranite (5–50 cm × 5–50 cm) (Figs. 3.24, 3.25 and 3.26).

Fine-grain syenogranite

The fine-grain syenogranite, in the shape of small stock or vein, intruded into medium-coarse and coarse-medium-grained syenogranite and medium-grain (biotite) monzonitic granite (Fig. 3.27) or the Huangjian Formation volcanic strata. The intrusive contact borderline is clear. Rocks are pink–light pink, mainly composed of K-feldspar

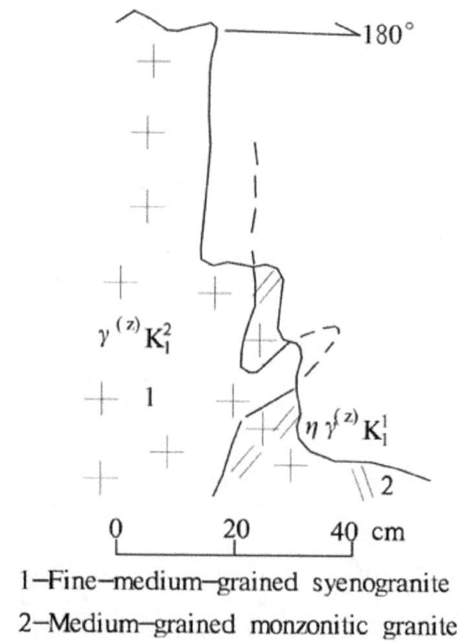

Fig. 3.24 Fine-medium-grained granite. It intrudes in medium-grained monzonitic granite

(45–50%), plagioclase (10–15%), quartz (30–35%), and a handful of biotite, with grain size of 0.5–1.5 mm.

Medium-grained syenite

The fourth stage medium-grained syenite is pink–light pink, mainly composed of K-feldspar (80–90%), plagioclase (5–10%), and a handful of quartz and biotite, with grain size of 1–5 mm.

3. **Wall-rock Alteration**

The Wushanguan pluton was intruded mainly along the contact zone between the Early Cretaceous Huangjian Formation volcanic rocks and the Early Paleozoic sedimentary rocks. The Huangjian Formation volcanic strata are main wall rocks, which suffered strong silicification. In the eastern part of the pluton locally distributed the Cambrian strata. The xenolith from the carbonatite and siliceous mudstone of the Yangliugang, the Xiyangshan, and the Yinzhubu Formations, occurred in the northern part of the pluton. These xenoliths underwant strong skarnization and silicification,

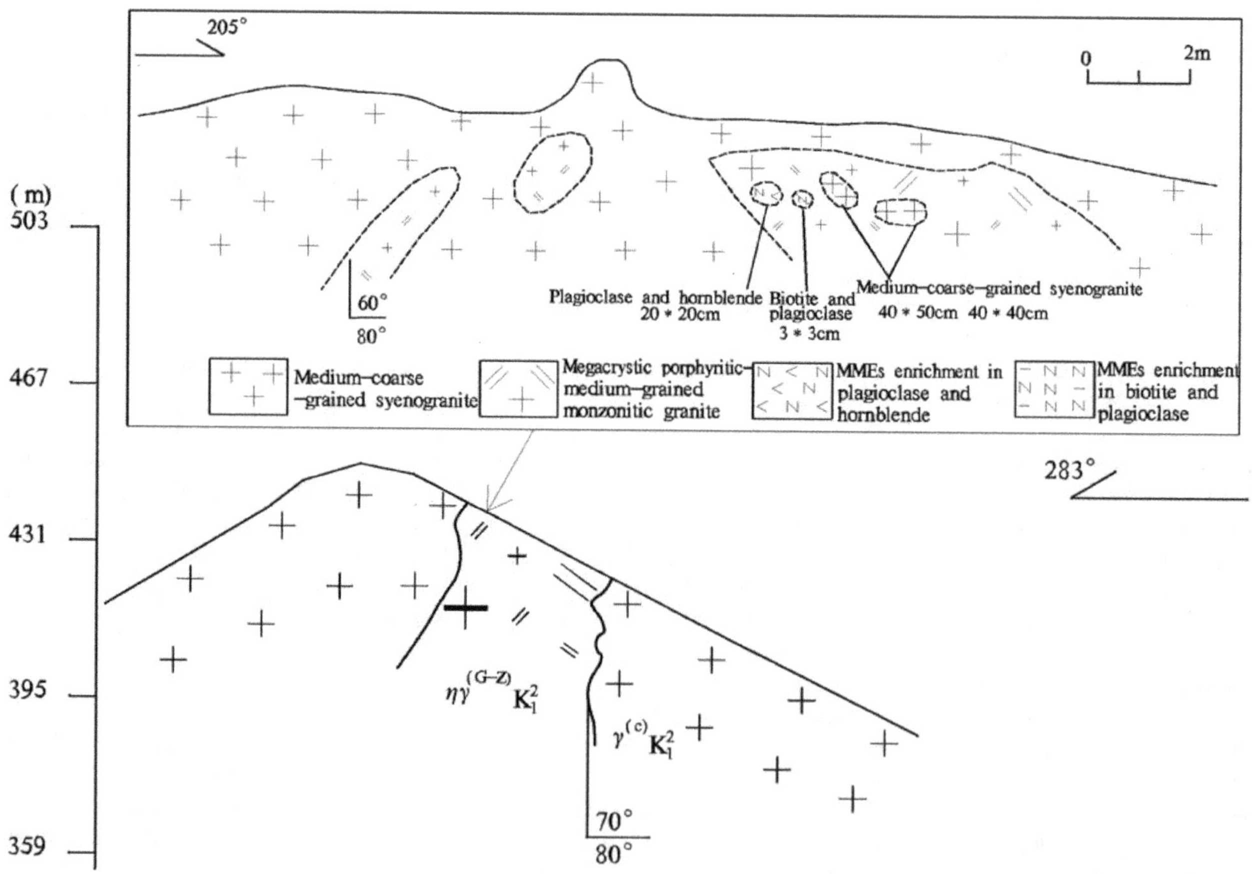

Fig. 3.25 Sketch showing the internal features of megacrystic porphyritic-medium-grained monzonitic granite and intrusive medium-coarse-grained granite contact relationship of the Wushanguan pluton

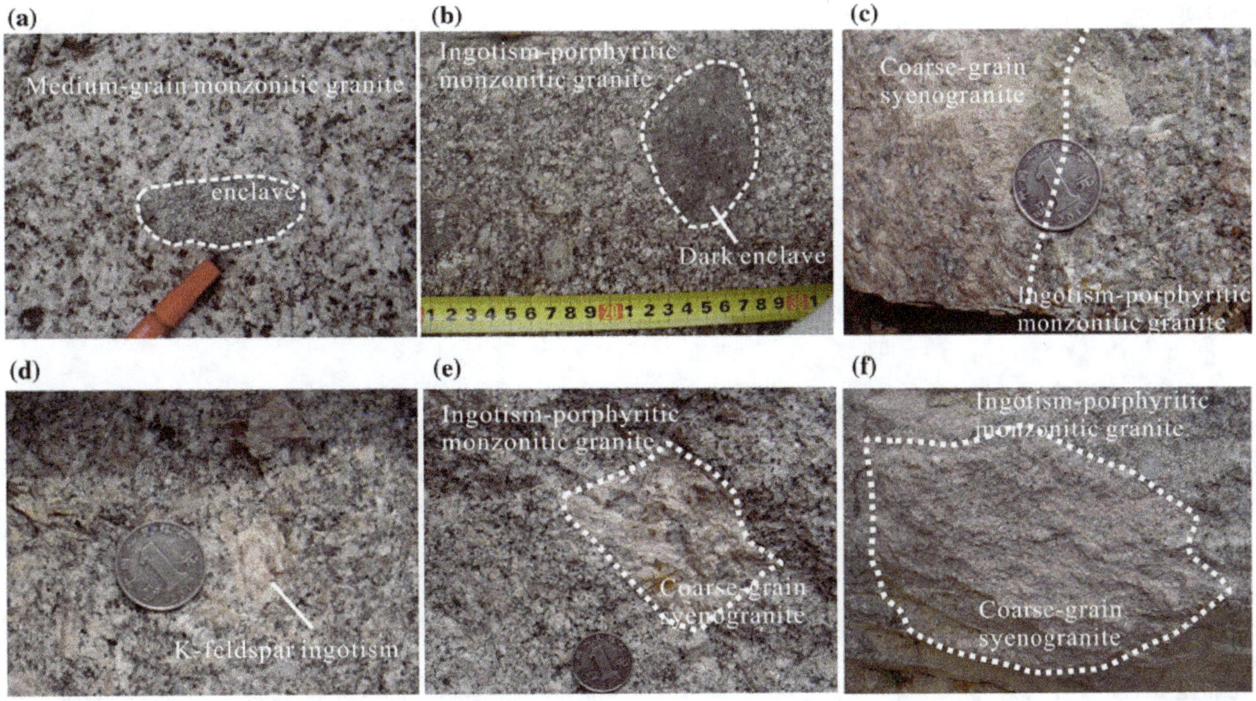

Fig. 3.26 Contact relationship between medium-coarse-grained syenogranite and megacrystic porphyritic monzonitic granite in the Wushanguan pluton. **a** MMEs (dark, biotite, and plagioclase bearing) developed in medium-grained monzonitic granite; **b** MMEs developed in megacrystic porphyritic monzonitic granite; **c** contact borderline of megacrystic porphyritic monzonitic granite and coarse-grained syenogranite; **d** K-feldspar phenocryst in megacrystic porphyritic monzonitic granite; **e** and **f** coarse-grained syenogranite wrapped in megacrystic porphyritic monzonitic granite

Fig. 3.27 Contact relationship of fine-grained syenogranite intrusion in medium-grained monzonitic granite at Zhaojiatang in the Wushanguan pluton. **a** fine-grained syenogranite intruded in early stage medium-grained monzonitic granite, with clear intrusion border line; **b** fine-grained syenogranite at the margin intruded as vein shape in early medium-grained monzonitic granite; **c** the early stage medium-grained monzonitic granite cognate xenolith in fine-grained syenogranite

and even formed skarn-type Fe–Pb–Zn polymetallic ore (mineralized) bodies. The Majiatang area, in the central part of the pluton, the xenolith of the tufaceous siltstone of the Nanhuaian Xiuning Formation is seen in the contact zone between the medium-grained monzonitic granite and the medium-coarse-grained monzonitic granite, with the size of 1.5×1 km^2, where the rocks developed strip-like strong hornfelsic alteration.

4. **Geochemical Features**

In the Wushangguan composite pluton, the medium-grained monzonitic granite, the medium-coarse and medium-grained and the fine-grained syenogranite are high in SiO_2 content (66.75–67.37%, 75.29–76.00%), enriched in alkali (K_2O + Na_2O = 7.49–7.62%, 8.20–9.14%), high in K_2O/Na_2O ratio (1.01–1.09, 1.16–1.33); MgO is 1.37%–1.74% and 0.10%–

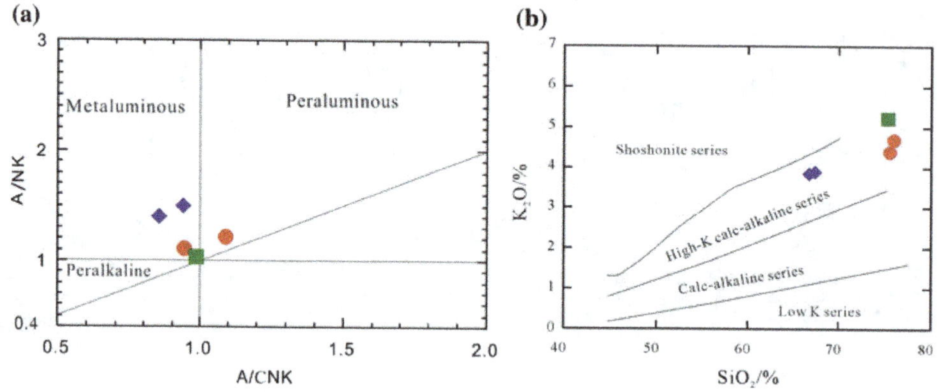

Fig. 3.28 A/CNK-A/NK diagram (**a**) and SiO_2-K_2O diagram (**b**) for the Wushanguan pluton (legends are shown in Fig. 3.29)

0.28%. P_2O_5 is 0.16%–0.17% and 0.01%–0.03%, and TiO_2 is 0.46%–0.53% and 0.07%–0.17%, respectively. A/CNK is 0.85–0.94 and 0.94–1.09, respectively. The medium-grained monzonitic granite is metaluminous, and medium-coarse and medium-grained and fine-grained syenogranite is metaluminious–peraluminous. Rittmann Index (σ) is 2.44–2.30 and 2.07–2.59, respectively, and the two types of the rocks are high K calc-alkaline (Fig. 3.28).

The medium-grained monzonitic granite is low in $\sum$REE content (154.36×10^{-6}–170.84×10^{-6}), LREE and HREE differentiated rather clear, LREE/HREE is 9.26–9.49. δEu is 0.66–0.75, showing weak negative Eu anomaly. The chondrite-normalized REE patterns show the feature of weakly dipping rightward; enrichment in K, Rb, Th, and U, weakly depleted in LILE such as Ba, Sr, and P; weakly depleted in HFSE such as Ti, Nb, and Ta (Fig. 3.29); $(^{86}Sr/^{87}Sr)_i$ is 0.70689, $\varepsilon Nd(t)$ is −4.46, and the T_{DM2} = 1.29 Ga. The medium-coarse and medium-grained, fine-grained syenogranite is high in $\sum$REE content (141.35×10^{-6}–252.09×10^{-6}), and LREE and HREE differentiated rather weakly, LREE/HREE = 4.56–9.01. δEu ranges in the range of 0.03–0.57, showing strong negative Eu anomaly. The chondrite-normalized REE patterns show the feature of weakly dipping rightward or "V" shape; enrichment in K, Rb, Th, and U, weakly depleted in LILE such as Ba and Sr; in terms of HFSE, weakly depleted in Nb and Ta, etc., and strongly depleted in P and Ti, etc.

5. **Isotope geochronology**

For the medium-grained monzonitic granite (D0029), the medium-grained syenogranite (D0021), the medium-coarse-grained syenogranite (D0022), and the fine-grained syenogranite (D0026), zircon is mostly in long and short strips, and less in grained, about 100–250 μm in length, the length-width ratio is mostly 2:1–3:1, less 1:1. Zircon can be seen magmatic oscillatory zoning (Fig. 3.30). Zircon's U content is 80×10^{-6}–655×10^{-6}, 98×10^{-6}–1241×10^{-6}, 28×10^{-6}–6659×10^{-6}, and 571×10^{-6}–1825×10^{-6}, respectively, Th content is 50×10^{-6}–1003×10^{-6}, 111×10^{-6}–1623×10^{-6}, 20×10^{-6}–3987×10^{-6}, and 289×10^{-6}–1524×10^{-6}, respectively, and Th/U ratio is 0.61–1.58, 0.69–1.48, 0.39–1.50, and 0.43–0.86, respectively, a typical feature of magmatism origin. Sample points tested are mostly projected on or near the concordant curve, $^{206}Pb/^{238}U$ weighted mean age is 136.3 ± 1.2 Ma (MSWD = 1.6), 133.9 ± 1.6 Ma

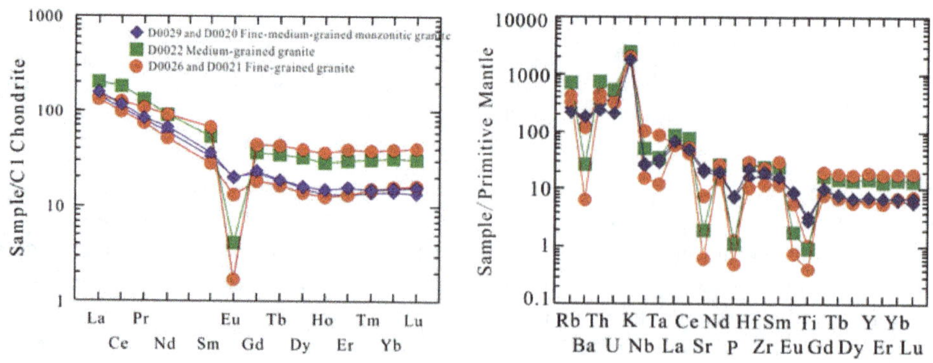

Fig. 3.29 REE chondrite-normalized distribution mode and primitive mantle-normalized trace element spider diagram for Wushanguan composite pluton (the normalized values of chondrite and primitive mantle come from Sun and McDonough 1989)

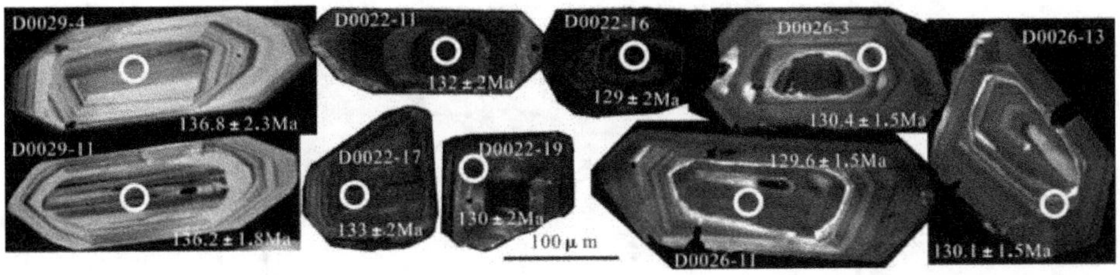

Fig. 3.30 Main Zircon CL images, dating location, and ages of the rocks in the Wushanguan composite pluton

(MSWD = 0.88), 131.1 ± 1.5 Ma (MSWD = 0.88), and 128.1 ± 0.82 Ma (MSWD = 1.5), respectively (Fig. 3.31), representing the crystallization age of the Wushanguan composite pluton, the early stage is medium-grained monzonitic granite (136.3 Ma), the middle stage is medium-coarse–medium-grained syenogranite (133.9–131.1 Ma) and the later stage is fine-grained syenogranite (128.1 Ma).

3.1.2.5 Tonglizhuan Composite Pluton

1. **Geological Features**

The Tonglizhuang composite pluton, of which the outcropping is small in the survey area, is located in the southeastern corner of the Hanggai Mapsheet. The pluton is nearly EW strike, distributed at the contact zone between the Cambrian–Ordovician siliceous rock, siltstone, carbonatite, and the Cretaceous volcanic tuff (Fig. 3.32), and intruded along the Luocun anticline axial, about 5 km long from east to west and 0.5–2 km wide from south to north. The pluton is mainly composed of the early stage medium-grain monzonitic granite inside, the middle stage fine-grain syenogranite at the northern margin, and the later stage medium-grain quartz orthophyre at the southern part and eastern margin. The granite, apilite, and sillite veins occurred in NW, NE or nearly SN strike inside the pluton.

2. **Petrological Features**

Medium-grained monzonitic granite is gray–light gray, medium-fine-grained and coarse-medium-grained subhedral granular texture, mainly composed of plagioclase (35–45%), K-feldspar (25–35%), quartz (20–30%), biotite (5–6%), and hornblende (<5%), with the grain size of 0.2–3 mm and bigger than 5 mm for a handful. Plagioclase is subhedral plate or column, slight sericitization on the surface, and growth Na-feldspar bicrystal and zoning structure; K-feldspar is mostly striped feldspar, irregular plate, with plagioclase replacement and inclusion structure; quartz, is anhedral granular, unevenly distributed among feldspar grains; biotite is dark brown, platy or laminated, partly alterated into chlorite along joints; hornblende is light green and long column (Fig. 3.33a, d, e).

Fine-grained syenogranite is pink–light pink, fine-grained texture, is mainly composed of K-feldspar (40–50%), plagioclase (20–25%), quartz (20–30%), and a handful of biotite (5%), in the grain size of 0.2–1 mm. K-feldspar is mainly orthoclase, wide plate, growth Carlsbad bicrystal; plagioclase is hypidiomorphic wide plate and plate-column, growth Na-feldspar bicrystal; quartz is hexagonal bipyramid and in form of fine granular aggregation; biotite is reddish brown and plate or laminated, and its margin is often surrounded by fine biotite aggregation and irony points (Fig. 3.33b, f).

Medium-grained quartz orthophyre is pink–light pink and medium-grained porphyaceous texture. Phenocryst is mainly composed of K-feldspar (25–30%), plagioclase (5–10%), and a handful of quartz and biotite, in the grain size of 1–5 mm; while matrix is micro-grained texture, and mainly composed of K-feldspar (40–45%), quartz (5–10%), and a handful of plagioclase (5–10%) as well as accessory minerals such as undeterminated metallic minerals, apatite, and zircon, with the grain size of 0.05–0.1 mm; K-feldspar is mainly orthoclase while plagioclase is mainly oligoclase (Fig. 3.33c).

3. **Wall-rock Alteration**

Medium-grained monzonitic granite and fine-grained syenogranite suffered from strong alteration, and its inside shows albitazation, greisenization, and chloritization and wall rocks show hornfelsic, skarnization, silicification, pyritization, and marbleization. The medium-grained quartz orthophyre suffered from weak alteration and show slight albitazation.

4. **Geochemical Features**

The early medium-grained monzonitic granite is high in SiO_2 content (70.61%), enrichment in alkali (Alk = K_2O + Na_2O, 7.75%), high in K_2O/Na_2O ratio (1.27); and low in content of MgO (0.98%), P_2O_5 (0.10%), and TiO_2 (0.32%).

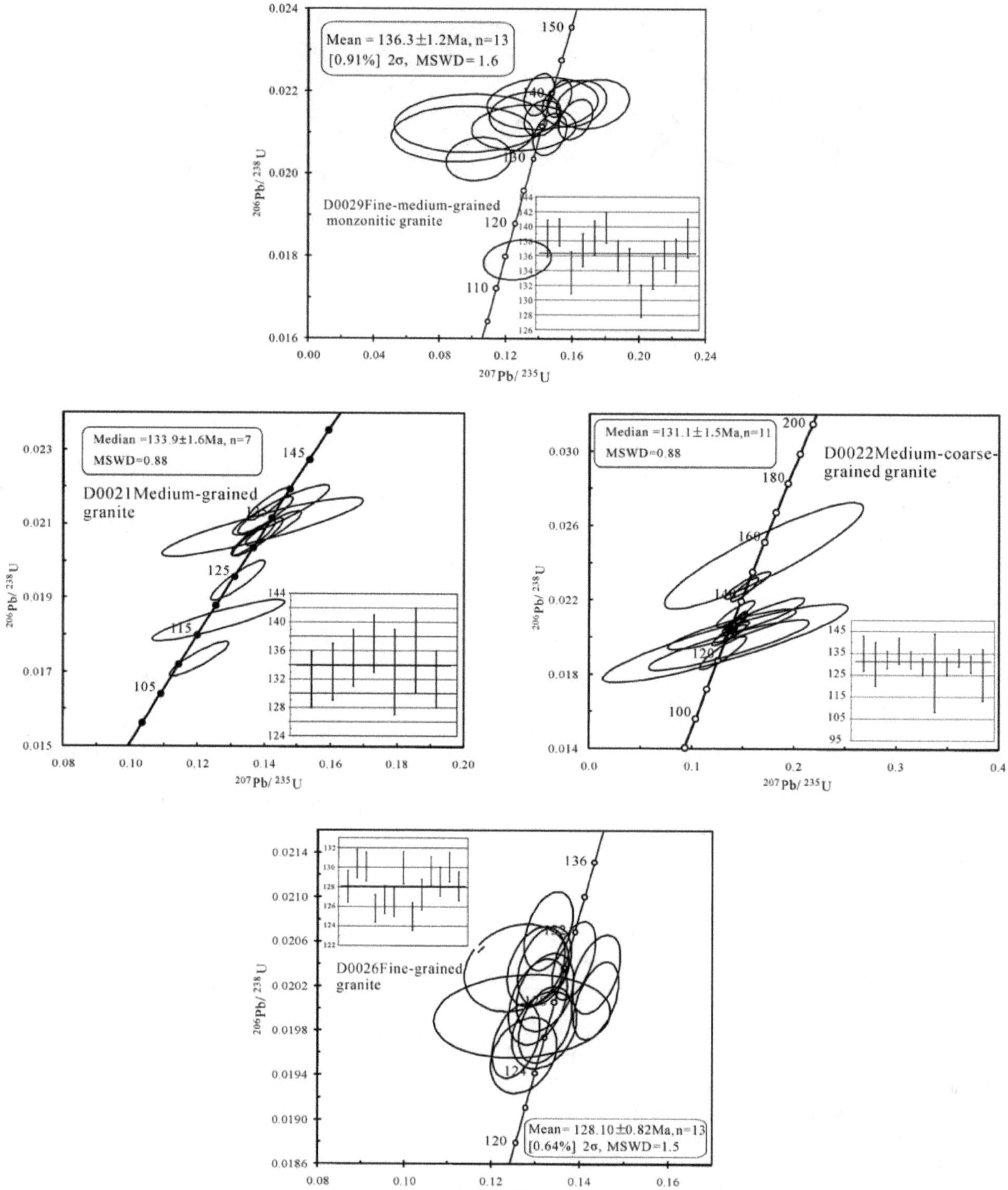

Fig. 3.31 Zircon U–Pb concordant curve of main lithologies of the Wushanguan composite pluton

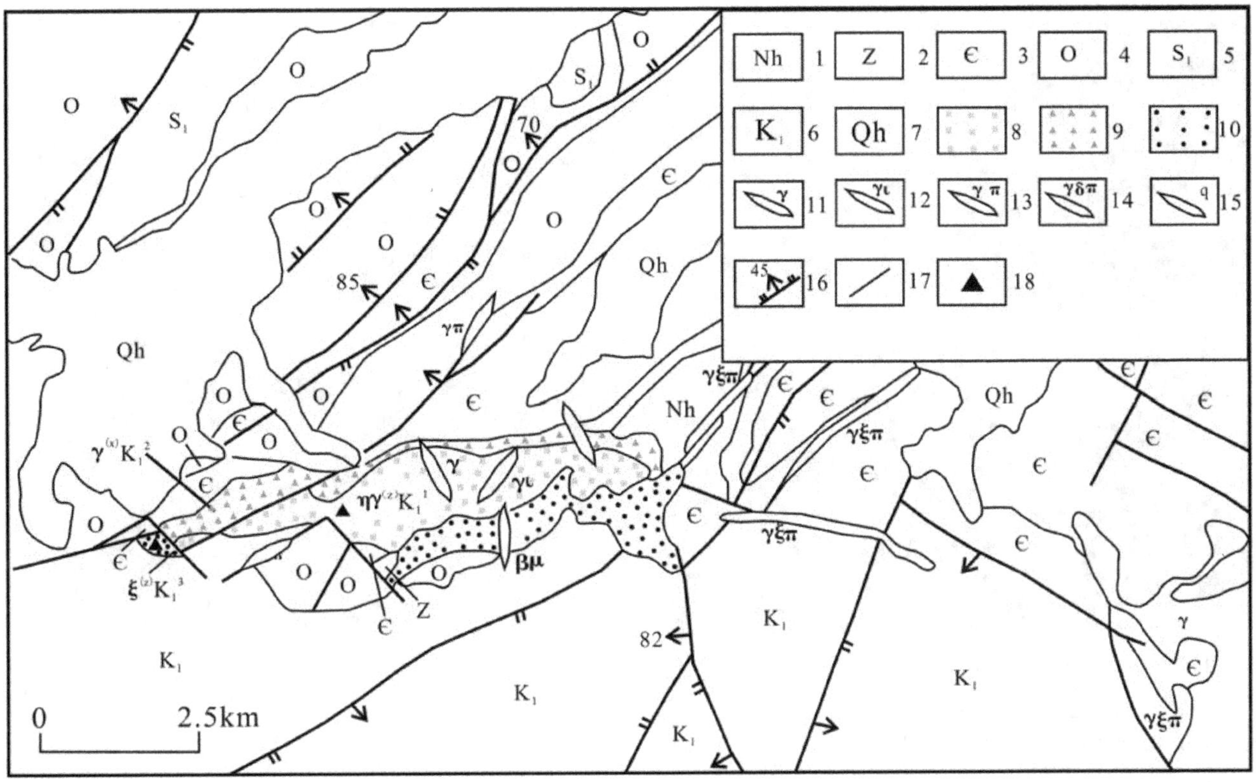

Fig. 3.32 Regional geological sketch of the Tonglizhuang composite pluton (based on the 1:200,000 Lin'an Mapsheet). 1. Nahuaian; 2. Sinian; 3. Cambiran; 4. Ordovician; 5. the lower Silurian; 6. the lower Cretaceous; 7. Quaternary; 8. medium-grained monzonitic granite; 9. fine-grained syenogranite; 10. medium-grained quartz orthophyre; 11. fine-grained granitic vein; 12. granitic aplite vein; 13. granitic porphyry vein; 14. granodiorite-porphyry vein; 15. quartz vein; 16. normal fault; 17. geological boundary; 18. sampling location

A/CNK is 1.00, Rittmann Index (σ) is 2.18, shown weakly peraluminous and high K calc-alkaline features (Fig. 3.34).

The later medium-grained quartz orthophyre is low in SiO_2 content (66.25%), enrichment in alkali (Alk = K_2O + Na_2O, 10.02%), high in K_2O/Na_2O ratio (1.66); low in content of MgO (0.67%), P_2O_5 (0.10%), and TiO_2 (0.37%). A/CNK is 0.94, Rittmann Index (σ) is 4.32 and is metaluminous and shoshonite features.

Both are low in $\sum$REE content (127.02 × 10^{-6}–183.02 × 10^{-6}), LREE and HREE differentiated rather clear, LREE/HREE = 11.71–9.12, La_N/Yb_N = 15.67–9.58, δEu = 0.72–0.79, showing weak negative Eu anomaly, the chondrite-normalized REE patterns show the feature of weakly dipping rightward; enrichment in K, Rb, Th, and U, weakly depleted in LILE such as Ba and Sr; weakly depleted in HFSE such as Nb, Ta, P, and Ti. Both have similar REE and trace elements features, indicating that both may have the same magma source.

5. **Isotope geochronology**

Zircons from the medium-grained monzonitic granite and medium-grained quartz orthophyre are mostly long and short strips, a small amount is granular, about 100–300 μm long, the length-width ratio is mostly 2:1–1:1, less 3:1. Zircons contain clear magmatic oscillatory zoning. Zircons have U content of (383–1635) × 10^{-6} and (80–1053) × 10^{-6}, respectively, Th content (329–1828) × 10^{-6} and (76–730) × 10^{-6}, respectively, and Th/U ratio 0.70–1.33 and 0.61–1.24, respectively, indicating typical magmatic origin zircon. Sample points tested are mostly projected on or near the concordant curve, and $^{206}Pb/^{238}U$ weighted mean age is 142.3 ± 1.4 Ma (MSWD = 1.5) and 125.3 ± 1.6 Ma (MSWD = 1.4), respectively (Fig. 3.35). For the medium-grained monzonitic granite's sample points D0032-16 and D0032-17, the $^{206}Pb/^{238}U$ ages are 758 Ma and 518 Ma, respectively, which indicate they are probably inherited zircons. The $^{206}Pb/^{238}U$ ages of a few sample

Fig. 3.33 Outcrop and micrograph images showing the lithological features in the Tonglizhuang pluton. **a** medium-grained monzonitic granite; **b** fine-grained syenogranite; **c** medium-grained quartz orthophyre; **d** K-feldspar (Kf) and plagioclase (Pl) in medium-grained monzonitic granite; **e** plagioclase (Pl) containing K-feldspar (Kf) core growth in medium-grained monzonitic granite; **f** plagioclase (Pl), K-feldspar (Kf), and quartz (Qtz) in fine-grained syenogranite

points (D0005-07, D0005-08, D0005-09, D0005-11) from the medium-grained quartz orthophyre are 148–137 Ma, possibly suggesting inherited zircons from the early stage monzonitic granite.

The Paleozoic strata and the volcanic rocks in Mount Tianmu are intrusive contact between the two sides of the Tonglizhuang composite pluton. The zircon U–Pb dating results indicated that crystallization age of the early medium-grained monzonitic granite is 142.3 ± 1.4 Ma. By comparing the lithological types and features of intrusive rocks in the survey area, the crystallization time of the middle stage fine-grained syenogranite was close to the fine-grained syenogranite in the Ma'anshan pluton and the Xianxia pluton (129.6–128.3 Ma). The forming time of the later stage medium-grained quartz orthophyre was 125.3 ± 1.5 Ma. Magmatism in the early and middle stages was likely to be related to the translational extensional fractures developed in the late Yanshanian in northern Zhejiang and southern Anhui, while the late-stage magmatism was likely to be related to regional volcanic eruption activities.

3.1.2.6 Zhinanshan-Dongkeng Volcanic-Intrusive Complexes

1. **Geological Features**

The Zhinanshan volcanic-intrusive complex, located in the central-south of the Chuancun Mapsheet, is NW-strike, about 1 km long and 700 m wide, with outcrop areas of 0.7 km^2. The complex intruded into the second member of the Huangjian Formation rhyolitic tuff lava, mainly consisting of fine-medium-grained quartz diorite. An early stage andesite pluton of 0.2 km^2 is presented in the east of the quartz diorite, and quartz diorite cemented andesite agglomerate breccia is visible at the margin of the intrusive contact zone (Fig. 3.36).

Dongkeng volcanic-intrusive complex is situated in the central of the Chuancun Mapsheet, about 3 km from the Zhinanshan volcanic-intrusive complex, nearly EW strike, ellipse, 2 km long from east to west and 1.5 km long from south to north, with outcrop areas of 1.2 km^2. The complex intruded at the border of the second member of the

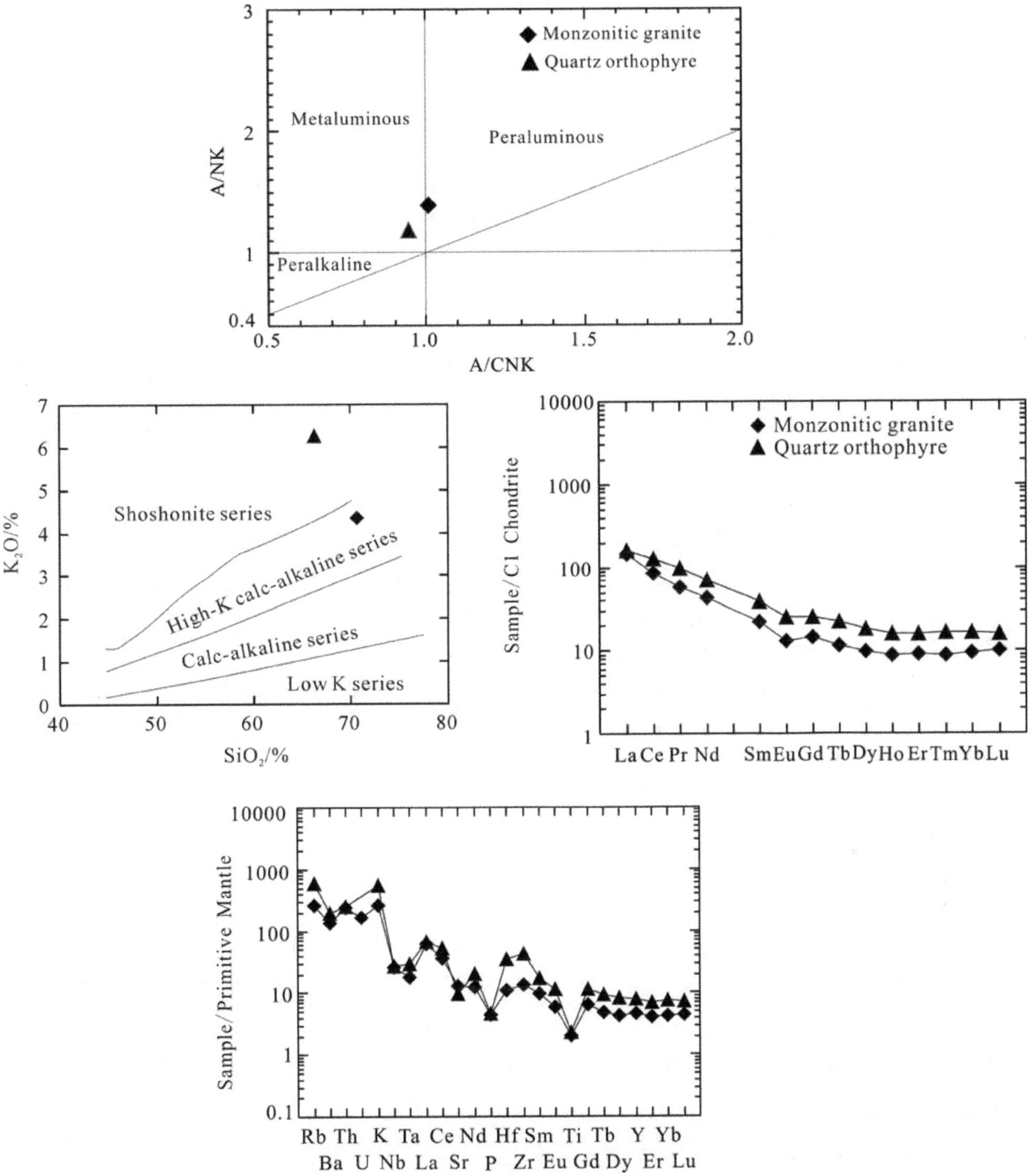

Fig. 3.34 REE chondrite-normalized distribution mode and primitive mantle-normalized trace element spider diagram for Tonglizhuang composite pluton (the normalized values about chondrite and primitive mantle come from Sun and McDonough 1989)

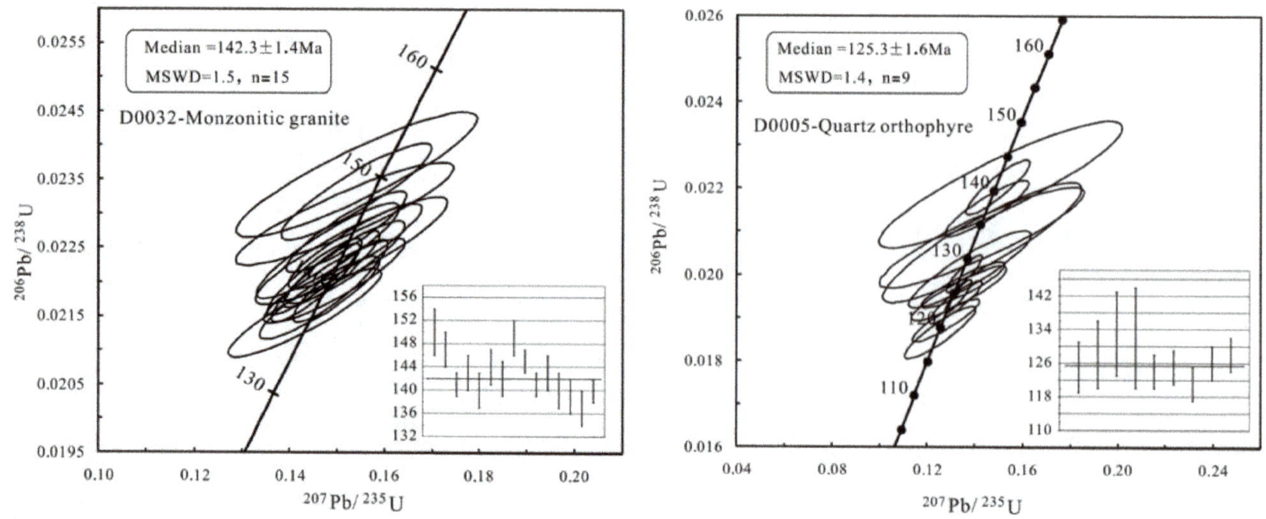

Fig. 3.35 Zircon U-Pb concordant diagrams for the Tonglizhuang composite pluton

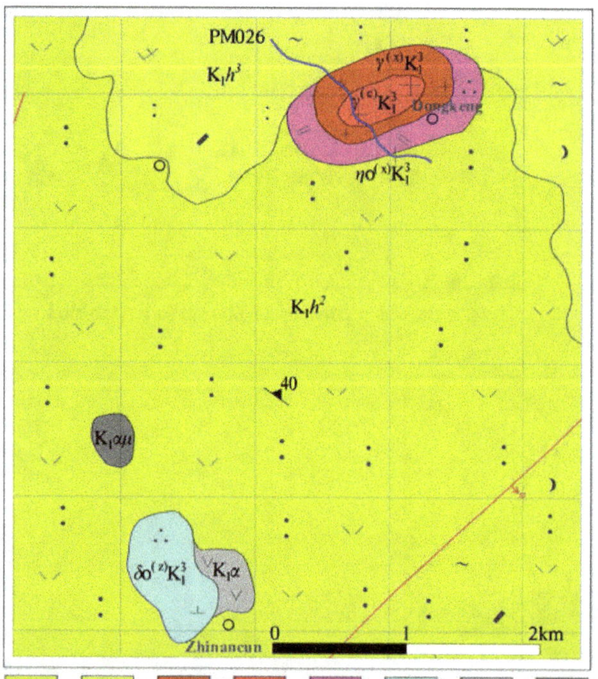

Fig. 3.36 Sketch geological map of the Zhinanshan–Dongkeng volcanic-intrusive complex. 1. the second member of the Huangjian Formation; 2. the third member of the Huangjian Formation; 3. fine-grained granite; 4. coarse-grained granite; 5. fine-grained quartz monzonite; 6. medium-grained quartz diorite; 7. andesite; 8. andesitic porphyrite

Huangjian Formation rhyolitic tuff lava and the third member of the Huangjian Formation rhyodacitic gravel-bearing vitroclastic ignimbrite and is composed of fine-grained quartz monzonite in the periphery, and medium-coarse-grained syenogranite and fine-grained syenogranite (porphyry) inside (Fig. 3.37). Grain sizes of the minerals in the Dongkeng complex were gradually larger from the edge to the center.

The south side of the Zhinanshan-Dongkeng complexes is widely in intrusive contact with the second member of the Huangjian Formation (K_1h^2) rhyolitic tuff lava, while the north side is in intrusive contact with the third member of the Huangjian Formation (K_1h^3) rhyodacitic gravel-bearing vitroclastic ignimbrite. Quartz diorite, granitic porphyry, felsite, and andesite veins are presented inside and outside of the contact zone between the complexes and the second–third member of the Huangjian Formation.

2. **Petrological Features**

Fine-medium-grained quartz diorite in the Zhinanshan complex is light gray, fine-medium-granular texture, mainly composed of quartz (15–22%), K–Na-feldspar (5%), plagioclase (50%), and hornblende (20%). The grain size is generally 0.4–1 mm but up to 1.5–3 mm for a handful, and K-feldspar and hornblende phenocryst locally seen with grain size up to 5–6 mm (Fig. 3.38). Quartz is granular, distributed pretty evenly. Plagioclase is plate-column-like, mostly weathered, mainly andesine-oligoclase based on the refractive index. Hornblende is column-like, distributed rather evenly. Rocks at the margin of the complex are cemented with early stage andesite agglomerate breccia, which content is 25–70%. The agglomerate is angular–subangular, and varies in sizes between 6 and 25 mm.

Based on the field contact relationship, from early to late stage, the Dongkeng volcanic-intrusive complex is medium-fine-grained (porphyritic) quartz monzonite → fine-grained syenogranite (porphyry) → medium-coarse-grained syenogranite.

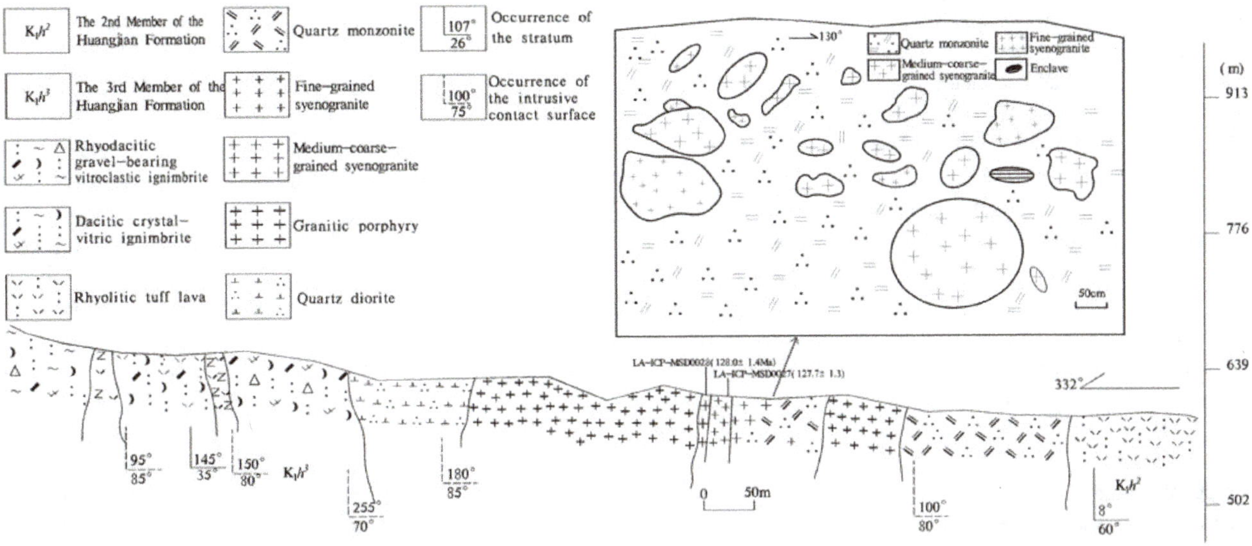

Fig. 3.37 Geological section map of the Dongkeng volcanic-intrusive complex

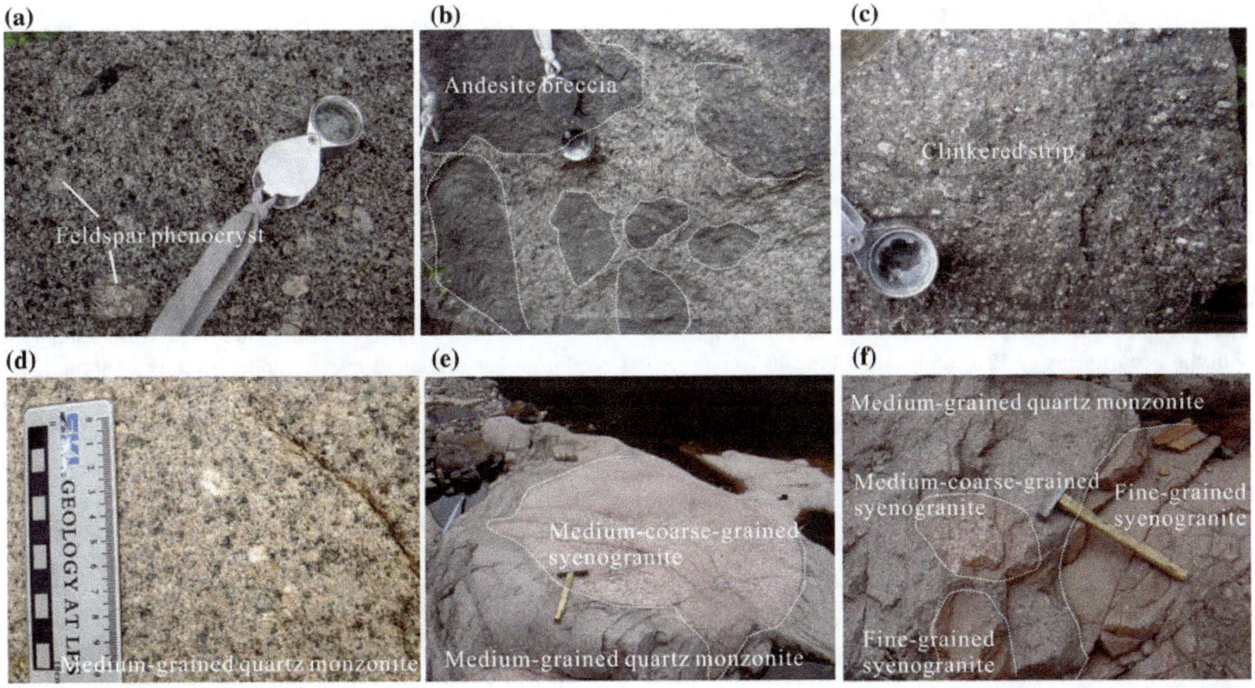

Fig. 3.38 Outcrop images showing the contact relationship between the main rock types in the Zhinanshan–Dongkeng volcanic-intrusive complexes. **a** Quartz diorite (porphyry) in the Zhinanshan complex; **b** quartz diorite(porphyry)-cemented andesite breccia in the Zhinanshan complex; **c** clinkdered strip occurred in rhyolitic tuff lava, which is the wall rocks of quartz diorite (porphyry) in the Zhinanshan complex; **d** the early stage medium-grained quartz monzonite in the Dongkeng complex; **e** the lump medium-coarse-grained syenogranite intruded in the early stage medium-grained quartz monzonite; **f** medium-coarse-grained, fine-grained lump syenogranite intruded in the early stage medium-grained quartz monzonite

Medium-fine-grained (porphyritic) quartz monzonite in the Dongkeng complex is light gray and medium-fine-grained (porphyritic) texture. Phenocryst contains plagioclase (20%), K–Na-feldspar (8%), hornblende (4%), and very little quartz. The grain size of the phenocryst is 1–3 mm, but a small amount up to 3–4 mm; and locally K-feldspar is 5–10 mm in size;

plagioclase is 1–4 mm in size, but a small amount up to 4–12 mm, and epidotization occurred somewhere; hornblende is 0.5–3 mm in size. Matrix contains quartz (12%), K–Na-feldspar (major), and plagioclase (minor) (55%) in the grain size of 0.2–0.5 mm. Some gray-green ellipse-like irregular enclaves are seen in the size of 1–8 cm × 1–8 cm. The light pink micro-grained to fine-grained irregular rhyolitic tuff lava xenoliths are also presented, in the size of 40 cm × 30 cm with no clear border. The coarse-medium-grained syenogranite and fine-grained syenogranite (porphyry) inclusions are developed in the medium-fine-grained quartz monzonite near the central of the complex, which are irregular and varies in size about 10–150 cm (length) × 10–150 cm (width). Few oval fluidal structure enclave is visible, about 40 cm × 30 cm in size. The border of syenogranite inclusions and quartz monzonite is irregular but clear, lack of burning-off or alteration, etc. Andesite veins, 1–2 m wide, intruded in quartz monzonite, the contact surface is irregular bending, with the occurrence of 100°∠80°.

Fine-grained syenogranite (porphyry) is light pink, granitic (porphyritic) texture, and matrix is micro-grained texture. Phenocryst mainly quartz (16%), K–Na-feldspar (10%), plagioclase (2%), and a handful of biotite, with the general size of 1–2 mm and very little of 2–6 mm, epidotization occurred to a handful of plagioclase, not evenly distributed. Matrix mainly contains feldspar (50%) and quartz (20%), about 0.1–0.2 mm in grain sizes; quartz is granular, feldspar is plate and mainly K–Na-feldspar with a handful of plagioclase. In the fine-grained syenogranite (porphyry), light pink micro-grained to fine-grained syenite veins and light pink coarse-medium-grained granite lumps can be seen locally. Two groups of joints are seen in outcrop of the complexes, with the occurrences of 322°∠50° and 138°∠40°.

Medium-coarse-grained syenogranite is light pink, medium-coarse-granular texture, mainly consisting of quartz (32%), K-Na-feldspar (56%), and plagioclase (12%), with the general grain size of 3–8 mm or a few of 8–12 mm. The large lump-like syenogranite intruded in fine-grained syenogranite and medium-fine-grained (porphyritic) quartz monzonite, while the lump varies in size ranging from 5–150 cm × 5–200 cm, with clear borderline and lack of clear burning-off alteration.

3.2 Geochemical Features

Quartz diorite in the Zhinanshan complex, and the medium-fine-grained quartz monzonite, the fine-grained syenogranite (porphyry), as well as the medium-coarse-grained syenogranite in the Dongkeng complex have similar geochemical features, possibly due to the differentiation evolution passes from the same magma origin.

Volcanic-intrusive rock is high in SiO_2 content (58.30–75.25%), enrichment in alkali (Alk = $K_2O + Na_2O$, 7.33–8.87%), high in K_2O/Na_2O ratio (1.20–1.97), variable in MgO content (0.29–3.47%), low in content of P_2O_5 (0.02–0.41%) and TiO_2 (0.10–0.99%); A/CNK is 0.81–1.04 and Rittmann Index (σ) is 2.44–3.51, having the features of metaluminous-weakly peraluminous and shoshonite features (Fig. 3.39).

All rocks in the Zhinanshan-Dongkeng complexes have medium $\sum$REE content (135.25×10^{-6}–202.71×10^{-6}), LREE and HREE differentiated rather obvious, LREE/HREE = 4.49–8.06, La_N/Yb_N = 3.42–9.24, δEu = 0.15–0.64, having strong negative Eu anomalies, the chondrite-normalized REE patterns show the feature of weakly dipping rightward, enrichment in K, Rb, Th, and U, weakly depleted in LILE such as Sr and Ba; depleted or weakly depleted in HFSE such as Ti, P, Nb, and Ta.

From fine-medium-grained quartz diorite → medium-fine-grained quartz monzonite → fine-grained syenogranite → medium-coarse-grained syenogranite, SiO_2 and A/CNK increased gradually, and contents of TiO_2, Al_2O_3, MgO, CaO, P_2O_5, MnO, LREE, Sr, and Ba as well as δEu decrease gradually. For fine-grained and medium-coarse-grained syenogranite, $(^{86}Sr/^{87}Sr)_i$ = 0.70554–0.70636, $\varepsilon Nd(t)$ = −5.73 to −4.19 and the T_{DM2} = 1.26–1.39 Ga.

3. **Isotope geochronology**

For the fine-medium-grained quartz diorite in Zhinanshan complex, the fine-grained syenogranite and medium-coarse-grained syenogranite in Dongkeng complex, zircons are mostly in long and short strips, and a few granular. Zircons have oscillatory growth zoning. Zircons in quartz diorite are about 70–120 μm long, its length-width ratio is mostly 1:1 and a few in 1.5:1, while in fine-grained syenogranite and medium-coarse-grained syenogranite, zircons is about 100–300 μm long, its length-width ratio is mostly 2:1–1:1, and a few 3:1.

Zircons in the quartz diorite of the Zhinanshan complex, U content is 120×10^{-6}–375×10^{-6}, Th is 118×10^{-6}–584×10^{-6}, and Th/U ratio is 0.81–1.66, showing a typical feature of magma origin. Sample locations mostly are projected on or near the concordant curve, $^{206}Pb/^{238}U$ weighted mean age is 130.5 ± 1.7 Ma (MSWD = 0.96) (Fig. 3.40), representing the crystallization age of Zhinanshan complex.

Zircons in the fine-grained syenogranite and medium-coarse-grained syenogranite of the Dongkeng complex, U content is 146×10^{-6}–558×10^{-6} and 85×10^{-6}–1445×10^{-6}, respectively, Th content is 101×10^{-6}–426×10^{-6} and 61×10^{-6}–978×10^{-6}, respectively, and Th/U ratio is 0.70–1.11 and 0.58–0.94, respectively, a typical feature of magma origin. Sample locations mostly are projected on or near the concordant curve, and $^{206}Pb/^{238}U$ weighted mean age is 127.6 ± 1.2 Ma (MSWD = 1.8) and 127.9 ± 1.3 Ma (MSWD = 1.7), representing the crystallization age of the Dongkeng complex.

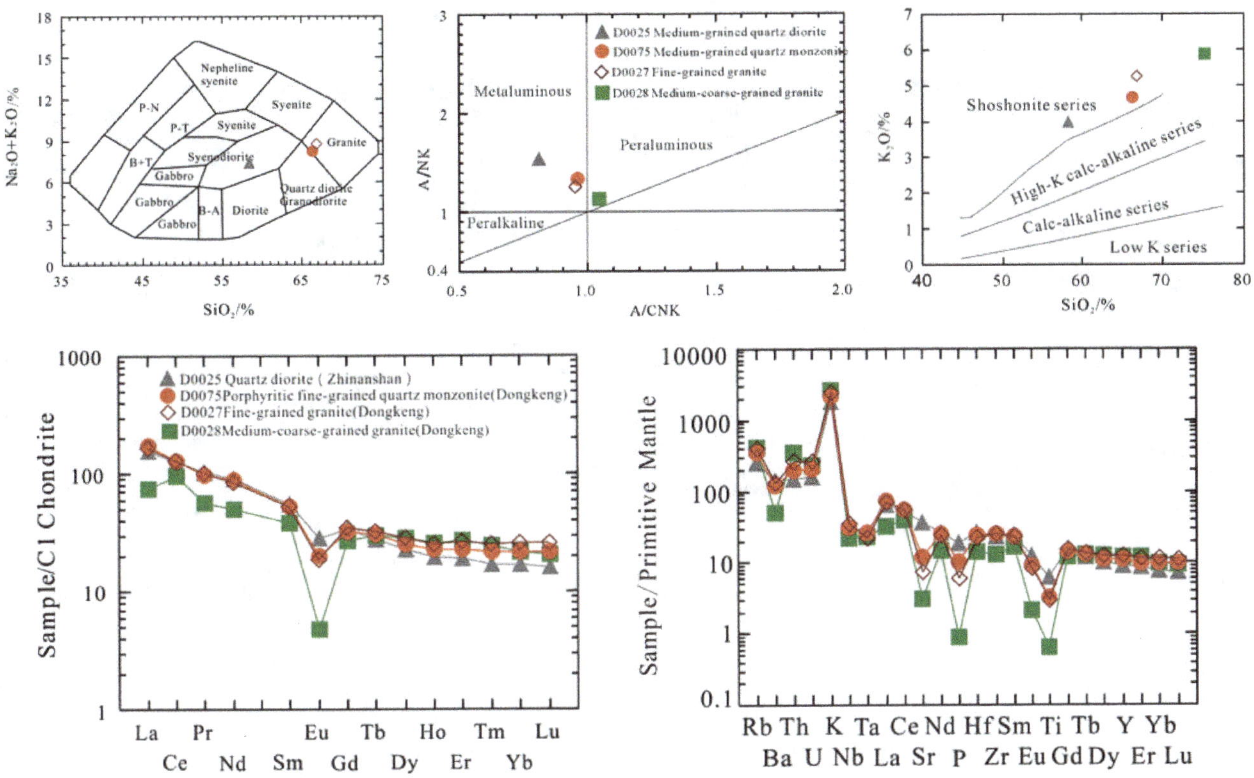

Fig. 3.39 SiO_2–Na_2O + K_2O, A/CNK-A/NK, SiO_2–K_2O, REE chondrite-normalized diagram, and trace element primitive mantle-normalized diagrams for the Zhinanshan–Dongkeng volcanic-intrusive complexes

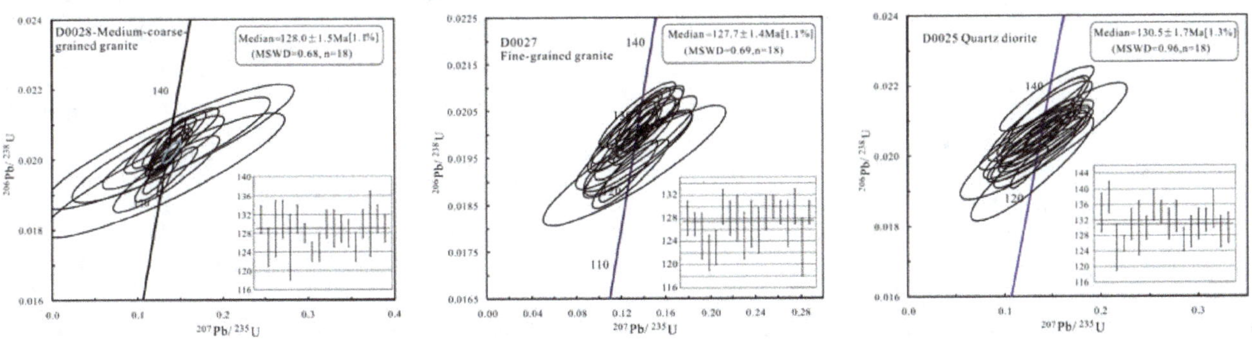

Fig. 3.40 Zircon U–Pb concordant curve of the main types of the rock from the Zhinanshan–Dongkeng volcanic-intrusive complexes

3.2.1 Origin, Evolution of Magma, and Its Relationship with Regional Metallogeny

3.2.1.1 Origin and Evolution Series of Magma

1. **Evolution Series of Magma**

According to zircon U–Pb ages, all intrusive rocks in the survey area were formed in the early stage of Early Cretaceous. Therefore, the intrusive magma activities can be divided into three stages and two series, including intrusive rocks and volcanic-intrusive rocks, in combination with lithological compositions, contact relationships, petrological, and geochemical features (Fig. 3.41).

(1) Intrusive rock series

Stage-1 (K_1^1): Zircon U–Pb age is 145.1–136.3 Ma. According to the contact relationship, the time sequence is

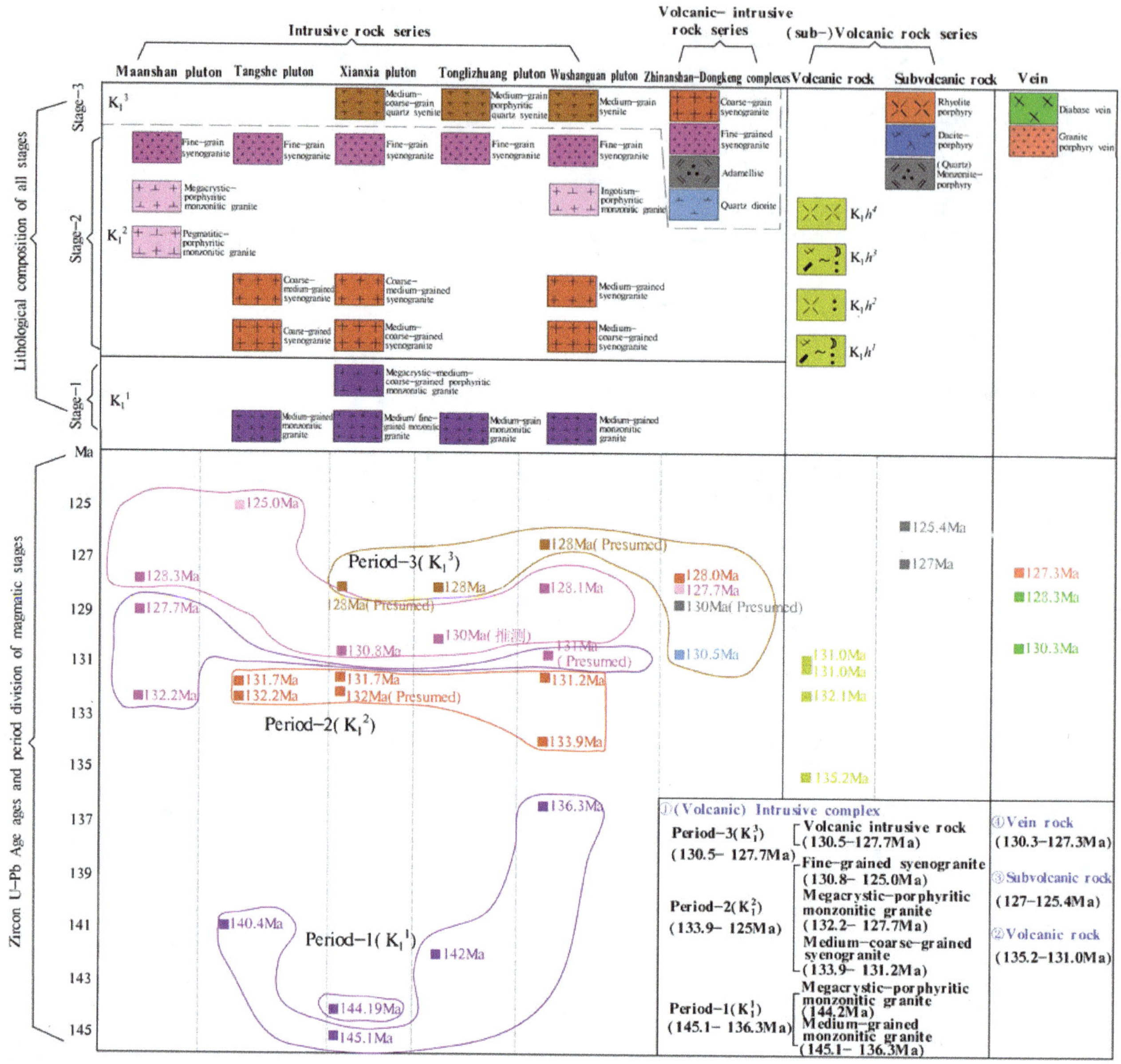

Fig. 3.41 Diagram showing lithological compositions and evolution series of magmatism in the Early Cretaceous

followed by medium-grained monzonitic granite (145.1–136.3 Ma) of Tangshe, Xianxia, Tonglihuang, and Wushanguan, and megacrystic porphyritic monzonitic granite (144.2 Ma) of Xianxia.

Stage-2 (K_1^2): Zircon U–Pb age is 133.9–125.0 Ma. According to the contact relationship, the time sequence followed by coarse-medium-grained to medium-coarse-grained syenogranite (133.9–131.2 Ma) of Tangshe, Xianxia, and Wushanguan; pegmatitic and megacrystic porphyritic monzonitic granite (132.2–127.7 Ma) of Ma'anshan and Wushanguan; fine-grained syenogranite (130.8–125.0 Ma) of Ma'anshan, Tangshe, Xianxia, Tonglizhuang, and Wushanguan.

(2) Volcanic-intrusive series

Stage-3 (K_1^3): Zircon U-Pb age is 130.5–127.7 Ma. Rocks are mainly smaller medium-grained quartz syenite in Xianxia, Tonglizhuang, and Wushanguan plutons, fine-medium-grained quartz diorite in Zhinanshan complex and medium-fine-grained quartz monzonite, fine-grained syenogranite and medium-coarse-grained syenogranite in Dongkeng complex.

2. **Petrogenetic types**

In the survey area, the time sequence of the intrusive rock series followed by medium-grained monzonitic granite → megacrystic porphyritic monzonitic granite → medium-coarse to coarse-medium-grained syenogranite → megacrystic-pegmatitic porphyritic monzonitic granite → fine-grained syenogranite to volcanic-intrusive rock sequences' fine-grained quartz diorite–medium-coarse-grained quartz syenite–medium-fine-grained quartz monzonite–fine-grained syenogranite–medium-coarse-grained syenogranite.

The intrusive rock series are generally characterized by high SiO_2 content, high alkali content, high FeO^T/MgO ratio, and high DI (79.79–93.53). From early to late stage, the rock series evolved from metaluminous to weakly peraluminous, from high K calc-alkaline to shoshonite (Fig. 3.42), alkali ($K_2O + Na_2O$) content and K_2O/Na_2O ratio increased. These geochemical features are similar to the highly differentiated I-type Fogang granite (82–94) in South China, and the Cayu granite (82–92) in east Gangdise, Tibet (Li et al., 2007). The $(K_2O + Na_2O)/CaO$–Al–Zr + Nb + Ce + Y diagram shows that from early to late stage, the undifferentiated granite gradually evolved into highly differentiated granite, which also indicated magma differentiation increases as time went on.

For the volcanic-intrusive rock series, SiO_2 content (58.30–75.25%), alkali content (7.33–10.02%), FeO^T/MgO ratio (1.81–4.38), and DI (62.05–94.58) are varied greatly, and mostly metaluminous and shoshonitic. However, all the features above increased gradually, followed by quartz diorite → quartz monzonite → syenogranite.

Some researches demonstrate that solubility of apatite is very low in the metaluminous–weakly peraluminous magma, and would be even lower when SiO_2 content increase during the process of magmatic differentiation. But in strongly peraluminous magma, the solubility of apatite increases with the increase of SiO_2 content; therefore, the changing trend of P_2O_5 and SiO_2 content with the evolution of magmatic differentiation can be used to distinguish between the I-type and A-type granites and S-type granitoid. In the survey area, the intrusive rocks have lower P_2O_5 content with an average of 0.04–0.17%. From early stage monzonitic granite to late-stage syenogranite, the P_2O_5 content decreased with the SiO_2 content increased, similar to the evolution trend of I-type granite. $(^{86}Sr/^{87}Sr)_i$ and εNd (*t*) also are similar to those of I-type granite in South China. As a result, intrusive rock in the survey area may be highly differentiated I-type granite. Together with A-type granite, both developed in the Early Cretaceous strata in northwestern Zhejiang and neighboring areas.

3. **Magma Sources**

Geochemical features of the major and trace elements of intrusive rocks in the survey areas indicate that the intrusive rock series from early stage monzonitic granite to late-stage syenogranite, volcanic-intrusive rock series from quartz

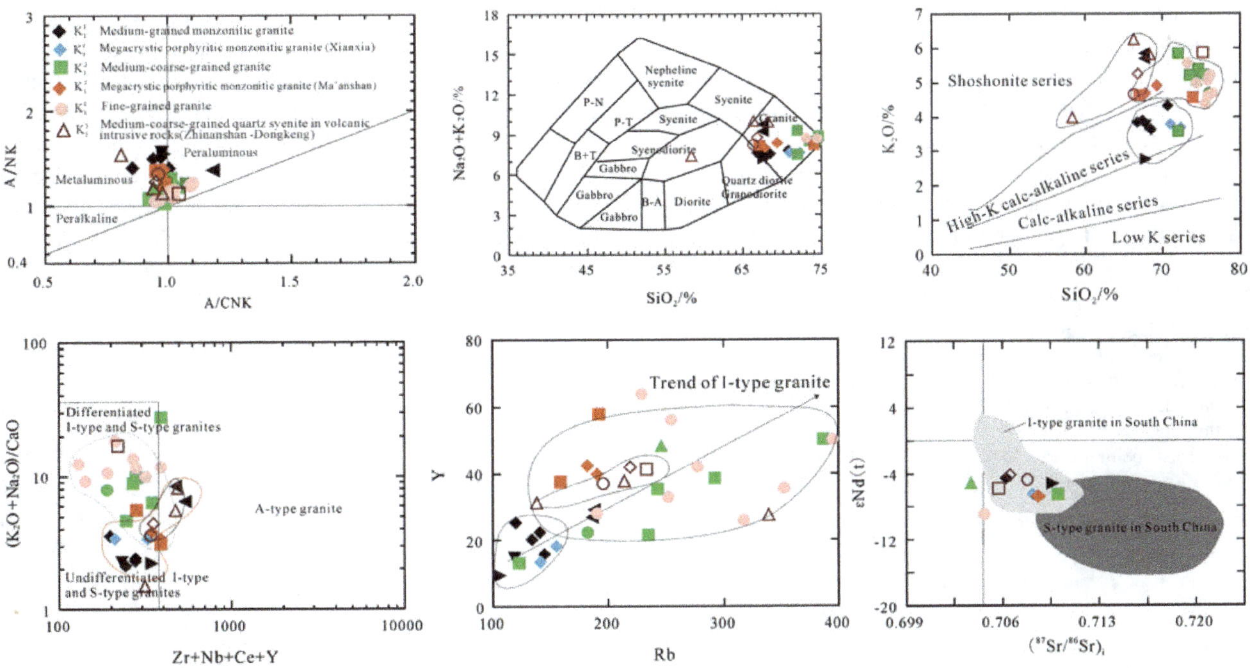

Fig. 3.42 Discrimination diagrams of intrusive rock and volcanic-intrusive rock types

diorite to syenogranite, the content of SiO_2, K_2O, and Rb, etc., increases gradually while the content of Al_2O_3, CaO, FeO^T, TiO_2, and P_2O_5, etc., decreases gradually, such linear relationships are obviously shown in the SiO_2 Hark diagrams. In the same way, REE and trace-elements distribution patterns are very alike, from early to late, REE total and Rb content tend to increase, $(La/Yb)_N$ ratio, Sr content and Sr/Y ratio tend to reduce, deficit in Ba, Nb, Ta, Sr, P, Eu, and Ti tends to enhance, indicating magmas are likely to have experienced highly fractional crystallization Ti-enrichment mineral facies (e.g., ilmenite and/or rutile), apatite and plagioclase or partial melting and remaining so that in late lithologies the content of minerals such as plagioclase and hornblende drops sharply; and the intrusive rock period-1 and the volcanic-intrusive rock period-3 are possibly dominated by differentiated crystallization while the intrusive rock period-2 is possibly dominated by partial melting (Fig. 3.43).

Rocks in the intrusive rock series and those in the volcanic-intrusive rock series have similar features of Sr–Nd isotopes. For rocks in the intrusive rock series, $(^{86}Sr/^{87}Sr)_i$ is 0.69215–0.70988 with average of 0.70536; $\varepsilon Nd(t)$ is −8.87 to −4.47 with average of −6.23, and two-stage Nd model age is 1.29–1.65 Ga and its mean value is 1.44 Ga. For rocks in the volcanic-intrusive rock series, $(^{86}Sr/^{87}Sr)_i$ is 0.70554–0.70768 and its mean is 0.70653; $\varepsilon Nd(t)$ is −5.73 to −4.19 and its mean is −4.90, $T_{DM2}(Nd)$ = is 1.29–1.39 Ga with average of 1.32 Ga. Rocks in both series have high $(^{86}Sr/^{87}Sr)_i$ and their $\varepsilon Nd(t)$ is higher than $\varepsilon Nd(t)$ (−8.12 to −9.06) of the lithospheric mantle at the north margin of Yangtze block; the Nd age in the stage-2 model is close to the peak (1.6–1.7 Ga) of model age of crust-source-type granite of Yanshanian in South China (Li 1993) but lower than Nd model ages of basement metamorphic rocks in southwest Yangtze block (2.0 Ga) and of Cathaysian orogenic system (1.8–2.2 Ga) (Chen and Jiang 1999), indicating that the magma source was dominated from reconstruction materials of in ancient crust, possibly with participation of a handful of mantle materials, and subsequently, mantle contributes more materials, which tallies with the geological fact that subsequently it is intruded by plenty of intermediate-basic veins. Therefore, intrusive rocks in the survey area is possibly produced from emplacement of magma under high-level crystallization segregation, which is formed by remelting crust materials in the east of the ancient Jiangnan Orogen, which is induced by magma of the lithospheric mantle of Yangtze block (Fig. 3.44), and this demonstrates that Mesozoic magma in South China succeeded Precambrian orogenic crust and mantle lithosphere materials, affected by paleo-Pacific plate subduction and rollback, with the reconstruction deformation, metamorphism, and melting of marginal materials from ancient continent under the new continental intraplate environment (Zheng et al. 2013).

Metamorphic basement in south Yangtze area (Li et al. 2003); MORB (mid-ocean ridge basalt) and OIB (oceanic island basalt) (Xia et al. 2004); lithosphere-enriched mantle cited from (Zhang et al. 2008); Yangtze lithospheric mantle, Yangtze lower crust, and crust in the east of Jiangnan orogen (Jiang et al. 2011; Wang et al. 2012); Yangtze upper crust (Wang et al. 2004); Tongcun Granodiorite porphyry (Zhu et al. 2014); Jinde granodiorite (Zhou et al. 2013); Mogan Mount granite (Zhang et al. 2012).

3.2.1.2 Tectonic Environment of Intrusive Rock

In Middle Jurassic, South China transformed from Tethys tectonic regime to Pacific tectonic regime. Many scholars investigated its tectonic settings profoundly and mainly the following ideas were obtained: (1) Lithosphere began extension in Middle Jurassic, and happened periodically until lithosphere regionally extended in Cretaceous (Chen et al. 2002; Li et al. 2007); (2) In Middle and Late Jurassic, subduction, compression, and orogenic environment of active continental margin in the Pacific tectonic regime (Xing et al. 2008; Zhang et al. 2009; Li et al. 2013) or intracontinent orogeny dominated by paleo-Pacific plate oblique-subduction (Mao et al. 2014), but extension also happens locally (Xing et al. 2008), and after collision orogeny the intracontinent

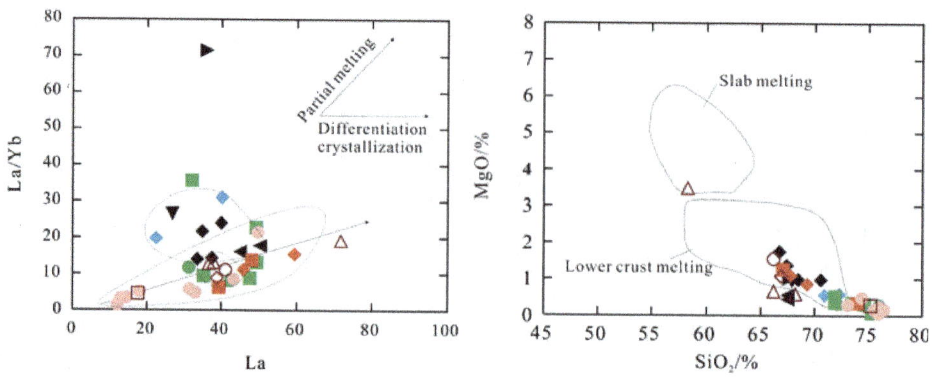

Fig. 3.43 La–La/Yb and SiO_2–MgO diagrams of intrusive rock and volcanic-intrusive rock (legends same as Fig. 3.42). Slab melting area (Zhu et al. 2009); lower crust melting area (Hou et al. 2004; Guo et al. 2007; Gao et al. 2010)

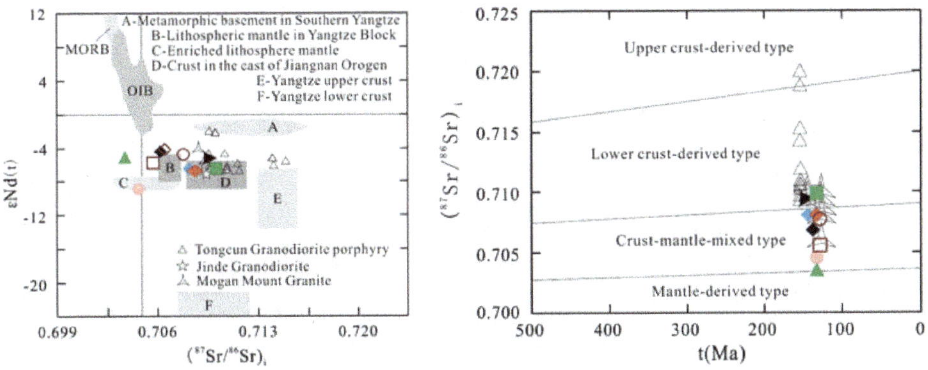

Fig. 3.44 $(^{87}Sr/^{86}Sr)_i$–εNd(t) and $(^{87}Sr/^{86}Sr)_i$–t(Ma) diagrams of intrusive rock and volcanic-intrusive rock (legends same as Fig. 3.42)

extension took place in the beginning of the Early Cretaceous (Li et al. 2013; Mao et al. 2014); (3) from Middle and Late Jurassic to the beginning of Early Cretaceous, multiple blocks compressed strongly intracontinent orogeny from multiple directions (165 ± 5 Ma–136 Ma), in the beginning of Early Cretaceous, extensional collapses and the lithospheric thinning (135–100 Ma), in the Late–EarlyCretaceous, weak compressional deformation took place (100–83 Ma) (Dong et al. 2007), stressing constraints of interactions of multiple blocks (Wang et al. 2013). Comprehensive research results show that in the northwest Zhejiang and even in South China, in Middle Jurassic (165 ± 5 Ma–145 ± 5 Ma), the tectonic settings were generally in an compressional environment, and transformed into extensional environment in the beginning of Early Cretaceous (145 ± 5 Ma–125 Ma), which was triggered by the oblique and shallow subduction direction of the Izanagi plate replaced by the straight and steep subduction of the Pacific plate (Zhu et al. 2010).

Similar to intrusive rocks in northwestern Zhejiang and southern Anhui, intrusive rocks and volcanic-intrusive rocks in the survey area fall within the post-collision granite area in the Rb–Y + Rb diagram (Fig. 3.45), and also, largely fall within the later orogeny and post-orogeny areas in the R_1–R_2 diagram, and combined geochemical features demonstrate that intrusive rocks and volcanic-intrusive rocks in the survey area are the product of intracontinent extension after subduction orogeny in Cretaceous.

3.2.1.3 Features of Magma Emplacement and Metallogenesis

1. Magma Emplacement Mechanism and Metallogenesis

The relationship between granite emplacement and metallogenesis is an important topic in geological study. In general, "active" emplacing (diapirism, dome, and balloon swelling) plutons are mostly produced in closed–semi-closed environments, adverse to exchange in materials and energy between plutons as well as formation of deposits; "passive" emplacing (e.g., cauldron subsidence, stoping and structure injection) plutons are produced in open environments, conducive to formation of contact metasomatic and other magmatic-hydrothermal deposits (Ma et al. 1994; Zheng et al. 2007; Feng et al. 2009). Intrusive rock series in the survey area, e.g., Ma'anshan, Tangshe, Xianxia, Tonglizhuang, and Wushanguan plutons are clearly under control of NE-strike fold core and fracture structures as well as wall-rock contact zones, thus having the feature of "passive" joint emplacement mechanism of stoping and structure injection. The Early Cretaceous mineralization types in the region are the contact metasomatic skarn type and medium-low temperature magmatic-hydrothermal type, which are different in mineralization characteristics and degree; mineralization associated with stoping emplacing pluton mainly happened at the contact zone of pluton which is in the shape of lenticle, balloon, and lamination, and of which the attitude always varies with the structure of the contact zone, for instance, skarn-type Pb–Zn polymetallic deposits are produced on the outer contact zone of Tangshe and Tonglizhuang plutons; mineralization associated with structure injection emplacing pluton mainly took place in the fractures of pluton and its wall rocks, pluton is often in the shape of stable and steep vein and lamination, the mineralization type is medium-low temperature magmatic-hydrothermal type, for instance, the magmatic-hydrothermal fluorite deposit formed at the margin of Ma'anshan megacrystic-pegmatitic monzonitic granite, and the quartz veinlet-stockwork W–Mo orebodies developed in the contact zone of Tonglizhuang-grained syenogranite.

In addition, magma ascent and emplacement is not only under control of multiple factors but also a process of formation, growth, and evolution itself. Change in the emplacement mechanisms in the process of ascent and emplacement, as well as resultant changes in physiochemical conditions of mineralization, may result in vertical zonation

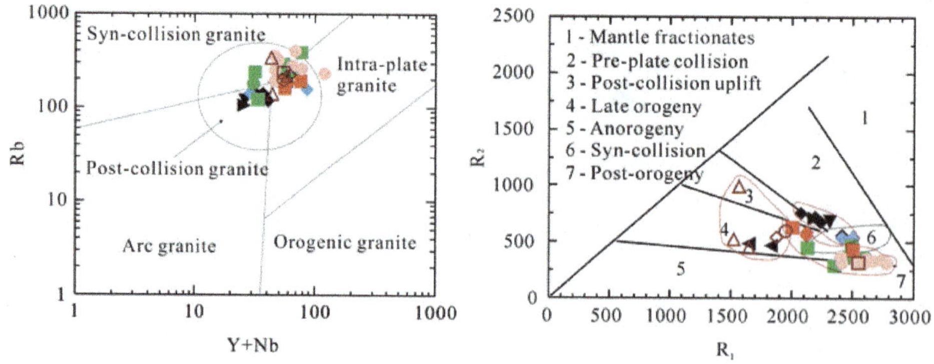

Fig. 3.45 Y + Nb–Rb and R_1–R_2 diagrams of intrusive rock and volcanic-intrusive rock (legends same as Fig. 3.42)

of ore deposits associated with the emplacement mechanism and the ore-controlling factors; metallogenic ability of granite is closely related to granite magma fractionate and evolve, in perspective of the continuity of composite pluton in the process of magmatic differentiation evolution from same source, it is often an intrusion body in a composite pluton series, and usually is a late intrusion body, because the more thorough magmatic differentiation the higher it evolves, the easier it is to form fluids that are enriched in metallogenic elements (Feng et al., 2009). This also demonstrates that some granite plutons' contact structure systems in the survey area may show a feature of horizontal zonation, for instance, Tangshe composite pluton, due to difference in distance to pluton and metallogenesis period, the mineralization type varies from quartz veinlet-stockwork type and skarn-type to structure alteration type or hydrothermal vein type, producing a rule of evolution from Wu–Sn–(Pb)–Ag to fluorite in terms of ore deposit types; geochemical researches indicate that, intrusive rock within the area are high-level fractionation, and differentiation enhances gradually from monzonitic granite to syenogranite in the intrusive rock series, showing a feature favoring Cu–Mo–W → W–Mo → Mo porphyry metallogenic evolution (Fig. 3.46), which is consistent with the feature that the known W–Mo and Mo deposits (ore occurrences) within the area are closely related to the distribution of fine-grained syenogranite, contact zones have often developed quartz fine-vein W–Mo and Pb–Zn polymetallic skarn, a main mineralization type in the survey area. But the prospecting potential for porphyry W–Mo deposits may exist in the deep.

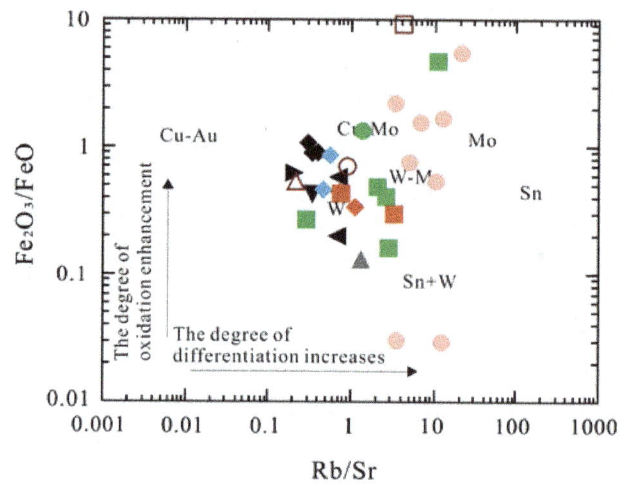

Fig. 3.46 Intrusive rock and volcanic-intrusive rock Rb/Sr–Fe_2O_3/FeO diagrams (legends same as Fig. 3.42) (Blevin 2003)

2. **Pattern of Emplacement Growth and Metallogenesis of Plutons**

Plutons of the same periods, even though they are basically identical in size, form, and geochemical components, differ greatly in whether their contain minerals and whether their mineralization is strong or weak, and their mineralization may vary from each other; the reasons are different emplacement mechanism will trigger different metallogenesis on one hand and different pattern of pluton emplacement and growth impact its metallogenic features and patterns on the other hand (Wu 1998; Rui et al. 1991). And the composite pluton's multiple pulsation emplacement process is the aggregation process the same magmatic source, in the ways of phasal accretion growth, emplace, and superimpose inside or around the intrusion body that emplaces initially, and finally a composite pluton in concentric-annular-belt, semi-annular-belt or irregular shape is produced; Here, granitic body in the concentric-annular-belt shape (including semi-annular-belt) is divided into "positive annular-belt (inner intrusion)" and "reverse annular-belt (outer intrusion)" and the former is common (Feng et al. 2009).

In the survey area, though there are similar emplacement mechanisms for Early Cretaceous composite plutons, finding a location at one off, a vast majority of granite pluton have gone through continuous and complex (very short between two explacements) pulsant emplacement growth process. Here, Ma'anshan pluton and Dongkeng volcanic-intrusive plutons show a positive annular-belt, the central unit (syenogranite) is later than the emplacement of the marginal units (monzonitic granite and quartz monzonite), the pluton accretes from outer to inner, "inner intrusion" (Fig. 3.47); Tangshe, Xianxia, Wushanguan, and Tonglizhuang plutons are composite plutons in the positive annular-belt—irregular shape, "inner intrusion" or "outer intrusion" accretion in some areas and randomly, no obvious rule in other areas (discrete) (Fig. 3.47). Within the area, the distribution features of known deposits (ore occurrences) related to magmatism also show the intrusive outer contact zones of composite plutons such as Tonglizhuang, Wushanguan, Tangshe, and the northeast section of Xianxia have developed skarn-type polymetallic deposits and quartz veinlet type Mo deposit, Ma'anshan pluton is mainly vein-type deposit but there is no mineralization in Dongkeng volcanic-intrusive pluton. Such metallogenic features indicate that in the condition of the same emplacement mechanism, for granitic pluton, which is formed through multiple period pulsation and owns the way of "outer intrusion" accretion, magma that emplace late may contact sedimentary wall rocks so as to exchange frequently materials and energy, it is prone to generate contact metasomatic deposit and hydrothermal deposits surrounding the pluton, and a vast majority of granite plutons in South China that are closely related to such ore deposit types are granite plutons with relatively small size and have reverse annular-belt structure and formed in the way of outer intrusion accretion; however, for inner intrusion granite plutons, since magma that late emplacement is unable to directly contact sedimentary wall rocks, it is not prone to generate contact metasomatic deposits and hydrothermal deposits. However, if late-stage magma contains enrichment high-temperature hydrothermal fluids, it can intruded upward and even puncture early plutons resulting in occurrence of metallogenesis, and thus pegmatite or porphyry (porphyrite) deposits, as well as greisen-vein-type deposits, are formed in or on the top of plutons (Feng et al. 2009).

3.3 Volcanic Rocks

3.3.1 Overview

In the survey area, volcanic rocks are mainly located in the east half of Xianxia Mapsheet and Chuancun Mapsheet with outcropping of 556 km^2, distributed a normal trapezoid. Its northern side, bordering with Baofuzhen, Anji–Wushanguan, Yuhang District, is in angular unconformable contact with Paleozoic sedimentary strata and in intrusive contact with Wushanguan composite pluton; its southern side, bordering with Majiafan–Gaolingcun–Linbeicun, Lin'an City, is in angular unconformable or fault contact with Paleozoic sedimentary strata; its western side, bordering with Dengcun, Lin'an City–Zhangcunzhen, Anji

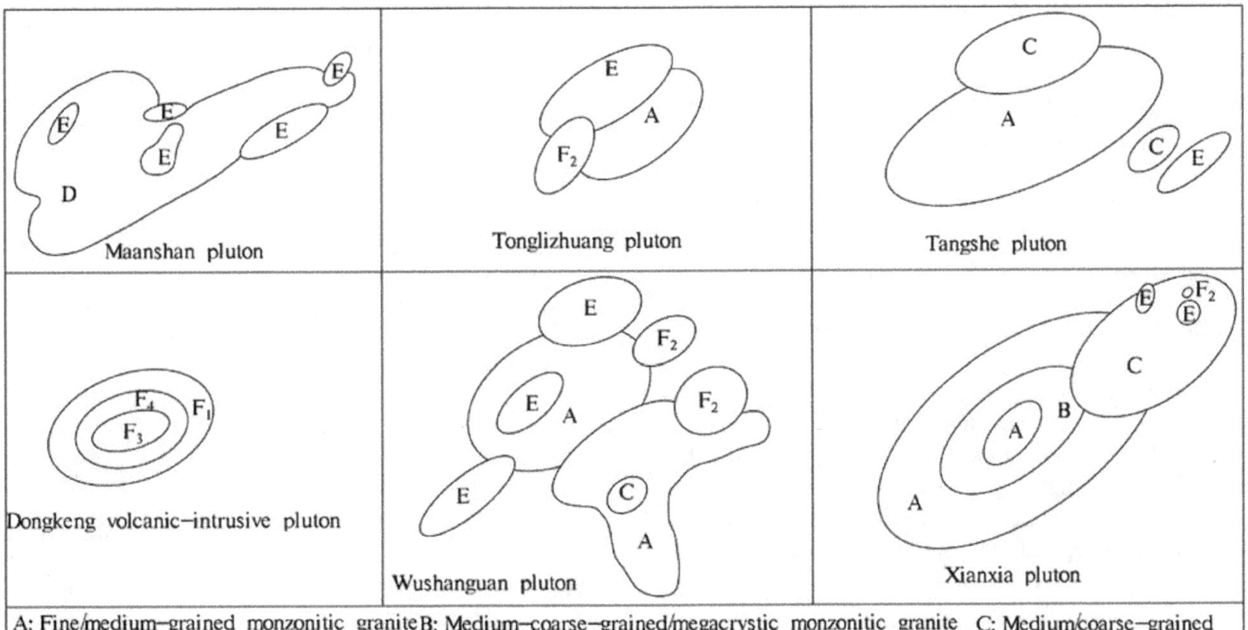

Fig. 3.47 Sketch showing the model for accretion growth ways of pluton in the survey area

County, in NE-strike fault contact with Paleozoic sedimentary strata an Xianxia composite pluton; it eastern side, bordering with Xianbaikengcun–Langjiacun, Yuhang District, is in intrusive or NW-strike fault contact with Wushanguan composite pluton.

In the survey area, volcanic rock belongs to Yianmu Mount–Mugan Mont volcanic structural depression of Tingzi Mount–Tianmu Mount–Mogan Mount sunzone of the active volcanic zone in northern Zhejiang. With many types of rocks and complex lithologies, volcanic rocks are characterized by great changes in lithofacies combinations, developed volcanic apparatus and multiple eruptions and migrations. On the basis of volcanic rock's distribution features, lithological and lithofacies combination as well as regional comparison, in the survey area, volcanic rockfalls within Mesozoic Huangjian Formation, and can be divided into four eruption rhythms (stratum unit at member level) by volcanic eruption from early to late and from strong to weak.

3.3.2 Rock Types and Features

In the survey area, volcanic rock has a number of rock types, can be mainly divided into four categories: volcanic clastic rocks, volcanic sedimentary rocks, lavas, and subvolcanic rocks, and rocks are named mainly by referencing to classification and naming schemes for volcanic clastic rocks in Research Report on Volcanic Structure—Lithology and Lithofacies—Volcanic Strata Mapping Methods, and other categories of rock are named as per this.

3.3.2.1 Volcanic Clastic Rock

In the survey area, there are various types of volcanic clastic rock with main lithology, including rhyolitic agglomerate breccia, rhyolitic agglomerate breccia tuff, rhyolitic/dacitic (breccia-bearing) crystal-vitric tuff, rhyolitic/dacitic (breccia-bearing) crystal-vitric ignimbrite, and rhyolitic/dacitic vitric ignimbrite, etc.

1. **Rhyolitic daciticacitic tuff (ignimbrite)**

It is the main lithology of the Member #1 (K_1h^1) and #3 (K_1h^3) of Huangjian Formation and the lithology consists of rhyolitic dacitic breccia-bearing crystal-vitric ignimbrite, rhyolitic dacitic agglomerate-breccia-bearing crystal-vitric ignimbrite, rhyolitic dacitic crystal-vitric ignimbrite, and rhyolitic dacitic crystal-vitric tuff, etc. Hereunder, describes its main lithological features:

(1) Rhyolitic dacitic (breccia-bearing) crystal-vitric ignimbrite

Mostly gray and offwhite, (breccia-bearing) ignimbrite texture, and massive structure; rock outcropping is obviously coarser that the lava type, and undeveloped joints structure; crystal pyroclast is mainly light-offwhite plagioclase (5–15%), K–Na-feldspar (5–10%) and a handful of quartz (1%), in the grain size of 0.5–1.5 mm; debris breccia is dark gray–light gray, with content of 1–5%, in subangular or subrounded shape, mainly rhyolite, dacite, andesite, and ignimbrite, etc., in the size of 0.5–1.5 mm for debris and 2–8 mm breccia. Plagioclase mostly in the angular and subhedral column shapes, some have Na-feldspar bicrystal and annular-belt structure; K-feldspar is mostly in the angular shape, and less in the shape of subhedral plate and column; quartz mostly in the angular shape and less in the shape of hexagonal bipyramid; dark minerals are mainly hornblende, biotite or augite, mostly replaced by chlorite and carbonatite, etc. Plastic vitroclastic, rock debris, and volcanic ash are 74–89% in content, plastic vitric is mostly in the rod-like, earthworm-like or irregular shape, and largely arranged directionally in rock; plastic rock debris in the shape of long strip, ripped, and lenticles, and it has been devitrified to be felstic and crystal-particle-like felsic minerals.

(2) Rhyolitic dacitic agglomerate-breccia-bearing (vitric) ignimbrite

Offwhite and gray, breccia-bearing crystal-vitric-plasticized tuffaceous texture and false-rhyolitic structure; agglomerate is about 5% in content and 6–10 cm in size, while breccia is 15% in content and 2–3 cm in size, agglomerate and breccia are mainly in the angular–subangular shape, mainly are crystal-vitric ignimbrite, vitric ignimbrite, and crypto-crystal siliceous, etc.; crystal pyroclast is mainly quartz (1–2%), K-feldspar (10–15%), and plagioclase (5–10%), some rock's crystal pyroclast is up to 30% in content, 0.2–1 mm in size, and up to 1.5 mm for less. Rocks have a handful of gray-green magmatic-fragment strips which are intermittently arranged directionally, and stripe is 5–30 mm (length) × 1–5 mm (width).

2. **Rhyolitic ignimbrite**

It is the main lithology of Member #2 of Huangjian Formation (K^1h^2) and the lithology consists of rhyolitic (breccia-bearing) crystal-vitric ignimbrite, rhyolitic agglomerate-breccia-bearing crystal-vitric ignimbrite, and rhyolitic crystal-vitric ignimbrite. The features of rhyolitic (breccia-bearing) crystal-vitric ignimbrite are:

Rock varies greatly in color, dark red, light purple, gray purple, gray, offwhite, and gray green, (breccia-bearing) ignimbritic texture and false-rhyolitic structure. In the rock, breccia is 5–10% in content, containing mainly gray green, dark gray and gray yellow ignimbrite, tuff or cryptocrystalline, some breccia are directionally arranged, in the size of 2–15 mm for most, 20–30 mm for some and up to 80 mm for a few (agglomerate), in the subangular–subrounded shape; crystal pyroclast is mainly light pink K-feldspar (5–10%), quartz (5–10%), and plagioclase (5%), in the size 0.5–2 mm, and 3–4 mm for a few, cemented with tuffaceous. K-feldspar is most in the angular shape, less in the shape of plate-like and column and occasionally Carlsbad twin visible; plagioclase is mostly in the shape of subhedral plate and column, Na-feldspar bicrystal often visible and annular-belt structure occasionally visible; quartz is mostly in the shape of angle and hexagonal bipyramid, occasionally in corrosion curved shape at the margin; dark minerals are mainly hornblende and biotite, which has been altered to be clay minerals. In the rock, plastic vitric pyroclast and volcanic ash are 60–82% in content, plastic vitric pyroclast is in the shape of bar, rod, strip, and fine stripe, and has been devitrified to be felsitic felsic minerals; plastic debris content is 1–10%, in the shape of lenticles, ripped, and striation, after devitrification it becomes felsitic texture and crystal micro-granular texture, K-feldspar, plagioclase, and quartz phenocryst; plastic vitric pyroclast and debris are clearly directionally distributed forming rhyolitic structure.

3. **Rhyolitic tuff**

It is a lithology of the lower part of Member #2 of Huangjian Formation (K_1h^2) and the lithology consists of rhyolitic agglomerate-bearing (breccia-bearing) crystal-vitric tuff, rhyolitic breccia-bearing crystal-vitric tuff, and rhyolitic crystal-vitric tuff. Taking the lithology at the lower part of Member #2 of Huangjian Formation (K_1h^2) East Tianmu Mount Scenic Area as an example, hereunder describes its main lithological features:

(1) Rhyolitic agglomerate-bearing breccia(-bearing) crystal-vitric tuff

Purple red, agglomerate-breccia-bearing tuff texture and massive-laminated structure, in the thickness of 1–5 m; breccia is about 20% in content and 1–6 cm in size (15%), and 2–10 mm for a few (5%); agglomerate is about 10% in content, 6–20 cm in size (8%), and 20–40 cm for a few (2%); agglomerate breccia is in the subrounded shape, not clear in sorting, in general, the content of agglomerate breccia is low in the upper and lower parts, more in the mid-part; agglomerate breccia has consistent component which is rhyolitic crystal-vitric ignimbrite. In rocks, crystal pyroclast is 30% in content, locally become rhyolitic crystal pyroclast tuff, crystal pyroclast is mainly feldspar (15%) and quartz (15%), in the size of 1–2 mm; cemented with tuffaceous. The mid of the stratum has developed interbeddings of purple-red siltstone (10 cm) and gray-green tuffite (about 1 m thick), whose attitude is 90°∠40° and 50°∠50°.

(2) Rhyolitic agglomerate-bearing breccia

Purple red and dark red, agglomerate-bearing breccia texture, and massive structure; agglomerate and breccia are mainly rhyolitic crystal-vitric ignimbrite; agglomerate is 10% in content and 6–25 cm in size, in the round and subangular shape, up to 60–70 cm for a few; breccia is 50–60% in content and 10–50 cm in size; crystal pyroclast is mainly quartz (4%), K–Na-feldspar (20%), and plagioclase (2%), and a handful of biotite, in the grain size of 0.2–1.5 mm; cemented with tuffaceous.

3.3.2.2 Volcanic Sedimentary Rock

It is the main lithology at the upper part of the Member #1 of Huangjian Formation's (K_1h^1), additionally, a handful of it is distributed at the lower part of the Member #1 of Huangjian Formation (K_1h^1) and the Member #3 of Huangjian Formation (K_1h^3) in the form of interbeddings (for instance, Tianhuangping and Lichanglong Mount), and lithology includes tuffaceous glutenite, tuffaceous (gravel-bearing) packsand-siltstone, and tuffite, etc.

1. **Tuffaceous Glutenite**

It is mainly distributed at Dashulin area of Dengcun Village, Lin'an District, and Hangzhou City. It is gray, gravel-bearing coarse-sand texture and mainly contains gravel and sand. Gravel is 30–35% in content, in the size of 2–500 mm varying greatly, generally in the round, subangular, subrounded, long-flat, and irregular shapes, contains complex and various components, mainly sandstone (20%), quartz (5%) and volcanic rock (75%), etc.; sand is 70% in content, in the size of 0.1–2 mm, in the round, subangular, and subrounded shape, mainly debris (65%), quartz (20%) and felsdpar (15%); demented with tuffaceous. Rock is 70–150 cm thick per layer and inside a handful of gravel is arranged in oblique rows.

2. **Tuffaceous (gravel-bearing) fine-siltstone**

It is mainly distributed at Shenxi Village Canyon and Tianhuangping area, Baofu Town, Anji County.

Rock at Shenxi Village Canyon is purple red, medium–thin-laminar structure, has minute rhythmic bedding, gravel-bearing coarse-fine-silt texture, in the interbedded shape; purple-red tuffaceous gravel-bearing gritstone stratum, purple gritstone stratum, purple packsand stratum, and dark purple silty mudstone are 10–20 cm, 10–30 cm, 5–10 cm, and 5–8 cm in thickness, respectively, with the attitude of 180°∠20°. Rocks contain a handful of plagioclase whose content is 3–5% and size is 1–3 mm; a handful of dark-black and gray-green angular tuff breccia locally seen, in the content of 2–3% and size of 3–10 mm; the lithology at the top is gray-black carbonaceous argillaceous siltstone. Terrigenous clast is 80–85% in content and mainly is silty-fine sand (65–70%), argillaceous (15%), and irony (5%). Silty-fine sand is mainly quartz, and a handful of debris, in the angular shape, evenly distributed; argillaceous is cryptocrystalline; irony is limonite, in rendered shape. Volcanic clastic is 10–15% in content, mainly crystal pyroclast (mainly plagioclase, and a handful of quartz), and a handful of volcanic cinder, pretty evenly distributed in the terrigenous clastics.

At Tianhuangping area, tuffaceous gravel-bearing packsand or siltstone is interbedded with sedimentary breccia crystal-vitric tuff, purple red and brown, medium–thin lamellar, locally interlined with medium–thin-lamellar tuffite, about 5–30 cm thick per layer; in the shape of intermittent long strips and bubbles, tuffaceous packsandy texture, composed of volcanic clastics (25%), silt and argillaceous (75%), volcanic ash and cemented with argillaceous. Gravel is in the subangular shape and in the grain size of 0.1–0.25 mm, mainly feldspar and secondly quartz.

3. **Tuffite**

It is distributed at the lower part of the Member #1 of Huangjian Formation (K_1h^1) and its Member #3 (K_1h^3) in the form of interbedding, for instance, tuffite at Tianhuangping area, light offwhite, weakly weathered, with the surface in the shape of smooth ellipse; composed of volcanic clastics (75–80%), silt and argillarceous (20–25%). Sand is fine grained, mostly 0.01–0.03 mm in size, and is mainly quartz, etc.

3.3.2.3 Lava

It the main lithology of the Member #2 of Huangjian Formation, there are complex and various rocks, mainly massive rhyolitic porphyry, rhyolitic tuff lava, rhyolitic agglomerate breccia lava, porphyritic rhyolitic, and bubble rhyolite.

1. **Massive rhyolitic porphyry**

It is mainly distributed at Shenxi Village Canyon, Baofu Town, in Xianxia Mapsheet, and based on features of texture and structure, and it can be divided into (breccia-bearing) vitric massive rhyolitic porphyry, felstic massive rhyolitic porphyry, and felsitic massive porphyritic rhyolitenevadite.

(1) (breccia-bearing) vitric rhyolitic porphyry

Gray and gray black, breccia-bearing porphyritic texture, and massive structure. Phenocryst is 10–15% in content and 1–2 mm in size, but up to 3 mm for few, mainly K–Na-feldspar (10%) and plagioclase (5%), a handful of biotite, in the granular or crushed granular shape. Breccia is 2–3% in content, in the angular–subangular shape, 2–15 mm in size, purple, gray, and black cryptocrystalline. Matrix is dark gray vitric lava.

(2) Felsitic massive rhyolitic porphyry

Light gray and gray, porphyritic texture, and massive structure. Phenocryst is 10–30% in content, mainly light pink K-feldspar and a handful of plagioclases, in the size of 1–3 mm and in the short column, granular, and irregular shapes. Matrix is light gray felstic lava cementation. Outcropping locally, K-feldspar porphyroclast is indistinctly seen on the weathered surface, NE60°-strike (330°∠10°) directional arrangement and rhyolitic structure, and K-feldspar porphyroclast featuring that it can be collaged.

(3) Felsitic nevadite

Gray, porphyritic texture, and massive structure. Phenocryst is 30–50% in content, and mainly light pink K-feldspar (25–40%) and a handful of offwhite plagioclase (5–10%), in the grain size of 1–3 mm generally, 4–5 mm for a few and up to 6–10 mm for individual K-feldspar phenocryst; K-feldspar is in the shape of plate and column, and Carlsbad twin occasionally seen; plagioclase is mostly in the shape of subhedral plate and column, Na-feldspar bicrystal often seen, a few like porphyritic; under the microscope, biotite and hornblende have been altered into chlorite and carbonate minerals; quartz is mostly in hexagonal bipyramid shape. Matrix is 50–70% in content, imbedded-crystal and felsitic, microcrystallite and micro-granular textures, etc., mainly cryptocrystalline felsitic or felsitic felsic minerals, in the grain size of 0.005–0.05 mm.

2. **Porphyritic (bubble) rhyolite**

It is main lithologies of the Member #2 of Huangjian Formation (K_1h^2), East Tianmu Mount–Shimen Village, Lin'an

District, Hangzhou City; also, those of the Member #4 of Huangjian Formation (K_1h^4) along West Tianmu Mount–Longwang Mount, and along Shifosi, Shanchuang Town, and lithology is mainly porphyritic rhyolite and bubble rhyolite.

Porphyritic rhyolite, light gray, porphyritic texture, bubble-rhyolitic structure, the content of phenocryst varies greatly, more rocks in the central, and less in the margins, mainly K-feldspar (5–20%), and a handful of plagioclase (1–2%) and biotite (1–2%). K-feldspar and plagioclase are in granular and short strip shapes, in the size of 1–3 × 1–2 mm, some of it has been epidotized; biotite is in the form of schistose agglomerate, in the size of 1–3 mm × 1–3 mm. Rocks at margins have developed bubble in the content of 5–10% and in the size of 3–15 mm, and rocks in the central have less or not bubble, Rocks have developed rhyolitic strip structure and eddy-like structure. K-feldspar and plagioclase are in the shape of subhedral plate and column, the former surface is argillization of different degrees, the latter surface has altered into sericite and some has Na-feldspar bycrystal; quartz is in the shape of hexagonal bipyramid and some show cataclastic phenomenon; dark minerals have been fully decomposed into chlorite, sericite, and carbonatite, etc., but on the basis of their features such as preserved appearance and transection, it is known they are mainly hornblende, biotite, or augite. Matrix is 75–93% in content, cryptocrystalline, felsitic, and crystal micro-granular textures, micro-graphic, spherulitic, and cataclastic textures sometimes, mainly felsic minerals, in the partizle size of 0.005–0.05 mm.

Bubble rhyolite, of which the most developed is at the mountain top of Shifosi, Shanchuang Town, and bubble is 50–60% in content and in the size of 2–15 mm, but 15–80 mm for a few.

3. **Rhyolitic tuff lava**

It is mainly distributed at Gaoling Village–Zhinanshan Village, Lin'an District, Hangzhou City, with bigger outcropping and thickness. Rocks are light pink, pink and light gray with porphyritic texture, and matrix is of felsitic-minute-grained texture; phenocryst is mainly K–Na-feldspar (15–30%), plagioclase (5–10%), and biotite (1–3%), and very few quartz, in the grain size of 1–2 mm in general and 2–2.5 mm for a few, and a few of K-feldspar is in the shape of long column, about 8 mm long and 2 mm wide. Matrix is feldspar (20–30%) and quartz (37–49%), in the grain size of 0.02–0.04 mm.

4. **Rhyolitic agglomerate-breccia-bearing lava**

Only seen in Yangjiashan volcanic vent, light purple gray and gray green, etc., breccia agglomerate lava texture, massive structure, mainly breccia (3–5%), agglomerate (5–10%) and crystal pyroclast (<5%), etc., breccia and agglomerate are gray-green andesite, in the subangular–subrounded shapes, and in the size of 5–50 cm; crystal pyroclast is mainly K-feldspar, plagioclase, quartz, and a handful of biotite, in the grain size of 0.05–1 mm, but up to 1.5–3 mm for a few. Since it is in the volcanic vent, it has universally developed alterations such as silicification, chloritization, carbonatization, and pyritization.

3.3.2.4 Subvolcanic Rock

It is distributed in Shenxi Village Canyon pitching up Tianpingshan ding–Dongling, Tianhuangping Reservoir, and Liwaichanglongshan–Linjiatang, and its lithology is mainly (quartz) monzonitic porphyry, dacite–porphyrite, and rhyolitic porphyry.

1. **(Quartz) monzonitic porphyry**

It is mainly distributed at Shexi Village Canyon Tianpingshan mountain top–Dongling, and the east peak of Tianhuangping Reservoir. Rocks are light pink and pink, of porphyritic texture and massive structure; phenocryst is mainly K-feldspar (5–10%), plagioclase (3–6%), and quartz (2–3%). K-feldspar is light pink, in the shape of strips, in the size of 4–10 × 3–5 mm (3–4%), and 10–20 × 5–10 mm for a few (1%), and it is seen that a handful of K-feldspar wraps plagioclase's growth margin; plagioclase is light green, developed epidotization, in the size of 2–10 × 2–5 mm; quartz is colorless and transparent, in the size of 1–5 mm, granular, and cataclastic but spliceable. Matrix is mostly gray and gray green, in felsitic–micro-granular shape, mainly alkaline feldspar and quartz microcrystalline, in the grain size of 0.1–0.2 mm. Accessory minerals such as zircon and apatite are very little.

2. **Dacite–porphyrite**

It is mainly distributed in Liwaichanglongshan–Linjiatang. Gray green and dark gray, porphyritic texture, phenocryst (10–30%), unevenly distributed, concentrated speckles locally seen, phenocryst is 0.1–1 cm in size, mainly milk-white plagioclase (10–20%), and a handful of augite (4%) and hornblende (2%), and a few quartz locally; plagioclase has pretty intact crystal form, in the shape of subhedral plate and column, Na-feldspar bicrystal often seen and annular-belt structure occasionally seen; K-feldspar is in the shape of plate and column, and has Carlsbad twin; quartz mostly in the shape of hexagonal bipyramid, showing cataclastic phenomenon; dark mineral have been fully decomposed into chlorite and carbonate, in the shape of hexagon and rhombus transection, occasionally in the shape of column, which should be hornblende. Matrix is of micro-grained granitic, spherulitic and micro-graphic textures, of which micro-grained granitic texture is mainly K-feldspar and quartz, in the anhedral granular shape, in the grain size of 0.05–0.1 mm; spherulitic texture is composed of fiber-like felsic microcrystalline arranged in radial shape or fan, and the center of some is fine feldspar; micro-graphic texture is composed of feldspar and quartz.

3. **Rhyolite (porphyry)**

It is mainly distributed at Tianhuangping Reservoir. Rocks are gray green and brown, of porphyritic texture and rhyolitic structure; phenocryst is mainly light pink K-feldspar (10%) and a handful of plagioclase (5%), in the size of 0.5–2 mm. Matrix has developed flow structure with attitude of 300°∠80° and 220°∠20°–60°. At the margin, there are a handful of gray-green cryptocrystalline breccia in the size of 5–10 mm, and 10–30 mm for a few. K-feldspar is in the shape of plate and column, and Carlsbad twin occasionally seen; plagioclase is in the shape of subhedral plate and column, and Na-feldspar bicrystal and concentrated speckle texture seen on some; quartz is in the shape of hexagonal bipyramid, often corroded and irregular; dark minerals have been fully replaced by chlorite and carbonate, its appearance is like column and sheet, which shall be hornblende and biotite, etc. Matrix is of cryptocrystalline felsitic texture, felsitic texture, and micro-graphic texture, etc., mainly cryptocrystalline felsic minerals (K-feldspar and quartz), in the grain size of 0.005–0.05 mm.

3.3.3 Facies of Volcanic Rocks

3.3.3.1 Division of Volcanic Rock Facies and Their Main Features

There are a number of previous schemes to divide volcanic rock facies. By referencing "Research Report on Volcanic Structure—Lithology and Lithofacies—Volcanic Strata Mapping Methods" and "Detailed Rules of Zhejiang on Regional Geological Survey", according to volcanic eruption type, volcanic material transfer way, orientation environment and state, volcanic rocks in the survey area, can be divided into eight main facies: explosive facies (fallout facies, clastic flow facies and surging facies), volcanic debris-flow facies, eruptive sedimentation facies, extrusive overflow facies, eruptive spill facies, explosive spill facies, volcanic vent facies and subvolcanic rock facies, and ten subfacies (Table 3.3).

1. **Explosive Facies**

(1) Fallout accumulative facies

In the survey area, fallout accumulative facies is the most typical in the mid-part of the Member #1 of Huangjian Formation (K_1h^1) of Shexi Village Canyon, Baofu Town and is the result of sedimentation and accumulation of volcanic debris under gravity after they are carried overhead by explosive airflow and migrated by wind. Lithology is mainly gray purple rhyolitic dacitic breccia(-bearing) crystal-vitric tuff, a handful of breccia crystal pyroclast tuff, breccia-bearing crystal-vitric weak ignimbrite, and locally interbedded with purple tuffite. This suite of rocks has developed better rhythmic beddings (Fig. 3.48), in medium–thick-lamellar structure, and strata vary greatly in thickness, 10–50 cm thick per layer; breccia is 10–20% in content, angular–subangular, mainly light pink rhyolite and gray-green ignimbrite, etc., in the grain size of 5–15 mm; crystal pyroclast is 15–25% in content, mainly K–Na-feldspar (10–18%), plagioclase (5–7%), and a handful of biotite, in the grain size of 2–3 mm; in this area, the attitude of strata is 45°∠15° at the lower part, and the content of breccia and crystal pyroclast becomes less

Table 3.3 Volcanic rock lithology and lithofacies division in the survey area

Stratigraphic unit	Lithofacies		Lithology
Huangjian Formation (K_1h)	Explosive facies	Fallout facies	Rhyolitic/dacitic (bearing-) crystal-vitric tuff, breccia (-bearing) crystal-vitric tuff, and breccia(-bearing) crystal pyroclast tuff
		Clastic flow facies	Rhyolitic/dacitic crystal-vitric ignimbrite, breccia(-bearing) crystal-vitric ignimbrite, breccia(-bearing) vitric ignimbrite, breccia(-bearing) crystal pyroclast ignimbrite, and debris ignimbrite
		Surging facies	Its lower part is interbedding of rhyolitic crystal pyroclast tuff and tuffaceous sandstone, and its upper part is rhyolitic crystal pyroclast tuff
	Volcanic debris-flow facies		Thick-lamellar–massive rhyolitic agglomerate breccia tuff, rhyolitic/dacitic agglomerate rock, agglomerate breccia, and breccia
	Eruptive sedimentary facies		Tuffaceous glutenite, tuffaceous gravel-bearing sandstone, tuffaceous sandstone, and tuffite
	Extrusive overflow facies		Vitric (breccia-bearing)rhyolitic porphyry, felsitic massive rhyolitic porphyry, and felsitic massive nevadite
	Eruptive spill facies		Bubble rhyolite (porphyre) and fluidal rhyolite (porphyre)
	Explosive spill facies		Rhyolitic tuff lava
	Volcanic vent facies		Agglomerate breccia lava
	Subvolcanic rock facies		Monzonitic porphyry, granite porphyry, diorite porphyry, and dacite–porphyrite

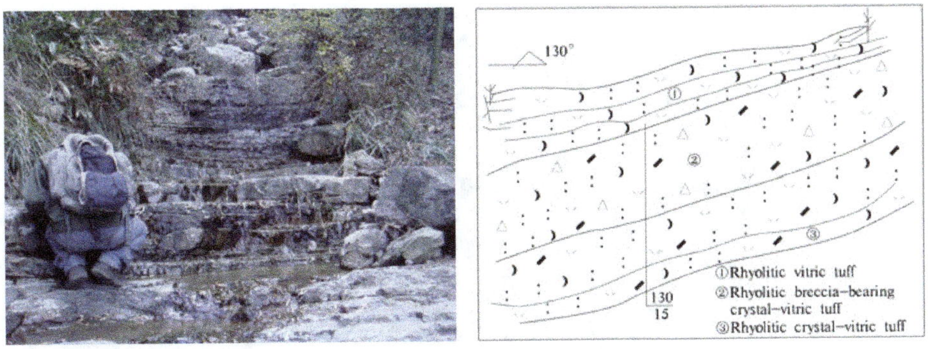

Fig. 3.48 Thin-medium-layered dacitic tuff in the mid-part of the Member #1 of Huangjian Formation (K_1h^1), Nanwu Forest Farm, Zhangcun

gradually upward, but increases locally, the attitude becomes 195°∠27° and the stratum attitude is gentle in general.

(2) Debris-flow accumulative facies

In the survey area, the debris-flow accumulative facies mainly developed in the lower part of the Member #1 of Huangjian Formation (K_1h^1) and its Member #2 (K_1h^2) at Baofu Town and Zhangcun Town, and the Member #3 of Huangjian Formation (K_1h^3) at Tianhuangping, where volcanic debris with small kinetic energy mainly move in the form of stratified flow in the process of plumed moving outward. Lithology is mainly a suite of rhyolitic dacitic breccia-bearing crystal-vitric ignimbrite and vitric ignimbrite, etc., at different clinkering degrees. Plastic vitric and magmatic fragments are widely developed in rocks, the feature of plastic deformation is very clear so as to form pseu-flowage structure (Fig. 3.49), and it is often seen that pseu-flowage structures flow around crystal pyroclast and breccia, and fold-like quasi-rhyolitic structures are seen locally, indicating the ancient flow direction of debris flow.

With the stable thickness and mostly massive structure, strata in these areas have developed hexagonal, pentagonal, and quadrangular columnar joints and massive joints (e.g., the peak of East Tianmu Mount Scenic Area). Where volcanic debris-flow cooling unit or flow unit is completely developed, their facies sequence structures are: the lower part is rocks of surging accumulative facies (In the survey area, it is only seen at Xikou, Lin'an District, as mentioned below), the mid-part is rock of debris-flow accumulative facies (ignimbrite), and the upper part is rocks of fallout accumulative facies, the debris-flow accumulative facies is in gradually transitional relationship with the surging accumulative facies while in abrupt change relationship with the fallout accumulative facies with clear borderlines (e.g., Shenxi Canyon).

(3) Surging accumulative facies

In the survey area, the surging accumulative facies is only seen at Xikou Village, Lin'an District, with outcropping locally, in which volcanic debris mainly move in beddy or turbulent form in plume and is not fully homogenized in the cloud formation and consumes energy lightly faster, which is the front of dense, high-concentration volcanic debris flow, one of the sequences of ignimbrite flow units and is in gradually transitional relationship with rhyolitic ignimbrite of the debris-flow accumulative facies.

For this suite of rocks, its lower part is interbeddings of medium–thick-layered light gray rhyolitic crystal pyroclast tuff and dark gray tuffaceous siltstone, gradually transitioned to be offwhite tuffaceous sandstone upwards, and it has wedge-shaped staggered bedding, wavy bedding, oblique bedding, and horizontal bedding, etc., and the borderline for each stratum is irregular (Fig. 3.50). It is in gradually transitional relationship with its overlying rhyolitic crystal pyroclast ignimbrite (debris-flow accumulative facies).

Rhyolitic crystal pyroclast tuff, light gray, tuffaceous texture, and medium–thick-layered structure. Crystal pyroclast is mainly light pink K-feldspar (15–20%), colorless transparent quartz (15%), plagioclase (10%), and a handful of biotite (5%), all in the grain size of 0.5–1.5 mm, where the latter developed chlorite alteration; rocks contain a handful of dark green breccia (1–2%). Rocks vary greatly in thickness, about 8–70 cm thick per layer. A single layer has developed reverse and normal graded bedding structures (Fig. 3.51), interlined with dark gray ignimbrite (1–2 cm) locally.

Tuffaceous siltstone, dark gray, tuffaceous silty texture, and medium–thick-lamellar structure, about 5–40 cm thick per layer; mainly composed of normal sediments and volcanic debris, argillaceous and volcanic ash cements. Normal sediments are 55–60% in content mainly silty and argillaceous, etc. The silty is finer (0.03–0.05 mm) and contains feldspar and quartz, not evenly distributed, more at local places. The volcanic debris is 40–45% in content, mainly vitric fragment, a handful of crystal pyroclast and volcanic ash, etc. Inside a single layer, normal graded bedding structure is developed, and grain size is transitional gradually.

Tuffaceous sandstone, offwhite, mainly sand grains, and argillaceous, etc. Sand is in the angular shape and in the

Fig. 3.49 1. Dacitic breccia-bearing crystal-vitric strong ignimbrite of the Member #3 of Huangjian Formation (K_1h^3) at the peak of East Tianmu Mount. 2. Dacitic tuff crystal-vitric strong ignimbrite in the lower part of the Member #1 of Huangjian Formation (K_1h^1). 3 and 4. Rhyolitic magmatic-fragment crystal-vitric strong ignimbritein the lower part of the Member #2 of Huangjian Formation (K_1h^2), Shenxi Canyon, Baofu Town

Fig. 3.50 Lower part at Xikou Village, Lin'an District, is interbedding of medium–thick-layered light gray rhyolitic crystal pyroclast tuff and dark gray tuffaceous siltstone (upward, gradually become offwhite tuffaceous sandstone, and rhyolitic crystal pyroclast tuff in a single layer has developed reverse and normal graded)

grain size of 0.1–0.2 mm, mainly feldspar and quartz, etc. The volcanic debris is less than normal sediments in content, about 40%, mainly vitric fragment, a handful of crystal pyroclast and volcanic ash, etc. Based on analysis on debris, such rocks experienced short movement and are very likely to accumulation at near places.

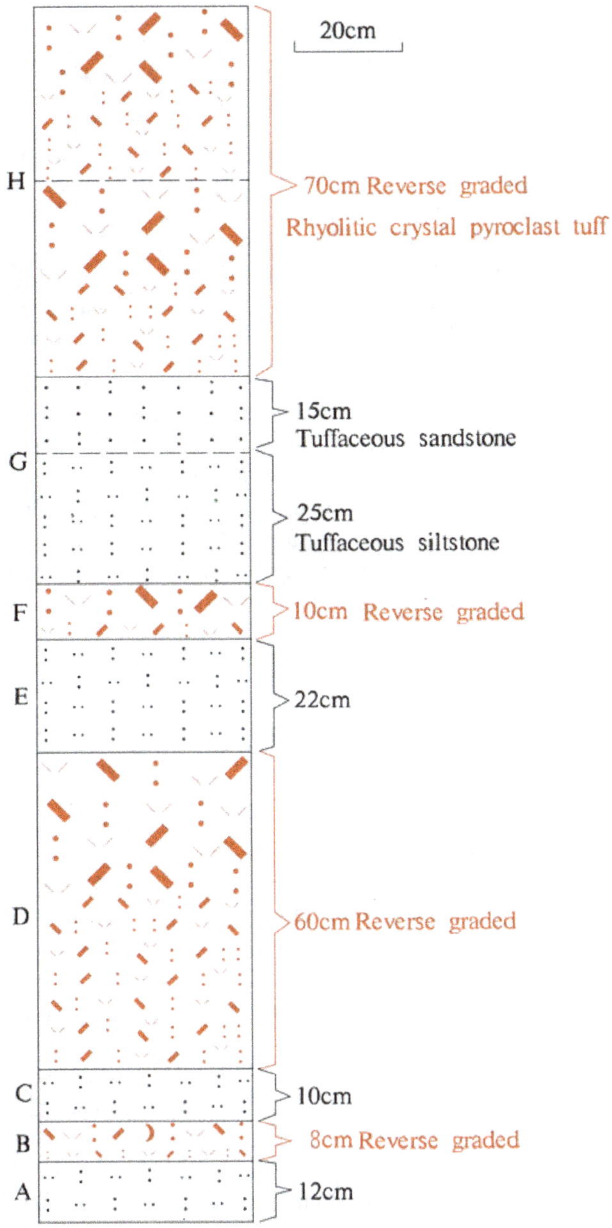

Fig. 3.51 Grade change of volcanic rocks of Surging facies at Xikou, Lin'an

For the surging facies at Xikou Village, Lin'an, the ratio of interbedding of both lithologies in the lower is 6:4, the contact surface is slightly wave, in general, the stratum at the lower is thicker and the one at the upper gets thinner; parallel bedding developed at the upper and oblique bedding at the lower; stratum's attitude is 62°∠42°.

2. **Volcanic debris-flow accumulative facies**

In the survey area, the volcanic debris-flow accumulative facies is only developed in the lower part of the Member #2 Huangjian Formation (K_1h^2) at East Tianmu Mount Scenic Area, in fault contact with the Member #3 Xiyangshan Formation ($\mathrm{\epsilon O}x^3$), produced by coarse-graded breccia, agglomerate and rock blocks quickly falling and accumulative near crater under gravity after volcanic debris were brought overhead by explosive gas; however, possibly under action of ephemeral drainage concurrent with volcanic eruption, these volcanic explosive accumulations move volcanic products down along the volcanic slope to form accumulation of volcanic debris-flow facies; its sediment in the lower part is thick-lamellar–massive rhyolitic agglomerate(-bearing) breccia, becomes rhyolitic agglomerate-bearing breccia tuff and tuffaceous gravel-bearing sandstone upward, locally interbedded with thin-lamellar tuffite and tuffaceous siltstone. Features of the suite of rocks are as follows.

(1) Volcanic debris varies greatly in size, volcanic ash coexists with agglomerate, in general, since debris-flow accumulation there is a certain sorting from bottom to top but the sorting is poor in a single stratum.
(2) Agglomerate and breccia had been partly rounded in the transfer process, angular agglomerate coexists with subangular, subrounded, and rounded coarse debris (Fig. 3.52), and on the surface of some agglomerate breccia various impressions due to friction and collision are visible.
(3) Most of the accumulations are volcanic debris and cement is mainly tuffaceous.

3. **Eruptive Sedimentary Facies**

In the survey area, the eruptive sedimentary facies is developed in the top of the Member #1 of Huangjian Formation (K_1h^1) at Baofu Town–Zhangcun Town, and in the top of the Member #3 of Huangjian Formation (K_1h^3) at Tianhuangping and Dongkeng Village, etc., which is formed by volcanic eruptive materials falling in water bodies, or being sedimented in water bodies through denudation and transportation, rather continually distributed at the margin of volcanic depression and caldera, or produced in volcanic rock in the form of interbedding, an important part of volcanic depressions or calderas. Sediments are volcanic debris sedimentary rock (tuffaceous glutenite, tuffaceous gravel-bearing sandstone–siltstone, and tuffite) and sedimented volcanic clastic rock (sedimentary breccia tuff), with

Fig. 3.52 Bottom feature of Huangjian Formation's Member #2 (K_1h^2) at East Tianmu Mount Scenic Area

clear horizontal bedding developed (Fig. 3.53). and sedimentary rhythm, and there are interlaced, wavy, and involuted beddings, etc., indicating a turbulent shallow lake facies.

4. **Extrusive overflow facies**

In the survey area, the extrusive overflow facies are mainly developed in the Member #2 of Huangjian Formation (K_1h^2) in Shengou Village Canyon, Baofu Town, and its lithology is mainly massive rhyolitic porphyry. It is a geological body produced by accumulative and cooling viscous and dense magma that is slowly extruded out of surface along volcanic vent or fractures beside volcanic craters. Usually, its shape is kupola, spine, rock monument, and rock ridge, etc., in different sizes and scales, its occurrence is steep, planarly in the shape of ellipse and round, and often in transitional relationship with rock of the eruptive spill facies.

There exists gradually transitional relationship between various facies zones in the massive rhyolitic porphyry of

Fig. 3.53 Interbeddings of tuffaceous gravel-bearing packstone and tuffaceous siltstone in the top of the Member #1 of Huangjian Formation, Nanfu Forest Farm

Shenxi Village Canyon, and the zonation phenomenon is clear. Horizontally, it is divided into vitric (breccia-bearing) massive rhyolitic porphyry at the margin, felsitic massive rhyolitic porphyry in the central and felsitic block nevadite, and meanwhile vein-like intrusion of Early Cretaceous sub-quartz monzonitic porphyry in the center (Yantianping). The contact relationship between block rhyolitic porphyry and the ignimbrite at the margin is irregular, at some local places steeply dipping vein-like intrusion occurs (see Yangtianping revived caldera for details) and at other local places it dips gently and overlies ignimbrite; affected by later NE-strike fault, it is in fault contact with rhyolitic tuff lava of the magmatic-liquation-type explosive spill facies in the south-eastern, and it is presumed that they were in gradually transitional relationship during early eruption stage.

Features of massive rhyolitic porphyry: (1) lithology is single, thick, no interbedding of other volcanic rock or sedimentary rock seen, and no developed rhyolitic structure; (2) the attitude of felsitic nevadite in inner facies zone is steep, with the dip angle of 40°–75° with oblique columnar joints structure developed (Fig. 3.54), and phenocryst content is 40–60%; (3) the felsitic rhyolitic porphyry of the transitional facies has developed massive joints, phenocryst content is 20–40%, locally rhyolitic structure indistinctly seen; (4) the vitric rhyolitic porphyry at the margin has developed horizontal plate-columnar joints, with gentle attitude and at the dip angle of 15°–30°, more breccia containing similar components are often seen, the border is clear, phenocryst content is 5–20% and phenocryst has the strong feature of collaging; (5) from central facies zones to transitional and marginal facies zones, phenocryst in rocks becomes more cataclastic, the matrix shows a trend of micro-grained texture–felsitic texture–cryptocrystalline texture; (6) the Early Cretaceous subvolcanic rock–rhyolitic porphyry–rhyolitic tuff lava are closely symbiotic, there were formed basically in the same or similar period, so as to create a symbiotic combined geological body integrating the subvolcanic rock facies, the extrusive overflow facies and the explosive spill facies, indicating they are product of differentiation evolution of comagma.

Fig. 3.54 Feature of massive rhyolitic porphyry at Shenxi Canyon

5. **Eruptive spill facies**

In the survey area, the eruptive spill facies is mainly developed in the mid of the Member #2 of Huangjian Formation (K_1h^2) of South Tianmu Mount and Member #4 (K_1h^4) of West Tianmu Mount–Longwangshan and North Tianmu Mount–Shifosi, which is lava of various types produced by later magma ascent from the deep underground via the volcanic vent to the earth surface and then spilling out of the crater. In the survey area, lithology is mainly porphyritic rhyolite and bubble rhyolite, one of the main rocks in dome-shaped volcanos, and rhyolitic structure is very developed, in West Tianmu Mount Scenic Area, there are often wrinkle-shaped rhyolitic structures (Fig. 3.55), and it also has developed bedding joints, columnar joints in the shape of embossment are developed at local places, rocks at the margin have developed bubble structure and slightly less phenocryst, and rocks in the mid has less bubble but more phenocryst. Surrounding Shifo Temple area, possibly since magma contains more volatile matters, bubble is very developed in rocks at the lateral of volcanic craters, and often seen concentratively distributed in layers.

6. **Explosive spill facies**

In the survey area, the explosive spill facies is mainly distributed in the upper part of Member #2 of Huangjian Formation (K_1h^2) of Gaoling Village–Dongkeng Village–Shuitaozhuang Village, Lin'an District, a transitional lithofacies due to eruption of volcanos in the form of boiling spill (between intensive explosion and quiet spray and spill), composed of volcanic debris and lava matrix, in overlying/underlying or fault contact relationship with the eruptive spill facies's porphyritic rhyolite in the mid and lower part of Member #2 of Huangjian Formation (K_1h^2).

Fig. 3.55 Porphyritic rhyolite in West Tiannu Mount–Longwangshan

Fig. 3.56 Rhyolitic tuff lava at Dongkeng Village, West Tianmu Mount, Lin'an District

Lithology is mainly rhyolitic (crystal pyroclast) tuff lava, its main difference from porphyritic rhyolite is: the former has no developed flow structure, but massive structure, with developed columnar joints (Fig. 3.56), in medium-fine-grained (quasi-porphyritic) texture, crystal pyroclast is high in content (30–70%) and matrix is felsitic–micro-grained; the latter has developed both flow structure and columnar joints, but mostly in sheet shape, porphyritic texture, with low phenocryst (5–20%), and matrix is cryptocrystalline.

7. **Subvolcanic rock facies**

The subvolcanic rock, very developed in the survey area, is mainly located in the center or surrounding caldera, a ultra-hypabyssal–hypabyssal intrusive body sharing the same source, the same time and the same space with volcanic rock, and it does not outcrop the surface but is located at the subsurface. Lithology is mainly quartz monzonitic porphyry and rhyolitic porphyry, most of them occur like

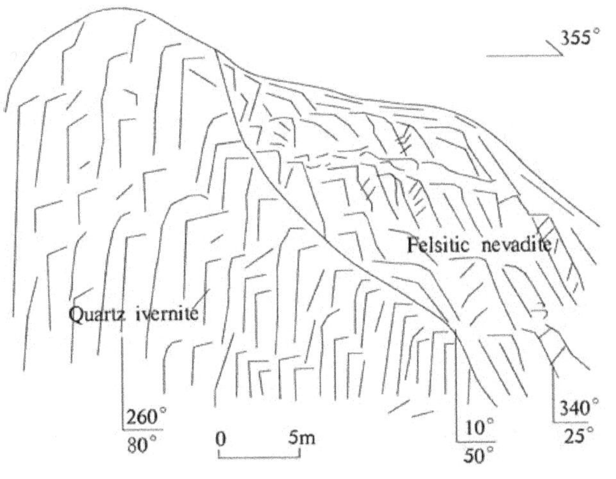

Fig. 3.57 Intrusive features of quartz monzonitic porphyry at Yangtianping

apophysis or boss, in the size of 1–20 km^2, for instance, the quartz monzonitic porphyry (Fig. 3.57) at Yangtianping caldera outcrops out of the center of the caldera, in irregular ellipse of which the long and short axes are 7 and 2 km, and it pulsed intruded into felsitic massive nevadite.

It is often filled in radial and annular fracture or fissures between strata, minerals in rocks are complex and many agglomerate breccias such as andesite wrapped inside.

3.3.3.2 Combination of Volcanic Facies

Combination of volcanic facies is the volcanic facies generated during one eruption in the history of a volcano and its generation sequence, and the type of combination of volcanic facies reflects comprehensively the features and rules of volcanic activities. There are temporal sequences and spatial superimposition relationships in lithofacies produced from multiple eruptions, research on combination types of volcanic facies help identify the source and direction of volcanic matters, determine the location of volcanic vent or volcanic eruption center, and restore the type and activity history of ancient volcanos. According to research on volcanic facies in the survey area, there are mainly three types of basic combinations as below.

1. **Explosive facies series (debris-flow accumulative facies–fallout accumulative facies)–eruptive sedimentary facies combination**

It is mainly seen at (the lower part) of Yangtianping revived caldera and Tianhuangping caldera, a common type of volcanic facies combinations in the survey area, reflecting that intense eruption and dormancy occur alternately during volcanic action, representing the mutual accumulation relationship between volcanic clastic facies and sedimentary facies. For instance, intense volcanic eruption occurs in the

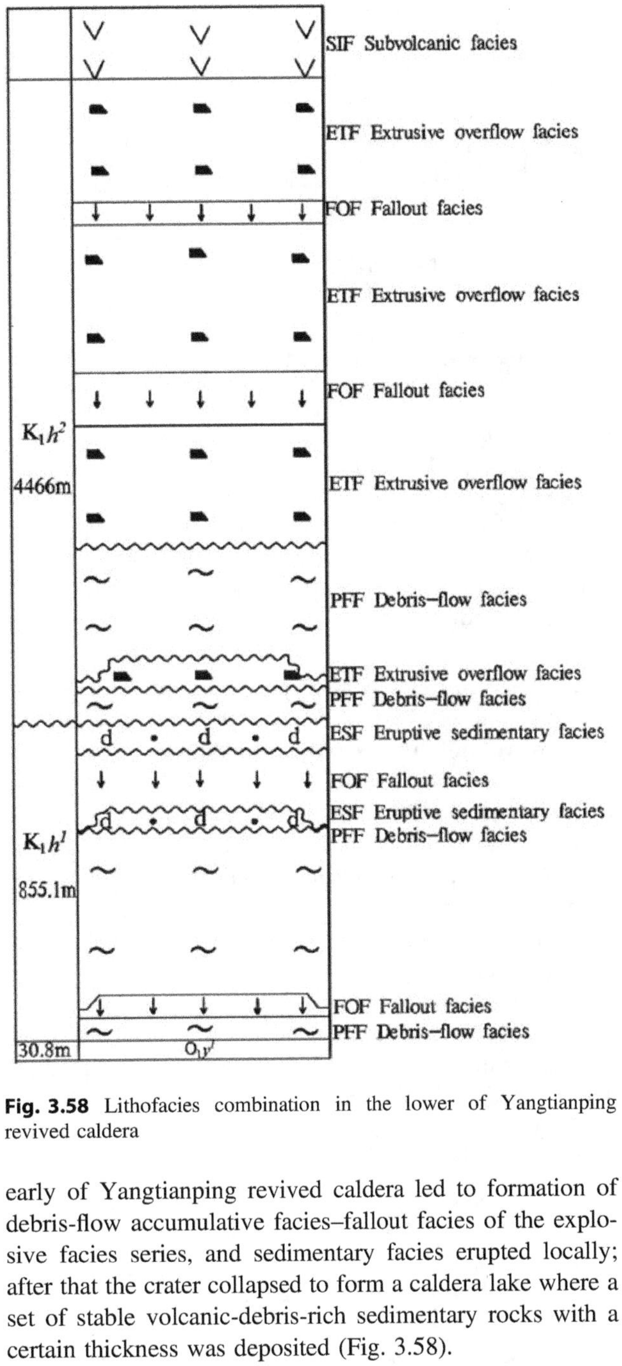

Fig. 3.58 Lithofacies combination in the lower of Yangtianping revived caldera

early of Yangtianping revived caldera led to formation of debris-flow accumulative facies–fallout facies of the explosive facies series, and sedimentary facies erupted locally; after that the crater collapsed to form a caldera lake where a set of stable volcanic-debris-rich sedimentary rocks with a certain thickness was deposited (Fig. 3.58).

2. **Explosive facies series (debris-flow accumulative facies–fallout accumulative facies)–extrusive overflow facies combination**

This series combination is mainly developed at (the upper of) Yangtianping revived caldera which, after collapse and sedimentation from early eruption, began reviving and erupting. Firstly, it erupted intensely for a very short time to

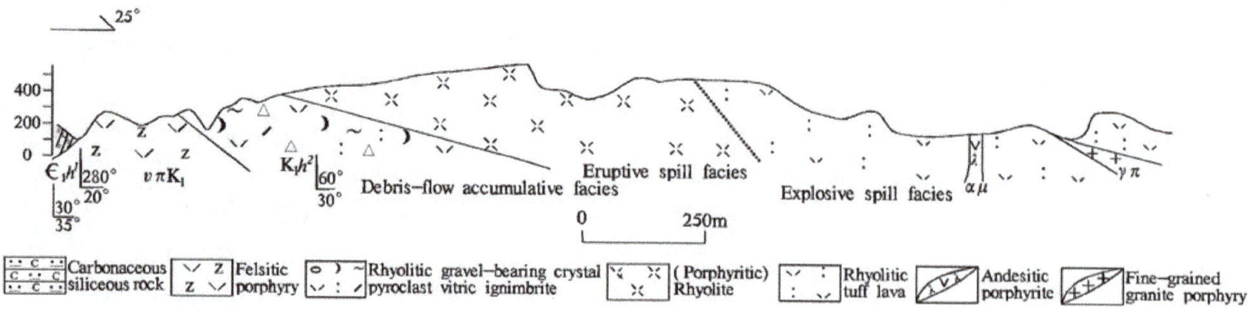

Fig. 3.59 Combination of debris-flow accumulative facies–eruptive spill facies–explosive spill facies in the Member #2 of Huangjian Formation in the western of East Tianmu Mount

form a set of rocks of debris-flow accumulative facies–fall-out accumulative facies, which is not table in thickness, thick at some local places (e.g., Changtanqiao), and thin at others (e.g., Shexi Village Canyon); secondly, the volcano erupted at a weak intensity and experienced extrusion and overflow in large scale, and finally it became intrusion of subvolcanic rock (Fig. 3.59).

3. **Explosive facies series (surging accumulative debris–flow accumulative facies–fallout accumulative facies)–eruptive spill facies–explosive spill facies combination**

It is mainly seen at East Tianmu Mount, Xikou Village, and Dongkeng Village, etc., also one of the important volcanic facies combinations in the survey area, and commonly seen in revived caldera (East Tianmu Mount–Caotanggang) and dome-shaped volcanos (South Tianmu Mount). East Tianmu Mount–Caotanggang revived caldera contains volcanic clastic rock of the explosive facies in its lower, porphyritic rhyolite (bubble rhyolite) of the eruptive spill facies in the mid and rhyolitic tuff lava of the explosive spill facies in the upper, reflecting that volcanic action changes from intense eruption in the early and quiet eruptive spill and explosive spill in the late (Fig. 3.60).

3.3.4 Volcanic Eruption Rhythms and Cycles

3.3.4.1 Volcanic Eruption Rhythm

Volcanic eruption rhythm means the cyclical changes of volcanic eruption, and such cyclical changes include regular changes in erupted material components, eruption intensity, eruption ways, and erupted thickness, etc. Generally speaking, a rhythm is composed of a or a few layers of rocks some of which could be very thin, just dozens of centimeters while others may be very thick, about hundreds of meters.

Based on volcanic eruption from strong to weak and its eruption ways, in the survey area, the Huangjian Formation volcanic rocks may be divided into four eruption rhythms (Fig. 3.61).

Eruption rhythm #1 is mainly the evolution of the debris-flow accumulative facies (locally interbedded with erupted sedimentary facies) → the fallout accumulative facies → the eruptive sedimentary facies, with the total thickness of > 855 m, mainly distributed along Zhangcun Town in the survey area.

Eruption rhythm #2 is mainly evolutions of the volcanic debris-flow facies and the surging accumulative facies (local) → the debris-flow accumulative facies → the extrusive overflow facies → the eruptive spill facies → the explosive

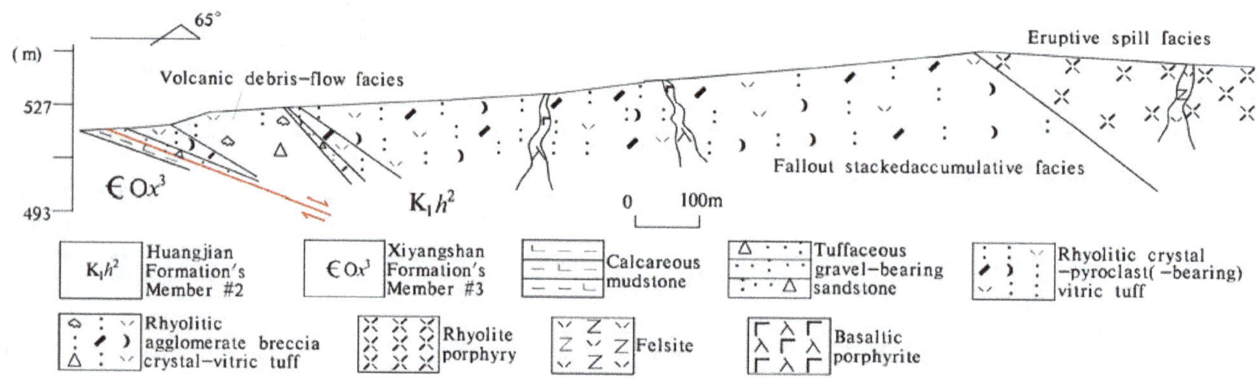

Fig. 3.60 Combination of volcanic debris-flow facies–fallout accumulative facies–eruptive spill facies in Huangjian Formation's Member #2 of Tianmu Mount Scenic Area

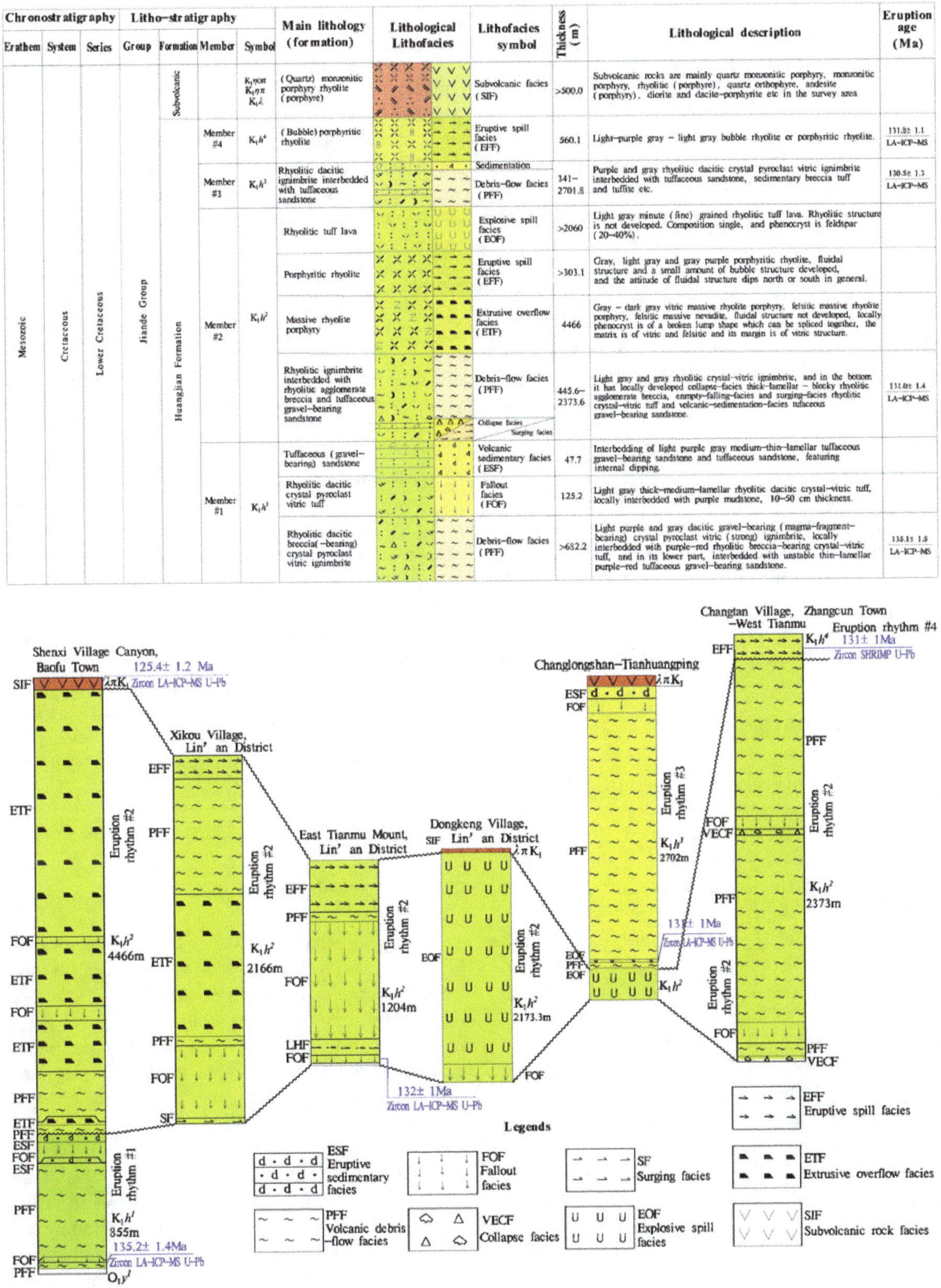

Fig. 3.61 Overall histogram on volcanic eruption rhythm and lithofacies formation in the survey area

spill facies → the subvolcanic rock facies, with total thickness of 7274–9202 m, an important part of volcanic rocks in the survey area. The combination of main rhythmic lithofacies varies in different areas, of these the Baofu Town–Zhangcun Town is mainly the evolution of the debris-flow accumulative facies → the extrusive overflow facies → the subvolcanic facies, the East Tianmu Mount is evolution of the volcanic debris-flow facies → the debris-flow accumulative facies → the eruptive spill facies → the explosive spill facies → the subvolcanic facies while Xikou Village is evolution of the surging accumulative facies → the debris-flow accumulative facies → the eruptive spill facies → the explosive spill facies.

Eruption rhythm #3 is mainly evolution of the debris-flow accumulative facies (locally interbedded with the eruption sedimentary facies) → the eruptive sedimentary facies → the subvolcanic facies, mainly developed in the north Tianhuangping–Linjiatang of the survey area, with the total thickness of > 341–2701 m.

Eruption rhythm #4, prettily single, mainly composed of the porphyritic rhyolite and bubble rhyolite in eruptive spill facies, is mainly developed at West Tianmu Mount–Longwangshan and South Tianmu Mount of the survey area, with total thickness >500 m.

3.3.4.2 Volcanic Eruption Time Limits and Cycles

1. **Time limit for eruption**

In order to limit the age of volcanic eruption, in the survey area, its top, bottom, and important volcanic rocks are analyzed for LA-ICP-MS zircon U–Pb chronology. In the survey area, in volcanic rocks, zircon mostly is irregular long strip, with the length of 100–150, 70–100, and 150–200 μm for a few, the length-width ratio is about 2:1, and zircon has developed zonal textures. Zircon in the Member #1–#4 of Huangjian Formation and volcanic rock, the U content is (130–569) $\times 10^{-6}$, (60–282) $\times 10^{-6}$, (24–332) $\times 10^{-6}$, (51–381) $\times 10^{-6}$, and (120–1447 and 112–1259) $\times 10^{-6}$, respectively, Th content is (103–755) $\times 10^{-6}$, (10–61) $\times 10^{-6}$, (3–61) $\times 10^{-6}$, (7–57) $\times 10^{-6}$, and (91–887 and 84–597) $\times 10^{-6}$, respectively, Th/U ratio is 0.49–1.41, 0.14–0.23, 0.12–0.25, 0.05–0.23, (0.59–1.25 and 0.47–1.48), a typical feature of magma zircon. Sample points tested are mostly projected on or near the concordant curve, and $^{206}Pb/^{238}U$ weighted mean age is 135.2 ± 1.4 Ma (MSWD = 0.93), 132.1 ± 1.0 Ma (MSWD = 1.16), 131 ± 1.0 Ma (MSWD = 1.06), 131 ± 1.0 Ma (MSWD = 0.92), 127 ± 2.0 Ma (MSWD = 1.06), and 125.4 ± 1.2 Ma (MSWD = 1.07), respectively (Fig. 3.62), indicating the time limit for volcanic rock eruption is 135.2–125.4 Ma, the beginning of Early Cretaceous, in the survey area.

2. **Eruption Cycles**

Volcanic cycles are intended to make divisions within a volcanic eruption area or a bigger range, it means formation of different volcanic eruption stages within the activity stage of a volcano and cyclic changes of volcanic products associated with a certain volcanic structures, and generally, there are a few of volcanic apparatus or volcanic groups (eruption basins).

By analyzing comprehensively such factors as features of volcanic activities, features of volcanic rock combinations and discontinuity of volcanic activities, in the survey area, the volcanic activity in Early Cretaceous is just one eruption cycle specifically based on the following evidences:

(1) In the survey area, the Huangjian Formation is the only stratum unit of volcanic rocks which has four eruption rhythms.
(2) Though the volcano erupts in multiple stages for a few times in the survey area, volcanic activities are basically continuous (135.2–125.4 Ma), with no regional structural unconformable surface.
(3) In the survey area, despite non-consistence in volcanic rock types and lithofacies combinations between eruption stages, chemical and geochemical features of the volcanic rocks are prettily similar (as mentioned below).
(4) In the survey area, the volcano acts in the Yianmu Mount–Mogan Mount volcanic depression where the structural environment is unchanged, and spatial distribution landscape and types of volcanic structures are pretty single.

3.3.5 Volcanic Structures

The distribution pattern of the Late Mesozoic volcanic rocks in Jixi County–Anji County, an area adjacent to the survey are in Zhejiang and Anhui Province indicates the NE-strike regional fracture plays a controlling role in volcanic activities. In the survey area, volcanic rocks are widely distributed, up to 140 km in length, the northwest border of the strip shaped Changhua–Tianmu Mount–Mogan Mount volcanic rock zone is under control of the NE-strike Jixi fracture zone and Xuechuan–Huzhou Fault to form a series volcanic eruption areas where plutonic intrusions probably are intermittently distributed (three eruption areas: Changhua, Tianmu Mount, and Mogan Munt), manifesting the relationships between volcanic activities and regional structures.

In the survey area, the main body of volcanic structures is a Tianmu Mount volcanic depression (basin) (Level-IV) in the shape of a normal trapezoid, and its south and north sides are in EW strike or NE strike fault or unconformable contact;

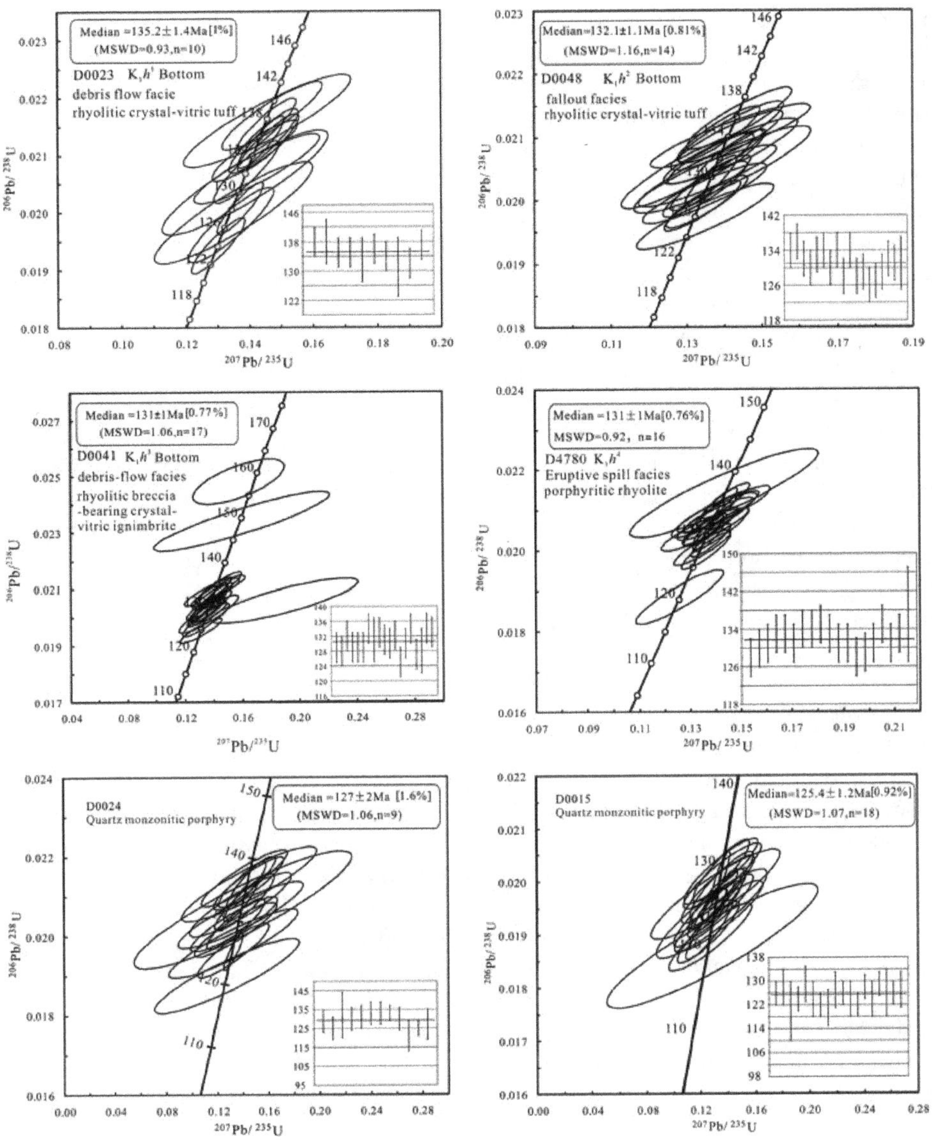

Fig. 3.62 Zircon U–Pb concordia diagrams and age histograms for mainly volcanic rocks in the survey area

its west side borders with NE-strike Maotan–Luocun facture and its east side is in fault or intrusive contact with Wushanguan pluton, which is the area accumulation products of Early Cretaceous volcanic activities in the regional fault-depressed structural basin, 30–37 km long from east to west and about 23 km wide from south to north; its inside developed the eruptive sedimentary rocks, which is intermittently periclinal and dips inward, rock strata are good in stratification, with gentle attitude, for instance at attitude is 95°–115°∠13°–15° for rock strata at Zhangcun Town in the west, 195°∠27° for rock strata at Shenxi Village Canyon in the north, and 50°∠35° for rock strata at East Tianmu in the south. The border of the volcanic depression is under control of nearly EW-strike buried faults and the NE-strike fault. The nearly EW-strike buried fault, possibly formed in Caledonian or earlier, mainly dipping southward, constituting the north margin of the volcanic rock basin; a large wide and gentle synclinorium formed in the Early Indosinian, i.e., regionally, buried syncline, forming the beginning shape of the volcanic basin, and forming the NE-strike Maotan-Luocun fault (Xuechuan–Huzhou fault) constituting the west margin of the volcanic basin; lifting occurs in Yanshanian so that the fault early formed at the basin border was activated, and the early syncline depression settled down along the border, and intense volcanic eruption and granite intrusion intruded.

Volcanic activity inside the volcanic depressions of Tianmu Mount is characterized by central eruption, the secondary volcanic structures are mainly central-type volcanic apparatus (Level V) and can be divided into five (revived) calderas and six dome-shaped volcanos based on eruptive and accumulation features. The calderas are included Yangtianping revived caldera, Tianhuangping, Changlongshan-Linjiatang, East Tianmu Mount-Caotanggang, and Yaotianfan caldera, while the dome-shaped volcanoes are included West Tianmu Mount-Longwangshan, South Tianmu Mount, Nanyushan, Shifo Temple, Dashulin, and Wuguishan, with other dome-shaped volcanoes in different sizes. Volcanic structures are characterized by obvious circular structures on 1:50,000 remote-sensing image, e.g., Yangtianping, East Tianmu Mount–Caotanggang, Tianhuangping (revived) caldera, West Tianmu Mount–Longwangshan, South Tianmu Mount, and Nanyushan dome-shaped volcano, with aeromagnetic anomalies developed in central and nearby areas of these volcanoes. All the characteristics of volcanic apparatus are shown in Fig. 3.63.

3.3.5.1 (Revived) Caldera

Caldera is a large volcanic apparatus formed after a volcano or a group of volcanoes collapsed, one of the main types of volcanic apparatuses in the survey area. Revived caldera means that after a caldera takes shape there are still volcanic clastic rock and lava erupting out of caldera, but mainly piedmont accumulation, collapsed accumulation at the caldera wall and lacustrine sediments; in some large caldera composed of acidic rocks volcanic activity often recurred to be developed into revived caldera; revived dome left some or all previously sunken fault blocks to uplift and ascent, so that rock strata previously horizontal or dipping inward possibly dip outward at the dip angle from a few degrees to dozens of degrees.

1. **Yangtianping revived caldera**

It is located around Shenxi Village Canyon, Baofu Town–Zhangcun Town, overall reflecting the whole process of intermittent eruption → eruption → intermittent → eruption → eruption and spill → subvolcanic rock intrusion. Its basic features are as follow:

(1) Morphological feature

Caldera spreads like an ellipse, about 120 km^2, nearly 10 km long from east to west and about 12 km wide from south to north, topographically like a protruding high mountain, planarly distributed like a ring, and on the remote-sensing image there developed clearly irregular circular structures, its margin developed NE-strike, NW-strike, and nearly EW-strike fracture structures, and its surroundings are associated with aeromagnetic anomaly in scale of 1:50,000.

(2) Lithological and lithofacies features

It is mainly composed of the Member #1 and #2 of Huangjian Formation, its outer strata are old and the inner ones are young, and it has the feature of pericline and dipping inward. The early lithofacies combination is the debris-flow accumulative facies interbedded with the volcanic sedimentary facies (rhyolitic dacitic crystal-vitric ignimbrite interbedded with tuffaceous siltstone) → the fallout facies (rhyolitic dacitic breccia crystal-vitric tuff) → the volcanic sedimentary facies (tuffaceous gravel-bearing siltstone), and the late one is the debris-flow accumulative facies (rhyolitic breccia-bearing crystal-vitric ignimbrite) → the extrusive overflow facies (rhyolitic porphyry) → the subvolcanic facies (quartz monzonitic porphyry) (Figs. 3.64 and 3.65); both combination's volcanic rocks are in unconformable contact relationship (Fig. 3.66a). The late rhyolitic porphyry in extrusive overflow facies is centered on quartz monzonitic porphyry at Yangtianping, which has relatively complete zonation, its margin features vein-like intrusion, and from inner to outer they are felsitic massive nevadite, felsitic massive rhyolitic porphyry, and vitric massive breccia-bearing rhyolitic porphyry (less speckles); for this suite of strata, its lower has developed horizontal bedding (190°∠20°) (Fig. 3.66b), its mid has developed horizontal and upright columnar joints and its upper part near the crater developed oblique columnar joints (50°∠35°).

(3) Structural features

In the margin of caldera, there are developed radial and circular fractures and the dipstrike for some of them is 250°∠80°, 80°∠70°, 170°∠30°, and 325°∠30°, and in the middle there are developed vein groups such as andesite and felsite, where there are plenty of developed andesite agglomerate breccia and veins inside quartz monzonitic porphyry subvolcanic rocks. The northern and western sides of the caldera have the tuffaceous gravel-bearing sandstone in volcanic sedimentary-facie developed early, etc., in the thickness of 47.7–100 m, having the feature of pericline and dipping inward, for instance, the volcanic rock strata are 95°–115°∠13°–15°, and 195°∠27° in dipstrike, respectively, at Zhangcun in its west and Shenxi Canyon in its north. There are no developed volcanic sedimentary strata in its eastern and southern sides due to late-stage fracture and volcanic eruption.

Active volcanic zone Level I	Active volcanic zone Level II	Eruption zone Level III	Volcanic–structural uplift or depression Level IV		Volcanic apparatus		Horizon lithological formation	Lithofacies combination	Subvolcanic rock (center)	Eruption age
Volcanic rockbelt in northwest Zhejiang	Active volcanic rock subzone in northwest Zhejiang	Changhua–Tianmu–Mount–Mogan–Mount volcanic eruption belt	Tianmu–Mount volcanic depression (basin)	Trapezoid; its south and north sides in EW–strike or NE–strike fault or unconformable contact, its west side borders with NE–strike Xuechuan – Xiaofeng fault and its east side is in fault or intrusive contact with Wushanguan pluton.	(Revived) caldera	Yangtianping	K_1h^1、K_1h^2 Rhyolitic dacitic ignimbrite, tuff; tuffaceous sandstone; rhyolitic porphyry	Debris–flow facies – fallout facies –volcanic sedimentary facies –debris–flow facies –extrusive overflow facies –subvolcanic facies	Quartz monzonitic porphyry	135–125Ma LA–ICP–MS
						Tianhuangping	K_1h^3 Rhyolitic dacitic breccia–bearing ignimbrite; tuffaceous sandstone; tuffite	Debris–flow facies – volcanic sedimentary facies –subvolcanic facies	Rhyolitic porphyry	-131Ma
						Changlongshan – linjiatang	K_1h^2、K_1h^3 Rhyolitic dacitic breccia–bearing(strong) ignimbrite; porphyritic rhyolite and bubble rhyolite	Debris–flow facies – volcanic sedimentary facies –subvolcanic facies	Dacite–porphyrite	-131Ma
						East Tianmu – Caotanggang	K_1h^2、K_1h^3 Rhyolitic dacitic breccia–bearing(strong) ignimbrite; porphyritic rhyolite and bubble rhyolite	Collapse facies (volcanic debris–flow facies) – debris–flow facies –eruptive spill facies–explosive spill facies–debris–flow facies		132–131Ma LA–ICP–MS
						Yaotianfan	K_1h^2 Interbedding of porphyritic rhyolite and rhyolitic crystal–vitric ignimbrite	Eruptive spill facies and fallout facies		-131Ma
					Dome–shaped volcano	West Tianmu Mount – Longwangshan	K_1h^2–K_1h^4 rhyolitic breccia–bearing ignimbrite; porphyritic rhyolite and bubble rhyolite	Debris–flow facies –eruptive spill facies		132–131Ma LA–ICP–MS
						South Tianmu Mount Nanyushan	K_1h^3、K_1h^4 rhyolitic dacitic ignimbrite; porphyritic (bubble) rhyolite	Debris–flow facies –eruptive spill facies		-131Ma
						Shifosi Temple Wuguishan	K_1h^3、K_1h^4 Rhyolitic dacitic ignimbrite; porphyritic (bubble) rhyolite	Debris–flow facies –eruptive spill facies		-131Ma
						Dashulin	K_1h^1、K_1h^4 Tuffaceous glutenite porphyritic rhyolite and bubble rhyolite	Volcanic sedimentary facies –eruptive spill facies		-131Ma

Fig. 3.63 Division of volcanic apparatus in the survey area

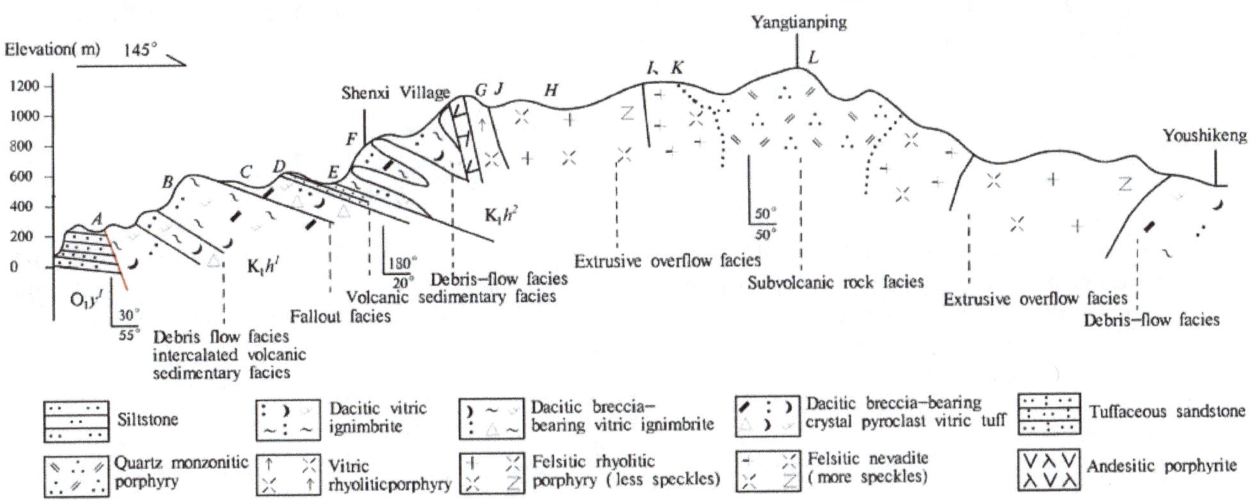

Fig. 3.64 Cross-section profile of Yangtianping revived caldera structure

Fig. 3.65 Images showing field features of Yangtianping revived caldera

(4) Time limit for eruption

Time limit for the caldera eruption was 135.1–125.4 Ma, and it experienced the process of fault depression → volcano eruption → volcanic collapse and sedimentation → reviving and eruption, magmatic extrusion and overflow → intrusion of subvolcanic rock in the beginning of Early Cretaceous (Fig. 3.67).

2. **Tianhuangping Caldera**

It is located around Tianhuangping Reservoir (Fig. 3.68), Tianhuangping Town, reflecting the process of eruption → intermittence → intrusion of subvolcanic rock, and its basic features are as follow:

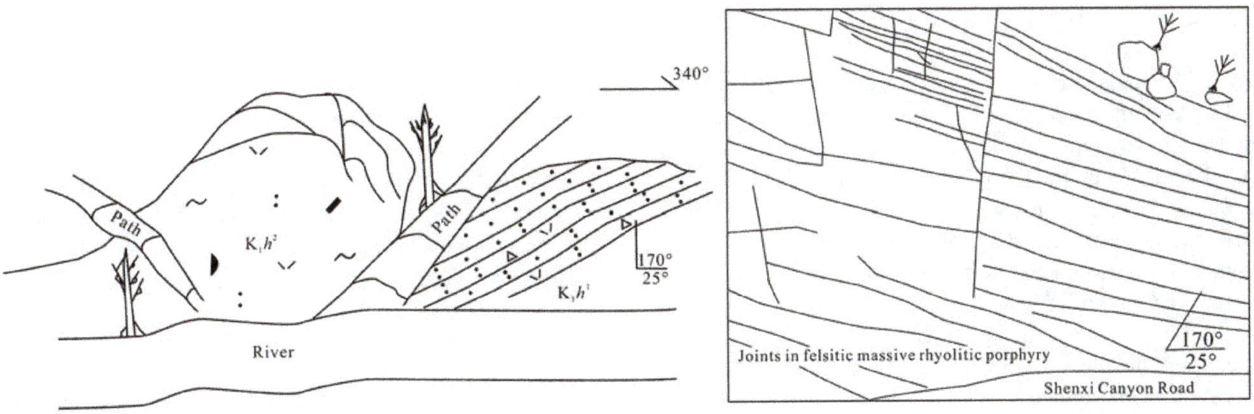

Fig. 3.66 Images showing (rhythmic) unconformable contact relationship between volcanic eruptions in two stages and horizontal joints developed in vitric rhyolitic porphyry in the lower part of late eruption

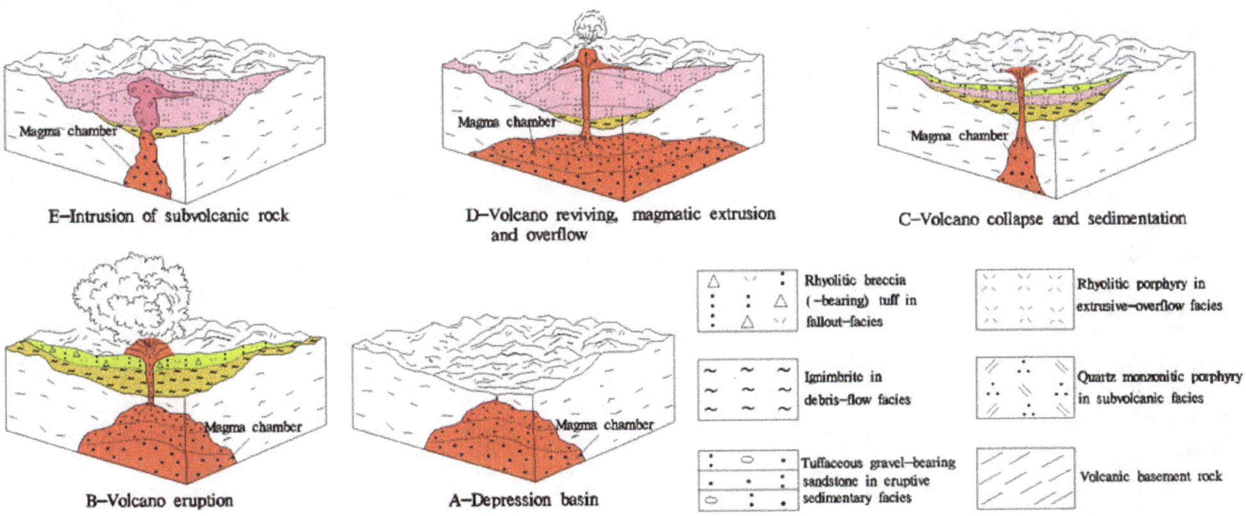

Fig. 3.67 Diagram showing evolution process of the Yangtianping revived caldera

(1) Morphological feature

Tianhuangping caldera spreads like a long-strip-like ellipse, about 2 km^2, nearly 1 km long from east to west and about 2 km wide from south to north, topographically like a protruding high mountain (hydropower station), and on the remote-sensing image there developed clearly ellipse-circular structures, and developed aeromagnetic anomaly in scale of 1:50,000.

(2) Lithological and lithofacies features

It is composed of strata in the Member #3 of Huangjian Formation, its outer strata are old and the inner ones are young, and it has the feature of pericline and dipping inward, and also planarly distributed like a ring; the lithofacies combination is the debris-flow accumulative facies interbedded with tuffaceous gravel-bearing sandstone (rhyolitic dacitic breccia-bearing crystal-vitric ignimbrite, locally interbedded with tuffaceous-powder gravel-bearing sandstone) → the volcanic sedimentary facies (interbeddings of tuffaceous gravel-bearing siltstone, sunken breccia tuff, and tuffite, etc.) → the subvolcanic facies(rhyolitic porphyry) (Fig. 3.69).

(3) Structural features

Perclinal and dipping-inward volcanic sedimentary rocks are developed in the periphery of the caldera, in its south side rhyolitic agglomerate-breccia-bearing crystal pyroclast ignimbrite is of lamination-like structure with the attitude of 330°∠25°; its west side has developed a large suite of combination of sedimentary breccia tuff, tuffaceous gravel-bearing sandstone, and tuffite, with the attitude of

120°∠20°, and its lower part has developed fold structures and the attitude of both wings of folds are 186°∠40° and 20°∠60°, indicating that volcanic sedimentation was affected by contemporaneous structural movement; its north has developed rhyolitic gravel-bearing crystal-vitric ignimbrite, interbedded with purple-red thin-lamellar tuffite and tuffaceous gravel-sandstone, with the attitude of 210°∠20°. The rhyolitic porphyry in late subvolcanic facies is in intrusive contact in the south and north sides, with the attitude of the contact surface of 330°∠70° and 200°∠60°.

Within 1–3 km in the south and east sides of the caldera there are developed subvolcanic rocks such as irregular rhyolitic porphyry and monzonitic porphyry, and veins such as andesite, felsite, and diorite porphyrite.

(4) Time limit for eruption

It is presumed that the time limit for volcano eruption at the caldera is 132–131 Ma and it generally experienced a process of volcano eruption → collapse and sedimentation → intrusion of volcanic rock.

3. **Changlongshan–Linjiatang Caldera**

It is located in Linjiatang village, Gaohong town, Lin'an District, with the following features:

(1) Morphological feature

Changlongshan–Linjiatang caldera spreads like a long strip in NW-strike, about 10 km^2, some 5 km long from northwest to southeast and 2 km wide from northeast to southwest, topographically in the shape of protruding high mountain. In the remote-sensing image, NW-strike circular structure developed, with 1:50,000 aeromagnetic anomaly, which is cut off by a later NE-strike fracture.

(2) Lithological and lithofacies features

Its west side is composed of rhyolitic dacitic breccia-bearing crystal-vitric ignimbrite of the Member #3 of Huangjian Formation, and its east side is distributed with the Member #2 of Huangjian Formation rhyolitic tuff lava and dacite–porphyrite in center's subvolcanic rock lithology (Figs. 3.70 and 3.71).

(3) Structural features

In the northwest and southwest sides of the caldera, a handful of tuffaceous glutenite and tuffaceous gravel-bearing siltstone, which have features of being slightly oblique and dipping inward outcrops about 1.5 m thick, the tuffaceous gravel-bearing sandstone in the lower part is 5–20 cm per

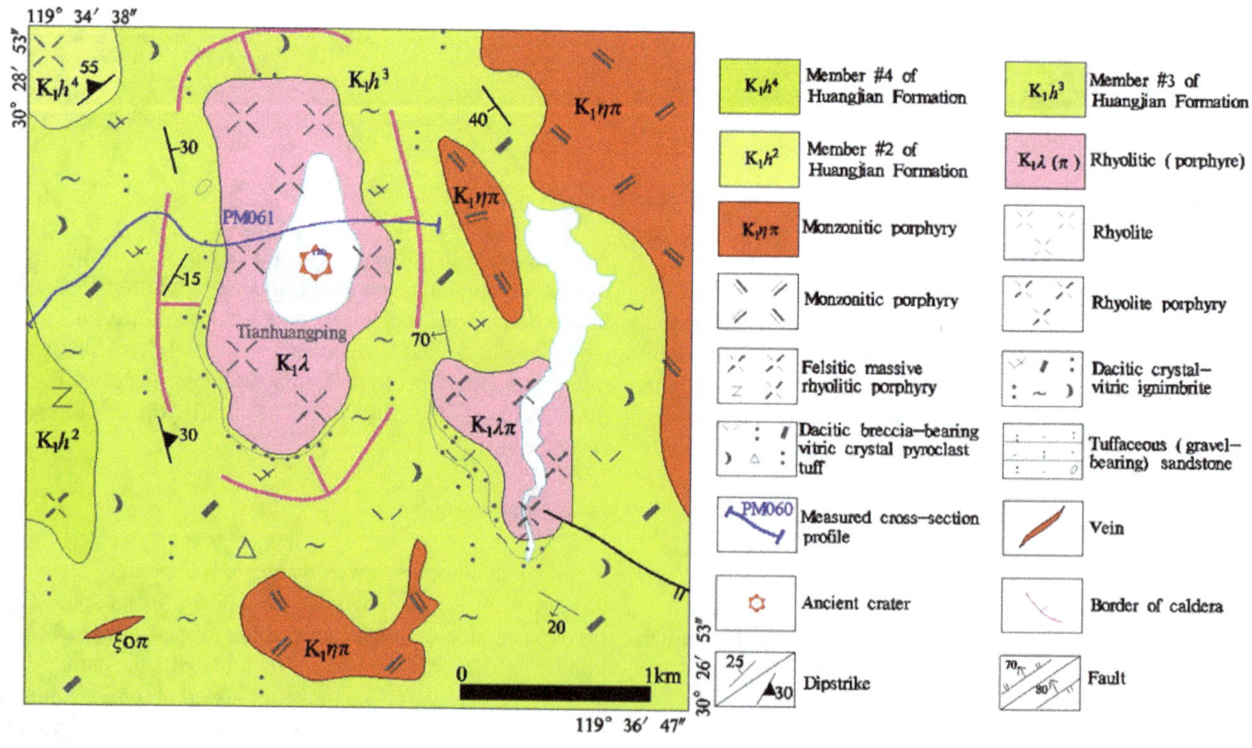

Fig. 3.68 Structural geological map of the Tianhuangping caldera

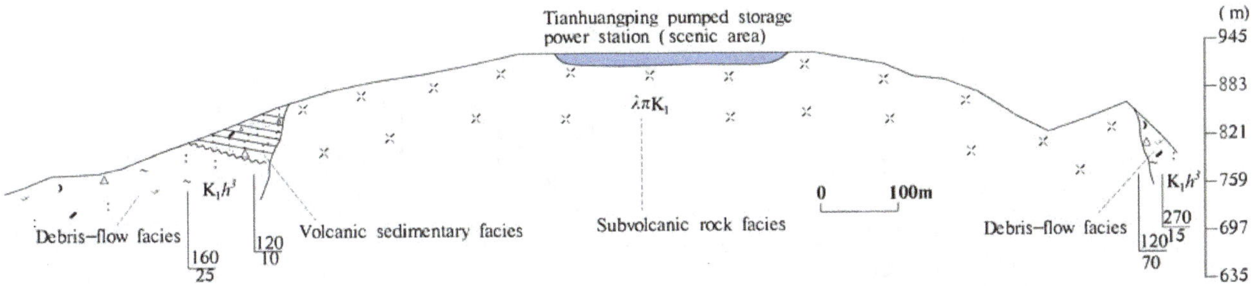

Fig. 3.69 Tianhuangping caldera cross-section profile

layer and the glutenite in the upper is strongly weathered, due to different components, gravel contour is clearly visible, gravel is 30–50% in content, about 1–8 cm in size generally, 10–20 cm for big ones, in the subangular shape, and a few gravel is 40–50 cm in size and in the shape of ellipse; their attitude is 140°∠25°, 125°∠15°, and 90°∠20°, respectively. In the late period, the caldera is cut off into two parts by intrusion of the Wushanguan composite pluton with fine-grained syenogranite apophysis, and silicification alteration developed at the margin of intrusive contact.

(4) Time limit for eruption

It is presumed that the time limit for volcano eruption at the caldera is about 131 Ma.

4. **East Tianmu Mount–Caotanggang Revived Caldera**

It is located in East Tianmu Mount, Lin'an District, in the south of the survey area, a well-preserved volcanic apparatus, in high elevation topographically, the peak of East Tianmu Mount is 1479 m above sea level and its main features are as follow:

(1) Morphological feature

Planarly it is ellipse, nearly 6 km long from east to west and 4–5 km wide from south to north, high and steep in the lower part and gentle in the peak; its south side is in angular unconformable contact with calcareous mudstone and marlstone in Cambrian–Ordovician Xiyangshan Formation, with the strata attitude of 50°∠40°, and in NW-strike fault contact in the East Tianmu Mount Scenic Area (Fig. 3.72). In the remote-sensing image, there is clearly developed ellipse-circular structure associated with 1:50,000 aeromagnetic anomaly in the mid and margin, which is cut off by a NE-strike fracture later.

(2) Lithological and lithofacies features

Strong in scale and intensity in the early, and it mainly reflects that erupted and collapsed acid magma washed by water flow and then sedimented and afterward it experiences eruptive and explosive spill so that it formed a set of lithology which was rhyolitic agglomerate breccia and crystal-vitric tuff in the lower, and porphyritic (bubble) rhyolite and rhyolitic tuff lava in the upper.

(3) Structural features

The early stage rock is largely lamellar, periclinal, and dipping inward, with steep dip angle, and the attitude of rhyolitic agglomerate breccia in the bottom is 50°∠35° and the attitude of rhyolite in the mid varies greatly, which is 110°∠75°, 220°∠30°, and 45°∠40°; there are two lateral craters in the west of Caotanggang. The late is mainly eruption of neutral-acid magma in small scale, only distributed in East Tianmu Mount, and its lithology is rhyolitic dacitic (breccia-bearing) crystal-vitric ignimbrite, especially inner rock which has the feature of strong clinkering, and the false-rhyolitic attitude is 55°∠15°, 30°∠25°, and 60°∠70° (Fig. 3.73).

(4) Time limit for eruption

On the basis of the zircon LA-ICP-MS U–Pb age of the early rhyolitic crystal-vitric tuff in the lower part (which is 132.1 ± 1.1 Ma), it is presumed that the time limit for the caldera is about 132 Ma; it generally experiences a process of eruption and sedimentation → magma overflow and explosive spill → reviving and eruption.

3.3.5.2 Dome-Shaped Volcano

Dome-shaped volcanic apparatus is a dome-shaped body formed by intrusive bodies and spilled lava. When

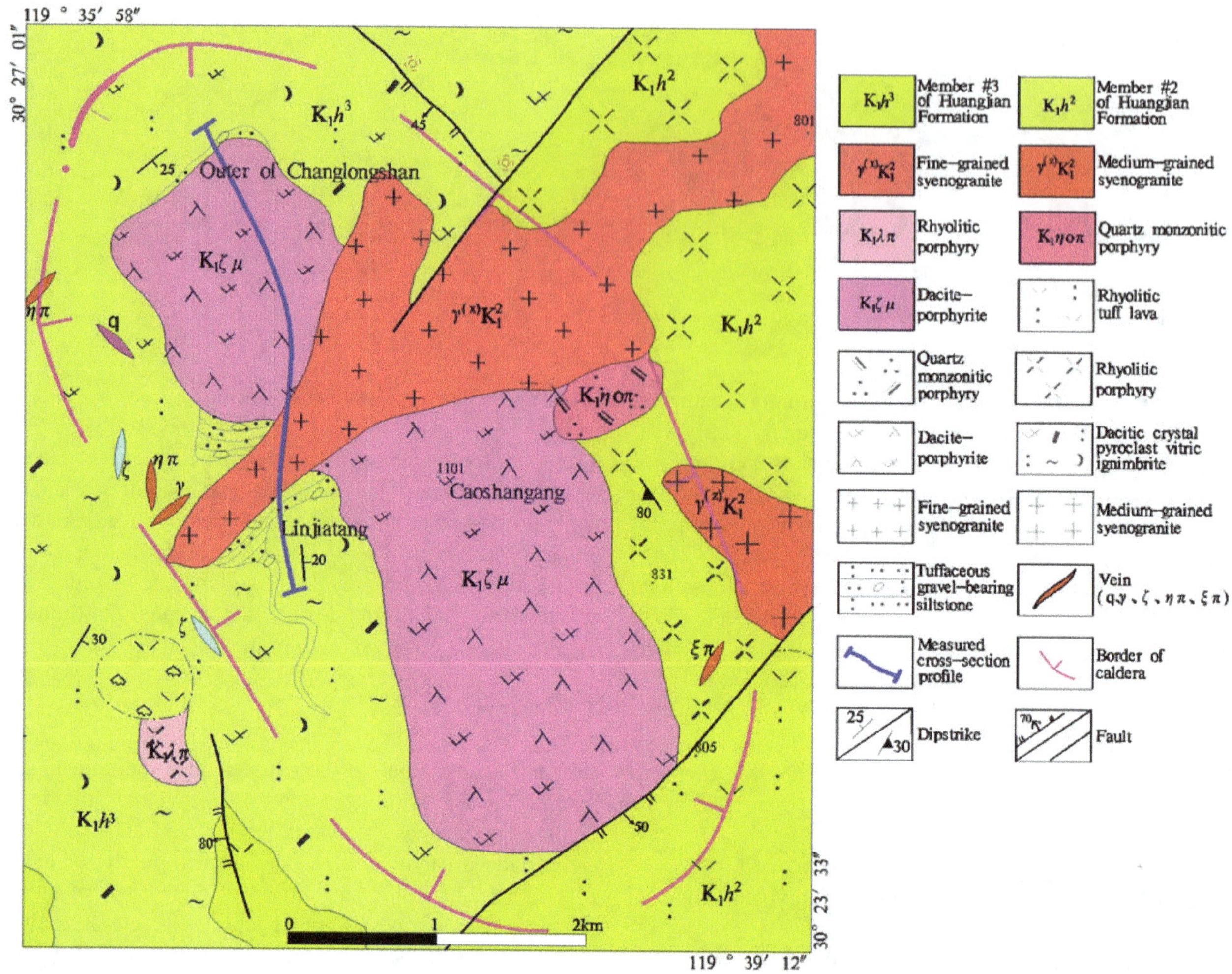

Fig. 3.70 Structural geological map of the Changlongshan–Linjiatang caldera

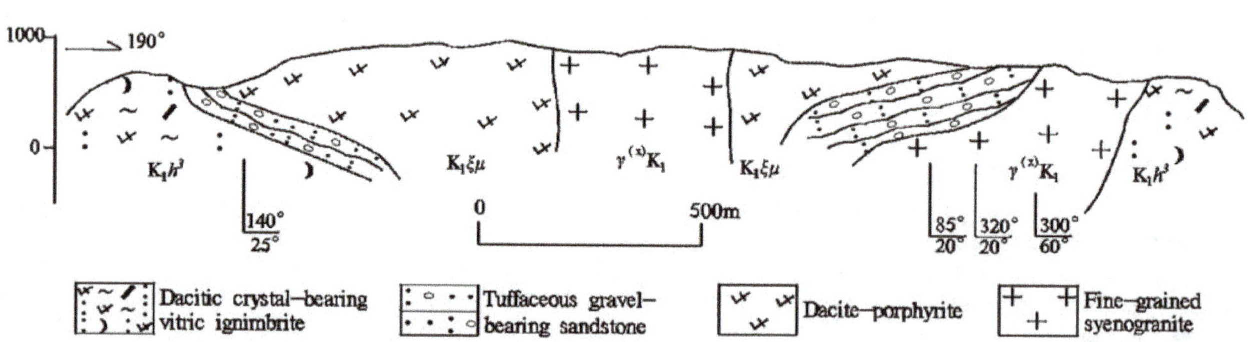

Fig. 3.71 Cross-section profile map of the Changlongshan–Linjiatang caldera

high-viscosity lava materials extruded out of the volcanic vent, they accumulated around or above the exit and do not overflow to form intrusive bodies, but at the same time lava that overflew slightly earlier is hunched up due to increased lava pressure so that it becomes the shape of a dome which is steep at the margin, gentle in the center and has cupola and lava dome in the peak. Main rocks constituting a dome-shaped volcanic apparatus are acid and neutral-acid rocks such as rhyolite and dacite, as well as trachyte and andesite.

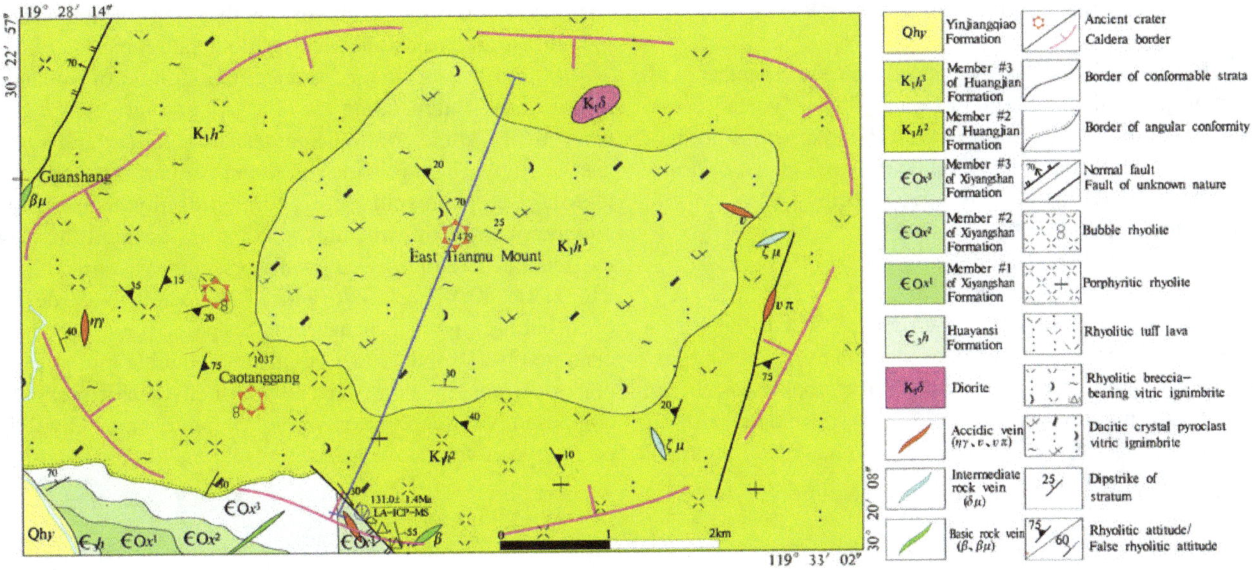

Fig. 3.72 Structural geological map of East Tianmu Mount–Caotanggang Revived Caldera

1. **West Tiannu Mount–Longwang Mount dome-shaped volcano**

The dome-shaped volcano situated at West Tianmu Mount–Longwang Mount national natural reserve in the south of the survey area, topographically with high elevation, and the highest elevation is 1505 m and 1587 m, respectively, at West Tianmu Mount and Longwang Mount. There mainly reflects eruption and spill of acid magma with the basic features as below:

(1) Morphological feature

It is in the NW-strike, nearly 8 km long and 2–5 km wide, steep at its margin, like a cliff at local places and rather flat at the upper. In the remote-sensing image there is clearly developed ellipse-circular structures, and NE-strike and NW-strike fracture structures at the margin.

(2) Lithological and lithofacies features

Lithology and lithofacies is mainly composed of the eruptive spill facies porphyritic rhyolite and bubble rhyolite of the Member #4 of Huangjian Formation (Fig. 3.74), porphyritic rhyolite in its lower part or margin has developed rhyolitic structure, phenocryst is 10–25% in content and has developed bubble structure with bubble in the size of 0.5–2 cm and content of 5–15%; its center is mainly rhyolite with more speckles, where phenocryst content is 25–70%.

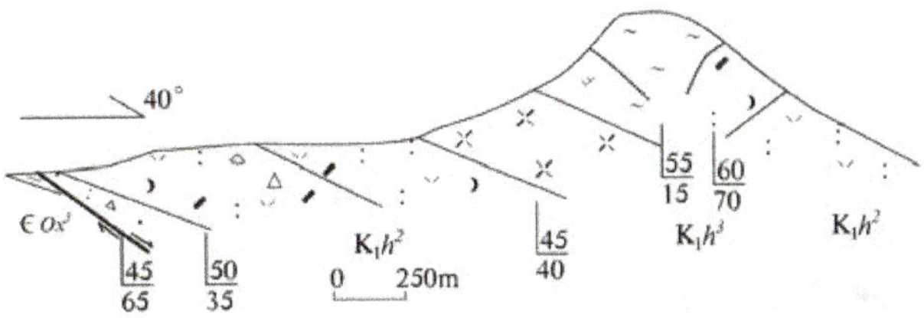

Fig. 3.73 Cross-section profile of East Tianmu Mount–Caotanggang Revived Caldera

(3) Structural features

Porphyritic rhyolitic in its western is in eruptive unconformable contact with the attitude of 110°∠50°, its east side is in NE-strike fault contact with Cambrian strata, with the fracture attitude of 300°∠80° (Fig. 3.75); rocks at the margin have developed rhyolitic structure with complex attitude, its attitude dips toward NE or East (80°–120°∠20°–40°) in general, and toward south locally (180°–210°∠15°–20°), there is no rhyolitic structure developed inner rocks but have developed round upright columnar joints structure, in the diameter of 20–100 cm per column; the features of rhyolitic attitude indicates it flows eastward in general.

(4) Time limit for eruption

Zircon LA-ICP-MS U–Pb age for porphyritic rhyolite is 131 ± 1 Ma.

2. **South Tianmu Mount Dome-shaped Volcano**

South Tianmu Mount dome-shaped volcano, together with Nanyushan and Shifo Temple dome-shaped volcanos, are distributed like bead streams in the EW strike, in the north of the survey area, also having relatively high elevation topographically, with the main features as below:

(1) Form and shape feature

Planarly in the shape of ellipse-round, 2–4 km long from east to west and 1–3 km wide from south to north, similarly it is steep in the lower part and flat at the mount peak. In the remote-sensing image, there is clearly developed ellipse-circular structure, with 1:50,000 aeromagnetic anomaly developed in the margin.

(2) Lithological and lithofacies features

Lithology in the lower part is mainly neutral-acid rhyolitic dacitic (breccia-bearing)crystal-vitric ignimbrite in the accumulation of debris flow early stage erupted, and locally has developed rock strata or interbedding of tuffaceous gravel-bearing sandstone and agglomerate breccia, about 20–100 cm per layer, for instance, the attitude of volcanic sedimentary interbedding is 140°∠45° and 290°∠25°, respectively, at the north of South Tianmu Mount and at the east of Nanyushan; lithology in the upper part is mainly the late-stage eruptive spill facies' porphyritic rhyolite, and the attitude of rhyolitic structures is complex, generally having the feature of dipping southward or southeastward (120°–213°∠30°–80°) (Figs. 3.76 and 3.77), locally have developed bubble rhyolite, in the bubble size of 0.3–5 cm, concentratively distributed like laminations, bubble rare at local places. Around the dome-shaped volcano, volcanic rocks are composed of quartz monzonitic porphyry and rhyolitic porphyry, as well as veins such as sillite and diorite.

(3) Time limit for eruption

The zircon LA-ICP-MS U–Pb age of porphyritic rhyolite in West Tianmu Mount Dome-shaped volcano is 131 ± 1 Ma, as the eruption time limit.

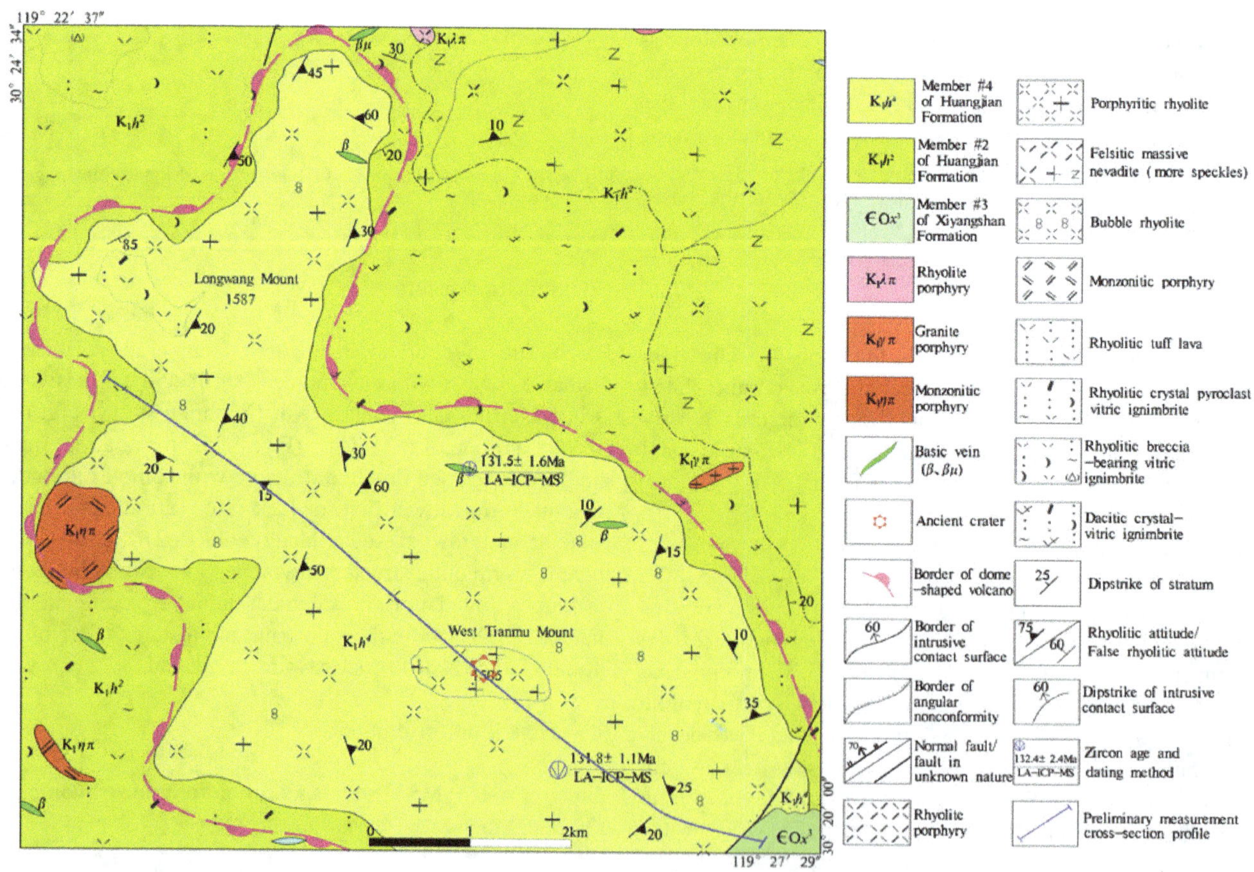

Fig. 3.74 Structural geological map of West Tianmu Mount–Longwang Mount dome-shaped volcano

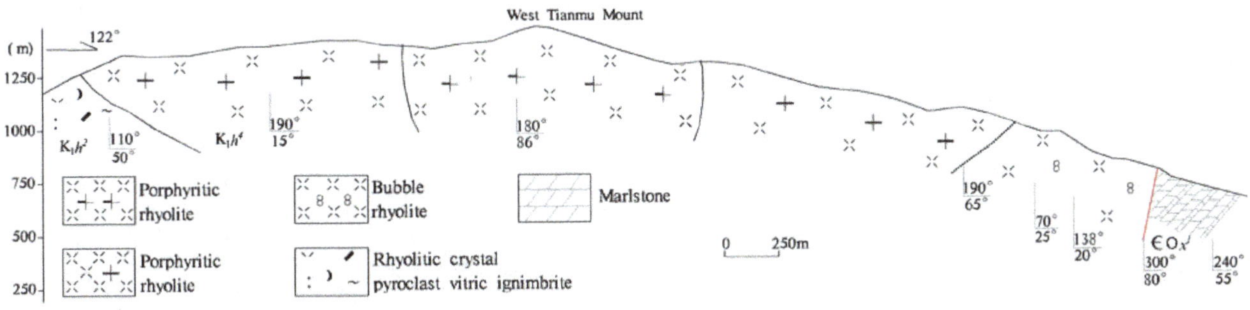

Fig. 3.75 Structural cross-section profile of West Tianmu Mount–Longwang Mount dome-shaped volcano

3. **Dashulin Dome-shaped Volcano**

Dashulin dome-shaped volcano is distributed Lin'am Yuanlinwukou at the southwest corner of the volcanic rock area of the survey area, with the following main features:

(1) Morphological feature

Planarly in the shape of long-strip ellipse, about 2 km long from south to north and 1.5 km wide from east to west, similarly it is steep in the lower part and flat at the mountain peak, extending south beyond the survey area (Fig. 3.78).

(2) Lithological and lithofacies structural features

Western of the volcano is in angular unconformable contact with Ordovician Changwu Formation siltstone strata, and locally in NE-strike fault contact. Around its bottom, there are developed tuffaceous glutenite and gravel-bearing sandstone, with gravel content of 20–25% and gravel size of 2–500 mm, varying greatly, which is generally rounded, subangular, subrounded, long-flat, and irregular, mainly sandstone (20%), quartz (5%), and volcanic rock (75%), 10–15 cm thick per layer, and a handful of gravel are arranged in oblique rows; of this suite of rock strata, the west and north sides are the most developed, with the attitude of 130°∠40° in the west and 230°∠30° in the north. The mid-part of local places in the north has a handful of sandwich of dacitic breccia-bearing crystal pyroclast vitric ignimbrite, and the attitude of rhyolitic-like structure is 230°∠30°. It the upper part, it is a suite of porphyritic rhyolite, the attitude of rhyolitic structure is 130°∠20°, 230°∠48°, 165°∠25°, and 220°∠20°, tilting southward overall (Fig. 3.79). Around the dome-shaped volcano, there are developed subvolcanic rocks and veins such as

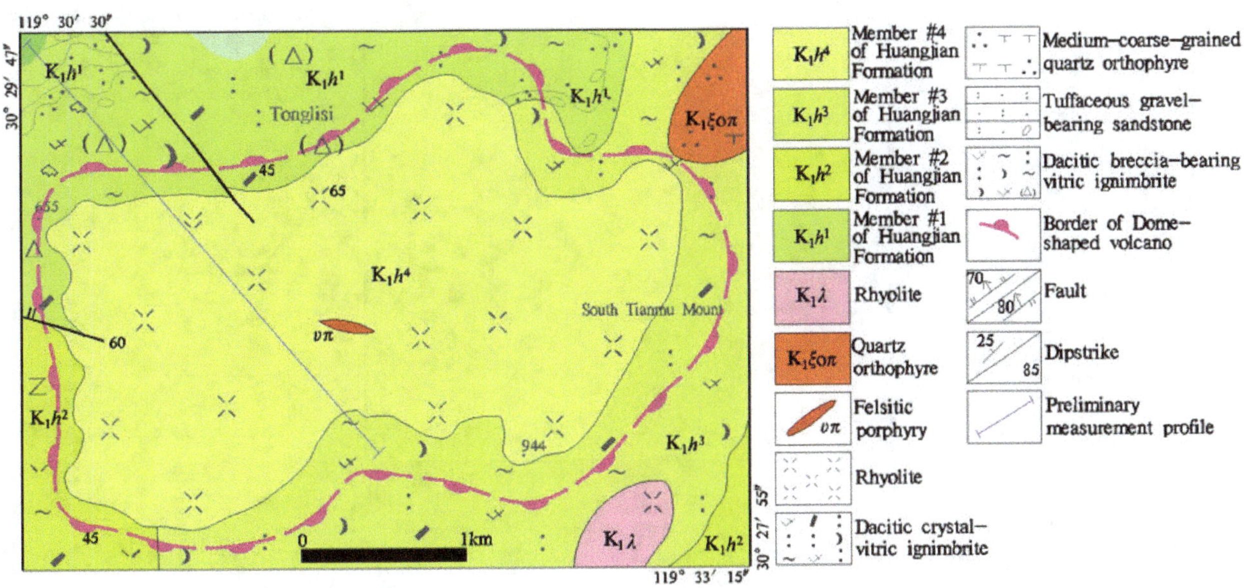

Fig. 3.76 Structural geological map of South Tianmu Mount–Longwang Mount dome-shaped volcano

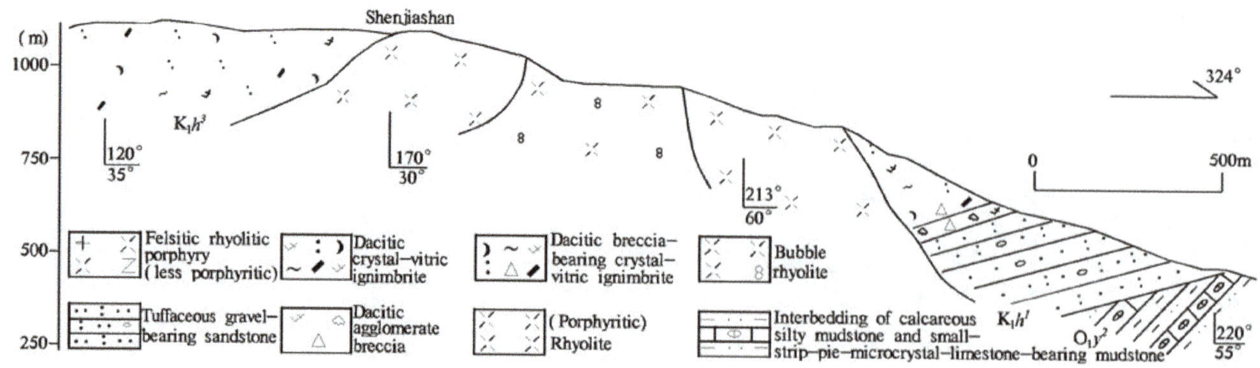

Fig. 3.77 Structural cross-section profile of South Tianmu dome-shaped volcano

granite porphyry, monzonitic porphyry, orthophyre, and dacite–porphyrite.

(3) Time limit for eruption

It is presumed that the time limit for its volcanic action is similar to that of West Tianmu Mount dome-shaped volcano (131 ± 1 Ma).

3.3.5.3 Regularity on Volcano Structural and Product Migration

In the survey area, volcanic apparatuses typically have the feature of bead-stream combinations, for instance: in the west side, West Tianmu Mount -Longwang Mount, South Tianmu Mount, and Nayushan dome-shaped volcanos, and Yangtianping revived caldera have clearly NE-strike arrangement feature; in the east side, South Tianmu Mount dome-shaped volcano, and Tianhuangping, Linjiatang–Changlong Mount and Yaotianfan calderas have the spatial combinations of NW-strike bead-stream-like arrangement; in the north side, South Tianmu Mount, Nanyushan and Shifo Temple dome-shaped volcanos have the feature of nearly EW-strike arrangement; in the south side, West Tianmu Mount–Longwan Mount and East Tianmu Mount–Caotanggang dome-shaped volcanos and Yaotianfan caldera also have the feature of nearly EW-strike arrangement, all

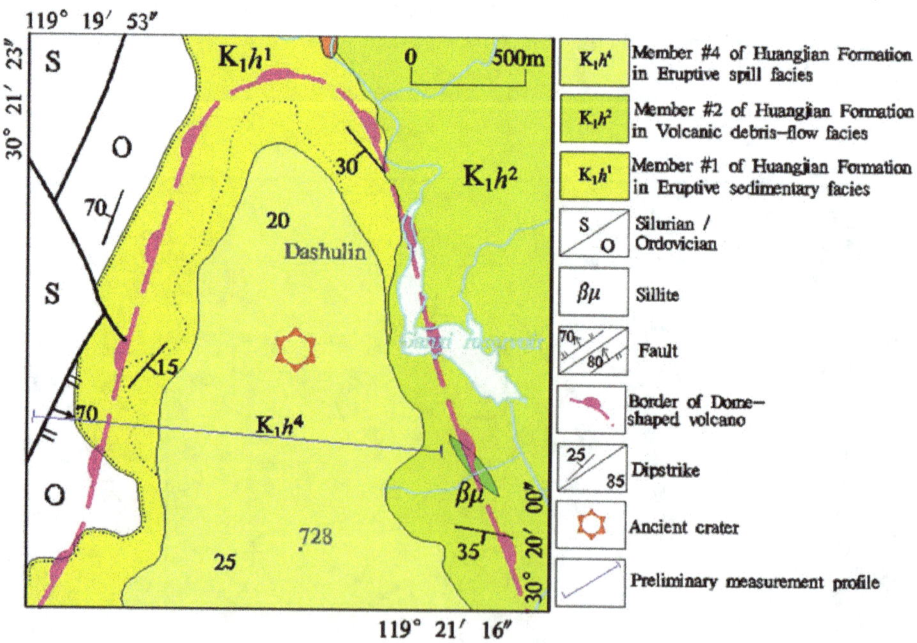

Fig. 3.78 Structural plane of Dashulin dome-shaped volcano

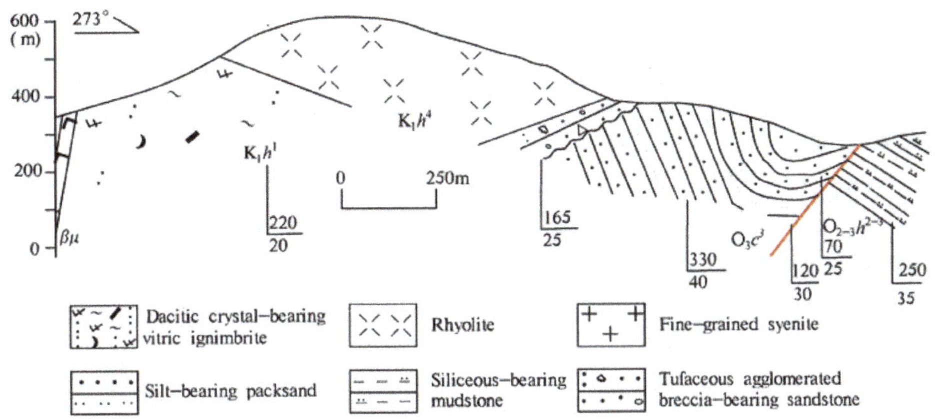

Fig. 3.79 Structural profile of Dashulin dome-shaped volcano

these indicate the controlling effect of the deep basement structure on volcanic apparatuses, constituting a normal-trapezoid volcanic basin. In general, in the survey area, volcanic structures feature migration from the central to east, north, and south, and the volcanism shows an evolution feature that is dominated by eruption in the early stage, extrusive overflow, eruptive spill, and explosive spill in the mid-stage, and eruption in the late stage.

3.3.6 Geochemical Characteristics and Tectonic Setting of the Volcanic Rocks

3.3.6.1 Geochemical Features

1. **Volcanic rocks**

In the survey area, volcanic rocks in the Huangjian Formation is high in SiO_2 content (66.68–76.90%), enrichment in alkali (Alk = $K_2O + Na_2O$, 7.15–9.93%) (Fig. 3.80), high in K_2O/Na_2O ratio (1.37–2.80%); low in MgO (0.24–1.14%), P_2O_5 (0.03–0.13%), and TiO_2 (0.14–0.45%); the DI index is high (81.04–93.21), similar to that of intrusive rocks. From the Member #1 to #4 of Huangjian Formation, the contents of SiO_2, K_2O, and TiO_2, and the K_2O/Na_2O ratio show a trend of gradual increase, and the contents of Al_2O_3, TFeO, MgO, CaO, and P_2O_5 show the trend of increase and then decrease. In general, Al_2O_3, CaO, MgO, FeO, TiO_2, P_2O_5, and Fe_2O_3 have a negative correlation with SiO_2, while K_2O has no clear correlation with SiO_2. All these features suggest that as differentiation evolution becoming more sufficient, lithologies evolve toward acidity and their alkalinity does not change greatly. A/CNK is 0.91–1.18, featuring evolution from metaluminous to weakly peraluminous (Fig. 3.81); Rittmann Index (σ) is 1.51–3.72 having the features of shoshonite series. TFeO/MgO ratio is 2.85–9.70, average value (6.17) is close to the I-type granite (2.27) (Whalen et al. 1987), but very different from A-type granite (13.4) (Turner et al. 1992).

Volcanic rocks are medium or high in $\sum$REE content (151.99×10^{-6}–296.83×10^{-6}), and the chondrite-normalized REE patterns show the feature of dipping rightward. The light and heavy rare-earths differentiate rather clear, La_N/Yb_N is 6.32–11.92 and δEu is 0.17–0.65, showing strong negative Eu anomaly. From early to late, the LREE and HREE differentiation weakens and the negative Eu anomaly gets slightly stronger. Enriched in K, Th, U, and Rb, strongly depleted in such LILE as Ba, and from early to late stage the depleted features weakens; weakly depleted in Nb and Ta, strongly depleted in HFSEs such as Sr, P, and Ti.

$(^{86}Sr/^{87}Sr)_i$ is 0.70029–0.70751, $\varepsilon Nd(t)$ is −6.24 to −4.69, and T_{DM2} is 1.31–1.43 Ga.

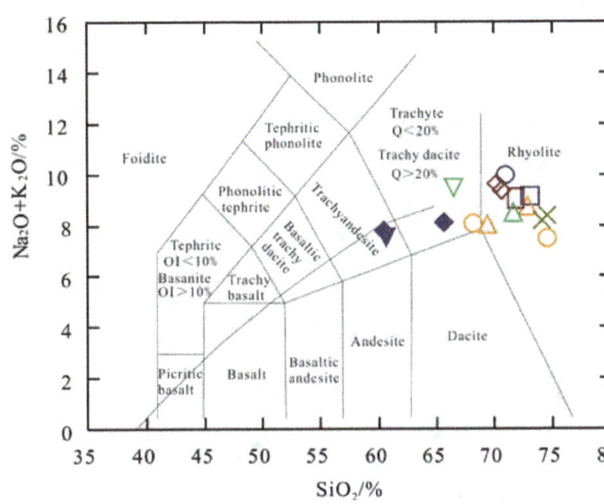

Fig. 3.80 SiO_2–$Na_2O + K_2O$ for volcanic rocks in the survey area

2. **Subvolcanic Rocks**

Subvolcanic rock is low in SiO_2 content (60.48–65.69%), enriched in alkali (Alk = $K_2O + Na_2O$, 7.46–8.09%), low in K_2O/Na_2O ratio (1.08–1.20%); high in MgO (1.44–2.55%), P_2O_5 (0.16–0.34%), and TiO_2 (0.58–0.83%); DI index is lower (67.26–76.94); A/CNK is 0.84–0.94, showing metaluminous. Rittmann Index (σ) is 2.88–3.41, and this means all have the features of shoshonite series.

Subvolcanic rock is high in content of $\sum$REE (202.73×10^{-6}–224.14×10^{-6}), and the chondrite-normalized REE patterns show a feature of weakly dipping toward right, light, and heavy rare-earth differentiate pretty obvious, La_N/Yb_N is 9.78–10.32 and δEu is 0.51–0.68, Eu showing medium negative anomaly. Enrichment in K, Th, U, and Rb, etc., and depletion in Ba, Sr, P, Nb, Ta, and Ti, etc., weaken compared with volcanic rock.

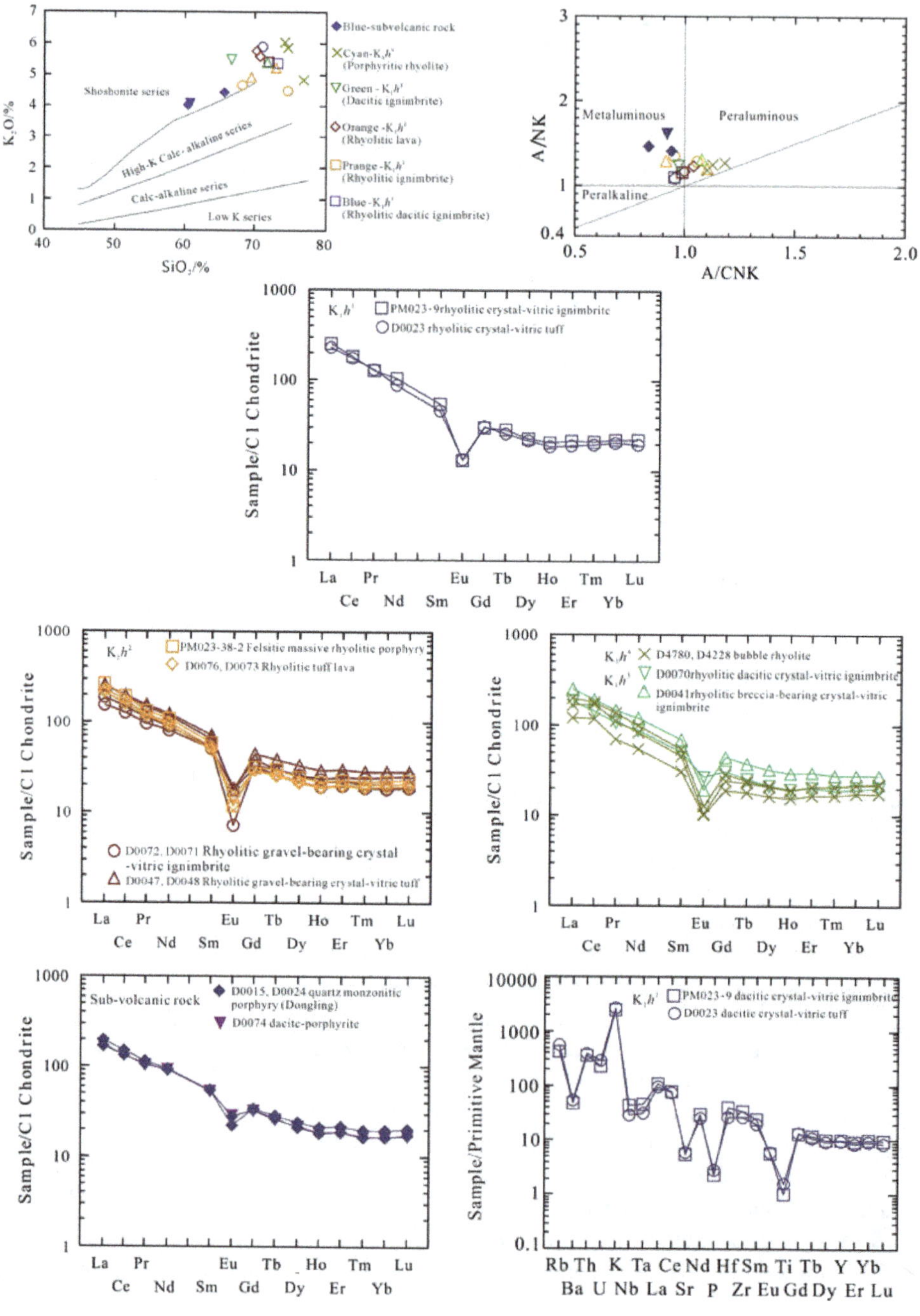

Fig. 3.81 REE chondrite-normalized diagram and trace element primitive mantle-normalized diagram for main volcanic rocks in the survey area

Fig. 3.81 (continued)

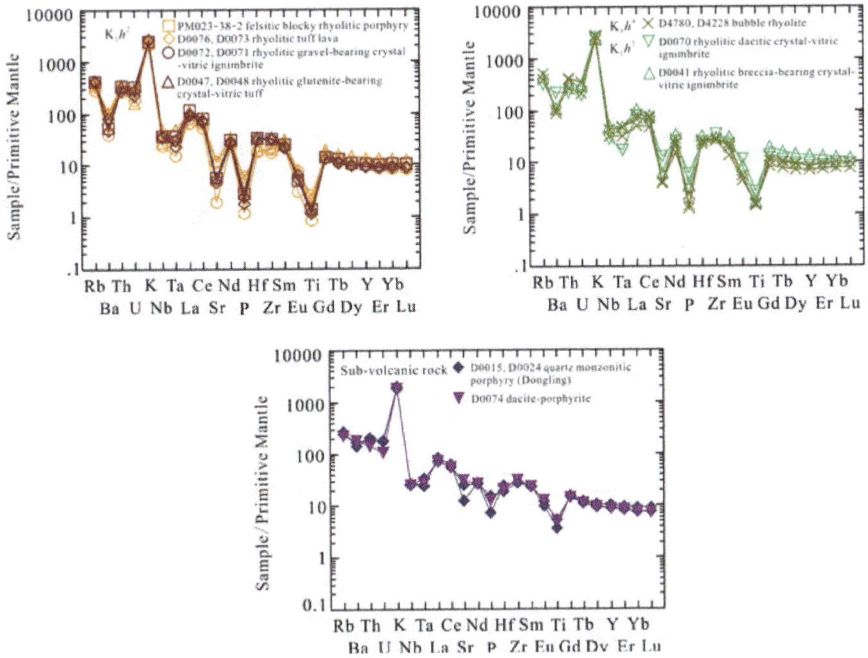

($^{86}Sr/^{87}Sr$) is 0.70733–0.70847, εNd(t) is −5.07 to −4.55, and T_{DM2} is 1.30–1.33 Ga, like other volcanic rocks in the survey area.

3.3.6.2 Magmatic Source and Tectonic Setting

Four eruptive rhythms can be divided into the survey area, based on the combination of volcanic lithologies and lithofacies as well as regional comparison. The geochemical features suggest that the early stage is a suite of shoshonite-series rhyolitic volcanic clastic rock and lava with high Si content (66.68–76.90%), low Sr (35.05×10^{-6}–229.7×10^{-6}), and strong negative Eu anomaly, the late stage is a suite of shoshonite-series volcanic rock with low Si (60.48–65.69%), relative high Sr (225.0×10^{-6}–551.4×10^{-6}) and medium negative Eu anomaly, and volcanic clastic rock and lava are relatively lower in content of Ti, Co, and Ni than subvolcanic rock, from early volcanic clastic rock and lava eruption to late intrusion of volcanic rock, the components between both change apparently from high Si, low Sr content and high differentiation to low Si, relatively high Sr and low differentiation (Fig. 3.82), which may reflect that differentiation evolution of magma in the upper horizon is lower than that in the bottom horizon, in Tianmu Mount, component zonation exists in magma chamber, highly differentiate magma concentrate at the top of magma chamber to be erupted early, and such component zonation in magma chamber is pretty universal in Late Mesozoic volcanic rock in southeast coastal regions (Xing et al. 2009).

For (sub) volcanic rock series, ($^{86}Sr/^{87}Sr)_i$ is 0.70029–0.70847 with average value of 0.70566; εNd(t) is −6.24 to −4.55, with average value of −5.05, and T_{DM2}(Nd) is 1.30–1.43 Ga; εNd(t) is significantly higher than εNd(t) (−8.12 to −9.06) of lithospheric mantle in the north margin of Yangtze block (Xue et al. 2009); the T_{DM2}(Nd) also significantly lower than the peak age (1.6–1.7 Ga) of crust-source-type granite of Yanshanian in South China (Li 1993) and the T_{DM2}(Nd) of basement metamorphic rocks in southwest of Yangtze block (2.0 Ga) and Cathaysian orogeny (1.8–2.2 Ga) (Che and Jiang 1999), indicating the magmatic source area may be the place where the lithospheric mantle of Yangtze block and ancient crust materials are mixed because there are clear crust–mantle interaction and addition of juvenile mantle materials in volcanic rock of Tianmu Mount.

Similar to the intrusive rock series and the volcanic-intrusive rock series in the survey area, intrusive (sub) volcanic rock series in the survey area belongs to the post-collision granite in the Rb-Y + Rb diagram (Fig. 3.45), and also largely belongs to the later orogeny and post-orogeny areas in the R_1–R_2 diagram. The (sub) volcanic-intrusive rock series are the product of intracontinent stretching after subduction orogeny in Cretaceous epoch, according to the geochemical and chronological features.

3.3.7 Volcanic Activity and Metallogenesis

In the survey area, volcanism and intrusion are developed well, and metallogenesis related to volcanic activity is much

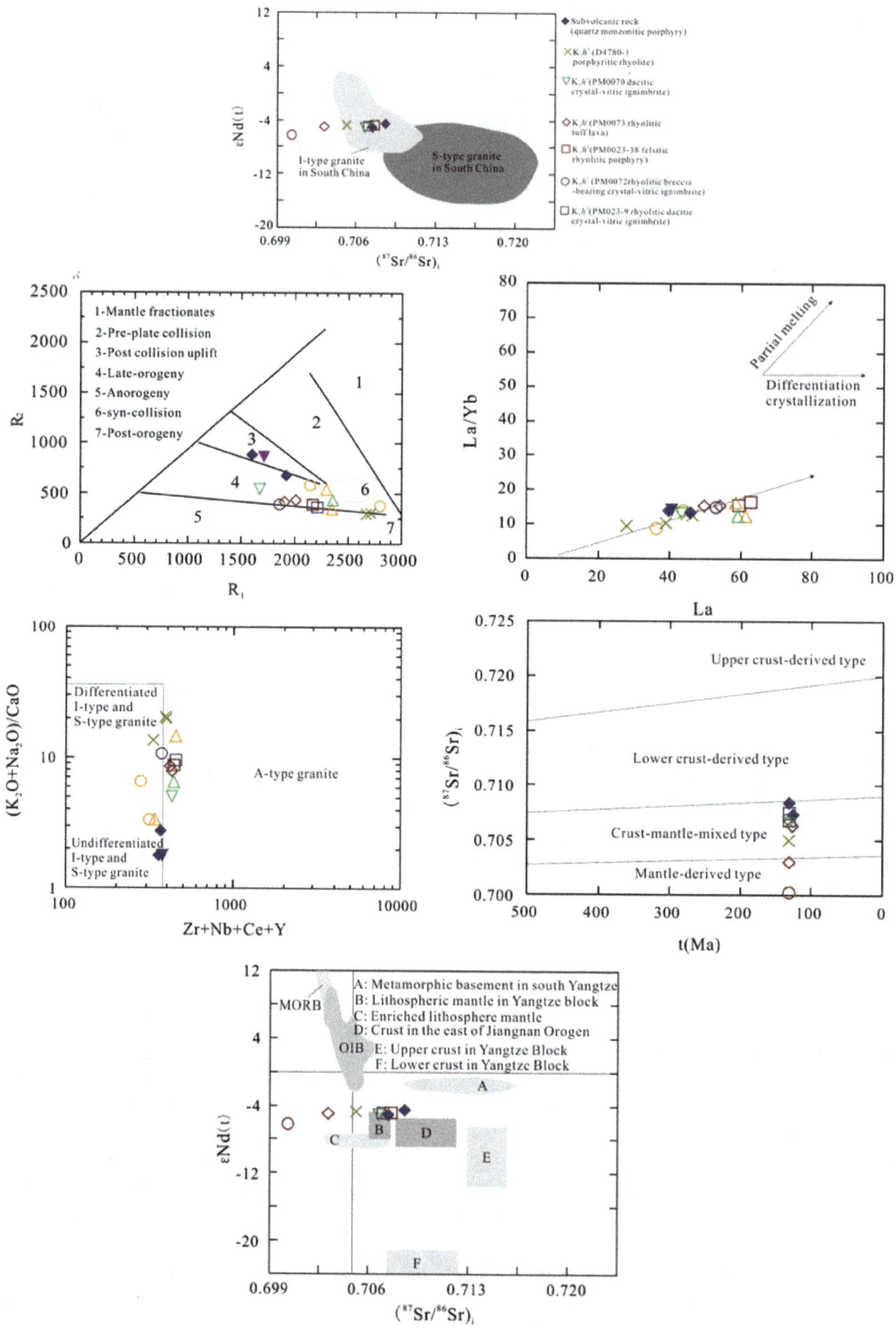

Fig. 3.82 Discrimination diagrams of volcanic rock types and tectonic setting in the survey area

weaker compared to intrusive activities. The reasons are that (1) volcanic activity mainly acts within the early depression basin where there is no wall rocks related to metallogenesis; (2) the product of volcanic eruption crystallizes fast, and there is no plenty of ore-forming fluids. Therefore, in the field survey, a handful of Zn–Pb and fluorite mineralization developed in fractures are casually seen. However, in the survey area, uranium ore occurrences are developed in rhyolitic porphyry of Mogan Mount, in the east periphery of the survey area. Volcanic structures and plenty of lava outcrops are developed, so in the future exploration and research should be focused on metallogenesis associated with volcanic structures. In addition, in the survey area, exogenous mineral resources associated with volcanic rock are mainly bentonite, kaolin, and clay minerals, which developed in ignimbrite and lava, in lamellar shape, and are worth prospecting.

3.4 Vein Rocks

3.4.1 Distribution Features

In the survey area, vein rocks are widely distributed, and areas for sedimentary strata, intrusive rocks and volcanic rocks have developed veins to different degrees, and there are various types of attitudes, mainly NW-strike and NW-strike, secondly NNE-strike, nearly SN strike and nearly EW strike, etc.; there are totally 194 veins, of which 59 are in sedimentary rock areas, 60 in intrusive rock areas, and 75 in volcanic rock area.

3.4.2 Rock Types

Veins have complex lithologies, followed by granite porphyry, granite, aplite, felsite, rhyolitic porphyry, andesite, andestic porphyrite, diorite, dacite, dacite–porphyrite, orthophyre, diabase, sillite, and lamprophyre, etc., which generally show the feature of transition from acid to neutral and basic. Main geological features of vein rocks are described in Table 3.4.

Granitic porphyry

Widely developed within Ma'anshan pluton, NE-strike intruded in monzonitic granite and syenogranite (Fig. 3.83), the dipstrike of the intrusive contact is 310°–330°∠50°–80°, borderline is straight, flat, and clear; veins are 1–10 km long and 10–100 m wide. Porphyritic texture, phenocryst unevenly distributed, mainly K-feldspar (3–10%), quartz (3–5%), plagioclase (3–5%), and a handful of biotite, generally in the grain size of 0.6–1.5 mm, less up to about 2 mm. Matrix: feldspar and quartz, mostly in the micro-grained shape, hidden graphic texture seen locally, generally in the grain size <0.5 mm.

Pegmatite

Granitic pegmatite, the vast distributed in Ma'anshan and Xianxia plutons, mainly lump and lentoid, the former diameter is just 50 cm, and the latter is generally not in a large scale, 30–50 cm long and 8–15 cm wide. The lentoid pegmatite in fine-grained syenogranite at Guocun transitions from coarse-grained and medium-grained syenogranite to fine-grained syenogranite, but there is no zonation on vein body itself, composed of pegmatitic-shaped Na-feldspar, quartz and a handful of biotite (Fig. 3.83), fine-grained syenogranite, feldspar quartz, and quartz zones can be seen in the outer contact zone of some vein bodies.

Dioritic porphyrite

Mostly inside Xianxia pluton, showing cyan-gray, a number of veins are arranged nearly in parallel and intruded as a high

Table 3.4 Main geological features of vein rocks in the survey area

Area	Lithology	Intrusive age	Lithological features
Ma'anshan	Granite porphyry	127.3 ± 1.8 Ma LA-ICP-MS Zircon U–Pb	NE-strike of 40°–60°, mostly dipping northwest, and a few dipping southeast, at the dip angle of 50°–80°; 1–10 km long and 10–100 m wide per vein. Porphyritic structure, phenocryst: quartz (10%), K-feldspar (10%), plagioclase (5%), and a handful of biotite, with grain size of 0.6–1.5 mm, and up to 2 mm for a few. Matrix: feldspar and quartz, mostly in the micro-gained shape, generally with the grain size <0.5 mm
Xianxia	Diabase	128.3 ± 2.4 Ma LA-ICP-MS Zircon U–Pb	Veins are prettily developed, granitic vein is the most common, vein strikes vary from NE to SW differently, more veins are NE-strike of 20°–60°, mostly dipping southeast, less dipping northwest, dip angle of 50°–80°; veins are mainly 1.5–3 m wide, a few up to 50 m wide; it is 0.2–1 km long per vein
West Tianmu	Diabase	130.3 ± 1.1 Ma LA-ICP-MS Zircon U–Pb	NW-strike or nearly EW strike intruded in porphyritic rhyolite, it is generally 10–100 m long, 0.1–1 m wide, dipping 0–10°, and at the dip angle of 75°–80°

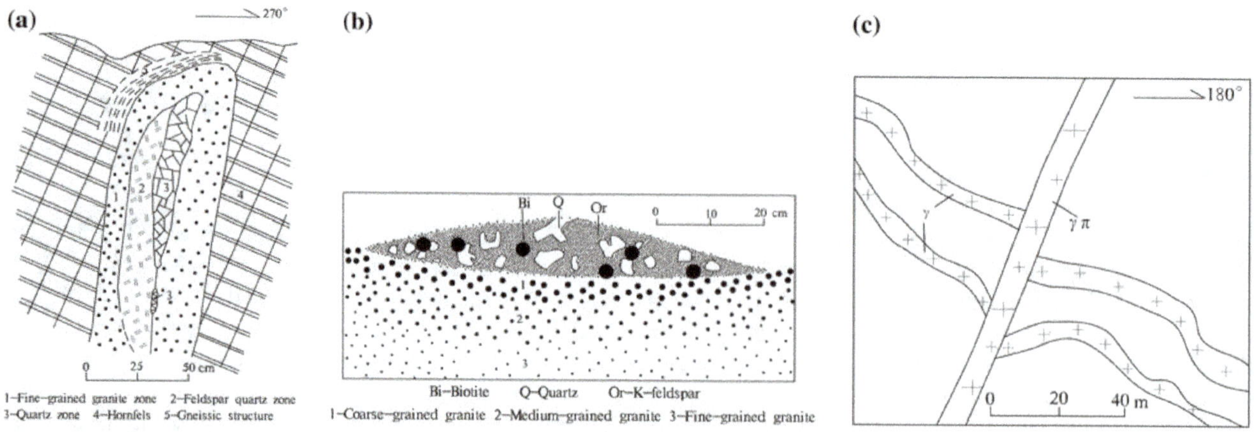

Fig. 3.83 **a**, **b** Outcropping sketch and zonation diagram of granite pegmatite inside Ma'anshan composite pluton; **c** fine-grained syenogranite vein cut off by granite porphyry inside Ma'anshan composite pluton

angle (15°∠75°) into monzonitic granite (Fig. 3.84). The rocks are strongly weathered, and spheroidal weathering can be seen. The weathering spheroid varies in size, generally 15 cm × 25 cm, in the shape of ellipse.

Diabase

Mostly developed in Xianxia pluton and volcanic rock, for instance, diabase vein in porphyritic rhyolite at West Tianmu Mount is about 10–100 m long, 10–100 cm wide and 10°∠85° in dipstrike. The diabase vein is wide in its east, and narrow in its west, and the contact surface is generally flat, straight, and clear. A very handful of plagioclase phenocryst is seen in rock locally, as with a few quartzs, and phenocryst is 1–3 mm in size. There are numerous pores on the vein surface, which are directionally arranged, in the size of 1–5 mm, with its dipstrike similar to the veins.

3.4.3 Geochronological Features

Zircon LA-ICP-MS U–Pb geochronological research is conducted on granitic porphyry vein (D0034) in Ma'anshan pluton, diabase veins in Xianxia pluton (D0013) and West Tianmu Mount porphyritic rhyolite (D4235). In granitic porphyry, zircon is mostly irregular long and short column, with the length of 50–200 μm, and the length-width ratio is about 3:1–1:1, showing obvious oscillatory zoning. In diabase,

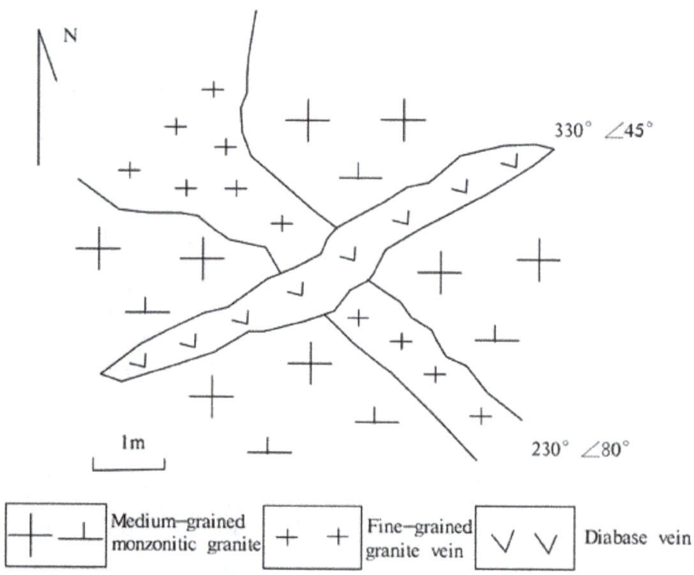

Fig. 3.84 Outcropping features of veins inside Xianxia composite pluton

①-Megacrystic-porphyritic monzonitic granite
②-Diorite porphyrite

zircon is mostly irregular long and short columns, with the length of 60–100 μm, the length-width ratio of 1.5:1–1:1, and also contains oscillatory zoning.

U content in zircon is 122×10^{-6}–836×10^{-6}, 83×10^{-6}–1717×10^{-6}, and 51×10^{-6}–336×10^{-6}, respectively, Th is 97×10^{-6}–349×10^{-6}, 11×10^{-6}–914×10^{-6}, and 8×10^{-6}–84×10^{-6}, respectively, and Th/U ratio is 0.54–1.62, 0.03–1.92, and 0.11–0.35, respectively, showing a typical feature of magmatic zircon. Sample locations mostly are projected on or near the concordant curve, 206Pb/238U weighted mean age is, respectively, 127.4 ± 1.8 Ma (MSWD = 0.77), 128.3 ± 2.4 Ma (MSWD = 2.8), and 130.3 ± 1.1 Ma (MSWD = 1.12) (Fig. 3.85), representing the overall intrusive age of veins in the survey area (130.3–127.4 Ma).

3.4.4 Geochemical Features and Origin of the Vein Rocks

1. **Geochemical Features**

Granitic porphyry

(1) The content of SiO_2, MgO, CaO, $Na_2O + K_2O$, P_2O_5, and TiO_2 is 76.74%, 0.21%, 0.40, 8.04%, 0.02%, and 0.07, respectively.
(2) Rittmann Index (σ) is 1.92, within the range of high K–Ca alkaline series.
(3) Peraluminous, A/CNK is 1.14.
(4) $\sum$REE is low, $\sum REE = 122.78 \times 10^{-6}$, LREE/HREE = 3.47, and the chondrite-normalized REE patterns show a feature of "V" shape. The light and heavy rare-earth differentiate unobviously, and Eu anomaly is obvious ($\delta Eu = 0.21$).
(5) The trace element spider diagram indicates enrichment in LILEs such as K, Th, U, and Rb, but strongly depleted in elements such as Ba; strongly depleted in HFSEs such as Sr, P, Nb, and Ti (Fig. 3.86).

Diabase

(1) The content of SiO_2, MgO, and CaO is 47.66–48.44%, 3.32–3.53%, and 5.41–6.19%, belonging to trachy basalt and tephrite in the TAS diagram, alkaline basalt in the Nb/Y–Zr/(10000TiO_2) diagram.
(2) $Na_2O + K_2O$ content is 5.49–6.87%, Rittman index (σ) is 5.54–10.13, within the area for shoshonite series.
(3) Metaluminous, A/CNK is 0.84–0.86, the content of P_2O_5 and TiO_2 is 0.41–0.54% and 2.14–2.23%.
(4) $\sum$REE is moderate, $\sum REE = 174.98 \times 10^{-6}$–$179.48 \times 10^{-6}$, showing enrichment in LREE, and LREE and HREE differentiate obviously, LREE/HREE = 7.93–7.95. The chondrite-normalized

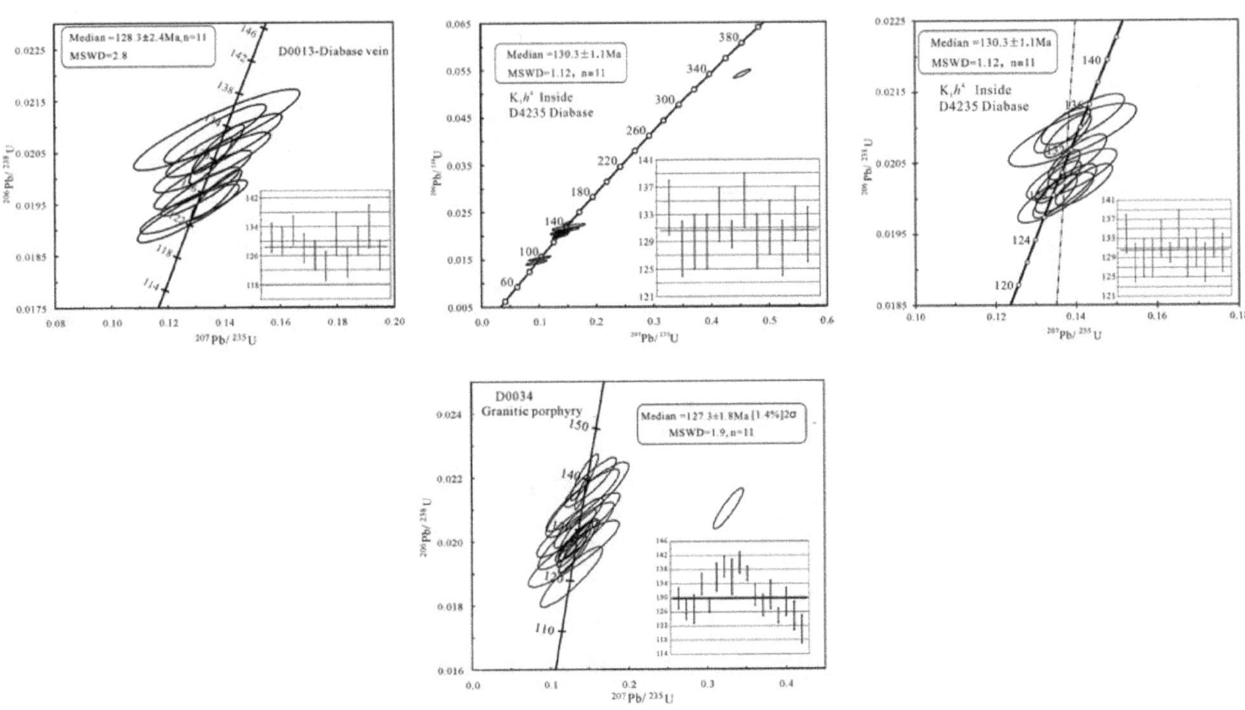

Fig. 3.85 Zircon U–Pb concordia diagrams and age histograms for vein rocks in the survey area

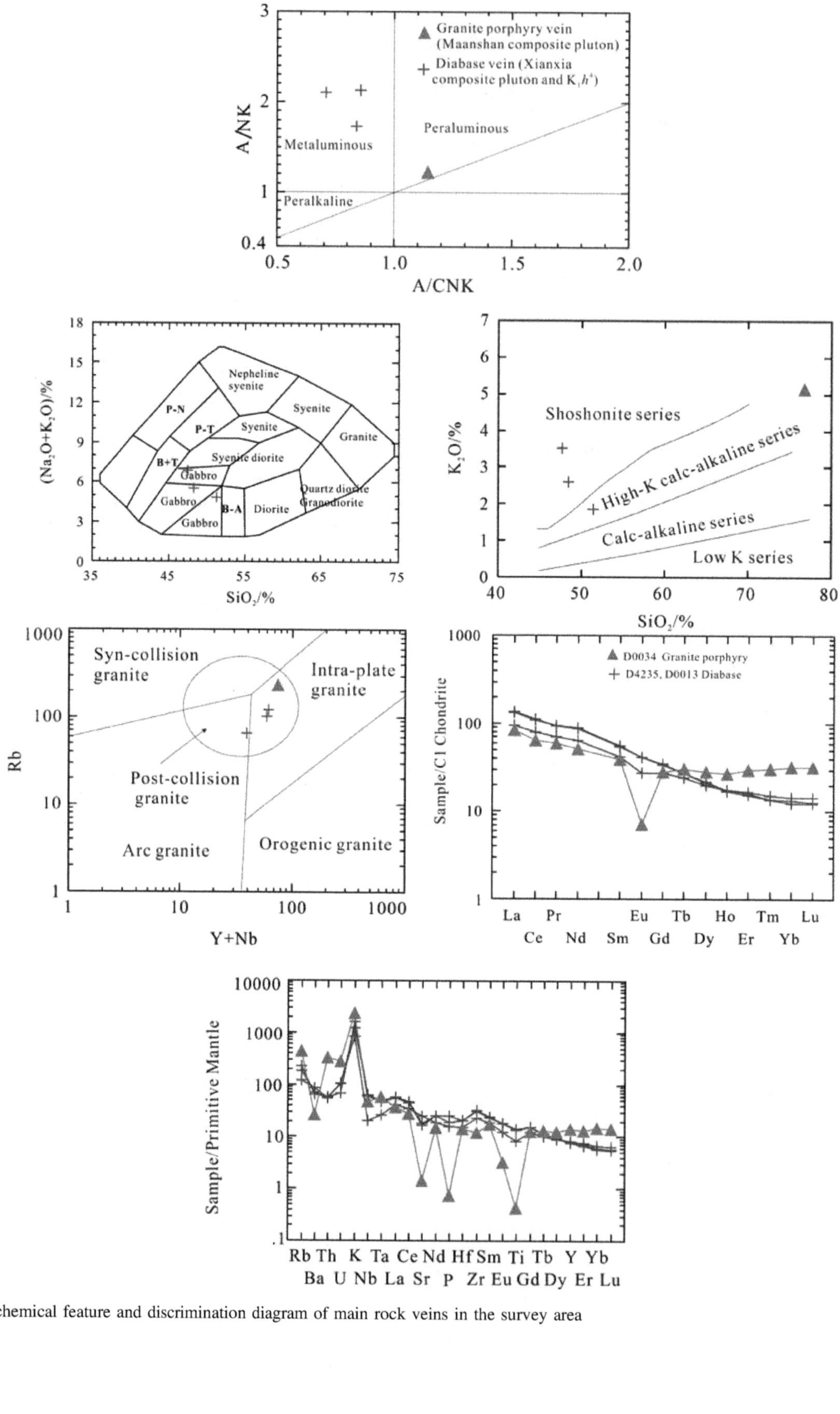

Fig. 3.86 Geochemical feature and discrimination diagram of main rock veins in the survey area

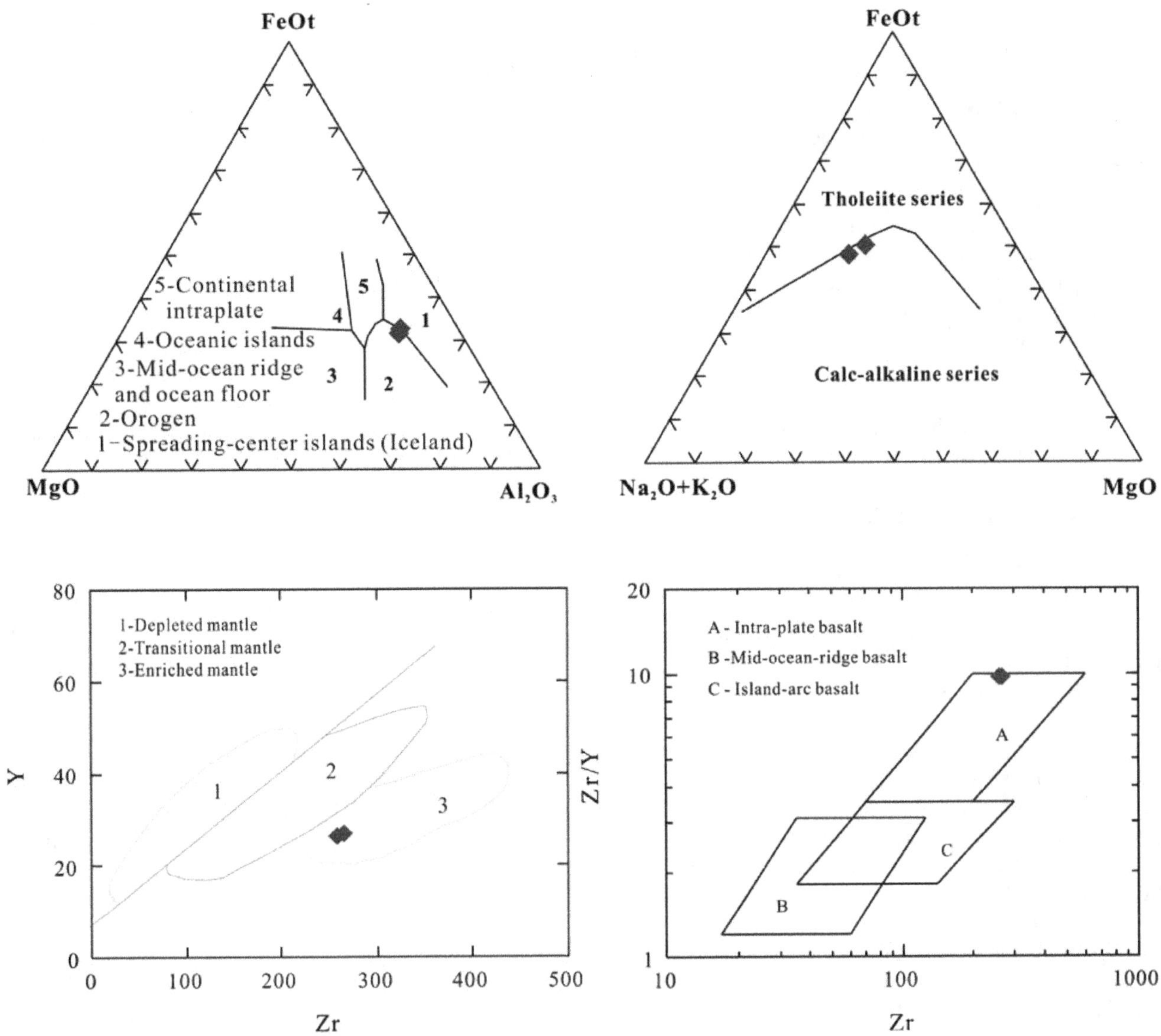

Fig. 3.87 Geochemical discrimination diagram of selective diabase veins in the survey area

REE patterns generally show the feature of dipping rightward, nearly no Eu anomaly (δEu = 0.91–0.95) and Ce anomaly (δCe = 0.95–0.96).

(5) The trace element spider diagram indicates enrichment in LILE such as Rb and K, and slightly weakly depleted in elements such as Ba, Th, U, and Sr.

2. **Origin of the vein rocks**

In the survey area, vein rocks formed close to or slightly later than the intrusion of volcanic rock and intrusive rocks, but all occurs in Early Cretaceous, and belong to the granitic area of post-collision plate in the Rb–Y + Rb diagram, products of intracontinent stretching after subduction orogeny in Cretaceous. Especially, diabase showing obvious relative HFSE deficit suggests that the enrichment of lithospheric mantle components is related to the modification of subducting slabs, and such enriching mantle-derived feature possibly records early modification of subducting slabs but also is the result of subduction of contemporaneous plates. The Zr/Y–Zr diagram shows that all diabases belong to the intraplate environmental area (Fig. 3.87), suggesting that diabase magma does not from island arc or active continental margin, and the enriching mantle-derived feature indicated by it possibly succeeds mainly the early modification record of subducting slabs.

Meanwhile, a series Early Cretaceous tensile red fault depression basins were developed in southeast China, indicating since the Early Cretaceous it had been regionally in the background of lithosphere expansion. In south China, from early to late stage (140–70 Ma) of the Late Mesozoic, diabase veins had been widespread for multiple periods, which also indicates the then background of intense crust tension, and such crust tension has the feature of episodic activity that is transitioned from local regions in the early stage (e.g., 140 Ma mainly happened in north Guangdong and west Fujian) to holistic tension in the late stage (since 94 Ma). Diabase vein rocks formed under the expansional background are widely distributed in west Fujian, coastal areas of Fujian, Hainan, Jiangxi, and Zhejiang, etc., a product from the joint melting of lithospheric mantle and convective asthenosphere.

3.5 Comparison Between Intrusion and Volcanism

3.5.1 Time and Space Relationship

According to the patterns of magmatism, magmatite in the survey area is divided into three series: intrusive rock, volcanic-intrusive rock, and (sub)volcanic rock, and their metallogenic time limits are 145.1–125.0 Ma, 130.5–127.7 Ma, and 135.2–125.4 Ma, respectively, all in the beginning of Early Cretaceous. Intrusive rocks are mainly developed at the margins of NE-strike faults, folds, and volcanic depressions, Ma'anshan pluton is mainly under control of the NE-strike, Tongkengcun–Qiguancun fault and Tangshe pluton are mainly under control of the NE-strike, Wangjia-Tangshecun anticline fold while Xianxia, Tonglizhuang, and Wushanguan plutons intruded around the margin of volcanic depressions, and their formation is closely related to basin-control structures in volcanic depressions. Volcanic-intrusive rock is mainly developed at the margin of early intrusive rock and volcanic depressions as well as in volcanic rocks; volcanism–subvolcanism is mainly under control of volcanic depressions. Overall, in the survey area, intrusion is a little earlier than volcanism, featuring migration from west to east, and there is close time-space relationship in terms of rock-control and basin-control structures.

3.5.2 Origin Relationship

In the survey area, product of intrusive rock is mainly acidic, followed by monzonitic granite–syenogranite; the volcanic-intrusive product is mainly intermediate acidic, followed by quartz diorite–quartz syenite–quartz monzonite–syenogranite; (sub) volcanic product is mainly intermediate acid–neutral, followed by rhyolitic dacitic-rhyolitic ignimbrite, lava, quartz monzonitic porphyry, and rhyolitic porphyry, etc. Evolution processed showing as acidic → intermediate acid → neutral and metaluminous → weakly peraluminous → (maluminous → metaluminous) from early to late stage; products of these three rock types show the feature of high differentiation evolution, and only the DI index of the volcanic rock is lower. The detailed characteristics of the Early Cretaceous magmatic rocks is listed in Table 3.5.

3.5.3 Comparison with Magmatite in Neighboring Areas

Based on the reliable dating data in recent 10 years, it is known that Yanshanian intrusive rocks in northwestern Zhejiang and neighboring areas were mainly formed in Middle Jurassic–Early Cretaceous (168–124 Ma), concentrated in three stages: 168–163 Ma, 150–147 Ma, and 142–124 Ma. The age trends to become young toward northeast, and in Late Cretaceous, small-scale magmatic activities occurred at 118–115 Ma locally.

In general, rocks were formed earlier in western Zhejiang, for instance, granodiorite at Tongcun formed at 162.1 Ma (Zhu et al. 2014), and the magmatic source is dominated by the lower crust, and crystallization differentiates weakly. However, in plutons located in neighborhoods in Zhejiang and Anhui, the mantle-derived materials increase, which belongs to crust-mantle-mixed area where crystallization differentiates strongly (Fig. 3.88). From Middle and Late Jurassic to Early Cretaceous, the tectonic setting of intrusive rocks evolved from collision-arc granite to intraplate granite. The diversity and changes in magmatic source also demonstrated the changes in subduction angle of paleo-Pacific-plate, transformation from subduction and extrusion orogeny in the Middle–Late Jurassic to expansion and tension in the beginning of Early Cretaceous, which resulted in different contributions of lithospheric mantle of Yangtze block and crust material source areas in Jiangnan Orogen, and differentiation of magma crystallization.

For volcanic basins developed in Late Mesozoic in northwestern Zhejiang and its surrounding, the main body of its volcanic rock strata is the lower volcanic rock series, and a handful of upper volcanic rock series, and bordering with the nearly EW-strike Changhua-Hangzhou Fault, it can be divided into two volcanic rock areas: Shunxi-Huzhou and Changshan-Tonglu. In Shunxi-Huzhou volcanic rock area, outcropped volcanic rock is mainly Huangjian Formation (K_1h) and a handful of Laocun Formation (K_1l) of lower volcanic rock series Jiande Group, and on the basis of zircon U–Pb age analysis on volcanic basins of Tianmu Mount, Huangjian Formation is erupted in 135–125 Ma, the beginning of Early Cretaceous. In Changshan–Tonglu volcanic rock area outcropped strata are mainly Laocun Formation (K_1l), Huangjian Formation (K_1h), Shouchang Formation

Table 3.5 Feature comparison of magmatism of the Early Cretaceous in the survey area

Series	Intrusive rock series	Volcanic-intrusive rock series	(Sub) volcanic rock series
Location and rock-control structure	Paleozoic stratum area; NE-strike fault, folds, and the margin of volcanic depressions	Margin of early intrusive rocks and volcanic depressions, as well as inside of volcanic lava	Inside of volcanic depressions, and fracture at the basin margin
Lithological combination	Monzonitic granite – syenogranite	Quartz diorite–quartz syenite–quartz monzonite–syenogranite	Ignimbrite, lava, quartz monzonitic porphyry and rhyolitic porphyry
Major elements geochemical features	① Acid ② Metaluminous ③ High K calc-alkaline series ④ Low → high differentiation	① Intermediate acid ② Weakly peraluminous ③ Shoshonite series ④ Moderate differentiation	① Intermediate acid → acid ② Weakly peraluminous ③ Shoshonite series ④ Low-high differentiation
Trace element geochemical features	Medium and high ∑REE, weakly dipping-right "V"-shape rare-earth curve, negative Eu anomaly. Enrichment in K, Th, U, and Rb, and deficit in Ba, Sr, P, Nb, Ta, and Ti, etc. From monzonitic granite to syenogranite, ∑REE, Rb content, and negative Eu anomaly trend to increase, and $(La/Yb)_N$ ratio, Sr/Y ratio and Sr content trend to decrease while deficit in Ba, Nb, Ta, Sr, P, Eu, and Ti, etc., trend to enhance	Medium and high ∑REE, weakly dipping-right "V"-shape rare-earth curve, negative Eu anomaly. Enrichment in K, Th, U, and Rb, and loss in Ba, Sr, P, Nb, Ta, and Ti, etc. From quartz diorite to syenogranite, ∑REE, Rb content, and negative Eu anomaly trend to increase, and $(La/Yb)_N$ ratio, Sr/Y ratio, and Sr content trend to decrease while deficit in Ba, Nb, Ta, Sr, P, Eu, and Ti, etc., trend to enhance	Medium and high ∑REE, weakly dipping-right rare-earth curve, negative Eu anomaly. Enrichment in K, Th, U, and Rb, and deficit in Ba, Sr, P, Nb, Ta, and Ti, etc. From volcanic rock to subvolcanic rock, ∑REE, Rb content, negative Eu anomaly, and loss in Ba, Nb, Ta, Sr, P, and Ti, as well as $(La/Yb)_N$ ratio, all trend to lower
Type of formation causes	I-type granite	I-type granite	I-type granite
Magmatic source	Mainly remodification of ancient crust materials, and a handful of lithospheric mantle materials of Yangtze block	Mixture of materials in lithospheric mantle of Yangtze block and ancient crust	Mixture of materials in lithospheric mantle of Yangtze block and ancient crust
Magmatic evolution	Differentiation crystallization → partial melting	Partial melting	Partial melting
Formation time limit (Ma)	145.1–125.0	130.5–127.7	135.2–125.4
Metallogenesis	Skarn-type Zn-Fe polymetallic and fluorite; quartz fine-vein tungsten; hydrothermal vein-type fluorite, Sb		Volcanic-subvolcanic hydrothermal Zn-Pb, fluorite mineralization, Uranium?

(K_1s), and Hengshan Formation (K_1hs) of lower volcanic rock series Jiande Group, and a handful of Zhongdai Formation (K_2z) of upper volcanic rock-system Qujiang Group, and in Jiande Group, the zircon U-Pb age is 134–115 Ma (Li et al. 2011). Overall, the period of lower volcanic rock series in Tianmu Mount and Shouchang, Jiande, northwest Zhejiang, is similar to the time limit for eruption of volcanic rock in its northwest neighborhood area, the middle and lower Yangtze River, (which is 135–123 Ma in Early Cretaceous) (Zhou et al. 2011).

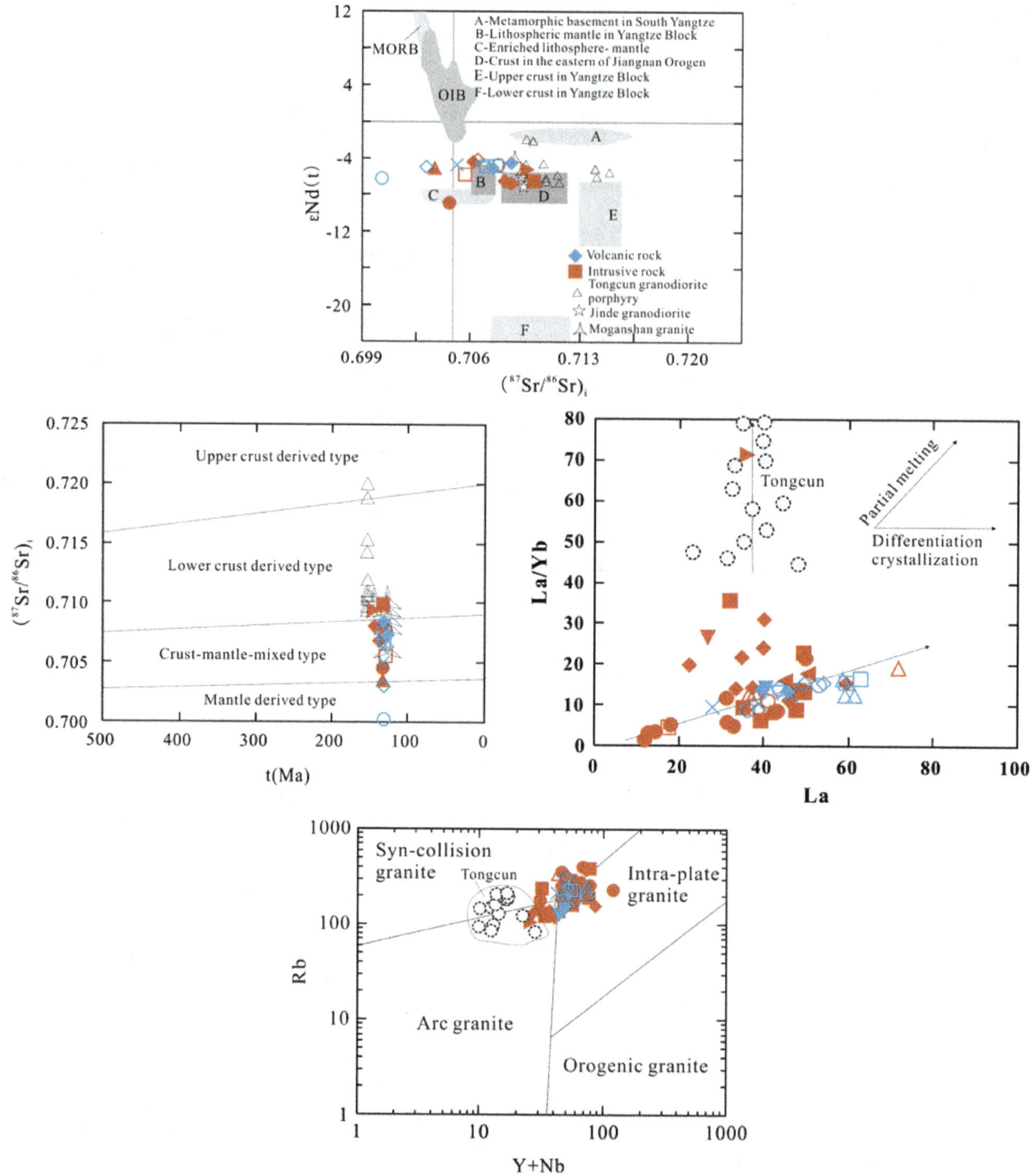

Fig. 3.88 Geochemical features and discrimination diagrams for magmatite in the survey area and neighboring areas

In Zhejiang Province, the lower and upper volcanic rock series can be divided into three volcanic activity cycles, the lower volcanic rock series corresponds to cycle #I and #II while the upper volcanic rock series corresponds to cycle #III (Yu et al. 2001; Yan et al. 2005). According to existing statistic chronological data, the period of the eruption of the lower volcanic rock series in Zhejiang is the beginning of Early Cretaceous (Fig. 3.89). In northwest Zhejiang,

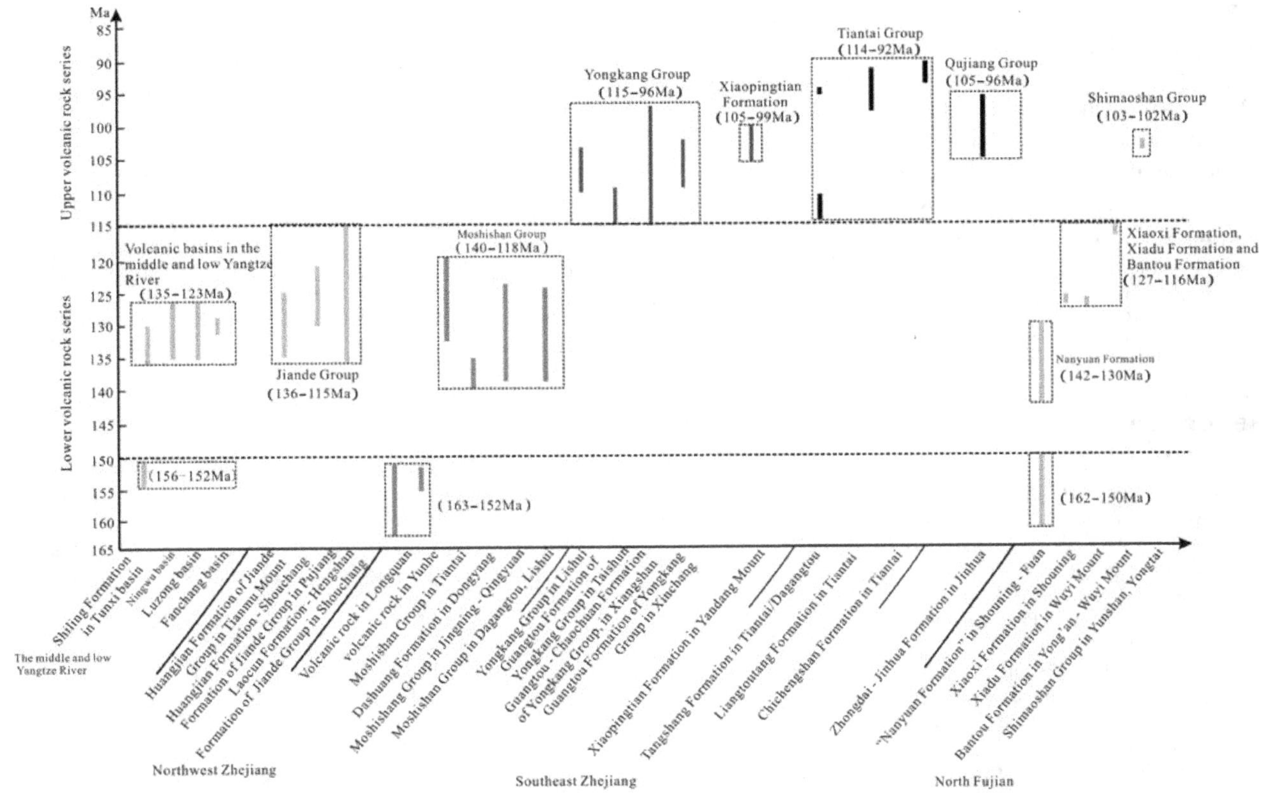

Fig. 3.89 Distribution of eruption time limits of volcanic rocks in Zhejiang and neighboring areas

volcanic rock horizons are Laocun Formation and Huangjian Formation of Cycle #I, and Shouchang Formation and Hengshan Formation of Cycle #II, of Jiande Group, for which the time limit of eruption concentrates at 136–115 Ma. In southeast Zhejiang, volcanic rock horizons are Dashuang Formation, Gaowu Formation and Xishantou Formation of Cycle #I, and Chawan Formation, Jiuliping Formation, and Zhucun Formation of Cycle #II, of Jiande Group, for which the time limit of eruption concentrates at 140–118 Ma. Volcanic rocks of Yongkang Group, Qujiang Group, and Tiantai Group of the upper volcanic rock series are erupted in 115–92 Ma, the end of Early Cretaceous–the beginning of Late Cretaceous.

In Fujian Province, of the lower volcanic rock series of Late Mesozoic, Nanyuan Formation volcanic rock is the most widely distributed rock types, with large era span, i.e., the end of Late Jurassic–the beginning of Early Cretaceous (Regional Geology in Fujian 2011), and the representative age of volcanic rock is 162.3–149.8 Ma (Late Jurassic) and 142.3–130.1 Ma (Early Cretaceous) (Xing et al. 2008), and the age of volcanic rock in Xiadu Formation, Bantou Formation, and Xiaoxi Formation of the lower volcanic rock series is 127–116 Ma; the age of volcanic rock in Shimaoshan Group of the upper volcanic rock series is 103–102 Ma. Regarding the tectonic setting of Late Jurassic volcanic rock in Nanyuan Formation, previous research determined that it formed under a different tectonic setting compared that under which the Early Cretaceous volcanic rock and both are products, respectively, before and after the end of great transformation of the tectonic regime in later Mesozoic (Xing et al. 2008; Li et al. 2009), and it is unsuitable to include the Late Jurassic volcanic rock in the lower volcanic rock series (Duan et al. 2013). Recently, some Late Jurassic volcanic rock strata were also discovered in Zhejiang and Anhui, for instance, Shiling Formation in Tunxi, south Anhui (154.7 Ma, Yu et al. 2016; 156–152 Ma, Tang et al. 2016), Huangjian Formation in Shouchang, north Zhejiang (150.9 Ma, Li et al. 2011), Dashuang Formation in Dagangtou, Lishui, southeast Zhejiang (155–152 Ma, Zhejiang Institute of Geological Survey 2013), and in Longquan (162.7–148.7 Ma, Zhejiang Institute of Geological Survey 2017). All of these ages indicate a volcanic eruption event happened in South China in Late Jurassic (163–150 Ma) but with a small scale, which may be a prelude of a large-scale volcanic activities of Early Cretaceous, or product of volcanic activities under different tectonic settings, which should be further researched in the future.

By combining comparative analysis on regional strata and chronology, it is known that Late Mesozoic volcanic activity in Zhejiang and Fujian should begin in 162 Ma (Mid to Late Jurassic); the Early Cretaceous lower volcanic rock series erupts in some 142–118 Ma in large scale, reaching peak in 135–125 Ma, which is a set of high K calc-alkaline volcanic eruptive rocks, most of which are intermediate acidic–acidic. While the upper volcanic rock series volcanic activities weaken significantly in 115–92 Ma, a set of basalt-rhyolite dual-peak combination and red layer sedimentary rocks; the regional unconformable surface between the lower and upper volcanic rock series (i.e., Minzhe movement) should have formed in about 118–115 Ma.

References

Blevin PL (2003) Metallogeny of granitic rocks. Gemoc Macquarie University. In: Magmas to mineralisation: The Ishihara symposium. Geoscience Australia, Canberra, Australia, pp 1–4

Chen JF, Jiang BM (1999) Nd, Sr, Pb isotope tracing and crustal evolution in southeastern China. In: Zheng YF (ed) Chemical geodynamics. Science Press, Beijing, pp 262–287 (in Chinese with English abstract)

Chen PR, Hua RM, Zhang BT et al (2002) Early Yanshanian Post-Orogenic Granitoids in the Nanling Region: petrological Constraints and geodynamic settings. Sci China (Ser D) 45 (8):756–768

Dong SW, Zhang YQ, Long CX, Yang ZY, Ji Q, Wang T, Hu JM, Chen XH (2007) Jurassic tectonic revolution in China and new interpretation of the Yanshan Movement. Acta Geol Sin 81 (11):1449–1461 (in Chinese with English abstract)

Duan Z, Xing GF, Yu MG, Zhao XL, Jin GD, Chen ZH (2013) Time sequence and geological process of Late Mesozoic volcanic activities in the area of Zhejiang—Fujian Boundary. Geol Rev 59 (3):454–469 (in Chinese with English abstract)

Feng ZH, Wang CZ, Wang BH (2009) Granite magma ascent and emplacement mechanisms and their relation to mineralization process. J Guilin Univ Technol 29(2):183–194 (in Chinese)

Fujian Institute of Geological Survey (2011) Fujian Regional Geology (in Chinese). National Geological Archives of China [distributor], 2018. http://www.ngac.org.cn/Data/FileList.aspx?MetaId=43FBD60C736B3FF1E05341015A0AECA0&Mdidnt=d00134772

Gao S, Qiu Y M, Ling W L, McNaughton NJ, Groves DI (2001) SHRIMP single zircon U–Pb dating of the Kongling high-grade metamorphic terrain: Evidence for >3.2 Ga old continental crust in the Yangtze craton. Sci China (Ser D), 44(4):326–335

Gao YF, Yang ZS, Santosh M et al (2010) Adakitic rocks from Slab melt-modified mantle sources in the continental collision zone of Southern Tibet. Lithos 119(3–4):651–663

Guo ZF, Wilson M, Liu JQ (2007) Post-collisional adakites in South Tibet: products of partial melting of subduction-modified lower crust. Lithos 96(1–2):205–224

Hoskin PWO, Schaltegger U (2003) The Composition of Zircon and Igneous and Metamorphic Petrogenesis[J]. Rev Miner Geochem 53:27–62

Hou ZQ, Gao YE, Qu XM et al (2004) Origin of adakitic intrusives generated during mid-Miocene east–west extension in Southern Tibet. Earth Planet Sci Lett 220(1–2):139–155

Jiang YH, Zhao P, Zhou Q et al (2011) Petrogenesis and tectonic implications of Early Cretaceous S- and A-type granites in the northwest of the Gan-Hang rift, SE China. Lithos 121:55–73

Li XH (1993) Geochronological framework and isotope system constraints of crustal growth and tectonic evolution in South China. Bull Mineral Petrol Geochem 3:113–115 (in Chinese with English abstract)

Li ZN, Quan H, Li ZT et al (2003) Mesozoic, Cenozoic igneous rocks and their deep processes in eastern China. Geological Publishing House, Beijing, pp 1–351(in Chinese)

Li XH, Li ZX, Li WX et al (2007) U–Pb zircon, geochemical and Sr–Nd–Hf isotopic constraints on age and origin of Jurassic I and A-Type granites from Central Guangdong, SE China: a major igneous event in response to foundering of a subducted flat–slab? Lithos 96:186–204

Li LM, Sun M, Xing GF et al (2009) Two late Mesozoic volcanic events in Fujian Province: constraints on the tectonic evolution of southeastern China. Int Geol Rev 51:216–251

Li XH, Chen SD, Luo JH, Wang Y, Cao K, Liu L (2011) LA-ICP-MS U-Pb isotope chronology of the single zircons from early Cretaceous Jiande Group in Western Zhejiang, SE China: significances to stratigraphy. Geol Rev 57(6):825–836 (in Chinese with English abstract)

Li ZL, Zhou J, Mao JR, Yu MG, Li YQ, Hu YZ, Wang HH (2013) Age and geochemistry of the granitic porphyry from the northwestern Zhejiang Province, SE China, and its geological significance. Acta Petrol Sinica 29(10):3607–3622 (in Chinese with English abstract)

Ma CQ, Yang KG, Tang ZH et al (1994) Magma dynamics of granite—theory and methodology as well as analysis of granite in East Hubei. China University of Geosciences Press, Wuhan (in Chinese)

Mao JR, Li ZL, Ye HM (2014) Mesozoic tectono–magmatic activities in South China: retrospect and prospect. Sci China Earth Sci 57:2853–2877

Rui ZY, Zhang HT, Li N, Wang LS (1991) On the multiple emplacement of the granitic magma and the multiple minerogenetic models. Acta Petrol Mineral 10(2):97–103 (in Chinese with English abstract)

Sun SS, McDonough WF (1989) Chemical and isotopic systematics of oceanic basalts: implication for mantle composition and processes. In: Saunder AD, Norry MJ (eds) Magmatism in the ocean basins. Geol Soc Spec Publ 42:313–345

Tang S, Xu XB, Yuan YM (2016) Geochemistry and geochronology of the volcanic rocks from Tunxi basin in southern Anhui and their tectonic significance. Acta Petrol Mineral 35(2):177–194 (in Chinese with English abstract)

Turner SP, Foden JD, Morrison RS (1992) Derivation of some A-Type magmas by fractionation of basaltic magma: an example from the Padthaway Ridge, South Australia. Lithos 28:151–179

Vavra G, Schmid R, Gebauer D (1999) Internal morphology, habit and U–Th–Pb microanalysis of amphibolite-to-granulite facies zircons: geochronology of the Ivrea Zone (Southern Alps). Contrib Miner Petrol 134:380–404

Wang YL, Wang Y, Zhang Q, Jia XQ, Hang S (2004) The geochemical characteristics of Mesozoic intermediate-acid intrusives of the Tongling area and its metallogenesis-geodynamic implications. Acta Petrol Sinica 20(2):325–338 (in Chinese with English abstract)

Wang XL, Shu XJ, Xu XS et al (2012) Petrogenesis of the early Cretaceous adakite–like porphyries and associated basaltic andesites in the eastern Jiangnan orogen southern China. J Asian Earth Sci 61:243–256

Wang YJ, Fan WM, Zhang GW et al (2013) Phanerozoic tectonics of the South China Block: key observations and controversies. Gondwana Res 23:1273–1305

Whalen JB, Currie KL, Chappell BW (1987) A-Type granites: geochemical characteristics, discrimination and petrogenesis. Contrib Mineral Petrol 95:407–419

Wu LS (1998) Type of magmatic emplacement in relation to mineralization in Jiujiang Ruichang District, Jiangxi Province. Min Deposits 17(1):36–45 (in Chinese with English abstract)

Xia LQ, Xia ZC, Xu XY, Li XM, Ma ZP, Wang LS (2004) Carboniferous Tianshan igneous megaprovince and mantle plume.

Reg Geol China 23(9–10):903–910 (in Chinese with English abstract)

Xing GF, Lu QD, Chen R, Zhang ZY, Nie TC, Li LM, Huang JL, Lin M (2008) Study on the ending time of Late Mesozoic tectonic regime transition in South China——comparing to the Yanshan Area in North China. Acta Geol Sin 82(4):451–463 (in Chinese with English abstract)

Xing GF, Chen R, Yang ZL, Zhou YZ, Li LM, Jiang Y, Chen ZH (2009) Characteristics and tectonic setting of Late Cretaceous volcanic magmatism in the coastal Southeast China. Acta Petrol Sinica 25(1):77–91 (in Chinese with English abstract)

Xue HM, Wang YG, Ma F, Wang C, Wang DE, Zuo YL (2009) The Huangshan A-type granites with tetrad REE: constraints on Mesozoic lithospheric thinning of the southeastern Yangtze craton? Acta Geol Sin 83(2):247–259 (in Chinese with English abstract)

Yan TZ, Yu YW, Chen JF, Xu XM, Wang JG, Cai ZH, Dong YH (2005) Nd–Sr isotope features of Cretaceous volcanic rocks in northwestern Zhejiang. Chin Geol 32(3):417–423 (in Chinese with English abstract)

Yu YW, Xu BT, Chen JF, Dong CW (2001) Nd isotopic systematics of the Late Mesozoic VOLCANIC Rocks from Southeastern Zhejiang Province, China: implications for stratigraphic study. Geol J China Univ 7(1):62–69 (in Chinese with English abstract)

Yu XQ, Chen ZW, Liu X, Ji X, Zhou SZ, Yang XP (2016) LA-ICP-MS zircon age of volcanic rock of Shiling Formation in Tunxi area, south Anhui, and redetermination of its epoch. Geol Bull China 35 (1):175–180 (in Chinese with English abstract)

Zhang HF, Goldstern SL, Zhou XH et al (2008) Evolution of subcontinental lithospheric mantle beneath eastern China: Re–Os isotopic evidence from mantle xenoliths in Paleozoic kimberlites and Mesozoic basalts. Contrib Mineral Petrol 155:271–293

Zhang JJ, Shi GH, Tong GS, Zhang ZY, Liu H, Wu RT, Chen L (2009) Geochemistry and geochronology of copper and polymetal-bearing volcanic rocks of the erhuling formation in Xujiadun, Zhejiang Province. Acta Geol Sin 83(6):791–799 (in Chinese with English abstract)

Zhang BT, Wang KX, Ling HF, Wu JQ (2012) Zircon U–Pb and whole-rock Rb–Sr chronology, Sr–Nd–O isotopes and petrogenesis of the mogaoshan granite pluton in the Zhejiang Province. Bull Mineral Petrol Geochem 31(4):347–353 (in Chinese with English abstract)

Zhejiang Institute of Geological Survey (2013) The regional geological survey of Shuangxi Town, Lishui City, Dagangtou Town and Zhangcun map sheets on a scale of 1,50,000 in Zhejiang (in Chinese). National Geological Archives of China [distributor], 2018. http://ngac.org.cn/Data/FileList.aspx?MetaId=43FBD60C7B413FF1E05341015A0AECA0&Mdidnt=d00135959

Zhejiang Institute of Geological Survey (2017) The regional geological and mineral resources survey of Xuanhu and Ruiyang map sheets on a scale of 1,50,000 in Zhejiang (in Chinese)

Zheng JM, Xie GQ, Chen MH, Wang SM, Ban CY, Du JL (2007) Pluton emplacement mechanism constraint on skarn deposit: a case study of skarn Fe deposits in Handan-Xingtai area. Min Deposits 26 (4):481–486 (in Chinese with English abstract)

Zheng YF, Xiao WJ, Zhao GC (2013) Introduction to tectonics of China. Gondwana Res 23:1189–1206

Zhou TF, Fan Y, Yuan F, Zhang LJ, Ma L, Qian B, Xie J (2011) Petrogensis and metallogeny study of the volcanic basins in the Middle and Lower Yangtze metallogenic belt. Acta Geol Sin 85 (5):712–730 (in Chinese with English abstract)

Zhou J, Jiang YH, Zeng Y, Ge WY (2013) Zircon U–Pb age and Sr, Nd, Hf isotope geochemistry of Jingde pluton in eastern Jiangnan orogen, South China. Geol China 40(5):1379–1391 (in Chinese with English abstract)

Zhou J, Jiang YH, Ge WY (2014) High Sr/Y jingde pluton in the Eastern Jiangnan orogen, South China: formation mechanism and tectonic implications. Acta Geol Sin 88(1):53–62 (in Chinese with English abstract)

Zhu DC, Mo XX, Wang LQ et al (2009) Petrogenesis of highly fractionated I–Type granites in the Chayu Area of Eastern Gangdese, Tibet: constraints from Zircon U–Pb geochronology, geochemistry and Sr–Nd–Hf isotopes. Sci China (Ser D: Earth Sci) 52(9):1223–1239

Zhu G, Niu ML, Xie CL et al (2010) Sinistral to normal faulting along the Tan–Lu Fault Zone: evidence for geodynamic switching of the East China continental margin. J Geol 118:277–293

Zhu YD, Ye XF, Zhang DH, Wang KQ, Wang CS, Yin XB (2014) Petrochemistry, SHRIMP dating and Sr–Nd isotopic constraints on the origin of the Kaihua porphyry Mo (Cu) deposit, Zhejiang Province. Earth Sci Front 21(4):221–234 (in Chinese with English abstract)

4

Geological Makeup and Structural Development

4.1 Overview

4.1.1 Geotectonic Location

The investigation area is located in the southern and eastern parts of the lower Yangtze block, sandwiched between the NE-trending Jixi-Ningguo fault and the Majin-Wuzhen fault, and its southeastern region is adjacent to the Cathaysia orogenic system along the Jiangshan-Shaoxing butt-joint belt. The whole investigation area is characterized as a structural transition belt. The investigation area is obliquely crossed by the NE-trending Xuechuan-Huzhou fault. Since Proterozoic, the investigation area has experienced multiple phases of tectonic activities including observable Caledonian movement, Indosinian movement, and Yanshanian movement to form complex geological structures, which are accompanied by other geological processes such as sedimentation, magma, metamorphism, and mineralization. All these geological processes together constitute the main structural framework of the investigation area (Fig. 4.1).

According to the results of previous regional geological investigations in the Xuancheng map (1:250,000), bounded by Xuechuan-Huzhou fault, the northwestern side of the investigation area belonging to Jixi-Ningguodun anticlinorium is located between Jixi-Ningguo fault and Xuechuan-Huzhou fault and extends NE 50° with the NE-trending Dakengkou-Renli-Ningguodun main fault developed. While the southeastern side of the area that belongs to Yuqian synclinorium is located between Xuechuan-Huzhou fault and Majin-Wuzhen fault and extending NE 40°.

4.1.2 General Features of Regional Structure

Based on the structural study, the tectonic evolution of the investigation area has experienced passive continental margin orogeny during the Chengjiang-Caledonian time, intracontinental orogeny in the Early Mesozoic, Mesozoic active continental margin stage, and Cenozoic tectonic uplift. The structural traces of strata and rocks show that the main deformation phases in the investigation area were Indosinian and Yanshanian, with strong structural expression and clear traces, while other deformation phases were relatively weak. In the Caledonian, deformation was mainly characterized by periodic uplift to forming relatively close NEE-SWW-trending anticlines and synclines; however, deformation in this episode was generally weak. After the subsequent superimposition, the anticlines became close but still not pronounced in the area. In the Early Indosinian, the compressional stress changed into SE-NW to form large-scaled broad and gentle anticlines and synclines trending nearly NE 60° accompany with a series of relatively small-scaled secondary fold groups, which outlines a general layout of folds in the area. In the Late Indosinian small-scaled NW-trending folds were developed, which superimposed on the Early Indosinian folds and interference with the early secondary folds formed dome and basin structures or nose structures. Deformation in the Yanshanian age was characterized by intense magmatic activity and large-scaled thrusting, extensional detachment and fault depression, resulting in the complex tectonic landscape in the region. In the Himalayan age, the tectonic stress is the tension shear action mode, and the region is dominated by the uneven and differential up-down motion, which is manifested as slow uplift, fluvial incision, and formation of multilevel terraces.

4.1.3 Structure Layers and Their Main Features

A structure layer is a set of rock assemblage with different deformation features formed in a period of the tectonic unit during the tectonic evolution stage. According to stage features of structural evolution history of the investigation area, contact relationship of disconformity and unconformity of stratigraphic sedimentation, differences of structural types, strength, structural style and spatial distribution, differences

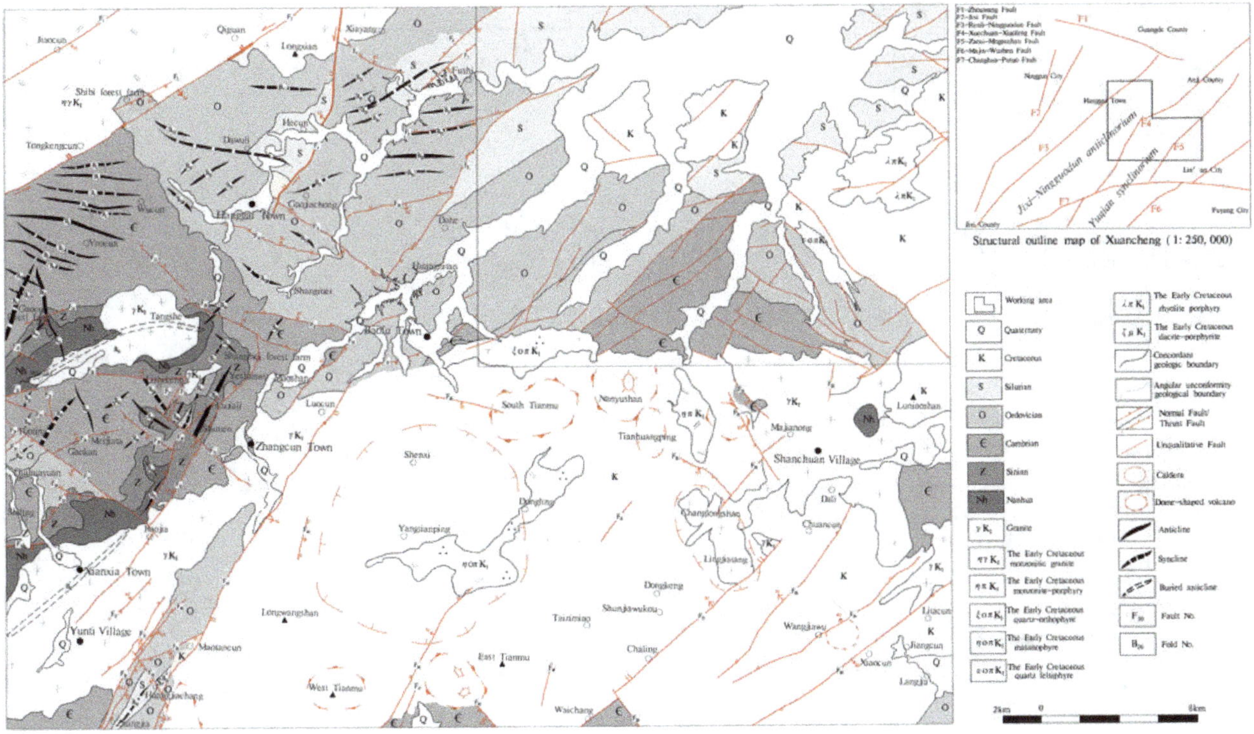

Fig. 4.1 Structural outline map of investigation area

in regional metamorphism, magmatic activity and mineralization, etc., the structural layers in the investigation area are divided as follows (Table 4.1).

4.1.3.1 Chengjiang-Caledonian Structural Layer

The Chengjiang-Caledonian structural layer is divided into the Nanhuaian substructural layer, the Sinian substructural layer, the Cambrian substructural layer, the Ordovician substructural layer, and the Silurian substructural layer.

Nanhuaian Substructural Layer

After the consolidation of the metamorphic basement, the area was in the continental margin development stage, with Xiuning Formation sandstone sediments, Nantuo Formation glacial sediments, and other stratigraphic sediments of a comparable period. During the Chengjiang movement, crust uplifts to form the disconformity between Nantuo Formation and Xiuning Formation in the early stage and the disconformity between the Sinian Lantian Formation and the Nantuo Formation in the late stage. The mudstone weathering crust at the top of the Nantuo Formation and the manganese dolomite deposit at the bottom of the Lantian Formation were formed. On the whole after the Chengjiang movement (about 680 Ma), the investigation area was changed to a relatively stable marine deposition environment.

Sinian Substructural Layer

The Sinian substructural layer is composed of the Lantian Formation and the Piyuancun Formation as a set of dolomites, dolomite limestone, limestone dolomite, siliceous siltstone, argillaceous siltstone, and siliceous shale intercalated with siliceous mudstone in the investigation area, showing the facies features of the open sea deposition and the shoal deposition in platform edge. A sedimentary disconformity developed between the Sinian substructural layer and the underlying Nanhuaian glacial diagenetic system. In terms of sedimentary formations, this disconformity belongs to jump discontinuity in sequence.

Cambrian-Silurian Substructural Layer

The Cambrian consists of the Hetang Formation, the Dachenling Formation, the Yangliugang Formation, the Huayansi Formation, and the Xiyangshan Formation, as a set of carbonaceous siliceous mudstone, shale, argillaceous limestone, and micrite, showing the facies features of siliceous argillaceous carbonate rock formation depositing in abysmal basin and continental shelf far away from terrestrial sources. It was of a silty abysmal basin sedimentary envi-

Table 4.1 A brief division of structural layer in investigation area

Tectonic period	Structural layer		Mechanical layer		Basin type	Sedimentary formation	Tectonic movement (Age/Ma)
Himalayan	Cenozoic structural layer		Weak layer			Diluvial sediment, alluvial sediment, slope wash	Himalayan movement
Yanshanian	Yanshanian structural layer	Missing					(65) Late Yanshanian movement
		Lower Cretaceous tectonic sublayer (K_1)	Competent layer		Volcano-sedimentary tectonic basin	Volcanic debris formation	(>96) Early Yanshanian movement
		Missing					(>145) Indosinian movement
Indosinian	Indosinian structural layer	Missing					(205) Caledonian movement
Chengjiang—Caledonian	Chengjiang—Caledonian structural layer	Silurian sub-structural layer (S)	Xiaxiang Formation	Weak layer	Peripheral foreland basin	Continental shelf sediment	(410)
		Ordovician sub-structural layer (O)	Changwu Formation-Wenchang Formation	Weak and hard interacted layer		Abyssal facies, continental shelf facies,tidal-flat facies varicolored composite terrigenous clastic formation.	
			Hule Forrmation-Huangnigang Formation	Weak layer	Passive marginal basin	Abyssal basin facies siliceous rock, carbonate rock, mudstone formation.	
			Yinzhubu Formation-Ningguo Formation	Weak layer		Continental shelf, abyssal basin, platform facies siliceous argillaceous flysch formation, carbonate rock formation.	Chengjiang movement
		Cambrian sub-structural layer (€)	Hetang Formation-Xiyangshan Formation	Competent and weak interacted layer		Abyssal basin, continental shelf facies, slope facies non-terrigenous siliceous argillaceous carbonate rock formation.	
		Sinian sub-structural layer (Z)	Lantian Formation	Competent and weak interacted layer		Continental shelf, abyssal basin facies P-bearing non-siliceous argillaceous and siliceous shale formation.	
		Nanhuaian sub-structural layer (Nh_{1-2})	Nantuo Formation	Weak layer		Littoral, bathyal sea, ice water continental shelf facies molasses.	680
			Xiuning Formation	Weak layer	Rift basin	Composite terrigenous clastic form non-terrigenousation, ice water volcano petromictic formation.	Jinning movement

ronment in this area in the Early Cambrian, a carbonate basin facies sedimentary environment in the Middle Cambrian, and the carbonate shelf inner margin facies sedimentary environment in the Late Cambrian.

Ordovician strata are composed of the Yinzhubu Formation, the Ningguo Formation, the Hule Formation, the Yanwashan Formation, the Huangnigang Formation, the Changwu Formation, and the Wenchang Formation. It is dominated by flysch formation, followed by the formation of siliceous argillaceous carbonate rocks away from terrestrial sources. The sedimentary facies are mainly Early Ordovician argillaceous shelf facies, Middle and Late Ordovician siliceous abyssal–subabyssal basin facies and open basin facies, reflecting the evolutionary law of transgression–regression–transgression–regression oscillation in the Ordovician.

The Silurian strata in the investigation area have a small distribution area and stable lithofacies, only formed by the Xiaxiang Formation. It is continental shelf facies of varicolored composite terrigenous clastic formation and littoral-shallow sea clastic rock formation. The middle and lower parts are mainly composed of mudstone, siltstone, silty shale, and sandstone, and the upper part is mainly feldspar–quartz sandstone. During Silurian, the sea area gradually shrank and the depth of seawater became shallow, which resulted in the deposition thickness thinned toward the east.

Throughout the entire Early Paleozoic sequence in the investigation area and in the northwest of Zhejiang, there are obvious differences in the features of stratigraphic sedimentary formation between the Early and Late Katian, which indicates that the Early Paleozoic was the beginning of the Caledonian crustal movement in southern China. The Caledonian movement resulted in the collage of the Cathaysian orogenic system and the Yangtze block into a unified continent, while the investigation area was uplifted into a land. The Upper Paleozoic strata are not exposed in the investigation area. According to the data of the investigation area in the Xuancheng map (1:250,000), the Devonian strata are generally unconformable to the underlying Silurian strata with a small angle after the Caledonian. In consequence, deformation in the investigation area was dominated by uplift in the Caledonian, showing less strong folding but the only formation of an overall broad bend in a differentiated way.

4.1.3.2 Indosinian Structural Layer

The whole structural layer of the Indosinian was missing. Judging from previous data and structural features around the investigation area, tectonic movements have occurred in the Indosinian, and deformation directly superimposed on the Caledonian and its early structural layers.

4.1.3.3 Yanshanian Structural Layer

The Yanshanian structural layer has a large distribution area in the investigation area and is composed of the Early Cretaceous Huangjian Formation. It is mainly distributed in the area of West Tianmu Mountain–East Tianmu Mountain–Shanchuan Village, generally extending NE. The Huangjian Formation lithologically is made of intermediate-acid, acid, and slightly alkaline continental volcanic rock and volcanic composite terrigenous clastic formation. The volcanic eruption rhythm is obvious, and the cycle is strong. The lithology and lithofacies of the eruptive centers changed little. This structural layer is also sporadically exposed in the southern margin of Xuanguang Basin in the northern edge of the investigation area, which is mainly composed of the Zhongfencun Formation and the Guangde Formation that are dominated by acid rhyolitic volcanic formation. Formation and development of the volcanic rock zone in the investigation area are closely related to the basement faults and are mainly controlled by the NE-trending structures.

4.1.3.4 Himalayan Structural Layer

Due to the absence of the Tertiary sediments in the investigation area, the Himalayan structural layer is merely composed of the Quaternary unconsolidated sediments. The Holocene series is distributed in the current river valley and the alluvial plains on both sides, making first to second level terrace landscape. It is diluvial–alluvial clastic formation.

4.2 Structural Deformation Features

4.2.1 Fold

The structural deformation is strong in the investigation area. Multiple phases of deformation produced folds with different deformed degrees and trends. According to their formation ages, four phases of folding can be identified: the near EW-trending folds in the Caledonian, the NE-trending folds in the Early Indosinian, the NW-trending folds in the Late Indosinian, and the basin structures in the Yanshanian (Table 4.2). The NE-trending folds in the Early Indosinian are the most important one in the area, which forms the layout dominated by NE-trending structures in the area.

4.2.1.1 The Caledonian Roughly EW-Trending Folds

According to the data of regional geological investigations including Xuancheng map (1:250,000), there exists a phase of folding showing approximately NE 70° trend by the orientations of axial plane cleavages, intersection lineation, and hinges of the folds of in the investigation area. However, this phase of folds is not shown as their original ca EW trend in the investigation area at present. With the core dominated by the Nanhuaian, it is speculated that this early phase of folds was originally formed with relatively tight geometry and rough EW trend and shows nowadays ca NE trend after compressive deformation and faulting modification in the later phase. The current investigation area exposes only of the residual traces of this phase of folds, represented by the Wangjia-Tangshecun anticline and Xianxia-Lujiali anticline. Since the axial traces of the anticlines are currently shown as S-shaped curves, as well as most of the residual anticlines have been deformed to the NE-trending folds and then become a component of the Indosinian NE-trending anticlinorium, this phase of folds will be described in detail in the following section about the Indosinian NE-trending folds.

4.2.1.2 The Early Indosinian NE-Trending Folds

The NE-trending folds are exposed in a large range in the investigation area. Bounded by Maotan-Luocun fault (F43) (Xuechuan-Huzhou fault), the northwestern side of the area is widely developed large-scaled NE-trending short-axial anticlinorium and synclinorium. It mainly consists of the Hanggai-Fushi synclinorium, the Shiling-Shangmeilinchang anticlinorium, and other small anticlines and synclines. Most of the southeastern side is covered with volcanic rocks, and only a few monoclinal structures are exposed at the eastern edge of the investigation area. Major NE-trending folds are summarized as follows.

Hanggai-Fushi Synclinorium

The Hanggai-Fushi synclinorium is located in the north of the investigation area. Its original anticlinorium has been successively faulted by the Tongkengcun-Qiguancun fault (F_3), the Hengshanjiao-Hulingjiao fault (F_4), the Changqingwu-Dalingjiao fault (F_9), the Hanggaizhen-Xiaolingshang fault

Table 4.2 A list of fold systems

Phase		Basin, complex fold and secondary fold		
Indosinian	Late	Dongshe syncline (B22), Xihekou anticline (B32), Xikengkou syncline (B33), Yankengkou anticline (B12), Gaocunlinchang anticline (B15), Shangmeilinchang-Baijiulinchang anticline (B36), Meijiata anticline (B42), Pingtoushan anticline (B45), West Lizhikeng syncline (B44), Huangshiyan-Damiao anticline (B50)		
	Early	Hanggai-Fushi synclinorium	Fushicun synclinorium	Xiaohangkeng syncline (B23), South Xiaohangkeng anticline (B24), North Haogouwu syncline (B25), Haogouwu anticline (B26), Feitoushan anticline (B27), Wanjiawu syncline (B28), Tianqinggang anticline (B29), Longtoushan syncline (B30)
			Gaojiachong synclinorium	Gaojiachong secondary synclinorium (B31)
			Hecun-Dawuli synclinorium	Liujia syncline (B1), Cishanmiao anticline (B2), Beichekeng syncline (B3), Wucun anticline (B4), Yinshanjian syncline (B5), Beichecun anticline (B6), Yujia syncline (B7), Changpucun anticline (B8), Jinjiabian-Wanshiwu syncline (B9), Chizikeng-Longwangmiao anticline (B10), Dayankeng syncline (B11), Laozhichang syncline (B17), Yuanxiwu syncline (B18), Qiaotou syncline (B19), Songkengwu anticline (B20), Siwukeng anticline (B21)
		Shiling-Shangmeilinchang anticlinorium	Wangjia-Tangshecun anticline (B16)	
			Xianxia-Lujiali anticline	Lutangchachang syncline (B37), Qingmingshan syncline (B47), Baishawu anticline (B46), Shimen-Yeshanwu anticline (B48), Taoshuwan-Shanwuli anticline (B51), Xianxia-Shenwulinchang anticline (B52)
			Guihuayuan-Lizhikeng syncline	Waizhang syncline (B38), Yutangshanwu anticline (B39), Tianuowu-Jiumushan syncline (B40), Tuotinglinchang syncline (B41), Tangwu syncline (B43), Huangshiyan syncline (B49)
		Other NE–trending folds	Zhangjia-Huangjiachang syncline (B54), Qishuping-Xiajiabeihou syncline (B53)	
Caledonian		The roughly EW-trending axial traces of folds have been completely modified in the area		

(F_{11}) into three parts: the Fushicun synclinorium, the Gaojiachong synclinorium, and the Hecun-Dawuli synclinorium.

Fushicun synclinorium

The Fushicun synclinorium is located in the Fushi Reservoir area, with the core exposed as the Silurian Xiaxiang Formation (S_1x) and the two limbs mainly formed by the Ordovician Changwu Formation (O_3c) and Wenchang Formation (O_3w) and older strata. The hinge of synclinorium plunges to the northeast and extends eastwards beyond the area. The synclinorium is cut off by large NE and NW striking faults to the west and south, respectively. There are many parasitic folds developed in the synclinorium. Based on the field data, attitudes of the parasitic folds are projected on the stereographic plots (Fig. 4.2a–g) to present the characteristics of the parasitic folds inside the synclinorium, as shown in Table 4.3.

As can be seen from the features of its internal parasitic folds, most folds are upright plunging and short-axial folds, with few being basin or dome structures, which reflect that the parasitic folds in this region may be affected by tectonic superimposition.

In a multilevel composite fold group (system), the establishment of overall stratigraphic attitudes, the description of the outline of a major fold, and the spatial distribution pattern of fold group (system) are all gradually determined in the process of establishing different levels of enveloping surfaces. Since the above folds are of similar sizes and ranks and basically contain the same rock layer (or formation), the hinges of the above-stated folds were analyzed and fitted with the enveloping surfaces (Fig. 4.2h).

Based on the outcrop strata, folds B23, B24, B25, and B26 can be seen as alternating anticlines and synclines in the north limb of the Hanggai-Fushi synclinorium. The attitude of the northern limb of the synclinorium is roughly fitted to be 122°∠46°. Folds B27, B28, B29, and B30 are displayed as alternating anticlines and synclines in the southern limb of the synclinorium. The attitude of the southern limb of the synclinorium is roughly fitted to be 22°∠30°. Thus, the limb angle of the synclinorium is calculated to be about 122°, the hinge trending about

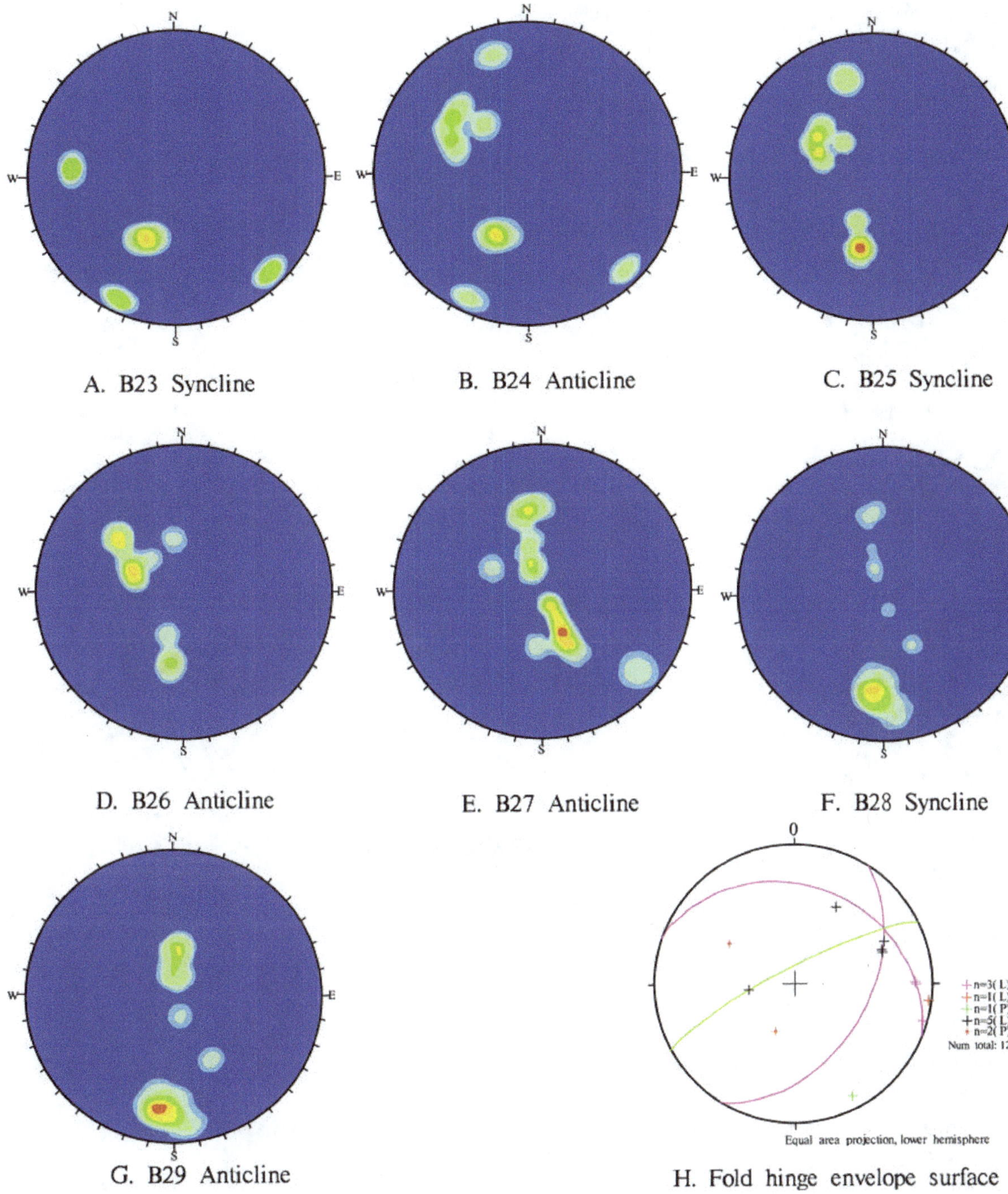

Fig. 4.2 Contour diagram of attitudes of the Hanggai-Fushi Synclinorium and its enveloping surface solution

58°∠25°, an axis surface of 333°∠80°, and the synclinorium is a inclined plunging fold. The structural styles of the parasitic folds that make up the synclinorium are shown in Fig. 4.3.

Gaojiachong synclinorium (B31)

The core of the synclinorium is exposed as the Xiaxiang Formation, and the limbs are composed of the Ordovician

Table 4.3 Parasitic fold features inside the Fushicun synclinorium

No.	Fold name	Length/km	Length–width ratio	Stratigraphic composition		Attitude		Fold type
				Core	Limb	Axial plane/°	Hinge/°	
B23	Xiaohangkeng syncline	2	5:2	O_3w	O_3c	316∠69	29∠37	Inclined plunging
B24	South Xiaohangkeng anticline	2	3:2	O_3c	O_3w	339∠82	65∠29	Upright plunging
B25	North Haogouwu syncline	>2	3:2	O_3w	O_3c	158∠89	69∠32	Upright plunging
B26	Haogouwu anticline	2.5	3:1	O_3c	O_3w	156∠83	70∠32	Upright plunging
B27	Feitoushan anticline	4.5	2:1	O_3w	O_3w	4∠72	89∠12	Inclined plunging
B28	Wanjiawu syncline	5	5:2	O_3w	O_3c	0∠88	90∠12	Upright plunging
B29	Tianqinggang anticline	3.5	2:1	O_3c	O_3w	187∠72	97∠1	Inclined horizontal
B30	Longtoushan syncline	5.5	2:1	O_3w	O_3c	195∠82	106∠3	Upright horizontal

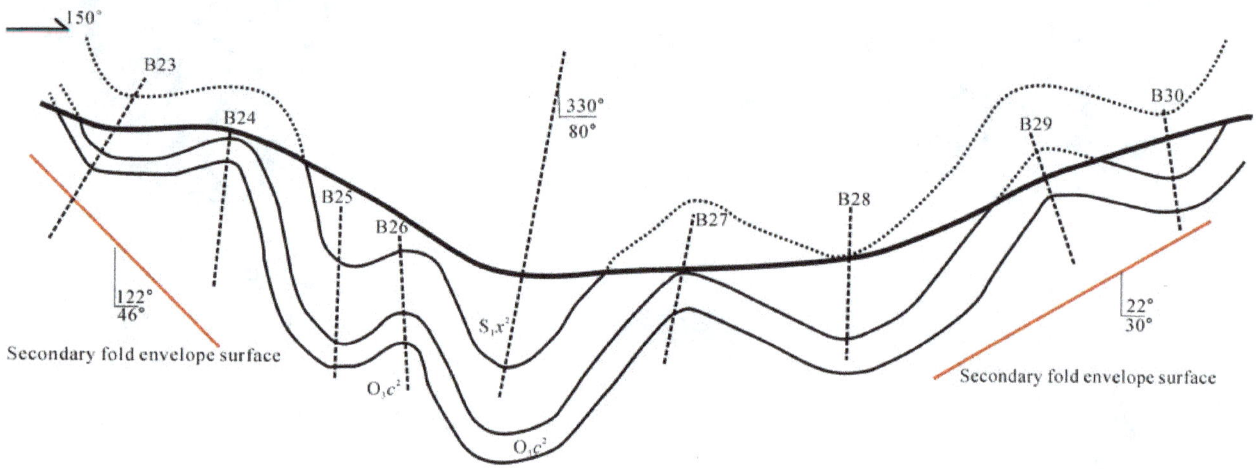

Fig. 4.3 Combination diagram of the parasitic folds of the Fushicun secondary synclinorium

Wenchang Formation, Changwu Formation, Huangnigang Formation, Yanwashan Formation, Hule Formation, and Ningguo Formation. Attitudes of the two limbs were calculated and analyzed by the stereographic plots. The results show that the representative attitudes of the two limbs are 105°∠30° and 0°∠70°, respectively, with a hinge of 79°∠28° (Fig. 4.4), a limb angle of 100°. The axis surface attitude is 155°∠66°. Consequently, this is an inclined plunging fold.

Hecun-Dawuli synclinorium

The center of the synclinorium is somewhere around the Hecun and the Dawuli. Its hinge generally plunges NEE and is cut off by NNE-NE striking Hengshanjiao-Hulingjiao fault (F4) to the east, adjacent to the Fushicun secondary synclinorium. Bounded by the Tongkengcun-Qiguancun fault (F3) in the north, the synclinorium is adjacent to the Maonanshan composite pluton.

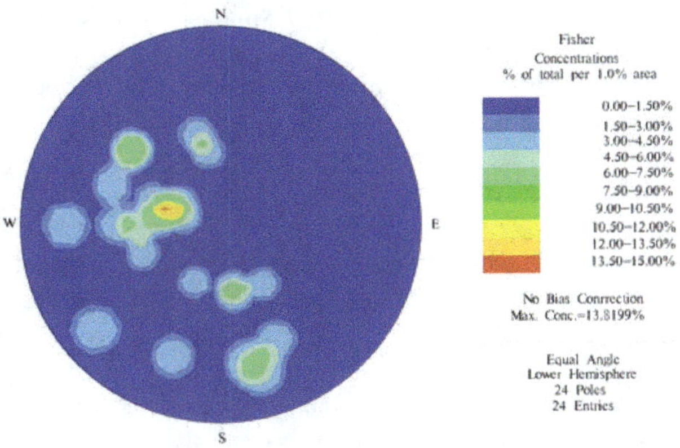

Fig. 4.4 Contour diagram of the limbs of the Gaojiachong secondary synclinorium (B31)

Strata of the core of the synclinorium belong to the Silurian Xiaxiang Formation, with both limbs exposed as the Ordovician Wenchang Formation, Changwu Formation, Yinzhubu Formation, and the Cambrian Xiyangshan Formation. A large number of parasitic folds were developed in the synclinorium. Based on field observations, the stereographic plots of the two limbs of the major folds (Fig. 4.5) display more obvious features of these parasitic fold features (Table 4.4).

The major parasitic folds inside the Hecun-Dawuli synclinorium are short-axial folds that are upright plunging with eastward plunging hinges. In the extension of B20 anticline hinge, an open anticline of a similar scale was formed by the Ordovician Wenchang Formation, of which the hinge is trending 243°∠4°. The hinges of the two anticlines are

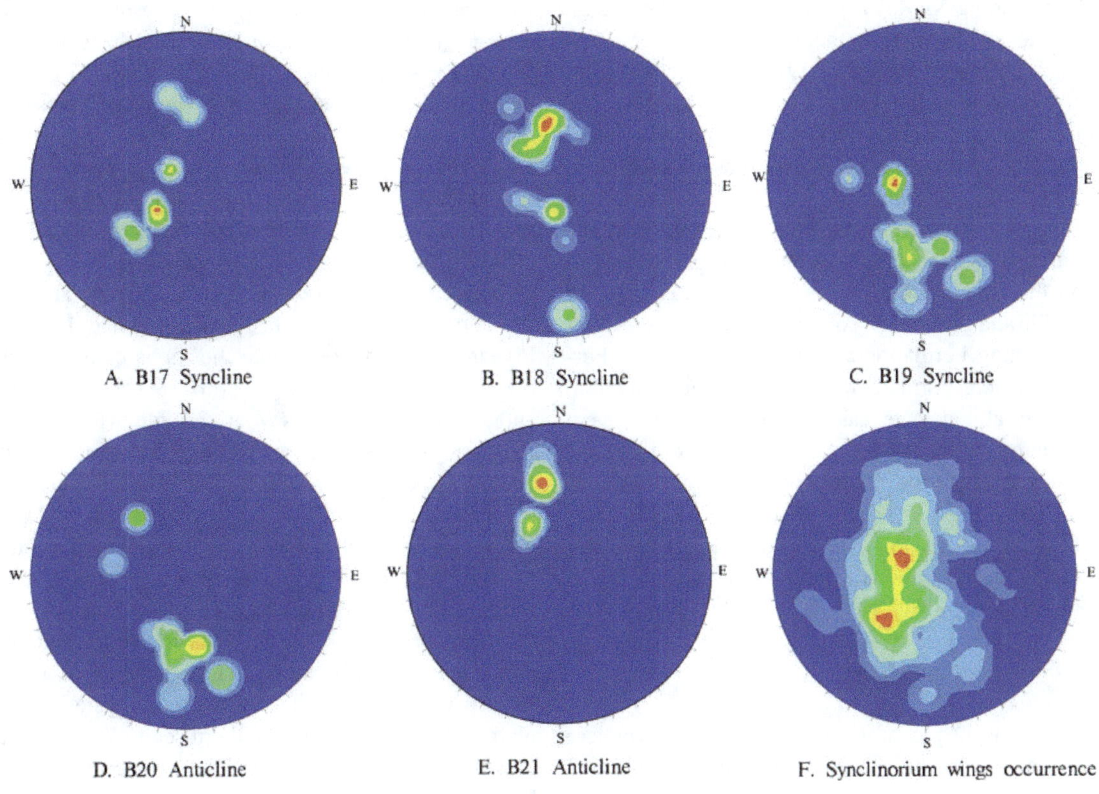

Fig. 4.5 Contour diagram of the limbs of the Hecun-Dawuli synclinorium and the parasitic folds

Table 4.4 Parasitic folds in the Hecun-Dawuli synclinorium

No.	Fold name	Length/km	Length–width ratio	Stratigraphic composition		Attitude		Fold type
				Core	Wing	Axial surface/°	Hinge/°	
B17	Laozhichang syncline	4	8:3	O_3w	O_3c	198∠83	110∠13	Upright plunging
B18	Yuanxiwu syncline	4.5	7:2	O_3w	O_3c	357∠83	86∠4	Upright horizontal
B19	Qiaotou syncline	2	3:1	O_3w	O_3c	162∠66	81∠20	Inclined plunging
B20	Songkengwu anticline	2.5	3:1	O_3c	O_3w	164∠88	75∠27	Upright plunging
B21	Siwukeng anticline	2	2:1	O_3c	O_3w	161∠51	96∠27	Inclined plunging

basically located on the same straight line but plunging oppositely, which may be related to the NW-trending fold superimposition in the later phase. The 573 groups of strata attitude data the two limbs of the secondary synclinorium were selected for projection statistics (Fig. 4.5f), which indicates the representative attitude of the limbs of the synclinorium are 122°∠19° and 40°∠44°, respectively. The hinge of the synclinorium is trending 110°∠19°, with a limb angle of 145° and an axial surface of 194°∠7°. This synclinorium is an open inclined plunging fold.

Shiling-Shangmeilinchang Anticlinorium

The Shiling-Shangmeilinchang anticlinorium regionally belongs to the Paleozoic Xuechuan-Baishuiwan anticlinorium. It is located between the NE striking Tongkengcun-Qiguancun fault (F3) and the Maotan-Luocun fault (F43), extending NE 45° and plunging NE. The overall geometry of the anticlinorium in the investigation area is relatively broad and gentle and formed by one syncline sandwiched between two anticlines as it appears. However, the anticlinorium is in fact formed by a syncline confined by a big radian curve of a closed anticline, showing as a whole as an anticlinorium. The anticlinorium is composed of the Wangjia-Tangshecun anticline, the Xianxia-Lujiali anticline, and the Guihuayuan-Lizhikeng syncline. The parasitic anticline inside the anticlinorium is relatively close, while the syncline is wide and gentle. The cores are exposed as the older Nanhuaian Xiuning Formation and the Sinian Lantian Formation and Piyuancun Formation. Based on field investigation and collection of attitude data, calculations and analysis were done by the stereographic plots of the well-exposed folds, and thus the fold features of the Shiling-Shangmeilinchang anticlinorium are shown in Table 4.5. Take the Wangjia-Tangshecun anticline, the Xianxia-Lujiali anticline, and the Guihuayuan-Lizhikeng syncline as examples. Their descriptions are given as follows.

Wangjia-Tangshecun anticline

The Wangjia-Tangshecun anticline is located somewhere around the Dingjiawan, the Wangjia, the Zhongtiancun, and the Shengli Reservoir, with the hinge extending rough NE 60° and plunging NE. The anticline is not completely exposed. The strata of its core are the Sinian Xiuning Formation in the Nanhuaian, Lantian Formation, and Piyuancun Formation. Its limbs were formed by the Cambrian Hetang Formation and Dachenling Formation. Tangshe composite pluton is exposed along with the anticline core. The anticline hinge turns to roughly 310° NW-trending after the intrusion of the Tangshe composite pluton (Shangmeilinchang-Baijiulinchang anticline (B_{36})). Southeastern limb of the Wangjia-Tangshecun anticline is mostly intruded by rocks and cut through by faults (F_{21}, F_{20}), while northwestern limb is cut through by the small NE striking fault (F_{14}) and observably superimposed by a NW-trending fold (Gaocunlinchang anticline (B_{15}).

Xianxia-Lujiali anticline

The Xianxia-Lujiali anticline is located somewhere around the Xianxia, the Shenwulinchang, and the Lujiali, generally extending NE 30° and plunging NE. The anticline is not completely exposed. Southeastern limb of the anticline is cut through by the NE striking Yuntixiang-Zhanglicun fault (F_{32}) that is parallel to the anticline axial trace and also cut by the NW striking Futangkou-Shiliangting fault (F_{31}), Dalingtou fault (F_{29}), and Yinshanwu-Shenwulinchang fault (F_{33}) into four parts that are named as the Shimen-Yeshanwu anticline (B_{48}), the Pingtoushan anticline (B_{45}), the Taoshuwan-Shanwuli anticline (B_{51}), and the Xianxia-Shenwulinchang anticline (B_{52}). Due to the severe damage by late faulting and rock intrusion, the Xianxia-Lujiali anticline is incompletely preserved. Here merely take the Shimen-Yeshanwu anticline (B_{48}) as an example to describe as below.

Shimen-Yeshanwu anticline (B48) is located along Shimen, Shuijiali, and Yeshanwu, generally extending with a NE 30° trend. The anticline was damaged later and became incomplete. It is crossed by faults in the core and limbs. The strata of the core are the Sinian Lantian Formation and Piyuancun Formation, while the limbs consist of the Cambrian Hetang Formation (Fig. 4.6). Southwestern end of the anticline is cut off by a NW striking fault (F_{31}). According to statistics and calculation of the attitudes of the two limbs strata, in the northern part of the fold near Yeshanwu, the hinge is oriented in 207°∠18°; limb angle is 84°; and the attitude of the axial surface is 122°∠70°. In the southern part of the fold around Shimen, the hinge of the anticline is oriented in 35°∠18°; limb angle is 128°; and the attitude of the axial surface is 123°∠82°. These attitude and orientation data classify the anticline into an inclined fold with its axial surface verging to NW. The northern segment of the hinge plunges to SW while the southern segment plunges to NE in the north, which shows an undulating geometry of the hinge. In addition, the strata at the core and two limbs of the anticline expanded and contracted along the NE-trending, indicating the later superimposition of folding nearly perpendicular to the hinge (NW trending).

Guihuayuan-Lizhikeng syncline

The Guihuayuan-Lizhikeng syncline belongs to the Shiling-Shangmeilinchang anticlinorium, which is sandwiched by two anticlines. The syncline is generally broad and gentle, with the northeastern end rising and the internal

Table 4.5 Brief parasitic fold features for the Shiling-Shangmeilinchang anticlinorium

NO.	Fold name	Size		Stratigraphic composition		Attitude		Fold type
		Length/km	Length–width ratio	Core	Wings	Axial surface/°	Hinge/°	
B13	Shibantang syncline	>3	3:2	ϵ_2y	ϵ_1d–ϵ_1h	310∠87	220∠5	Upright horizontal
B14	Shizishan anticline	2	5:2	Nh_2n	$Z_{1-2}l$–Z_2p	116∠81	206∠4	Upright horizontal
B16	Wangjia-Tangshecun anticline	>6	>3:1	Nh_2n–Z_2p	ϵ_1h–ϵ_2y	NE striking	Trending 60	–
B38	Waizhang syncline	>1.2	>2:3	ϵ_1d	ϵ_1h	103∠70	183∠24	Inclined plunging
B41	Tuotinglinchang syncline	>3	3:2	ϵOx	ϵ_3h–ϵ_1d	322∠68	234∠6	Inclined horizontal
B40	Tianluowu-Jiumushan syncline	>10	2:1	O_1y	ϵOx–ϵ_2y	325∠80	50∠33	Inclined plunging
B34	Changjiaokeng anticline	>3	3:2	Nh_2n	$Z_{1-2}l$	127∠89	37∠29	Upright plunging
B35	Dukengwu anticline	>1.5	>1:1	ϵ_1d	ϵ_2y–ϵ_3h	304∠75	32∠8	Leant horizontal
B37	Lutangchachang syncline	>2	>2:1	Z_2p	$Z_{1-2}l$	283∠69	12∠3	Inclined horizontal
B43	Tangwu syncline	2	5:2	ϵ_2y	ϵ_1d–ϵ_1h	Striking 60	Trending 60	Tectonic basin
B47	Qingmingshan syncline	>2	>2:1	ϵ_1h^2	ϵ_1h^1	133∠75	49∠22	Inclined plunging
B46	Baishawu anticline	>1.2	>3:2	Nh_2n	$Z_{1-2}l$	288∠85	17∠14	Upright plunging
B48	NE end of Shimen-Yeshanwu anticlilne	>3.8	>3:1	$Z_{1-2}l$–Z_2p	ϵ_1h	122∠70	207∠18	Inclined plunging
B48	SW end of Shimen-Yeshanwu anticlilne	>3.8	>3:1	$Z_{1-2}l$–Z_2p	ϵ_1h	123∠82	35∠18	Upright plunging
B49	Huangshiyan syncline	>2	>2:1	ϵ_1h^2	ϵ_1h^1	328∠85	58∠5	Upright horizontal
B51	Taoshuwan-Shanwuli monocline	>4	>2:1	Nh_1x – Nh_2n	ϵ_1h	Striking 35	Trending 35	–
B52	Xianxia-Shenwulinchang anticline	>8	>3:1	Nh_1x – Nh_2n	ϵ_1h	Striking 35	Trending 35	–

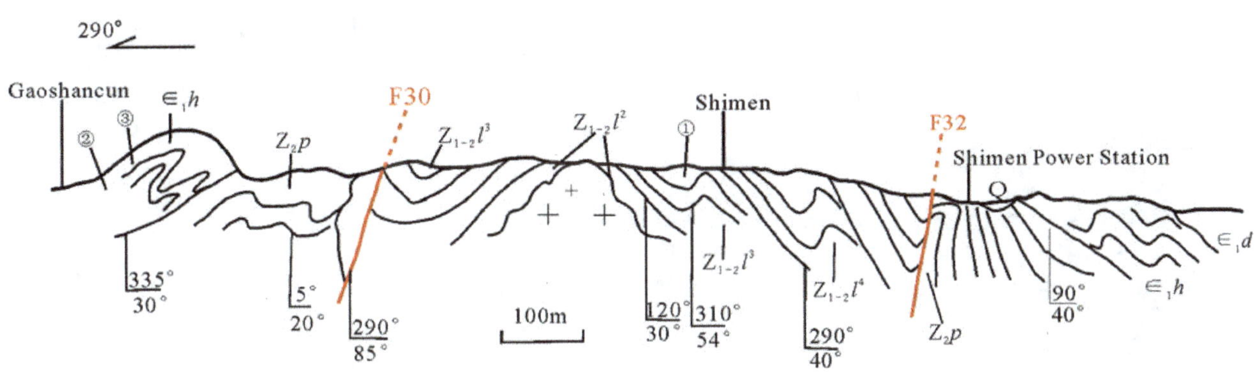

Fig. 4.6 Cross section of the southwestern end of the Shimen-Yeshanwu anticline

faults developing. In addition, the fold is superimposed by the NW-trending folds, and thus the fold geometry is extremely complex. Here, taking the relatively large-scaled Tianluowu-Jiumushan syncline (B_{40}) and the Tuotinglinchang syncline (B_{41}) as examples, the brief descriptions of the synclines are as follows:

The Tianluowu-Jiumushan syncline (B_{40}) is located somewhere in Tianluowu, Jiumushan, Longtingcun, and Shilingcun. Outcrop of the syncline is relatively wide and gentle in the area, with its core formed by the Ordovician Yinzhubu Formation that is damaged by the NE fault. The strata of the two limbs are the Cambrian Xiyangshan Formation, Huayansi Formation, and Yangliugang Formation. The southeastern limb is relatively complete exposure, while the northwestern limb is incomplete. The core part is damaged by faults F_{24} and F_{25}, resulting in the northwestern strata of the core partially overturned. According to the statistics of the 98 groups of attitude data of the syncline limbs (Fig. 4.7a), the hinge of the syncline is oriented as 50°∠33°; its limb angle is 145°; the attitude of the axial surface is 325°∠80°. All these show an open inclined plunging syncline with the axial surface verging to SE. The syncline has a limited length and a width of about 5 km in the investigation area. Considering the extension of the periphery of the investigation area, the axial ratio of the syncline is about 2:1, which indicates that this syncline is actually a small tectonic basin.

The Tuotinglinchang syncline (B_{41}) is located in and to the east of Tuotinglinchang. The core of the syncline is exposed as the Cambrian Xiyangshan Formation, while the limbs are formed by the Cambrian Huayansi Formation, Yangliugang Formation, and Dachenling Formation (Fig. 4.8). The syncline is bounded by several groups of faults. Its northwestern limb was damaged later. To the west of the syncline, several alternative anticlines and synclines develop trending parallel to the syncline. According to the statistics and calculation of the 47 groups of attitude data of the syncline limbs (Fig. 4.7b), the hinge of the syncline is oriented as 234°∠6°; the limb angle is 118°; and the axial surface has an attitude of 322°∠68°. All data indicate that it is an open inclined plunging syncline with the axial surface verging to SE.

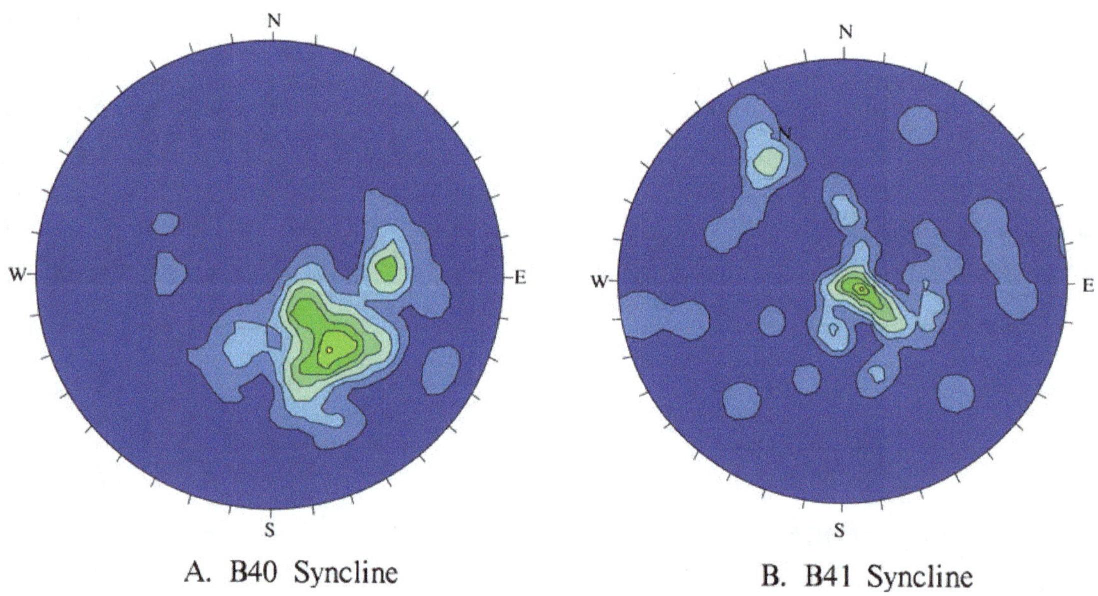

Fig. 4.7 Contour diagram of the limbs attitudes of the Guihuayuan-Lizhikeng syncline

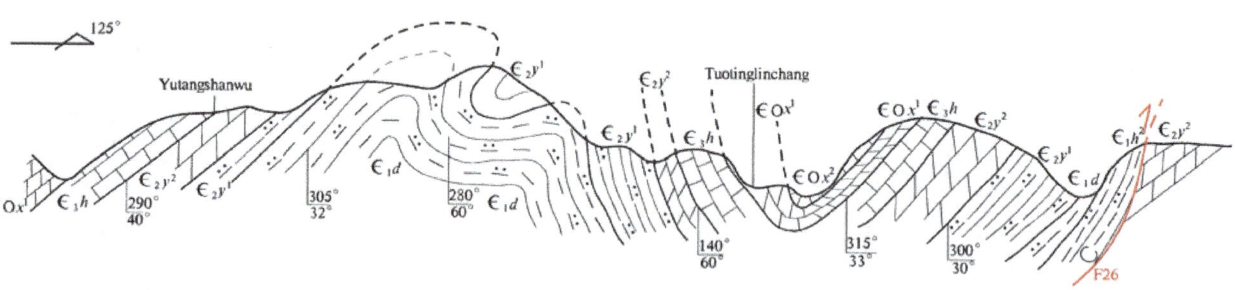

Fig. 4.8 Cross section of the northeastern end of the Guihuayuan-Lizhikeng syncline

According to the above analysis, the Guihuayuan-Lizhikeng syncline is a roughly 50° NE-trending inclined plunging syncline with the axial surface dipping to NW. Bounded by the NW-trending Yinshanwu-Shenwulinchang fault (F_{33}), the northeastern end of the fold hinge rises to NE, and the southwestern end rises to SW. The two ends of the hinge are dextrally displaced about 1 km along fault.

The Secondary Fold Bundle

This phase of folds is mainly characterized by the deformation of the Cambrian-Ordovician strata and distribute in a limited area. They are developed along the Yaocun-Wucun in the northern part of the investigation area. The outcrop is clearly exposed especially to the south of Ningguodun fault and in the western part of the Hanggai-Fushi synclinorium. The folds were mainly developed in the rising end of the regional Hanggai-Changxing synclinorium, forming roughly EW-trending parallel fold bundle.

In the Yaocun-Wucun area, the Early Indosinian fold bundle is roughly trending EW and is located to the south of the Tongkengcun-Qiguancun fault (F3) and the outer margin of the Hecun-Dawuli secondary synclinorium. The bundle of folds is composed of nearly parallel anticlines and synclines arranging at intervals, with hinges of extending 290° and generally plunging SEE. Based on the field survey, statistical calculation of stereographic plots was done for the attitudes of the folds. The features of the folds are described in Table 4.6.

The above folds were all developed in the Yangliugang Formation, the Dachenling Formation, and the Hetang Formation of the Cambrian. They are similar in geometry and alternative as synclines and anticlines, basically short-axial folds extending nearly east–west. Based on the stereographic plots of the above fold hinges in the area, the optimal orientation is obtained to be 100°∠2° and 280°∠2° (Fig. 4.91). The fold hinge is nearly horizontal. Assuming the parasitic fold hinges are coplanar in space, when changing the plunging direction from 280° to 100°, the enveloping surface attitude of the parasitic folds is roughly obtained to be 40°∠25°, which represents the attitude of one limb of the higher-order fold that the above folds belong to (Fig. 4.10). The assemble pattern of the folds is shown in the Fig. 4.11.

The set of near EW-trending fold bundle generally develops in the relatively older strata of the Hecun-Dawuli secondary synclinorium limbs, with most of them in the Xiyangshan Formation, the Huayansi Formation, and the Yangliugang Formation of the Cambrian-Ordovician. The parasitic fold is not well developed after it generally extends eastwards to the Xiyangshan Formation of the Hecun-Dawuli secondary synclinorium limbs and shows no EW extensibility in the investigation area. Taking the fold bundle enveloping surface fitting condition into consideration, the fold bundle basically belongs to the southern limb of the higher-order syncline, i.e., the southern limb of the Hecun-Dawuli synclinorium. Thus, the near EW-trending fold bundle is the parasitic fold of the

Table 4.6 Brief table of Yaocun-Wucun nearly EW-trending fold bundle features

No.	Fold name	Length/km	Length–width ratio	Stratigraphic composition		Attitude		Fold type
				Core	Wing	Axial surface/°	Hinge/°	
B1	Liujia syncline	3.5	5:1	ЄO*x*	$Є_3h$	345∠77	68∠24	Inclined plunging
B2	Cishanmiao anticline	5	>5:1	$Є_2y$	$Є_3h$–ЄO*x*	6∠88	96∠14	Upright plunging
B3	Beichekeng syncline	5.5	>9:2	ЄO*x*	$Є_3h$–$Є_2y$	183∠54	111∠23	Overturned inclined plunging
B4	Wucun anticline	4.2	6:1	$Є_2y$	$Є_3h$–ЄO*x*	186∠47	106∠10	Overturned inclined plunging
B5	Yinshanjian syncline	3	>3:1	ЄO*x*	$Є_3h$–$Є_2y$	205∠72	119∠12	Inclined plunging
B6	Beichecun anticline	>5.5	>3:1	$Є_2y^1$	$Є_2y^2$	214∠77	124∠6	Inclined horizontal
B7	Yujia syncline	5	3:1	$Є_2y^2$	$Є_2y^1$	193∠46	280∠2	Overturned inclined horizontal
B8	Changpucun anticline	3.8	4:1	$Є_1h$	$Є_1d$–$Є_2y$	192∠43	280∠1	Overturned inclined horizontal
B9	Jinjiabian-Wanshiwu syncline	3.5	4:1	$Є_2y^2$	$Є_2y^1$	18∠87	107∠1	Upright horizontal
B10	Chizikeng-Longwangmiao anticline	>2.5	>3:1	$Є_1h$	$Є_1d$–$Є_2y$	41∠84	310∠1	Upright horizontal
B11	Dayankeng syncline	>1.5	>3:1	$Є_2y$	$Є_1d$–$Є_1h$	185∠81	275∠8	Upright horizontal

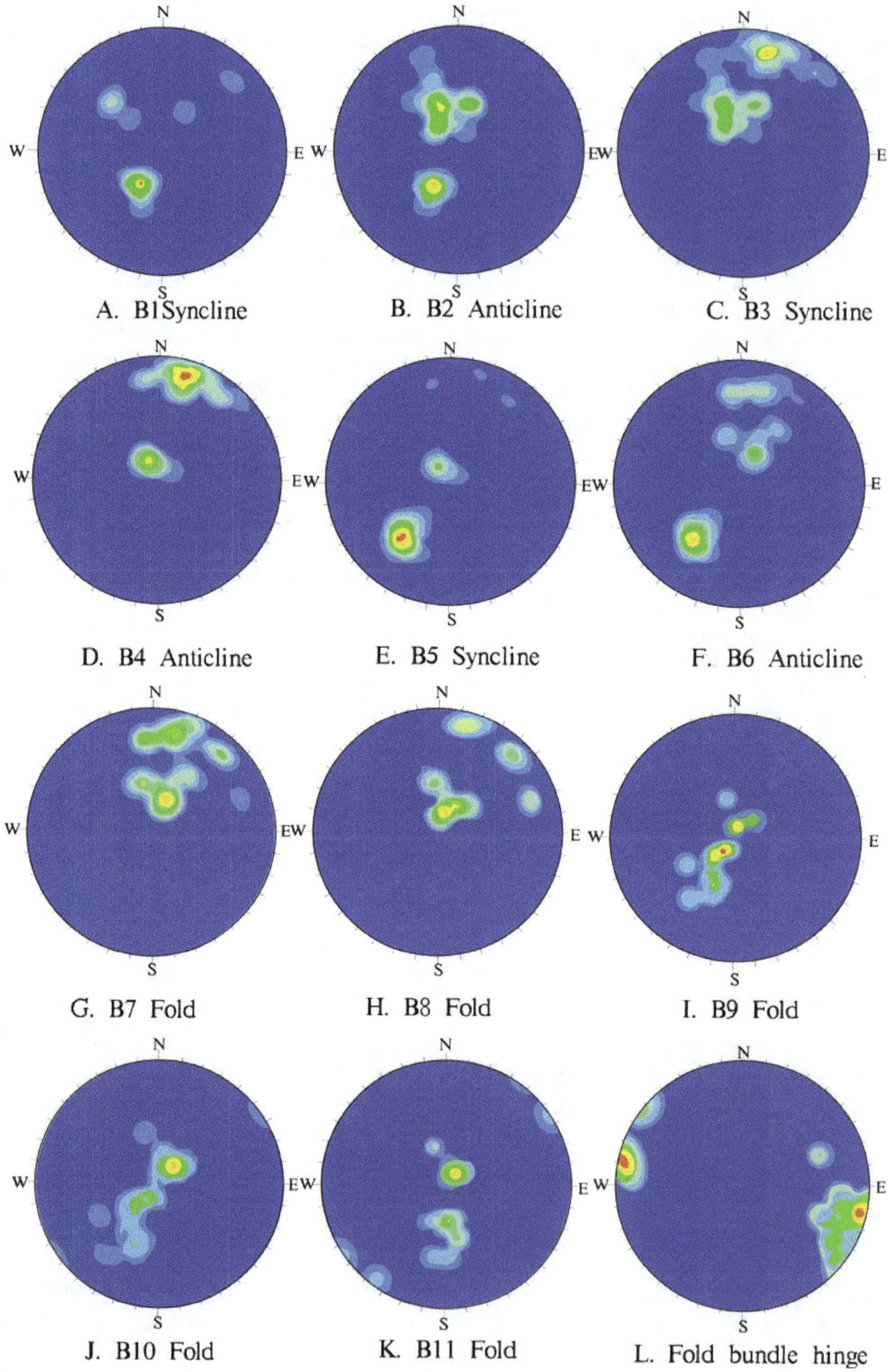

Fig. 4.9 Contour diagram of the attitudes of the Yaocun-Wucun roughly EW-trending fold bundle

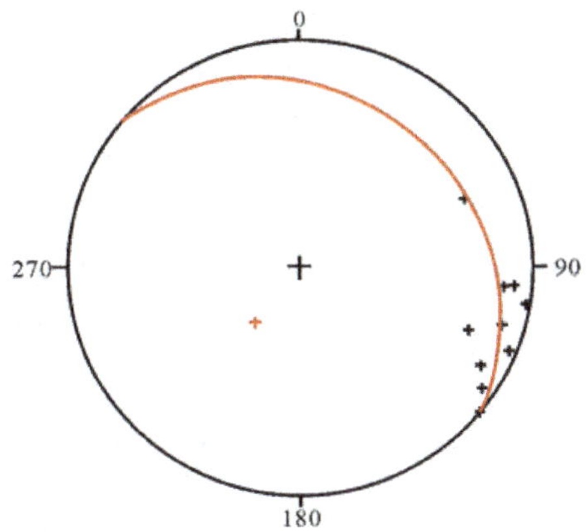

Fig. 4.10 Yaocun-Wucun fold enveloping surface fitting figure

Hecun-Dawuli secondary synclinorium, but not representing an individual regional tectonic superimposition.

Other NE-Trending Folds

Synclinal structures are developed between the southeastern side of the Xianxia composite pluton and the Maotan-Luocun fault (the Xuechuan-Huzhou fault). The southeastern part of this fault is basically covered by the Cretaceous volcanic or intruded rocks. Folds are not well observed. The Paleozoic strata are sporadically exposed. The NE striking monocline strata are merely exposed at the eastern edge of the investigation area.

Zhangjia-Huangjiachang syncline (B_{54})

It is located along the Zhangjia, Paifangwu Reservoir, Ma'anshan, and Huangjiachang. Its core is exposed as the Silurian Xiaxiang Formation, and limbs are mainly formed by the Ordovician Wenchang Formation and Changwu Formation. The anticline is extending to NE and cut off by the NW striking fault (F_{42}) (Fig. 4.12). Through the statistical calculation and analysis of the strata attitudes of the two limbs of the syncline, it comes out that the orientation of the syncline hinge of 59°∠11°, the limb angle of 88° and the axial surface attitude of 330°∠80°, which displays an inclined plunging fold with the axial surface verging to SE.

Back Qishuping-Xiajiabei syncline (B_{53})

It is located in Qishuping and extends NE. Its core is in the Silurian Xiaxiang Formation, with a NE striking fault developed. The two limbs are deformed by the Ordovician Wenchang Formation and Changwu Formation. The northwestern limb is damaged by fault (F_{41}) while the southeastern limb is unconformably covered by volcanic rocks. The syncline hinge is oriented as 205°∠6°. The limb angle is 111°. The attitude of the axial surface is 116°∠75°. All these data indicate an inclined plunging fold with the axial surface verging to NW.

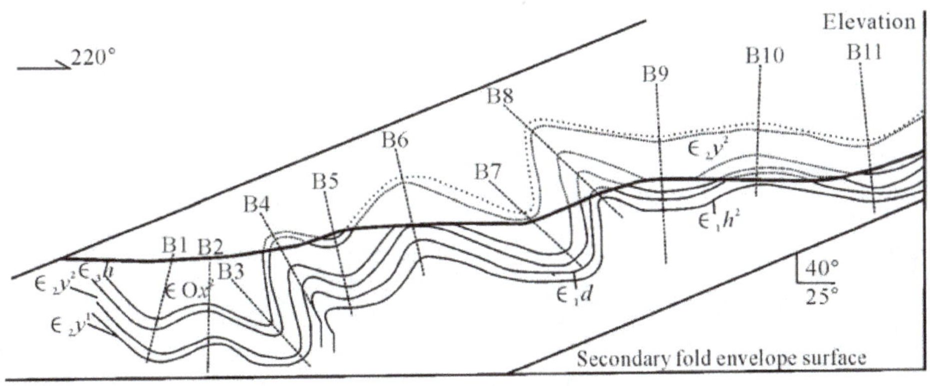

Fig. 4.11 Sketch of the assemble pattern of the Yaocun-Wucun nearly EW-trending parasitic folds

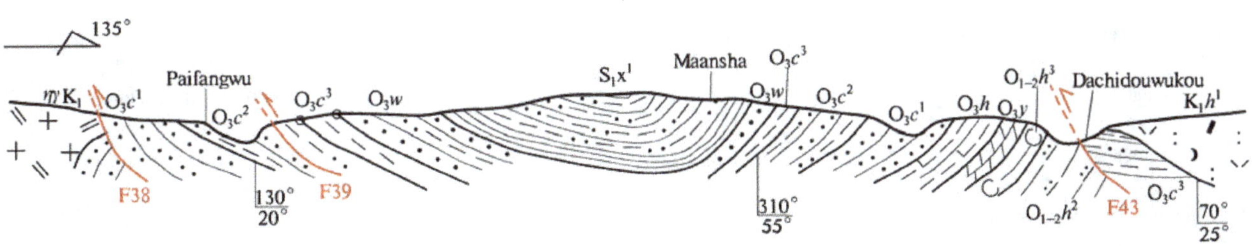

Fig. 4.12 Cross section of Zhangjia-Huangjiachagn syncline

4.2.1.3 The Late Indosinian NW-Trending Folds

In the investigation area, the NW-trending folds are relatively small-scaled folds that superimpose across on the early large NE-trending folds. Due to the affection from later faulting, these folds show complex geometries of nose-like structures and small dome and basin structures locally. Features of the NW-trending fold in the investigation area are shown in Table 4.7. Take the Shangmeilinchang-Baijiulinchang anticline and Xikengkou syncline as examples.

Shangmeilichang-Baijiulinchang anticline (B$_{36}$)

It is located near Shangmeilinchang and the hinge extends NW. The core strata are the Sinian Lantian Formation. The limbs are deformed by the Sinian Piyuancun Formation. The structure is located at the closure of the NE plunging Shiling-Shangmeilinchang anticline, forming an obvious dome and basin structure. After calculating the limbs attitudes through stereographic plots, the anticline hinge is oriented as 298°∠4°; the limb angle is 96°; and the axial surface has an attitude of 209°∠78°, which comes out an inclined horizontal anticline.

Xikengkou syncline (B$_{33}$)

It is located to the northwest of Shangmeicun and bounded by a NE striking fault in the southeast. The core of the anticline is composed the second member of the Indovician Yinzhubu Formation, while the wing is the first member of Indovician Yinzhubu Formation. After calculating the limbs attitudes through stereographic plots, the syncline hinge occurrence oriented as 314°∠8°; the limb angle is 101°; and the axial surface has an attitude of 42°∠80°, which comes out an upright horizontal syncline. An overturned anticline can be seen on the actual outcrop about 2.5 km to the southwest of the syncline (Fig. 4.13), of which the strata of its core is the Cambrian Dachenling Formation and the limbs are deformed by the Yangliugang Formation. This is a NW plunging overturned anticline with the hinge orientation of 280°∠15°.

The NW-trending folds in the investigation area are relatively small in scale and mostly controlled by the early large-scaled NE-trending fold system. The NW-trending folds overprint the NE-trending secondary folds to form dome and basin structures or nose folds locally. The overall orientation of the fold hinge is 320°–330° with a dip angle of 20°.

4.2.2 Fault

Fault structure is one of the main forms of orogeny, which develops at different degrees in each orogenic stage. In the investigation area, NE, NNE, and NW striking faults are well developed and show features of multiphase activity. On the whole, basement faults and regional large faults were formed in early stages and have been active in multiphase, and their mechanical properties changed as the changes of regional stress fields. In addition, a group of faults with the same direction can also experience multiple phases of

Table 4.7 Summary of NW-trending fold features in the investigation area

No.	Fold name	Size		Stratigraphic composition		Attitude		Fold type
		Length/km	Length–width ratio	Core	Wing	Axial surface/°	Hinge/°	
B22	Dongshe syncline	2	3:1	$O_{1-2}n$	O_1y^3	214∠83	125∠13	Upright plunging
B32	Xihekou ancline	1	5:3	$O_{1-2}n$	$O_{2-3}h$–O_3c	213∠77	298∠20	Inclined plunging
B33	Xikengkou syncline	3	3:2	O_1y^2	O_1y^1	42∠80	314∠8	Upright horizontal
B12	Yankengkou ancline	2.6	3:2	Z_2p	$Є_1h$ – $Є_1d$	240∠73	325∠17	Inclined plunging
B15	Gaocunlinchang ancline	2	2:1	Nh_2n	$Z_{1-2}l$–Z_2p	57∠83	328∠4	Upright horizontal
B36	Shangmeilinchang-Baijiulinchang ancline	1.8	2:1	$Z_{1-2}l$	Z_2p	209∠78	298∠4	Inclined horizontal
B42	Meijiata ancline	2	3:2	Z_2p	$Є_1h$ – $Є_1d$	253∠76	334∠29	Inclined plunging
B45	Pingtoushan ancline	>2.5	>2:1	$Z_{1-2}l$–Z_2p	$Є_1h$	44∠27	335∠11	Overturned tight inclined plunging
B44	West Lizhikeng syncline	1.8	1:1	$Є_1h^2$	$Є_1h^1$	249∠77	338∠2	Inclined horizontal
B50	Huangshiyan-Damiao ancline	2	1:1	Nh_2n–Z_2p	$Є_1h$	207∠84	291∠24	Nose structure

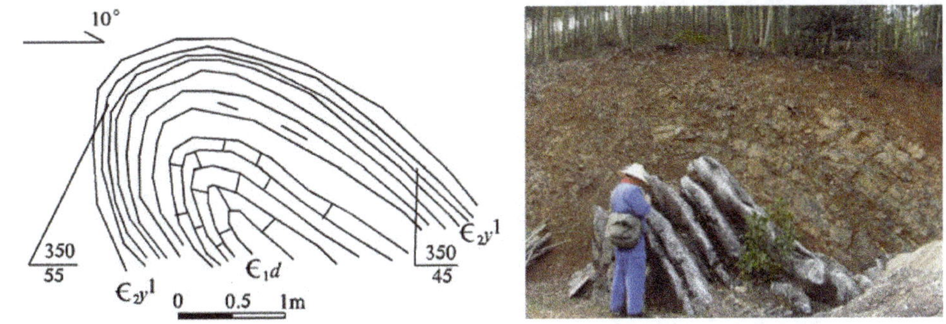

Fig. 4.13 Tangkengwu overturned anticline

activity. Therefore, the faults in different directions intersect, displace, and recombine with each other, resulting in a complex manner. The main faults are described in detail here and the general faults are briefly described as follows.

4.2.2.1 Major Fault

Tongkengcun-Qiguancun Fault (F_3)

The fault belongs to the Jixi-Ningguo fault zone that is known as the "Anhui, Zhejiang, and Jiangxi fault zone" and is an important regional boundary fault zone. In the investigation area, the Tongkengcun-Qiguancun fault starts from the southwest and extends toward northeast through Paomachang, Tongkengcun, Yonghe Forest Farm, and Qiguancun, with a general strike of NE 55° and a length of 18.5 km. Based on analyses of remote-sensing images, the location of the fault passing through appears to be a discontinuous linear trace in spatial morphology. The southwestern segment of the fault shows the continuous distribution of triangular facets, and the northeastern segment of the fault has formed cutting marks on the mountain edge, which are particularly obvious in the 3D images.

The fault surface in the investigation area is generally dipping to SE and consists of several left-lateral en echelon reverse faults that are parallel to each other. The Early Paleozoic strata lie to the southeast of the fault. The fault cuts through the Ma'anshan composite pluton extending toward NE. In the pluton, there are several faults with the same strike as this main fault and several granite porphyry veins with nearly parallel distribution, which roughly represent the clear demarcation line between intrusive rocks and sedimentary rocks in the investigation area.

The fault cut the basement of the investigation area, and it was formed in the Jinning Period. In Caledonian, the Jinning basement fault revived, which was a ductile-brittle fault that developed in the cover and had an obvious controlling effect on the Ma'anshan composite pluton. The fault was strongly active in the Yanshanian and characterized by several phases of activities of strong thrusting and left-lateral movement (Figs. 4.14 and 4.15), which caused a general counterclockwise rotation of fault strike from near EW to NE. Angular diagenesis with a width of 3–10 km developed along the southeastern sides of the fault, and compressive cleavage belt and lens isobaric structures with a width of 100 m developed near the fault. In the late stage of the Yanshanian, tensional normal faults developed in different directions, and the phenomena of rock fragmentation, silicification, and vein-rock filling were common, which implies that the mechanical characteristics of the Yanshanian changes from early compression to later extension.

There are obvious differences in structures, magmatic rocks, and mineral products between the two sides of the fault. The northwestern side belongs to Jixi-Ningguo fault zone. Within the Jixi-Ningguo fault zone, tectonic schistosities are stable (dipping to 130°–160°); several subparallel secondary faults developed; and en echelon and imbricate fault are commonly observed, which shows the characteristics of multiphase of activities and the same geological history of development and evolution. In the investigation area, the main body is the Ma'anshan composite pluton, in which there developed a series of parallel-distributed faults and dikes with the same strike of rough NE 55° that is consistent with the main fault strike. The southeastern side belongs to the Xuechuan-Huzhou fault zone, with large

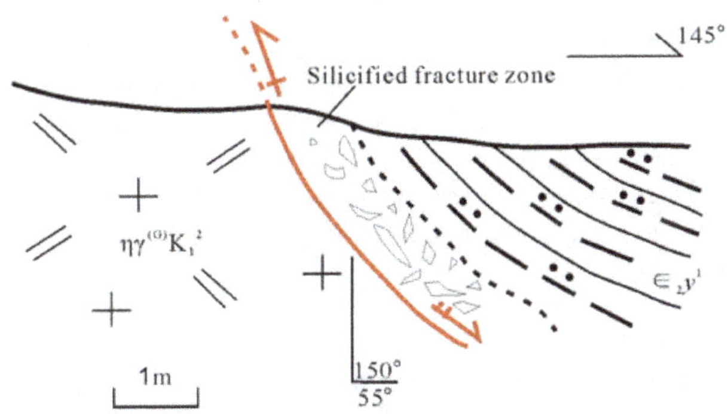

Fig. 4.14 Tongkengcun thrust fault

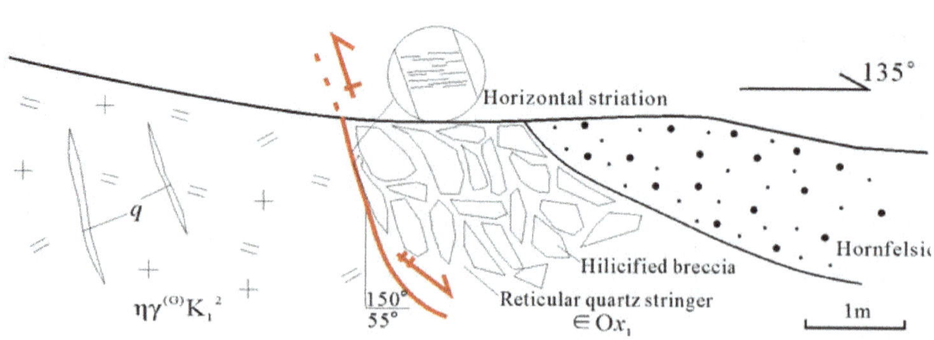

Fig. 4.15 Fayundong thrust fault

crustal activity and an obvious linearity of fold. A series of NE-NNE striking fault bundles developed within the zone, and the fault surface generally dips to SE with steep dip angle of more than 60°; it was cut by the late nearly EW striking fault.

This fault basically controlled the formation and distribution of the Ma'anshan composite pluton. The fault was formed in the Jinning Period and reactivated along basement faults in the later period, so that provided a good ascending channel for magmatic activity. In the early stage of the Late Yanshanian, the regional NW-SE extension that was perpendicular to this fault resulted in the most important magmatic activity period and ore-controlling and metallogenic period in the region during the Late Yanshanian (after 145 Ma). The multistage pulsating intrusive rock formed the composite pluton with different scales in the investigation area. During the later stage of the Late Yanshanian, there was a strong compression in NW-SE in the region. The southeastern block of the fault strongly thrust toward NW. So far, the distribution of the pluton and strata was basically controlled.

Maotan-Luocun Fault (F_{43})

The Maotan-Luocun fault (F_{43}) belongs to the Xuechuan-Huzhou fault. The main segment of the fault is located inside of Zhejiang Province. It goes from Xuechuan, Bainiuqiao, West Tianmu Mountain, Xiaofeng-Baishuiwan, Moganshan to Huzhou for about 140 km, with the general strike of NE (40°–45°). Fault surface dips to SE with an angle of 60°–80°. The fault is about 22 km long in the investigation area, as the boundary of aeromagnetic positive and negative anomalies and also the obvious boundary between sedimentary rocks and volcanic rocks in the area. The fault was formed in the Indosinian and strongly reactivated several times in the Yanshanian. The fault experienced intense extensions in the early stage and became the main channel of the Yanshanian magmatic activities to form the Tianmushan-Moganshan NE-trending volcanic-granite belt in the investigation area. In the middle stage, the fault thrust, sheared, and cut across the Paleozoic Shiling-Shangmeilinchang anticlinorium caused mostly loss of the eastern limb of the anticlinorium. In the late stage, NNE striking sinistral faults were intensely active and recombined with the early faults, enabling the fault zone to have complex multiphase activity features. The surficial features are shown as below:

The fault zone in the investigation area is composed of 3–4 roughly parallel and discontinuous NE striking faults, most of which developed in the lower Paleozoic strata. Individual faults show combined kinematics of thrusting, normal slipping, and shearing. The fault zone shows a series of thrusts moving from SE to NE (Figs. 4.16 and 4.17). Along the fault zone, multistage of monzonitic granite and granite intruded, and these intrusions were lately broken and filled with quartz veins. The fault zone shows ductile deformation in the volcanic rocks around Changtanqiao, probably due to thrust movement lifting up the deeper tectonic layers to the surface. Additionally, horizontal striations of left-lateral movement are observed on the fault surfaces of the fault zone, which indicates that intense sinistral slip has occurred in the late stage.

The rock-controlled and basin-controlled action of this fault is very pronounced in the investigation area. The faults were formed in the Early Indosinian and extend in NE direction. Three to five large-scaled faults developed parallel to the main fault, which together provided an ascending channel for the later magmatic activity. Rifting occurred during the Early Yanshanian, which reactivated the earlier boundary faults in the basin. Then, the earlier synclinal graben underwent subsidence along the boundary faults, and afterward intense volcanic eruption and granite intrusion occurred, determining the formation and distribution of the

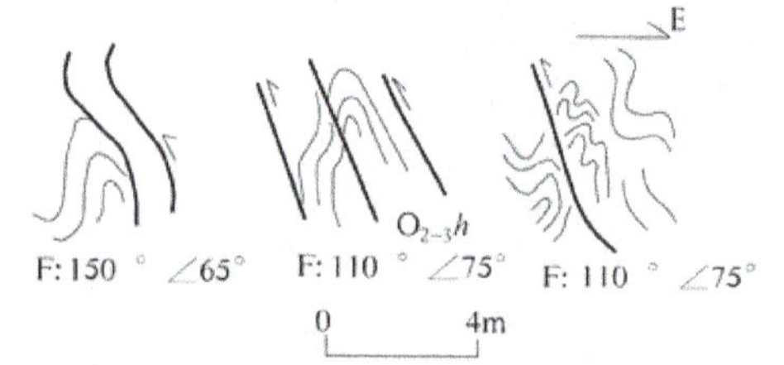

Fig. 4.16 NE-NNE-trending fault belt tectonic features (Anhui Institute of Geological Survey 2003)

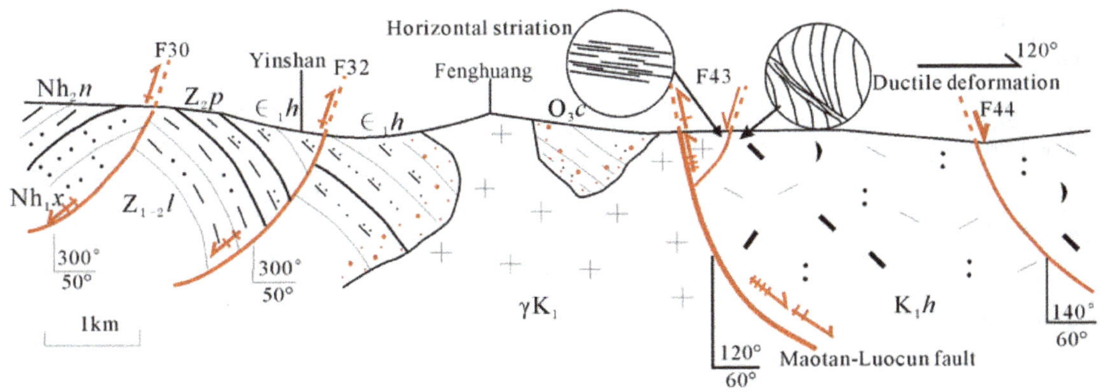

Fig. 4.17 Maotan-Luocun fault tectonic features

Xianxia composite pluton and the Tianmushan volcanic basin. In the middle stage of the Yanshanian, under a strong contraction regime, the Maotan-Luocun fault (Xuechuan-Huzhou fault) controlled the strong northwards and northwestwards thrusting of the volcanic rocks in the southeast, resulting in the uplift and denudation of extremely thick volcanic rock and plutons in the southeastern block of the fault. In the Late Yanshanian and the Himalayan, the stress relaxed, which caused that the preexisting basins dropped off again along the Maotan-Luocun fault (the Xuechuan-Huzhou fault) and thus controlled the volcanic basin fault depression.

Waichangcun-Chuancun Fault (F_{57})

Three to four faults developed subparallel to the Waichangcun-Chuancun fault in the investigation area, which belongs to the Zaoxi-Moganshan fault zone. The fault generally strikes NE 45° for about 15 km in the area, showing as a broken line trace of continuous distribution. Through remote-sensing image, it can be seen that the fault location is a tonal line where light and dark colors intersect. Relatively small-scaled fault triangular facets are seen along the fault. The fault surface is dominated by NW dipping and dip angle of 50°–70°. These faults are a series of the normal fault system.

4.2.2.2 General Fault

The fault structures in the investigation area were well developed, and there are certain orientation arrays in space, mainly including four groups of arrays: roughly EW striking, NE striking, NNE striking, and NW striking. In this area, faults in the same group of arrays also show features of multiple phases of activities, so the faults in different directions often cut, dislocate, and recombine each other, resulting in a complex manner.

NE-Trending Fault

There are a number of NE striking faults in the investigation area, which can be divided into early and late stages in terms of their scales and natures. Faults formed in the early stage are relatively large-scaled with a length of more than 10 km, featured by left-lateral compression-torsion. They mainly strike NE 30°–60°. Width of the fault zones range from several 10 cm to several 10 m, with tectonic lenses, cataclastic rocks, rock crumpled, cleavages, and fold cleavages developed throughout the zones. The attitudes of fault surfaces are mainly vertical and steep dipping to NW, with a few large faults dipping to SE. Formation time of the early NE striking faults is the same as that of the Early Indosinian folds. These faults are mainly striking NE 30°–60° with large scale and thrust domination.

The scale of the later formed NE striking fault is relatively small, which is characterized by extensional fault and is often related to volcanic eruption and magma intrusion. The late NE striking faults were formed as either reactivation of on the basis of the earlier fault or new ones They were especially intensely active during the Yanshanian, mainly characterized as: Intense extension in the early stage provided the main channel of magmatic intrusion in the Yanshanian to form NE extending volcanic-granite belts of composite plutons such as the Ma'anshan, Tangshe, and Xianxia in the survey area; strong thrusting in the middle stage cut through the Shiling-Shangmeilinchang anticlinorium to cause great loss of the southeastern limb of the anticlinorium; In the late stage, the stress relaxed and thus deformation was dominated by fault-block movements.

Features of the NE striking fault are listed in Table 4.8.

NNE Striking Faults

The NNE striking faults were less developed in the investigation area and are mostly characterized by sinistral

Table 4.8 Brief northeast striking faults in the investigation area

No.	Fault name	Strike/°	Attitude of surface/°	Nature	Length/km	Fault feature	Note
F1	Xiejiacun-Jiaocun	55	145∠70	First thrust, later normal fault	>7	Rock fragmentation, silicification, quartz vein body along the strike	Controlled by Tongkngcun-Qiguancun fault
F2	Shibilinchang-Lingxi	60	150∠80	Normal fault	4	Rock fragmentation, silicification, granite along the strike	
F6	North Tianfu Holiday Resort	45	135∠60	Thrust fault	>2.5	Rock fragmentation, strata loss, inconsistent stratigraphic occurrence on both sides	
F7	Tianfu Holiday Resort	45	135∠70	Thrust fault	>3.5	Rock fragmentation, fault gouge in fault rupture zone	
F9	Changqingwu-Dalingjiao	30	300∠70	Sinistral slip-thrust fault	>9	Multiphase activity, foliated cataclastic breccia, opposite strata occurrence	
F18	Shatianfan	40	310∠60	Thrust fault	4.5	Foliated splintering, fragmentation	
F27	Lizhikeng	30	300∠60	Normal fault	2.2	Extension, rock fragmentation, siliceous breccia	
F28	Baishawu	30	290∠70	Normal fault	2	Extension, quartz vein, striation	
F30	Shuijiali-Wuyunshan	30	300∠50	First thrust, later normal fault	8	Multiphase activity, foliated cataclastic breccia zone, multiphase activity	
F32	Yuntixiang-Zhanglicun	30	300∠50	First thrust, later normal fault	>16	Multiphase activity, foliated cataclastic breccia zone, multiphase activity	
F25	Tianluowu	30	310∠55	Thrust fault	>2	Rock fragmentation	
F35	Longtoukan	35	300∠50	Normal fault	2.5	Rock fragmentation, strong silicificationalteration	
F37	Puwutang	45	130∠60	Thrust fault	3.5	Rock fragmentation, pluton cutoff	
F38	Shinianwu	30	120∠60	Thrust fault	4.2	Rock fragmentation, pluton cutoff	
F39	Dengcun-Qingshanta	30	120∠65	First thrust, later normal fault	>8.5	Thrust, rock fragmentation, pluton cutoff, multiphase activity	
F40	Zhangjia-Huli	30	120∠60	First thrust, later normal fault	>8.5	Compression-shear, strata loss, silicification, fragmentation, multiphase activity	
F49	Xiangchunwu	30	300∠70	Normal fault	>8.5	Extension, rock fragmentation	
F51	Shiling	35	305∠60	Normal fault	2.5	Extension, rock fragmentation	

(continued)

Table 4.8 (continued)

No.	Fault name	Strike/°	Attitude of surface/°	Nature	Length/km	Fault feature	Note
F56	Laoyingke	40	310∠65	Normal fault	5	Extension, rock fragmentation, pluton cutoff	
F58	Longshangcun-Shaxishan	35	305∠60	Normal fault	11	Extension, rock fragmentation, pluton cutoff	A series of normal faults, controlled by the Waichangcun-Chuancun fault
F59	Wangjiawu-Lingcunwu	55	325∠60	Normal fault	11	Extension, rock fragmentation	
F60	Daxi-Litacun	50	140∠65	Normal fault	>13	Rock fragmentation	

Table 4.9 Brief NNE striking faults features

NO.	Fault name	Strike/°	Attitude of surface/°	Nature	Length/km	Fault features	Note
F4	Hengshanjiao-Hulingjiao	10–30	280∠65	Left-lateral-Normal fault	>10	Rock fragment zone, quartz vein along fault and fault effect are obvious	
F44	Longli	9	280∠60	Normal fault	2	1–2 m silicified fragment zone and felsite veins	
F46	Gaolingcun	20	290∠70	Normal fault	>2	Extensibility, fragmentation and felsite veins	
F47	Xiyoucun	18	288∠65	Normal fault	2.5	Fragment belt, Syenite porphyry veins	
F55	Wushanguan	7	277∠70	Normal fault	>2	Rock fragmentation	

strike-slip, and their mechanical properties are mainly left-lateral compressive torsion (Table 4.9). The fault zone is nearly vertical with a width ranging from 10 m to tens of meters, usually consisting of vertical cleavage bands. Striations were developed, and structural lens and structural breccia are seen in some faults.

NW Striking Faults

The NW striking faults in the investigation area are well developed and can be divided into two types according to their geneses. The first type is transverse faults associated with folds, of which the attitudes are generally affected by the structural positions where they are located within the folds, mainly striking NW 300°–330°. Mechanical properties of these faults are mainly tensile and tensional. They cut across the NE-trending folds and the axis parallel faults. With a large amount of throw, these faults show a significant effect of normal faulting and normal translation faulting. The type one fault was formed at the same time as the folds, beginning in the Caledonian-Indosinian. Formation of another NW striking faults is related to activities of the NE, NNE, and EW striking major faults, which were mostly formed in the late stage of the Early Cretaceous. Due to the strong thrusting from SE to NW in the investigation area, and under the effects of differential stress and boundary conditions, the NW striking faults were well developed to cut across and dislocate the major fault zone. They are hence of mainly tensional properties with a large scale and far in extension. Such faults are commonly formed by reactivation of the early NW striking transverse faults, showing complex features of multistage activities (Table 4.10).

Near EW-Trending Fault

Relatively less near EW striking faults developed in the investigation area, which can be roughly divided into two generations from the aspects of fault cutting relationship and relative scales. The first generation was formed as EW striking basement faults probably during the Caledonian, which is dominated by brittle–ductile deformation. They are mainly striking 70°–80° which is consistent with the trend of the Caledonian wide and gentle NEE-trending folds. In the investigation area, there may be EW striking basement faults developed in the position where the Tangshe pluton lies in. These basement faults extend eastwards to act as volcanic basin-controlling faults, which induced subsidence of the volcanic basin during later reactivation. Formation of the

Table 4.10 Brief NW striking faults

NO.	Fault name	Strike/°	Attitude of fault surface/°	Nature	Length/km	Fault feature	Note
F5	Xiaohangkeng	300	30∠60	Normal fault	3	Rock fragmentation, strong silicification alteration with inconsistent strata on both sides.	
F13	Xiaolingjiao	310	40∠65	Normal fault	3.5	Rock fragmentation, pluton cutoff.	
F11	Hanggaizhen-Xiaolingshang	300	210∠50	Left-lateral	7	Rock fragmentation, pluton cutoff.	
F16	Tangkengwu	295	25∠70	Normal fault	6	Thrust, rock fragmentation, pluton cutoff, multiphase activity.	
F15	Shizishan	300	30∠35	Thrust fault	2	Compression-shear, lost strata, silicification, fragmentation, multiphase activity.	
F23	Baiyangshan	290	200∠45	Normal fault	>2.5	Extensibility, rock fragmentation	
F33	Yinshanwu-Shenwulinchang	305	215∠70	Normal fault	>10	Extensibility, rock fragmentation.	
F34	Shenglijiawu	280	190∠65	Thrust fault, left-lateral	3.3	Extensibility	
F29	Dalingtou	305	215∠60	Normal fault	3.5	Extensibility, rock fragmentation, pluton cutoff.	
F31	Pingtoushan-Langcuncun	310	220∠60	Normal fault	5.8	Extensibility, rock fragmentation	
F42	Huangjiachang	325	235∠65	Normal fault, left-lateral	3.3	Extensibility, strata dislocation.	
F50	Xikou	320	50∠58	Normal fault	>2	Extensibility	
F52	Xianrentou	305	215∠45	Normal fault	3.5	Extensibility	
F54	Jinzhuping	330	240∠65	Normal fault	5	Extensibility, pluton cutoff.	
F53	Niutoujing	315	225∠75	Normal fault	2.5	Extensibility, pluton cutoff.	
F17	Paiwukou	282	12∠80	Normal fault	5.5	Extensibility, strata missing	
F45	Lantangli	280	10∠70	Normal fault	4	Extensibility	

second generation of EW striking faults was related to the Late Yanshanian fault-block activities, and most of them are brittle faults with shallow depth (Table 4.11).

Based on the above fault development features, it can be concluded that in the early stage, roughly EW or NEE striking faults were formed in the basement, which were dominated by thrusting faults. In the middle stage, NE striking faults were formed in the basement with main kinematics of thrusting. Most of these thrusts experienced reactivation later under the mechanical conditions changing from extensional normal faulting to compressive thrusting and then extensional normal faulting. NW striking faults were formed later or in the same age with the NE striking basement faults, which was dominated by the folds cutting transverse faults. In the late stage, NNE striking faults were formed with mainly left-lateral shearing. The shallow level roughly EW striking faults were formed latest, and most of them were brittle faults.

Table 4.11 Brief near EW striking faults

NO.	Fault name	Strike/°	Attitude of fault surface/°	Nature	Length/km	Fault feature	Note
F14	Gaocunlinchang	70	160∠60	Thrust fault	>3.0	Stratigraphic discontinuity on both sides, rock fragmentation, silicified belt, granitic veins	
F21	Wangjia	75	345∠40	Thrust fault	>2.0	Rock fragmentation	
F26	Gaokan	69	340∠85	Normal fault, left-lateral	3.5	Rock fragmentation, obvious fault effect	
F41	Laohuping	65	155∠85	Normal fault	3.5	Rock fragmentation	
F8	Shikengwu	78	170∠70	Thrust fault	>4.0	A few of schistosity, rock fragmentation	
F10	Dongxiwu	75	165∠80	Thrust fault	>5.6	A few of schistosity, rock fragmentation	
F19	Xiazhang	70	160∠60	Normal fault	2.5	Stratigraphic discontinuity, rock fragmentation	
F20	Wenkouao-Lizhikeng	95	5∠45	Thrust fault	5	Rock fragmentation, subangular breccia	
F36	Changwa-Zhaojia	80	170∠65	Normal fault	>2	Stratigraphic discontinuity, rock fragmentation	

4.3 Structure Hierarchy and Structure Sequence Division

Although structures of the investigation area are very complicated and diverse, however, if we look through the phenomenon and explore its essence, we will find that they have certain rules to follow, no matter in spatial arrangement and distribution or development sequence in time. In terms of the view that the various structures produced by a certain tectonic movement under a certain mechanism have a genetic relationship. According to the characteristics of structural development patterns, strain facies, unconformities, structural compound and dike groups, tectonic generations were divided and tectonic sequences were established (Table 4.12).

4.3.1 Chengjiang—Caledonian Deformation Cycle

4.3.1.1 General Uplifting in the Chengjiang Period (D_1)

After the consolidation of metamorphic basement of the Nanhuaian, the area was in a stage of passive continental margin development. Since the Nanhuaian Period, large-scale transgression has deposited a set of Xiuning Formation sandstone and its equivalent strata dominated by littoral–epicontinental marine clastic deposit formation. With the influence of the Chengjiang movement, the investigation area was uplifted and formed into a land, which suffered from the short time of weathering and denudation, resulting in the disconformity between the glacial strata of the Nantuo Formation and the Xiuning Formation. In the Early Sinian, the sealevel eustasy occurred frequently, resulting in the disconformity between the Lantian Formation of the Sinian and the Nantuo Formation of the Nanhuaian. In this period, the overall uplift was dominant and the fold deformation was weak.

4.3.1.2 Near EW-Trending Wide and Gentle Fold in the Caledonian (D_2)

The deformation of this period was mainly developed in the old strata during and before the Silurian age. The investigation area was under the stress field of near SN compression though, still carried on the main uplifting movement succeeding from those in the Chengjiang Period. At the same time, deformation of the rock strata was dominated by relatively weak plastic manner that produced relatively close anticlines and synclines of near EW-trending and possibly gave birth to nearly EW striking basement faults in the investigation area. For example, the features of residual multistage compression are seen in the anticline core in the Wangjia-Tangshecun anticline and in the formation fault in the northern boundary of the Tianmu Mountain volcanic depression.

Table 4.12 Tectonic complex events in the investigation area

Tectonic event							Magmatic event	Sedimentary event	Tectonic movement (Age/Ma)
Deformation cycle	Generation	Tectonic environment	Regime	Tectonic type	Movement direction	Deformation phase			
Himalayanian cycle	D_5	Continental margin active belt	Extension	Uneven rise and fall movement, faulted basins		Brittle shear fault deformation phase		Alluvial, diluvial, slope deposits	Himalayan movement
Yanshanian cycle	$D_{4\text{-}3}$		Contraction	NNE left-lateral faults, NW-lateral faults; NE striking boundary faults reactivation and strong thrusting from SE to NW	← NW	Brittle shearing, brittle failure		Missing	(65) Late Yanshan movement
	$D_{4\text{-}2}$	Continental margin active belt	Extension	Formation of volcanic basins and development of brittle normal faults in NE striking	NW-SE	Magma thermodynamic deformation phase	Volcanic eruption-magmatic intrusion	Terrestrial volcanic clastic	(>96) Early Yanshan movement
	$D_{4\text{-}1}$		Extension	Foreland compression basins, intermountain faulted basins	← NE	Rifting, and brittleness		Missing	(>145)
Indosinian cycle	$D_{3\text{-}2}$	Epicontinental sea basin	Contraction	NW-trending folds, fold superposition and nose structure	NE-SW	Elastoplastic longitudinal bending deformation phase			Late Indosinian movement (205)
	$D_{3\text{-}1}$		Contraction	NE-trending large anticlinal structures and secondary folds	← NW	Elastoplastic longitudinal bending deformation phase			(?) Caledonian Early Indosinian movement
Caledonian cycle	D_2	Peripheral foreland basin Passive marginal basin	Contraction uplifting	Relative tight and roughly EW and NEE-trending folds	← SN	Elastoplastic longitudinal bending deformation phase		Cinerous silty mudstone of Xiaxiang Formation Manganese-bearing dolomite deposit at the bottom of Lantian Formation	(410) Chengjiang movement
Chengjiang cycle	D_1	Continental rift basin	Uplifting	Upper and lower strata disconformity	↑	General uplifting		Argillaceous weathering crust at the top off Nantuo Formation	(680)

4.3.2 Indosinian Cycle

4.3.2.1 NE-Trending Wide and Gentle Folds in the Early Indosinian (D_{3-1})

Deformation in this period is the most important and largest scaled structural deformation in the investigated area, which has laid the basic structural pattern of the NE-trending folds in the investigation area. In the Early Indosinian, the investigation area was under the SE-NW compression and contraction regime, which was dominated by plastic deformation, forming large-scaled NE-trending short-axial folds that superimposed on the Caledonian wide and gentle NEE-trending folds in a roughly coaxial way. At the same time, the Nanhuaian Xiuning Formation and Nantuo Formation in the cores of the Caledonian wide and gentle folds are weak layers. The cores of the folds are easily tightened under compression. The Caledonian anticlines are close, and their axial traces show the bent and refolded features. The overall framework of the Caledonian folds is characterized by wide and gentle NE-trending synclines and relatively close and S-shaped bending anticlines. The prototype of Hanggai-Fushi synclinorium in the north of the investigation area, the Shiling-Shangmeilinchang anticlinorium in the middle, and the Yuqian synclinorium of volcanic basin in the southeast of the investigation area were formed. With further compressive deformation in the direction of SE-NW, the important boundary fault, the Maotan-Luocun fault (the Xuechuan-Huzhou fault) was formed.

4.3.2.2 NW-Trending Folds in the Late Indosinian (D_{3-2})

Deformation in this period is a relatively weak deformation in the Late Indosinian, subjected to the stress field of SW-NE compression (Fig. 4.18) and dominated by plastic deformation, which resulted in formation of NW-trending folds that overprinted the NE-trending folds of the Early Indosinian. This phase of folding did not produce large-scale dome–basin structures in the investigation area, but only locally small dome–basin and nose structures were formed by superimposition on the secondary folds of the NE-trending folds. There are also small tectonic basins and domes in the western periphery of the investigation area. Therefore, it could be speculated that the NW-trending folds were relatively weak deformation in the Late Indosinian, which is quite different from the scale of NW-trending folds formed in the Early Indosinian. For example, the NW-trending nose-shaped structures developed in Gaocunlinchang in the southwest of the Hanggai map and Damiao of the Xianxia map, and series of small dome–basin structures were superimposed and formed in Shangmeilinchang and Yaocun-Wucun where near EW-trending secondary fold were developed.

4.3.3 Yanshanian Cycle

The diversity of tectonic activities in the Yanshanian is the main feature of intracontinental orogeny in the investigation area, forming diagnostic faults, fold system, and magmatic evolution sequences, as well as fault uplift and basin structures. Therefore, the Yanshanian tectonic period had a profound influence on the region that deformation transformed from formation of marginal sea to formation of intracontinental basins, from ductile (brittle) to brittle manners, from folding dominant to fault-block lifting and translation shearing. Deformation in the Yanshanian played a basic role of the foundation of the modern tectonic features of the investigation area. According to the developmental conditions of the strata, intrusive rocks and volcanic rocks in the investigation, and combining with the regional geological evolution features, the Yanshanian was divided into the Early Yanshanian and the Late Yanshanian by using the Early Yanshanian movement (145 Ma) as the boundary.

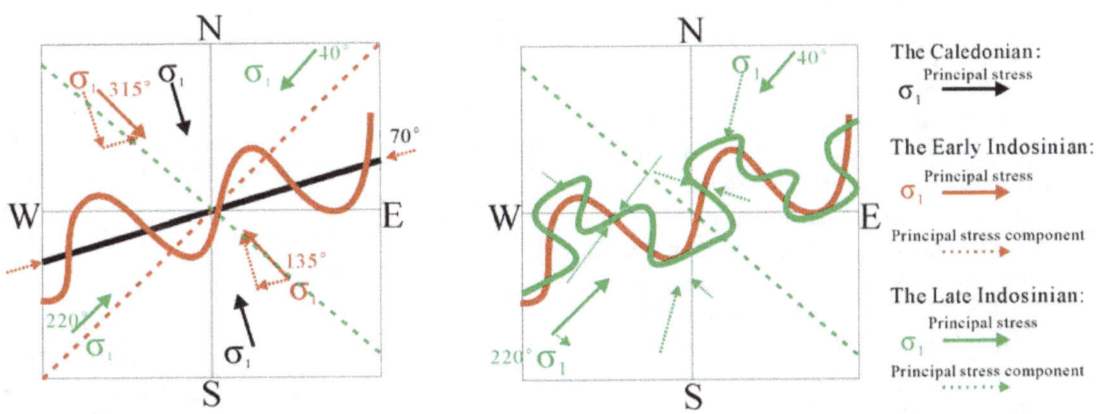

Fig. 4.18 Ideal model of the Caledonian-Indosinian anticlinal axes deformation in the investigation area

4.3.3.1 The Early Yanshanian (Jurassic)

In the Early Yanshanian, the investigation area came to a new stage of development of continental active margin, and the littoral Pacific domain dominated by the formation of terrestrial basin and mountain structures.

Rifting ($D_{4\text{-}1}$)

In the Early Yanshanian (Jurassic), the regional stress field changed from compression in NW-SE direction to extension. The cores of the anticlines and synclines ruptured to form foreland compression basins and intermountain faulted basins.

4.3.3.2 Late Yanshanian (Cretaceous)

The Late Yanshanian (Cretaceous) tectonic movement was complex, when volcanic sedimentation basins, large composite plutons, strong thrust nappe structures, and complex fault-block activities were formed. In the investigation area, it can be divided into three distinct generations: the early, the middle, and the late.

Early stage: Rifting, intermediate-acid volcanic sedimentation basin, granite intrusion ($D_{4\text{-}2}$)

In the Early Cretaceous, the rifting process was characterized by magmatic activity from intermediate-acid magma to intense volcanic eruption. Under the control of regional NE- and NNE-trending structures, broad and gentle volcanic domes and volcanic basins with short axes were formed; the imbalanced structure uplifting was occurred; the NE, NNE, and NW striking faults were reactivated; and the fold deformation was weak.

Middle stage: Thrusting structure ($D_{4\text{-}3}$)

Controlled by the littoral Pacific domain, the NE striking boundary fault zones were strongly active. Regionally, large-scaled thrust nappe moving from SE to NW developed, which strongly modified the early structures to form a series of ductile and brittle shear zones, compressive shear schistosity zones (130°–160°∠25°–60°), broken cleavage belts, imbricate thrust zones, overturned fold fragments, drag folds, and squeezed the lens. In the investigation area, the main features of deformation are characterized by the thrust shearing along the Maotan-Luocun fault formed in the Early Indosinian, which cut across the Shiling-Shangmeilinchang anticlinorium, as a result most of the southeastern wing of the large anticlinorium was missing. The large faults also deformed the early formed volcanic rocks nearby them. Secondly, the near-vertical Tongkengcun-Qiguancun main fault was thrusted along the NW in this period and coupled with left-lateral shear, the southeastern plate of this fault suffered from denudation. Stress relaxation induced the formation of dominant extensional faults and, bedding parallel gravity sliding, recombination of the newly formed NNE striking sinistral shearing faults and the NW and NS striking faults, and fault-block movement instead of fault-fold movement.

4.3.4 Himalayan Cycle

Himalayan fault effect (D_5)

In the Himalayan, the investigation area experienced a relatively tectonic quiet interval. The mode of tectonic stress gradually turns to tension shear and forms a series of faulted basins. It is mainly composed of uneven and differential lifting movements, accompanied by deep fault activities. Quaternary sedimentation is mainly distributed in the Xitiao River basin of the investigation area. Mountain glacier activity and river erosion shaped the modern topography.

4.4 Controlling Relationship Between Structure and Strata, Ore, and Basin

4.4.1 Fault-Controlled Rock and Ore

In the Late Yanshanian, spatial distribution of magmatic intrusion, volcanic eruption, and endogenetic ore deposits was obviously controlled by faults and folds.

4.4.1.1 Fault-Controlled Rock

The Late Yanshanian (after 145 Ma) is the most important magmatic activity period as well as ore-controlling and metallogenic period. The volcanic eruption was unconformable above the preformed basement rock. The multi-stage pulsating intrusive rocks formed the composite plutons with different scales in the investigation area. Controlled by basement structures or fault structures, the general tectonic lines of volcano-intrusive rock belts in the investigation area are trending in NE direction, being consistent with the trends of basement structures. For example, the distribution of the Ma'anshan pluton, the Xianxia pluton, the Yangtianping-Dongling pluton, and the Tonglizhuang pluton in the investigation area are obviously controlled by the NE striking faults. In the middle of the investigation area, the plutons in EW trending are distributed along the edges of volcanic basins extending to the Tangshe pluton. It is speculated that the volcanic basin margins in the investigation area may be controlled by near EW striking blind faults.

4.4.1.2 Fault-Controlled Ore

Faults of different scales play different roles in the process of mineralization. Those faults of regional scales generally control the distribution of metallogenic belts. In the middle of the investigation area, the dense distribution of EW-trending ore deposits and ore spots is particularly obvious. It is speculated that this distribution pattern is closely related to the intersection of the NE striking fault zones and the possible near EW striking basement fault zones.

There are three major groups of faulted structures in the investigation area: NE striking, roughly EW striking, and NW striking. The fault structures closely related to mineralization are mainly NE striking and roughly NW striking, and the NW striking faults are mostly ore-bearing structures. The endogenetic metal deposits are mainly controlled by NE striking and roughly EW striking fault structures, especially the intersection of faults with different strikes. For example, mineralization of the Gaocun antimony deposit occurred in the fractures of siliceous rocks of the Cambrian Hetang Formation in the northern part of the Tangshe pluton. The ore body is controlled by the NW striking normal faults. Hehuatang scheelite formed in the NNW 340° striking faults in the Tangshe monzogranite pluton, and the surrounding rock has experienced silicification and silky mica-quartzite.

4.4.2 Fold-Controlled Rock and Ore

Strong tectonic activity makes the strata in the investigation area strongly folded, providing favorable conditions for mineralization. In the process of fold formation, due to the different stresses in different places, folds of different scales and different curvatures are formed. It is easy to form layer-parallel slip and detachment in the core and limbs of a fold, as well as the folded layers are prone to be deformed by joints, fractures, and cleavages, which creates space for the migration, precipitation, and occurrence of ore-bearing hydrothermal fluid. The most obvious ore-controlling folds in the investigation area are those anticlines with the cores of the Nanhuaian and the Sinian strata. The deposit is mainly located in the Nanhuaian–Sinian strata where the structural positions are produced by the intersection of EW-trending and NE-trending structures. The main fault is 30°–35° NE striking, dipping to SE. In the early stage, it is compressive and torsional, and in the late tensile stage, it becomes ore-bearing structures containing, e.g., tungsten ore. Such as the Wangjia-Tangshecun anticline, its core is extensionally broken. Due to the influence of regional faulting in the later stage, the NE striking and roughly EW striking faults developed. The core of the faults was later intruded by the Tangshe pluton, and contact metasomatic deposits are often formed around the contact site between the pluton and the Sinian strata.

4.4.3 Structure-Controlled Basin

The basin types in the investigation area are relatively simple, with the volcanic tectonic basin most developed, which is clearly controlled by the structures and formed by the combined action of nearly EW and NE striking faults and fold deformation. For example, the volcanic basin of East Tianmu mountain in the investigation area is bounded by nearly EW striking blind faults and NE striking faults. The blind faults of near EW strike were formed early, possibly in the Caledonian or even earlier. It mainly dips to the south, forming the northern edge of the volcanic basin. In the Early Indosinian, a large wide and gentle synclinorium was formed, that is, the Yuqian synclinorium, which became the embryonic form of volcanic basin. The NE striking Maotan-Luocun fault (the Xuechuan-Huzhou fault) was formed and became the western edge of the volcanic basin. Formation of the Maotan-Luocun fault (the Xuechuan-Huzhou fault) plays a significant role in controlling the formation and development of the volcanic basin at Tianmu Mountain: rifting in the Early Yanshanian reactivated the boundary faults formed early. The early synclinal sag experienced subsidence along the boundary faults, strong volcanic eruption, and granite intrusion. In the Middle Yanshanian, under the strong contraction regime, the Maotan-Luocun fault controlled the strong northwards and northwestwards thrusting of the volcanic rocks in the southeast. While in the Late Yanshanian, the stress was relaxed to cause that the early basins again collapsed along the Maotan-Luocun fault.

4.5 Tectonic Development

The investigation area is located in the southeast margin of the lower Yangtze block and has undergone four tectonic evolution stages, namely Chengjiang-Caledonian epicontinental cover orogeny, Indosinian intracontinental orogeny, Yanshanian active continental margin, and Himalayan tectonic uplift (Fig. 4.19).

4.5.1 Tectonic Evolutionary Stage of Epicontinental Cover Orogeny in Chengjiang-Caledonian Continental Margin Caprock

4.5.1.1 Continental Margin Rift Basin Stage in the Early Nanhuaian Period

In the Early Nanhuaian Period, with the occurrence of the Rodinea supercontinent breakup, the rift basin developed in the southeastern margin of the Yangtze block (where the Proterozoic island arc zone was exposed). With the large-scaled transgression, the tidal flat-shallow shelf facies

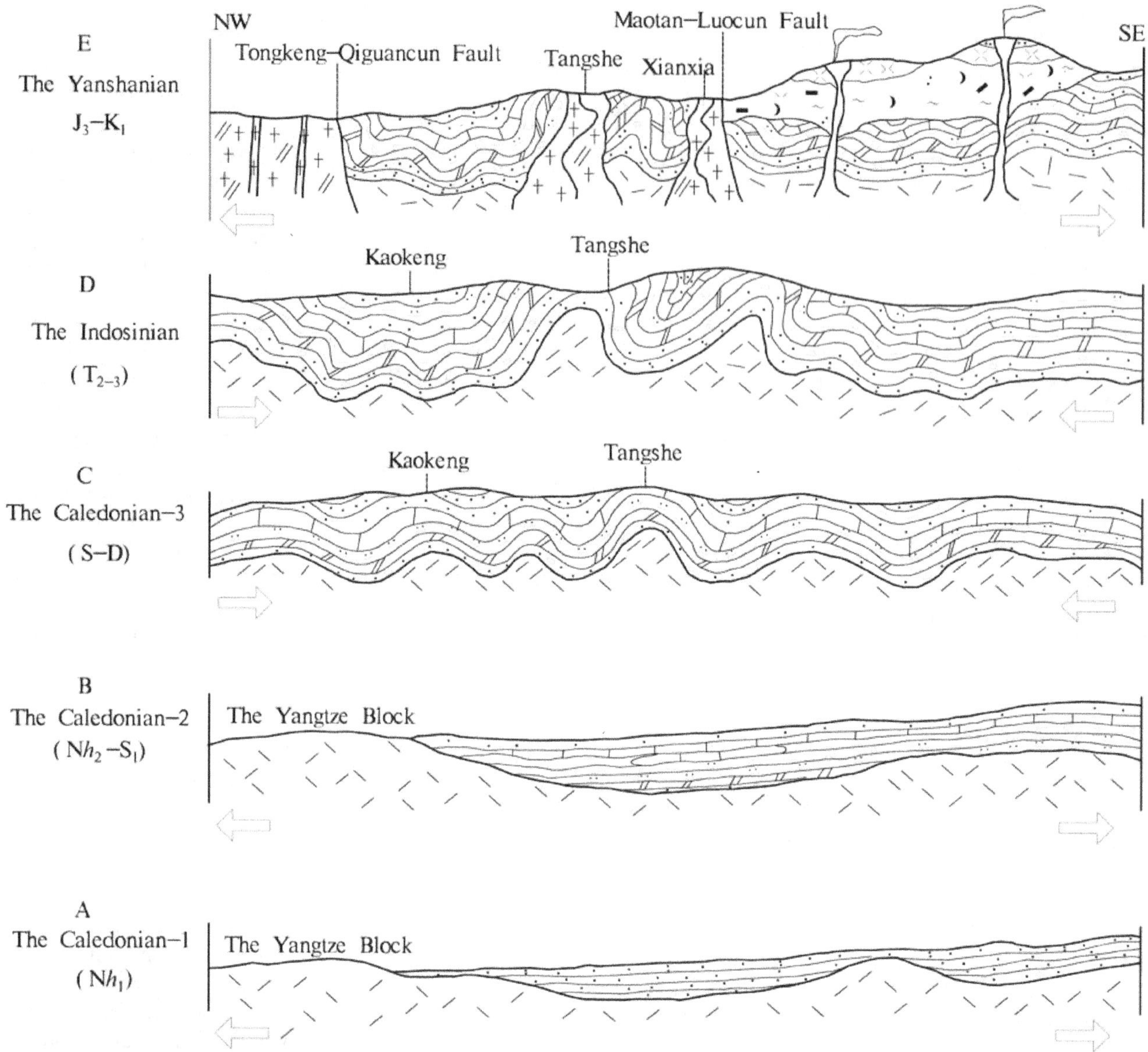

Fig. 4.19 Structure evolution pattern diagram of the investigation area

Xiuning Formation flysch was formed in Tangshe area of the investigation area. In the Early Xiuning Period, the glutenite and sandstone rock assemblage of the lower Xiuning Formation was deposited, which was missing later in the investigation area. Conglomerate in the bottom of the Xiuning Formation is unconformable on the underlying Neoproterozoic strata and pluton, representing existence of a global tectonic event. In the Middle Xiuning Period, with the further expanding transgression range and seawater deepening, sandstone and silt rock formations of tidal flat facies were formed in the Xiuning Formation, developing flaser bedding and current bedding. In the Late Xiuning Period, sandstone and mudstone formations were formed. A great amount of tuffite and tuff sandstone intercalations were generated during the Middle and Late Xiuning Periods, indicating that there were intermittent volcanic eruption activities.

4.5.1.2 Passive Marginal Basin in the Late Nanhuaian–Early Stage of the Late Ordovician

In the Nantuo Period of the Late Nanhuaian age, the climate became cold to form ice water deposits representing by a set of tillite-gravelly sand-mudstone. In the middle period, the climate warmed briefly and deposited siliceous, dolomitic and manganous mudstones, which indicates that sedimentation in the investigation area had entered the mixed deposit stage of passive continental margin and generally belonged to neritic shelf facies deposition.

From the Early Sinian to the early stage of the Late Sinian, the climate became warmer and the sea area expanded, forming a shallow sea shelf, carbonate platform and subabyssal basin facies deposition of the Lantian Formation. Due to the frequent sealevel eustasy, carbonate formation, siliceous mudstone-carbonaceous mudstone formation, carbonate formation, and argillaceous formation were successively formed from bottom to top. In the Late Sinian, the sea level firstly rose and then fell, which forms the subabyssal sedimentary facies siliceous rocks of Piyuancun Formation, siliceous mudstones, and neritic shelf facies carbonaceous siliceous mudstones of the Piyuancun Formation. The Ediacaran biota developed in the area during this period, showing the features of oxidation environment.

The Hetang stage in the Early Cambrian continued the paleogeographic pattern of the Sinian. Formation of black siliceous rocks, siliceous mudstones, and carbonaceous mudstones of the deep-subabyssal basin facies rich in sulfur, phosphorus, and other organic matters is due to subsistence of abundant organisms. Trilobites, brachiopoda, small crustaceans, sponges, and spicules were formed in the whole region; however, no fossils were found in the survey area. In the Early stage of the Middle Cambrian (the Dachenling Period), formation of carbonate rocks in the neritic continental shelf facies was mainly formed, and the argillaceous–siliceous deposits in the subabyssal basin facies were partly formed due to the change of sea level. In the Late stage of the Middle Cambrian (the Yangliugang Period), the sea area expanded further with frequent sealevel eustasy, which induced that the sedimentary features of the subabyssal basin deepwater shelf and shallow sea shelf facies occurred repeatedly, and carbonate formation and siliceous mudstone formation were formed. During the Huayansi-Xiyangshan stage of the Late Cambrian, the seawater became shallow and the muddy sediments increased, so that the investigation area entered the shallow sea shelf carbonate platform facies deposition, and thus sedimentary thin layer argillaceous carbonate rock formation and argillaceous rock formation were formed.

During the Yinzhubu-Ningguo Period of the Early Ordovician, sedimentary environment in the investigation area changed from the shallow sea to the sub-deep sea, which gave birth to deposition of argillaceous and silty mudstone formations of sedimentary shallow sea shelf facies and subabyssal basin facies. In the Ningguo Period, planktonic graptolites were the most abundant and could be divided into as many as ten graptolith zones, and two relatively complete graptoliths were found in the investigation area. Hule Period in the Middle Ordovician inherited the depositional environment of the Ningguo Period and formed the siliceous rock formation and siliceous mudstone formation of subabyssal basin-abyssal basin facies. Graptoliths also appeared on a large scale. Three graptoliths were found in the investigation area. In the Late Ordovician Yanwashan Period, the seawater gradually became shallower and was deposited as shallow marine shelf carbonate rocks, forming nodular marl formation, and containing a few conodonts.

4.5.1.3 Peripheral Foreland Basin of the Late Stage of the Late Ordovician–Middle Silurian

With the intensification of the Caledonian movement, the investigation area changed from passive marginal basin to peripheral foreland basin sedimentation.

During the Late Ordovician Yanwashan-Huangnigang Period, the sea level gradually became shallower, so that shallow sea shelf to deepwater shelf carbonate facies was deposited to form calcareous mudstone and nodular marl formations containing a large number of benthic trilobites, brachiopoda, and a small number of planktonic globular plectozoites. During the Changwu Period, it was composed of abyssal slope–abyssal shelf flysch–flysch-like deposits, forming silty mudstone and siltstone formation, and developing typical Bouma sequence rock assemblages such as AE, ABE, and ADE. The organisms are mainly planktonic graptolites, micropaleontology and benthic mollusks, and chitin worm is found in the investigation area. As a result of the intermittent volcanic activity in this period, as many as 47 layers of porphyry interlayer were found in the strata of the Huangnigang Formation and the Changwu Formation. In the Wenchang Period, the facies of neritic shelf and subabyssal basin formed sandstone formation, siltstone formation, and carbonaceous rock formation. The period experienced the Hernant global glacier event at the end of the Ordovician, presenting the characteristics of two main glacial periods and one interglacial period, when the *Normalograptus extraordinarius* graptolith belt, *Songxites–Aegiromenella* fauna, and *Normalograptus persculptus* graptolith belt were developed.

In the Xiaxiang stage of the Early Silurian, the seawater in the investigation area became shallow so that shallow sea shelf facies deposited, to form carbonaceous siliceous mudstone formation, silty mudstone formation, and sandstone formation. During the Middle and Late Silurian, the marine area further narrowed to cause all the investigation areas uplifted, and thus, no deposition was occurred.

Since the Early Paleozoic, the geological process of subduction-collision-collage of the Neoproterozoic island arc belt downward Yangtze block has ended. Consequently, the investigation area has become the stable block area and presented mainly passive marginal basin sedimentation. During the Caledonian, the Yangtze block subducted under the Cathaysian block along the Jiangshan-Shaoxing docking belt and the collision occurred. Affected by the subduction and collision and also the near EW-trending Changhua-Putuo fault zone (outside the area), the investigation area is featured by strong fold orogeny, volcanic tectonic events, and abrupt changes of sedimentary environment. The main evidences are listed as follows: (1) Under the effect of near NS compressive stress field, the Nanhuaian-Cambrian strata was deformed to be the Jurassic style folds with nearly EW-trending axes. The typical examples are the close anticlines formed by the Xiuning Formation–Nantuo Formation–Lantian Formation–Piyuancun Formation and exposed in Tangshe. (2) As many as 47 layers of bentonite (volcanic ash) were found in the strata of the Late Ordovician Huangnigang-Changwu Formation. (3) Thickness and deposition rate of the Changwu Formation have great changes compared with that of the Yanwashan Formation and the Huangnigang Formation. The deposition rate and thickness of the Changwu Formation are dozens of times as fast and thick as those of the Yanwashan and the Huangnigang Formations. At the same time, the features of the inverted grain order appeared in the Changwu Formation.

4.5.2 Continental Tectonic Evolution Stage in the Indosinian

The Caledonian Yangtze block and the Cathaysian orogenic system were joined together to form a unified continent. In the Mid-Late Silurian–Devonian the investigation area was uplifted into a land without any deposition. During the Early Carboniferous to the Early Triassic, the northwestern Zhejiang Province started the epicontinental sea deposition stage and formed a series of trough-like epicontinental seas trending NE. Due to the elevation higher than sea level, deposition did not occur in the investigation area during this period.

At the end of the Middle Triassic, under the effect of the NW-SE tectonic stress field, the NE-trending folds were formed and superimposed on the NW-trending Caledonian folds in the investigation area. The main structures in the area are represented by the Hanggai-Fushi large synclinorium and Shiling-Shangmeilinchang large anticlinorium with the hinges plunging NE. The large area between the investigation area and the Jiangshan-Shaoxing butt zone in this period is characterized by the development of the NE-trending close fold structures.

4.5.3 Active Continental Margin Stage in the Yanshanian

Due to the continuous subduction of the ancient Pacific Ocean to the South China block during the Yanshanian, the investigation area started to experience an active continental margin stage and characterized by intense magmatic activity.

In the early stage of the Early Cretaceous, bounded by the Maotan-Luocun fault zone, fault basins were formed under the NW-SE tensile stress field in the western side of the investigation area, with a large amount of overall subsidence, followed by volcanic magma activities. The northwestern side is mainly featured by magma intrusion, forming Ma'anshan, Tangshe, and Xianxia composite plutons. The Ma'anshan composite pluton intruded along the Tongken-Qiguancun fault zone. The Tangshe composite pluton intruded into the cores of the near EW-trending anticlines formed in the Caledonian. The Xianxia composite pluton intruded along the Maotan-Luocun fault zone. These composite plutons are spatially distributed in a zone of gradually narrowing from SE to NE; intrusion of the composite plutons shows time sequence of getting younger from SW to NE. In the southeast, there were strong continental volcanic eruptions and intermediate-acid magma intrusions. The volcanic activity in this period experienced a relatively complete volcanic eruption cycle in the period of the Huangjian Formation and four volcanic eruption rhythms. The typical volcanic structures were formed, such as the Yangtianping revived caldera, Tianhuangping caldera, West Tianmu-Longwangshan dome volcano. After an eruption of the volcanoes, volcanic magma intrusion developed along the northern and northeastern boundaries of the outlying volcanic rocks.

In the late stage of the Early Cretaceous, the investigation area was featured by a NW-SE compressive tectonic stress field, and thrusts showed movement from SE to NW. The Tongkengcun-Qiguancun fault zone is mainly characterized by a series of acidic dike intrusion and thrust uplift and denudation on the southeastern side. The Maotan-Luocun fault zone thrust and left-lateral sheared the southeastern limb of the Hanggai-Fushi large synclinorium, which causes the strata of this limb incomplete. At the same time, the volcanic rocks are in fault contact with Paleozoic strata and the Xianxia composite pluton.

4.5.4 Tectonic Uplift Stage in the Himalayan

In the Himalayan, the investigation area has experienced a relatively quiet interval. Deformation was dominated by differential vertical movement of ascending and descending. The 1400-m-high tectonic planation surface in Tianhuangping is the most typical case.

Reference

Anhui Institute of Geological Survey (2003) The regional geological survey of Xuancheng map sheet on a scale of 1:250,000[R] (in Chinese). National Geological Archives of China [distributor], 2018. http://ngac.org.cn/Data/FileList.aspx?MetaId=F5F68B3483841B60E0430100007F0760&Mdidnt=d00122329

5 Summary and Reflections

Under the care and support of the leaders and experts from the executing entity of the Nanjing Center, China Geological Survey, and the undertaking entity of Zhejiang Institute of Geological Survey, the project team of the *Regional Geological and Mineral Surveys of Hanggai Map sheet (H50E009022), Xianxia Map sheet (H50E010022) and Chuancun Map sheet (H50E010023) in Zhejiang (1:50,000)* has worked hard and assiduously for more than three years and successfully completed various tasks. Based on the systematic analysis of this survey and previous data, abundant geological data have been obtained about the strata, structures, intrusives, volcanic rocks, and mineral deposits, together with a number of research results. The most outstanding results among them were the establishment of the "Standard Cross Section of the Lower Yangtze Region in the Upper Ordovician Hirnantian" of Hanggai Town, Anji County, Zhejiang, the establishment of ten graptolithic zones in the Ordovician-Silurian, the first discovery of the sponge fossils in the Late Ordovician Hirnantian, and the development of the Late Ordovician volcanic events. These results not only provide basic and detailed original data, but also improved the research level of related professional fields in the survey area, Zhejiang Province and South China. It would provide abundant evidences of biological evolution and volcanic events for further stratigraphic study and the Caledonian tectonic event studies.

The project results have been converted and applied in geological relic planning and protection, scientific research, geological science popularization, college teaching practice, mineral exploration guidance, and other aspects and have achieved significant social benefits. The project achievements were rated as the top ten geological achievements of the "12th Five-year plan" in Zhejiang Province. The "standard cross section" has been listed as a provincial geological relic for protection by the Anji Municipal Land Resources Bureau and People's Government of Hanggai Town. From April 2013 to now, researchers from the Nanjing Institute of Geology and Paleontology, Chinese Academy of Sciences, Key Laboratory of Resource Stratigraphy and Paleogeography, Chinese Academy of Sciences, the National Museum of Wales, Academician and Expert Workstation of Nuclear Engineering Jingxiang Construction Group Co., Ltd., have repeatedly investigated and studied the sedimentary stratigraphic sections in the surveyed area. The School of Earth Sciences, Zhejiang University, has organized students to conduct field internships in the investigation area many times and obtained good practice teaching achievements. The Hangzhou Daily, Xiaoshan Daily, and Huzhou Aishan Primary School have organized several batches of primary and secondary school students to carry out geological science popularization activities in typical sections and important fossil sites found in the project and greatly improved the general public's awareness of geological work. Up till now, the "standard cross section" has become an important base for geological science popularization education of the Hangzhou Newspaper Group.

5.1 Innovative Results

5.1.1 Established the "Standard Cross Section of the Lower Yangtze Region of the Upper Ordovician Hirnantian" in Hanggai, Anji County, Zhejiang

Through the comprehensive study and analysis of multiple strata, including lithostratigraphy, sequence stratigraphy, biostratigraphy, chronostratigraphy and chemical strata, and the discussion of sedimentary environment, as well as the comprehensive comparison with stratotype section, the standard cross section of the lower Yangtze Region of the Upper Ordovician Hirnantian and the typical section of the boundary between the Ordovician and the Silurian are established for the first time. It provides a reference for the study of regional stratigraphic correlation and greatly improves the level of stratigraphic research in the lower Yangtze Region.

In November 2015, the All China Commission of Stratigraphy organized relevant academician experts to conduct field investigation and demonstration of the declared "Standard Section of the Lower Yangtze Region in the Upper Ordovician Hirnantian" of Hanggai, Anji County, Zhejiang. At the same time, the Commission advised Zhejiang Provincial Department of Land and Resources, the People's Government of Anji County, and Zhejiang Institute of Geological Survey to incorporate the protection and research of Hanggai "Standard Section" in Anji County into the planning of geological environment protection and ecological construction, and further strengthen the protection and utilization of the section and scientific research. The Commission believed that the following innovative results had been achieved in the section:

1. There was a complete outcropping of the Hirnantian and Ordovician-Silurian boundary strata in the Hanggai "Standard Section" of Anji, Zhejiang Province. Cyclic deposition of graptolite shale and fine-grained clastic rock with a thickness of more than 360 m developed and the chemical stratigraphic studies of carbon isotopes have been carried out. No structural deformation is developed near the main boundaries, and the geological events and biological sequences of the Hirnantian have been completely recorded, which meet the requirements of establishing the standard stratigraphic section of the Hirnantian and the Ordovician-Silurian boundaries in the Lower Yangtze region.
2. The section is rich in well-preserved fossils. Identified in the Upper Ordovician Katian-Hirnantian and the Lower Silurian Rhuddanian strata were six graptolite zones, including *Dicellograptus complexus, Paraothograptus pacificus, Normalograptus extraordinarius, Normalograptus persculptus, Akidograptus ascensus, Parakidograptus acuminatus*, and 1 (one) crustacean fauna *Songxites–Aegiromenella*, as well as diverse graptoliths, Chitinozoans, sponges, trilobites, gastropods, brachiopoda, cephalopods, and other phylum fossils.
3. Rich sponge fauna (more than 10 genera and species) was first discovered in the Hirnantian, opening an important window for further understanding the global biosphere appearance of the Hirnantian and the evolution of sponge organisms after the Cambrian life explosion.

5.1.2 Fossils of Sponges in the Upper Ordovician Hirnantian Have Been Discovered for the First Time

Throughout the field geological investigation, a large number of sponge fossils were found in the special lithologic layer of black carbonaceous shale in the middle of the Wenchang Formation of the Upper Ordovician Hirnantian in the survey area. It is the first discovery of this contemporaneous horizon, which has important research significance in the following three aspects. (1) It is the initial discovery in the strata during this period, with great research value. (2) These sponges are of the Burgess Shale type, which was previously believed to have died out in the first major extinction event at the early stage of the Late Ordovician Hirnantian, while the sponges in the survey area were symbiotic association with the *Normalograptus persculptus* graptolith belt at the end of the Hirnantian, showing that it lived till the end of the Hirnantian and did not die out during the glacial period. (3) These sponges are generally growing in an oxygen-rich environment, while the sponges in the survey area are found in the carbonaceous shale that indicated an oxygen-deficient environment. It is suggested that the survey area was in a very special paleogeographic environment during that period, having important research significance in paleoecology, bio-paleogeography, and evolutionary paleontology.

Through the follow-up study of the Nanjing Institute of Geology and Paleontology, Chinese Academy of Sciences, more than 75 species of sponge fossils in the section have been identified so far, significantly exceeding the total number of sponge genera collected over 100 years in the famous Cambrian burgess shale in Canada. It is the most abundant and diverse sponge fauna found during the geological history, revealing that the residual seafloor after the great catastrophe was not as quiet and desolate as previously thought, but still had abundant biological reproduction.

5.2 Other Results

5.2.1 Strata

(1) Stratigraphic unit framework is more accurately determined

Through the geological survey and combining with the foundation of predecessors, the stratigraphic units in the survey area were determined systematically. A total of 45 stratigraphic units were divided, including 19 formations and 26 members. The fundamental lithostratigraphy in the stratigraphic units was comprehensively studied, in combination with biological, sequence, geochronological and event stratigraphy, and the stratigraphic division and correlation of the Ordovician-Silurian strata were mainly carried out, which were divided into 8 second-order sequences and 25 third-order sequences.

(2) Ten graptolithic biobelts of the Ordovician-Silurian have been identified

The Ordovician-Silurian biostratigraphy has made great progress and breakthrough, and 10 graptoliths, and 1 fauna have been identified from bottom to top. Among them:

two graptolite zones of the Ningguo Formation (*Acrograptus ellesae* belt and *Nicholsonograptus fasciculatus* belt).
two graptolite zones of the Hule Formation (*Pterograptus elegans* belt and *Hustedograptus teretiusculus* belt).
two graptolite zones of the Changwu Formation (*Dicellograptus Complexus* belt and *Paraothograptus pacificus* belt).
two graptolite zones of the Wenchang Formation (*Normalograptus extraordinarius* belt and *Normalograptus persculptus* belt); in addition, 1 Chitinozoans—Wuningensis Songxites, Metacrinus, etc. (*Songxites* + *Aegiromenella*) fauna.
two graptolite zones of the Xiaxiang Formation (*Akidograptus ascensus* belt and *Parakidograptus acuminatus* belt).

(3) Fossils of other creatures from the Cambrian-Ordovician have been discovered

In the Late Cambrian–Early Ordovician Xiyangshan Formation, a *Lotagnostus americanus* was found. Conodonts were found in the Late Ordovician Yanwashan Formation, and 8 trilobites of genus and species were found in the bottom of the Late Ordovician Huangnigang Formation. The discovery of these fossils provides paleontological evidence for the exploration of the sedimentary environment, as well as original data for the subsequent biological research in the northwestern Zhejiang Province.

5.2.2 Intrusives

(1) A systematic investigation and classification of intrusives

The investigation and classification of the complex plutons, such as the Ma'anshan, Tangshe, Xianxia, Tonglizhuang, Wushanguan, and Zhinanshan-Dongkeng, were carried out. The features of the size, spatial and temporal distribution, contact relationships and rock type of the intrusives were well identified. It is determined that the intrusives in the survey area are mainly composed of monzonite, syenogranite, quartz monzonite, quartz syenite (porphyry), and the vein rocks are mainly composed of granitic rocks, pegmatite, diabase, and diorite porphyrite.

(2) A framework of regional tectonic–magmatic–thermochronology was established

Based on the zircon U–Pb geochronology and the rock geochemical characteristics, the tectonic-magmatic-thermochronology framework of three magmatic stages of the Cretaceous in the area was established, followed by K_1^1 (145.1–136.3 Ma), K_1^2 (133.9–125.0 Ma), and K_1^3 (130.5–127.7 Ma). The intrusive age of the vein rocks was determinate as 130.3–127.4 Ma.

The intrusion ages of the Yanshanian intrusives in northwestern Zhejiang and its adjacent area are mainly Middle Jurassic to Early Cretaceous (168–124 Ma) and mainly concentrated at 168–163 Ma, 150–147 Ma and 142–124 Ma.

(3) The series of intrusive rocks–volcanic intrusive rocks–(sub) volcanic rocks are divided and compared

Through a comparative study of the petrological, geochemical, and geochronological features of the intrusives in each period, the intrusives in the survey area are divided into two series of intrusive rocks and volcanic intrusive rocks, which were compared synthetically together with subvolcanic rocks, in order to discuss the source of magma, tectonic environment, intrusive mechanism and mineralization.

It is proposed that the intrusive rocks, volcanic intrusive rocks, and (sub) volcanic rocks in the survey area were all formed in the extensional environment that followed subduction orogeny in the Early Cretaceous. The early magmatism is dominated by differentiation crystallization and the late stage may be dominated by partial melting. The magmatic source of the intrusive rocks is dominated by the reconstruction of crustal materials in the ancient Jiangnan orogenic belt. The magmatic source areas of volcanic intrusive rock series and (sub) volcanic rock series may involve some lithospheric mantle materials of the Yangtze block. The igneous rocks in the survey area may be emplaced and formed successively by high degree of crystallization of magma derived from the remelting of crustal material in the eastern part of the ancient Jiangnan orogenic belt induced by the ascent of lithospheric mantle magma of the Yangtze block in the extensional and tensile tectonic environment after subduction orogeny in the Early Cretaceous.

5.2.3 Volcanic Rocks

(1) Volcanic lithology and lithofacies are systematically divided

The Mesozoic volcanic rocks in the survey area were analyzed by the survey method of "volcanic strata—lithologic lithofacies—volcanic structure." The volcanic lithology can be divided into 4 categories and 13 subcategories, and volcanic facies can be grouped into eight types: explosive facies (fallout facies, clastic flow facies, surging facies), eruption-sedimentary facies, invasion overflow facies, explosive spill facies, eruptive spill facies, volcanic debris flow facies, volcanic vent facies, and subvolcanic rock facies. The field identification marks of volcanic facies were established, which provided preliminarily study of the Mesozoic volcanic rocks in northwestern Zhejiang Province.

(2) The evidence of rhythmic division and geochronology of volcanic eruption was determined

The survey area was divided into four volcanic eruption rhythms. The first eruption rhythm is mainly composed of the evolution of clastic flow accumulation facies (local sandwiched eruption-sedimentary facies) → fallout accumulation facies → eruption-sedimentary facies. The second eruption rhythm primarily consists of volcanic debris flow facies (local) → clastic flow accumulation facies → invasion overflow facies → eruptive spill facies → explosion spill facies → subvolcanic rock facies. The third includes clastic flow accumulation facies (locally intercalated eruption-sedimentary facies) → eruption-sedimentary facies → subvolcanic rock facies, while the fourth is mainly made of eruptive spill facies.

The zircon U–Pb isotope dating provides geochronological evidence for each volcanic eruption rhythm, restricting the volcanic eruption time of the Area to be in the range of 135.1–125.4 Ma in the Early Cretaceous.

(3) The volcanic edifice is divided

According to the control of regional faults on volcanic edifices and the features of volcanic edifice assemblage and volcanic lithofacies assemblage, the grades and types of volcanic structures are classified. The volcanic edifice in the region is divided into 1 Tianmu Mount volcanic depression, 5 (revived) calderas in Yangtianping, Tianhuangping, Changlongshan-Linjiatang, East Tianmu Mount-Caotanggang, and Yaotianfan, and 6 dome-shaped volcanoes in West Tianmu Mount—Longwangshan, South Tianmu Mount, Nanyushan, Shifo Temple, Dashulin and Wuguishan. A number of lateral craters of different sizes were also delineated in the dome volcano. The prioritized description was focused on the typical volcanic edifices of 4 calderas follow by Yangtianping, Tianhuangping, Changlongshan-Linjiatang, East Tianmu Mount-Caotanggang, and three dome-shaped volcanoes follow by West Tianmu Mount-Longwangshan, South Tianmu Mount and Dashulin.

5.2.4 Structure

(1) The structural traces of fold and fault are systematically investigated

Through the investigation of morphology, kinematics, and dynamics, the fold system in the survey area is divided into three stages: Caledonian, Early Indosinian, and Late Indosinian. In Caledonian, the archetype of the equally wide and gentle fold of the Wangjia-Tangshe anticline in the near EW trending was formed. In the Early Indosinian, a large baffle-type fold system was formed in the northeast direction, significantly controlling the main structural framework of the investigation area. In the Late Indosinian, a mainly straddling and relatively small and weak NW-strike fold was formed.

The fault system in the survey area is divided into four groups: EW trending, NE trending, NNE trending, and NW trending. The early basement thrust faults were near EW or NEE trending. In the middle stage, the NE-trending basement thrust faults were activated multistage in the later stage; NW-trending faults formed later or at the same time with NE-trending basement faults and most were transverse faults which cut folds. Late NNE-trending faults are dominated by left-lateral shear. The surface near EW-trending brittle fault was formed in the most recent period, mostly brittle fractures.

(2) The relationship between structure and formation of rocks, ore deposits, and basin is discussed

NE-trending basement faults control the outcrop and distribution of plutons in the survey area. The fractures closely related to ore mineralization are NE-trending and EW-trending faults, and NE-trending faults are mostly rock and ore conducting structures, while NW-trending faults are mostly ore-bearing structures. The large NE-trending fold system controls and forms the prototype of the Tianmu

Mount volcanic basin, which interacts with the near EW-trending concealed basement faults and NE-trending basement faults, resulting in the subsidence fault depression and formation of the volcanic basin.

(3) The development history of the structure is studied, and the evolution model is established

The survey area is located in the southeastern margin of the Yangtze block and has undergone four tectonic evolution stages, including the epicontinental overburden orogeny of the Chengjiang-Caledonian, the Indosinian intracontinental orogeny, the Yanshanian active continental margin, and the Himalayan tectonic uplift. In the Chengjiang-Caledonian, the Area has experienced three evolutionary stages: the Early Nanhuaian continental margin rift basin, the Late Nanhuaian–early stage of the Late Ordovician passive continental margin basin, and the late stage of the Late Ordovician–Early Silurian peripheral foreland basin, so that the Yangtze block and the Cathaysian orogenic system merged into an unified continent. In the Indosinian, the Area was uplifted and exposed, forming the NE blocking fold system and NE basement fault system, which greatly influenced and controlled the tectonic features of the surveyed area. Due to the continuous subduction of the paleo-Pacific plate into the South China Block, the Yanshanian survey area entered the active continental margin stage, characterized by intense magmatic activities and mineralization. In the Himalayan, the survey area was in a relatively quiet interval, dominated by uneven differential movement.

5.2.5 Database

A series of maps including geological map (1:50,000), geological and mineral resources map, volcanic lithofacies structure map, ore-bearing formations and structural map and metallogenic elements and metallogenic prediction maps of typical deposits have been prepared. By means of digital regional geological survey, the databases of the raw data have been established, including freehand field map library (1:25,000), total field map library, original data distribution map library and section database, stream sediment survey database (1:50,000), as well as the manuscript database and the result database. All the data formats conform to related specifications.

5.3 Problems and Suggestions

The geological structure of the survey area is complex. The strata, structure, igneous rocks, and mineral resources are all relatively developed. Due to the limited time and funds, there are still some basic and key geological problems that restrict the ore prospecting, which need to be further studied and solved:

1. Due to poor outcrop, strong alteration, and other reasons, the control accuracy of the Nanhuaian in this work is not enough.
2. The Ordovician-Silurian stratigraphic cross section in the survey area is an ideal research area for biostratigraphy. Although 10 graptoliths and 1 fauna have been found in this work, more graptoliths cannot be found without enough funds and time. Further research should be carried out and comparison with Hirnantian stratotype section should be made. It is suggested that these sections should be protected as typical sections of the Ordovician and Silurian.
3. The complex structure of the survey area experienced tectonic evolution during multistages of the Caledonian, the Early Indosinian, the Late Indosinian, the Yanshanian, and the Himalayan. Although this work was carried out the investigations and studies on the distribution pattern and mechanical properties of important fold and fault structures, there is still a lack of detailed research on the superposition and combination of structures, and further investigation should be conducted.
4. In this work, the important mineral source strata and igneous rocks in the survey area have been preliminarily divided, but the metallogenic strata in the stratigraphic unit have not been precisely determined. It is not clear about the difference of intrusives and structures on mineralization, which needs further investigation and study.
5. The survey area has a desired metallogenic prospecting. Due to the limitation of funds, this work is less invested in mineral exploration to ensure regional geological survey, and only the Hanggai stream sediment measurement was made. The results only show a very small number of important anomalies and mineral occurrences (spots) inspection with a preliminary inspection level. In the following work, regional geophysical and geochemical prospecting, mineral inspection and comprehensive research should be strengthened to provide theoretical and technical support for regional prospecting breakthroughs.

During the project, the survey team received strong support and help from many leaders and experts from the Nanjing Center, China Geological Survey and the Zhejiang Institute of Geological Survey. Prof. Rixiang Gong, Kongzhong Wang, and Zhongda Chen have guided the project team in the fieldwork for many times. In order to compile the report, the experts including Prof. Yanjie Zhang from the Nanjing Center, China Geological Survey, Prof. Yida Luo and Xiaoyong Wu from the Zhejiang Institute of Geological Survey put forward many valuable suggestions and opinions. The survey team also received thoughtful guidance and help in investigating and studying stratigraphic paleontology in the area from Academician Xu Chen, Prof. Yuandong Zhang, Prof. Peng Tang, Dr. Xiang Fang and Dr. Xuan Ma, of the Nanjing Institute of Geology and Paleontology, Chinese Academy of Sciences, Prof. Zhiqiang Zhou of the Xi'an Center, China Geological Survey, Joe Botting and Lucy Muir of the National Museum of Wales, as well as Prof. Guohua Yu of Zhejiang Institute of Geological Survey. In the effort, Academician Chen Xu, Prof. Yuandong Zhang, and Dr. Xuan Ma assisted in identifying the graptoliths, Prof. Peng Tang helped in identifying the chitin fossils, and Prof. Zhiqiang Zhou lent a support in confirming the trilobite fossils.

In addition, we extend gratitude to some colleagues for their participation in the project and generous guidance for this work, in no special order they include Jian Liu, Fenglong Liu, Zhen Wang, Ming Wu, Shengqiang Yu, Jinhua Chen, Shuanghui Xu, Zengcai Tang, Wenjie Hu, and Huaisheng Xie.

The authors sincerely appreciate the support and help of all the leaders, experts, and colleagues.

Due to the limited knowledge and time, there are inevitably some problems and deficiencies in this book, which we sincerely wish readers to criticize and correct.

Permissions

All chapters in this book were first published by Springer; hereby published with permission under the Creative Commons Attribution License or equivalent. Every chapter published in this book has been scrutinized by our experts. Their significance has been extensively debated. The topics covered herein carry significant information for a comprehensive understanding. They may even be implemented as practical applications or may be referred to as a beginning point for further studies.

The contributors of this book come from diverse backgrounds, making this book a truly international effort. We would like to thank all the contributing authors for lending their expertise to make the book truly unique. They have played a crucial role in the development of this book. Without their invaluable contributions this book wouldn't have been possible. They have made vital efforts to compile up to date information on the varied aspects of this subject to make this book a valuable addition to the collection of many professionals and students.

This book was conceptualized with the vision of imparting up-to-date and integrated information in this field. To ensure the same, a matchless editorial board was set up. Every individual on the board went through rigorous rounds of assessment to prove their worth. After which they invested a large part of their time researching and compiling the most relevant data for our readers.

The editorial board has been involved in producing this book since its inception. They have spent rigorous hours researching and exploring the diverse topics which have resulted in the successful publishing of this book. They have passed on their knowledge of decades through this book. To expedite this challenging task, the publisher supported the team at every step. A small team of assistant editors was also appointed to further simplify the editing procedure and attain best results for the readers.

Apart from the editorial board, the designing team has also invested a significant amount of their time in understanding the subject and creating the most relevant covers. They scrutinized every image to scout for the most suitable representation of the subject and create an appropriate cover for the book.

The publishing team has been an ardent support to the editorial, designing and production team. Their endless efforts to recruit the best for this project, has resulted in the accomplishment of this book. They are a veteran in the field of academics and their pool of knowledge is as vast as their experience in printing. Their expertise and guidance has proved useful at every step. Their uncompromising quality standards have made this book an exceptional effort. Their encouragement from time to time has been an inspiration for everyone.

The publisher and the editorial board hope that this book will prove to be a valuable piece of knowledge for students, practitioners and scholars across the globe.

Index